DSL Advances

Prentice Hall PTR Communications Engineering and Emerging Technologies Series

Theodore S. Rappaport, *Series Editor*

DOSTERT *Powerline Communications*

DURGIN *Space–Time Wireless Channels*

GARG *Wireless Network Evolution: 2G to 3G*

GARG *IS-95 CDMA and cdma2000: Cellular/PCS Systems Implementation*

GARG & WILKES *Principles and Applications of GSM*

HAĆ *Multimedia Applications Support for Wireless ATM Networks*

KIM *Handbook of CDMA System Design, Engineering, and Optimization*

LIBERTI & RAPPAPORT *Smart Antennas for Wireless Communications: IS-95 and Third Generation CDMA Applications*

PAHLAVAN & KRISHNAMURTHY *Principles of Wireless Networks: A Unified Approach*

RAPPAPORT *Wireless Communications: Principles and Practice, Second Edition*

RAZAVI *RF Microelectronics*

REED *Software Radio: A Modern Approach to Radio Engineering*

STARR, CIOFFI & SILVERMAN *Understanding Digital Subscriber Line Technology*

STARR, SORBARA, CIOFFI & SILVERMAN *DSL Advances*

TRANTER, SHANMUGAN, RAPPAPORT & KOSBAR *Principles of Communication Systems Simulation with Wireless Applications*

DSL Advances

Thomas Starr

Massimo Sorbara

John M. Cioffi

Peter J. Silverman

PRENTICE HALL
Professional Technical Reference
Upper Saddle River, New Jersey 07458
www.phptr.com

A CIP catalog record for this book can be obtained from the Library of Congress.

Editorial/production supervision: *Jessica Balch (Pine Tree Composition, Inc.)*
Cover design director: *Jerry Votta*
Art director: *Gail Cocker-Bogusz*
Interior design: *Meg Van Arsdale*
Manufacturing manager: *Alexis Heydt-Long*
Manufacturing buyer: *Maura Zaldivar*
Publisher: *Bernard Goodwin*
Marketing manager: *Dan DePasquale*
Editorial assistant: *Michelle Vincenti*
Full-service production manager: *Anne R. Garcia*

Publishing as Prentice Hall Professional Technical Reference
Upper Saddle River, NJ 07458

Prentice Hall books are widely used by corporations and government agencies for training, marketing, and resale.

For information regarding corporate and government bulk discounts, please contact: Corporate and Government Sales (800) 382-3419 or corpsales@pearsontechgroup.com

Printed in the United States of America
10 9 8 7 6 5 4 3 2 1

ISBN: 0-13-093810-6

Pearson Education Ltd., *London*
Pearson Education Australia Pty, Limited, *Sydney*
Pearson Education Singapore, Pte. Ltd.
Pearson Education North Asia Ltd., *Hong Kong*
Pearson Education Canada, Ltd., *Toronto*
Pearson Educación de Mexico, S.A. de C.V.
Pearson Education-Japan, *Tokyo*
Pearson Education Malaysia, Pte. Ltd.

To my wife Marilynn Starr, and my children Matthew and Eric.

—T.S.

To my wife Josephine Sorbara, and our children Kayla,
Christina, Mark, Stephanie, and Stacy.

—M.S.

To my dear wife Assia, and my children Robert, Christa,
Laura, and Jenna Cioffi.

—J.M.C.

To Pat Lane, and Aaron and Alyssa Silverman.

—P.J.S.

PREFACE

The title *DSL Advances* refers to the multitude of developments in digital subscriber line technology, network architectures, and applications since DSL entered the mass market in early 1999. The title also speaks of the evolution of DSL from an obscure network technology to a consumer service used by tens of millions of customers throughout the world. Many books, including *Understanding Digital Subscriber Line Technology* (Starr, Cioffi, and Silverman; Prentice Hall, 1999) described the fundamentals of DSL technology and its usage up to early 1999. DSL has transformed from a technology to a networked lifestyle enjoyed by millions of people.

This book is the second in a series of books on DSL. Chapters 1 and 2 of *DSL Advances* provide the reader with a brief recap of the fundamentals, and then continue the story with the developments during DSL's golden age. Although this book is useful on its own, it is best to have previously read *Understanding Digital Subscriber Line Technology.*

After 1998, the DSL flower bloomed and flourished. The recent developments are manifold. DSL services reached the one-million-customer mark by mid-2000, and accelerated to serve over 30 million customers by the end of 2002. Five percent of homes in the United States had broadband connections by the end of 1999.[1] The transition from low-volume, pilot service offerings to serious attempts to reach a mass market was stimulated by the rapid onslaught of data service competition from cable modems and the long needed stabilization of the regulatory environment. In the United States, as well as a growing list of other countries, regulatory developments provided the key elements for the advance of competitive local exchange carriers (CLECs): assured local loop unbundling, central office collocation, access to telco operations systems, and asymmetric DSL (ADSL) line sharing. Innovation was not limited to DSL equipment; the phone line infrastructure evolved to improve its ability to support DSLs. SBC Communications led the way by deploying many thousands of fiber-fed next generation digital loop carrier (NGDLC) terminals that shortened the

[1]As reported in the January 2001 issue of *Computer Telephony* magazine.

length of twisted-wire copper lines. Increasingly, customers more than three miles from the Central Office can receive DSL services. With approximately one-third of U.S. customers served via DLC and NGDLC in the year 2000, and projections for this number to grow to nearly 50% by 2004, the average loop length is shrinking.

Internationally, Asian markets have outpaced the United States in DSL deployments. Korean and Japanese DSL lines each outnumber total U.S. deployments, and China will soon pass the United States as well. A Japanese deployment of nearly 10 million 7–12 Mbps ADSL systems is expected by the end of 2003. British, French, and German ADSL deployments presently parallel the United States' in size, but are growing faster. Thus, DSL—paricularly ADSL—is fast becoming the world's choice for broadband Internet access.

DSL was a dark horse in 1988 when ISDN was stumbling out of the starting gate and chief technology officers were asking why bother to develop HDSL since fiber would be "everywhere" in a couple of years. Few would have imagined DSL advertisements on city buses, and multi-megabit per second DSL modems selling for about the same price as a 14.4 kb/s voice-band modem in 1988. However, DSL is an interim technology. Eventually, fiber-to-the-home/business and broadband wireless will capture the market. Some fiber-optimistic telephone companies expect that by the year 2010 DSL will have a minority market share in the top market areas. Yet major telecommunications technologies do not die quickly; DSL will see niche usage well past 2010. Will the market success of cable modems persist? Cable modem service is expected to grow, but the growth rate may slow if the upgrading of cable plant for two-way digital service stalls with the completion of upgrades for the most attractive areas. As the cable-modem take rate grows, cable companies will face the dilemma of pumping much more money into service with poor return or providing a service with poor performance. Cable modem service is available in few business areas; unlike DSL, where high-margin business customers help pay the way for the low-margin residential customers. DSL's strengths are its provision by many differentiated service providers and its ultimate availability in nearly all business centers and most residential areas.

Chapters 3 through 7 introduce the newest members of the DSL family and report on the maturation of some of the older family members. These chapters show that the various types of DSL share similarities in the underlying technology and network architecture. The recent DSL advances include symmetric DSLs, as well as the popular ADSL. The symmetric HDSL and SDSL technologies transformed into HDSL2 and SHDSL. HDSL2 conveys 1.5 Mb/s via two wires, whereas HDSL required four wires. SHDSL conveys the same data rates as SDSL, but on loops 2,000 to 3,000 feet longer. The newer symmetric DSL technologies will quickly make HDSL and SDSL obsolete because of the improved performance and two other factors vital for the success of any new DSL technology. HDSL2 and SHDSL are based on industry standards with multivendor interoperability. HDSL2 and SHDSL are designed to minimize crosstalk into other lines. In some ways, VDSL is a delayed opportunity. Very high bit-rate DSL (VDSL) conveys rates as high as 52 Mb/s via very short copper end-section that is usually served via an optical network unit (ONU) that connects to the network by fiber. The cost of the required multitude of ONUs complicates the VDSL business case, while radio frequency ingress and egress force major technical compromises. VDSL is searching for its best opportunity: Will the line code be DMT or single-carrier? Will the prime services be symmetric or asymmetric? VDSL's opportunities are diminished by the industry standards's inability to reach a single VDSL standard.

Chapters 9 and 10 warn that the continued success of DSL is not assured. DSL could become a victim of its own success if the crosstalk between DSLs in the same cable is not managed. This calamity will hopefully be averted by the T1.417 local loop spectrum management standard developed in the T1E1.4 Working Group. This standard is expected to be the technical foundation of telecommunications policy in the United States for the reliable use of unbundled loops. Other countries are developing spectrum management rules to address their unique environment. Traffic laws and police are needed for drivers to safely and efficiently share the road. Limited transmitted signal power and equipment certifications are required for service providers to effectively share the copper pairs in the access cables. Even with such existing static spectrum management, DSL crosstalk diminishes the data rates on other lines, in many cases by a factor of 3 to 10. Chapter 11, which discusses the new area of dynamic spectrum management (DSM), is an exciting first look at the tremendous future possibility as DSL management becomes more automated and sophisticated.

High-speed Internet access and the growing popularity of telecommuting stimulated the mass market for DSL, but in the spirit of "what is old is new," a "new" application for DSL was discovered: voice. DSL was not just for data. Although ADSL was always designed to carry one voice channel, Chapter 16 presents the various ways voice-over-DSL (VoDSL) carries many voice channels in addition to data. The additional service revenue from the same line was icing on an already delicious cake. Furthermore, the many voice calls on one pair of wires reduced the exhaustion of spare pairs in cables. The other new/old emerging application is video. Video-on-demand spurred ADSL in 1994 until the economic reality of competing with videotape rentals stalled ADSL deployment. In a cliff-hanger rescue, ADSL was saved by the Internet. The ADSL lines to millions of homes can connect to video servers as well as data servers. Video-on-demand has resurfaced, and this time it is cost competitive. Video-on-demand via ADSL will become especially attractive with 1.3 Mb/s MPEG4 video coding providing picture quality better than VHS videotapes.

Chapters 12 to 16 show the results from use of DSL in the real world, and reach beyond the physical layer core of DSL by discussing the end-to-end network that surrounds DSL. The customer end of the phone line is not the end of the story. Chapter 12 compares methods for DSL to connect to all networked devices within the customer's premises: dedicated LAN wiring, shared phone wiring, shared AC power wiring, and wireless. Both centralized splitter and distributed in-line filter premises wiring configurations are discussed. Chapter 13 presents the developments in the DSL Forum and elsewhere to remove the labor from DSL: automatic configuration of the customer's equipment and flow-through service provisioning. Chapter 14 explores and compares the diverse breadth of upper-layer communications protocols used by DSLs. Chapter 15 exposes a topic of great importance that is barely mentioned in most DSL books: security. Security risks and essential safeguards for every DSL vendor, service provider, and user are discussed.

The views expressed in this book are solely those of the authors, and do not necessarily reflect the views of the companies and organizations affiliated with the authors.

T.S.
M.S.
J.M.C.
P.J.S.

ACKNOWLEDGMENTS

A great many leading experts generously assisted the creation of this book. Grateful acknowledgment is given to Dr. Martin Pollakowski for his thorough review of the manuscript and many helpful suggestions. The authors also thank the following people for their valuable contributions: from Globespan/Virata: Amrish Patel, George Dobrowski, Ehud Langberg, Bill Scholtz, Patrick Duvalt, and Eli Tzanhany; from Centillium: Les Brown; and from Valo, Inc.: Gary Fox.

Chapter 3 benefited from the help of Syed Abbas of Centillium, Inc., Art Carlson, Howard Levin, Les Humphrey, Jim Carlo, Daniel Gardan, Krista Jacobsen, Louise Hoo, Evangelos Eleftheriou, Frederic Gauthier, Aurelie Legaud, Fabienne Moulin, Sedat Olcer, Giovanni Cherubini, Meryem Ouzzif, Frank van der Putten, Dimitris Toumpakaris, Wei Yu, Nick Zogakis, and Ranjan Sonalkar.

Chapter 7 benefited from the help of Jeannie Fang, Krista Jacobsen, Jungwon Lee, George Ginis, Wei Yu, Steve Zeng, Frank Sjoberg, Rickard Nilsson, Mikael Isaksson, Jacky Chow, Giovanni Cherubini, Evangelos Eleftheriou, Sedat Olcer, Behrooz Rezvani, Sigurd Schelestrate, Thierry Pollet, Vladimir Oksman, J.J. Werner, D. Mesdagh, C. Del-Toso, and K. Hasegawa.

Chapter 8 benefited from the help of Larry Brown of SBC, Inc.

Chapter 11 benefited from the help of G. Ginis, S. T. Chung, W. Yu, D. Gardan, S. Zeng, C. Aldana, J. Lee, A. Salvekar, J. Lee, W. Rhee, W. Choi, K.W. Cheong, J. Fan, N. Wu, B. Rezvani, Ioannis Kanellakopoulos, S. Shah, Michail Tsatsanis, R. Sonalkar, R. Miller, G. Cherubini, V. Poor, G. Zimmerman, G. Al-Rawi, K. Foster, T. Pollet, C. del-Toso, S. Schelesratte, N. Nazari, M. Alba, K. Kim, K. Hasegawa, N. Mihito, and Y. Kim.

Chapter 12 benefited from the help of Dave Alm, Don R. House, and Frederick J. Kiko of Excelcus, Inc.

Appendix A benefited from the help of Jeff Neumann and John Rozema of SBC, Inc.

Appendix B features Matthew and Marilynn Starr and benefited from the help of Marilynn Starr, who proofread many chapters.

The cover art is derived from a photograph of a tapestry named *Jeu d'espirit;* it was designed and woven by Gretchen Starr.

The DSL Forum is thanked for permission to use information from its Technical Reports, especially in preparing the content of Chapters 13 and 14.

CHAPTER 1

Introduction to DSL

Digital subscriber line (DSL) technology transforms an ordinary telephone line into a broadband communications link, much like adding express lanes to an existing highway. DSL increases data transmission rates by a factor of twenty or more by sending signals in previously unused high frequencies. DSL technology has added a new twist to the utility of twisted-pair telephone lines.

1.1 THE TELEPHONE LOOP PLANT

The twisted-wire pair infrastructure (known as the *loop plant*) connects customers to the telephone company network. The loop plant was designed to provide economical and reliable plain old telephone service (POTS). The telephone loop plant presents many challenges to high-speed digital transmission: signal attenuation, crosstalk noise from the signals present on other wires in the same cable, signal reflections, radio-frequency noise, and impulse noise. A loop plant optimized for operation of DSLs would be designed quite differently. Local-loop design practices have changed relatively little over the past 20 years. The primary changes have been the use of longer-life cables and a reduction in loop lengths via the use of the digital loop carrier (DLC). In recent years, primarily in the United States, many thousands of DLC remote terminals (DLC-RTs) have been placed in neighborhoods distant from the central office. Telephone and DSL service is provided directly from the DLC-RT. DSL performance is improved because the DSL signals traverse only the relatively short distance (generally, less than 12,000 feet) from the DLC-RT to the customer site. However, DSLs must cope with the huge embedded base of loop plant, some of which is 75 years old.

The term *loop* refers to the twisted-pair telephone line from a central office (CO) to the customer. The term originates from current flow through a looped circuit from the CO on one wire and returning on another wire. There are approximately 800 million telephone lines in the world.

The loop plant consists of twisted-wire pairs, which are contained within a protective cable sheath. In some parts of Europe and Asia the wires are twisted in four-wire units called "quads." Quad wire has the disadvantage[1] of high crosstalk coupling between the four wires within a quad. Within the CO, cables from switching and transmission equipment lead to a main distributing frame (MDF). The MDF is a large wire cross-connect frame where jumper wires connect the CO equipment cables (at the horizontal side of the MDF) to the outside cables (at the vertical side of the MDF). The MDF permits any subscriber line to be connected to any port of any CO equipment. Cables leaving the CO are normally contained in underground conduits with up to 10,000 wire pairs per cable and are called feeder cables, E-side, or F1 plant. The feeder cables extend from the CO to a wiring junction and interconnection point, which is known by many names: serving area interface (SAI), serving area concept box (SAC box), crossbox, flexibility point, primary cross-connection point (PCP). The SAI contains a small wire-jumper panel that permits the feeder cable pairs to be connected to any of several distribution cables. The SAI is at most 3,000 feet from the customer premises and typically serves 1,500 to 3,000 living units. The SAI is a wiring cross-connect field located in a small outside cabinet that permits the connection of any feeder wire pair to any distribution wire pair. The SAI predates DLC. The SAI contains no active electronics, and is located much closer to the customer than the carrier serving area (CSA) concept originally developed for DLC.

1.2 DSL REFERENCE MODEL

As shown in the generic DSL reference model in Figure 1.1, a DSL consists of a local loop (telephone line) with a transceiver at each end of the wires. The transceiver is also known as a modem (modulator/demodulator). The transceiver at the network end of the line is called the line termination (LT) or the transmission unit at the central end (TU-C). The LT may reside within a digital subscriber line access multiplexer (DSLAM) or a DLC-RT for lines fed from a remote site. The transceiver at the customer end of the line is known as the network termination (NT) or the transmission unit at the remote end (TU-R).

---

[1]Chapter 10 shows that such high crosstalk coupling can be an advantage when all four wires are used in a coordinated way by a single customer, but otherwise this characteristic is a disadvantage.

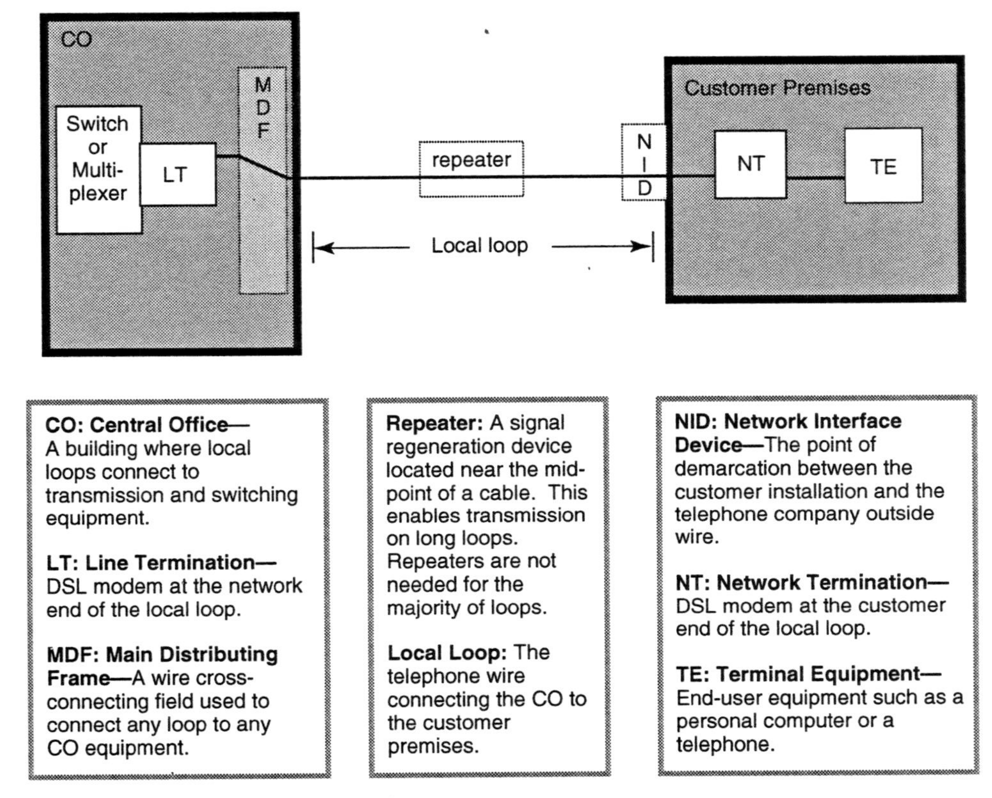

Figure 1.1 DSL Reference Model

The majority of DSLs are served via copper lines extending all the way from the central office to the customer's premises as shown in Figure 1.1. To address DSL's limited line reach, a repeater may be placed near the midpoint of the copper line to boost the signal on the line. However, installation of midspan repeaters has a high material and labor cost. More often, DSL service for customers in areas distant from a CO have DSL service enabled by the placement of a DLC-RT in their neighborhood as shown in Figure 1.2. The DSL signals traverse the relatively short copper wire between the customer and the DLC-RT. The link between the CO and DLC-RT is usually optical fiber. Alternatively, distant areas may also be served via a DSLAM located at a remote site; sometimes the DSLAM may be located within a building having many customers.

Figure 1.3 shows the network architectures for a voice-band modem, private-line DSL, and switched DSL services. The voice-band modem transmission path extends from one customer modem, a local line, the public switched telephone network (PSTN), a second local line, and a second customer modem. In contrast, the extent of DSL transmission link is a customer-end modem (TU-R), a local line, and a network-end modem (TU-C), with the digital transport through the remainder of the network being outside the scope of the DSL transmission path. The private-line service is not switched within the network and thus provides a dedicated

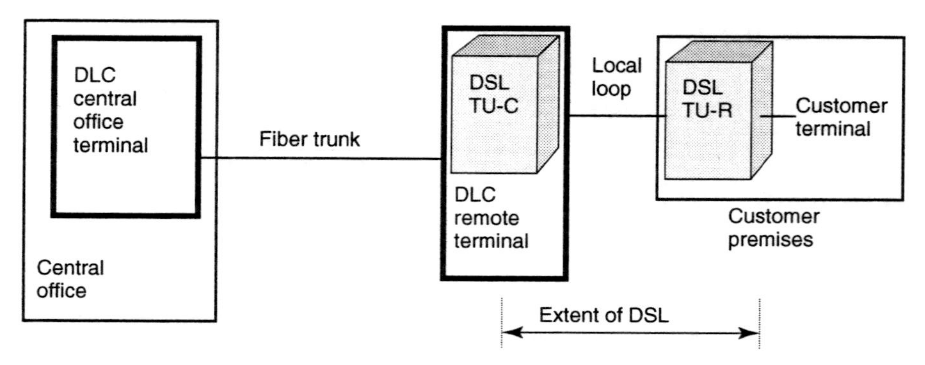

Figure 1.2 Digital Loop Carrier Reference Model

connection between two endpoints. The private-line architecture may contain a service-provisioned cross-connection within the central office. The switched DSL service permits the customer to connect to many end points simultaneously or sequentially. Often, the TU-C for the switched DSL service resides within a DSLAM. The switching or routing functions may reside within the DSLAM or in separate equipment.

1.3 THE FAMILY OF DSL TECHNOLOGIES

Several species of DSL have resulted from the evolution of technology and the market it serves. The earliest form of DSL, 144 kb/s basic rate ISDN, was first used for ISDN service in 1986, and then was later applied to packet mode ISDN DSL (IDSL), and local transport of multiple voice calls on a pair of wires (DAML: digital added main line). Basic rate ISDN borrowed from earlier voice band modem technology (V.34), and T1/E1 digital transmission technology (ITU Rec. G.951, G.952).

As shown in Figure 1.4, DSL transmission standards have evolved from 14.4 kb/s voice-band modems in the 1970s to 52 Mb/s VDSL in the year 2001. This has been an evolution, with each generation of technology borrowing from the prior generation.

High bit-rate DSL (HDSL) was introduced into service in 1992 for 1.5 Mb/s (using two pairs of wires) and 2 Mb/s (using two or three pairs of wires) symmetric transmission on local lines. HDSL greatly reduced the cost and installation time required to provide service by reducing the need for midspan repeaters and simplifying the line engineering effort. HDSL is widely used for private line services, and links to remote network nodes such as digital loop carrier remote terminals and wireless cell sites. In 2000, HDSL2 was introduced to accomplish the same bit-rate and line reach as HDSL but using one pair of wires instead of the two pairs required for HDSL. Both HDSL and HDSL2 operate over CSA (carrier serving area) length lines consisting of up to 12 kft[2] of 24 AWG wire, 9 kft of 26 AWG

[2]1 kft equals 1,000 feet.

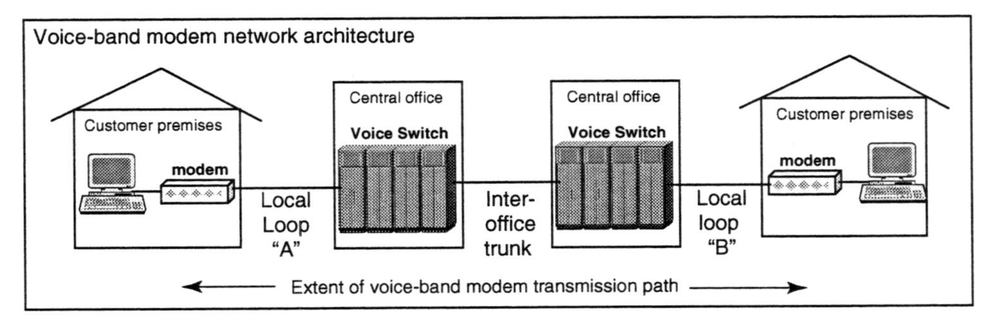

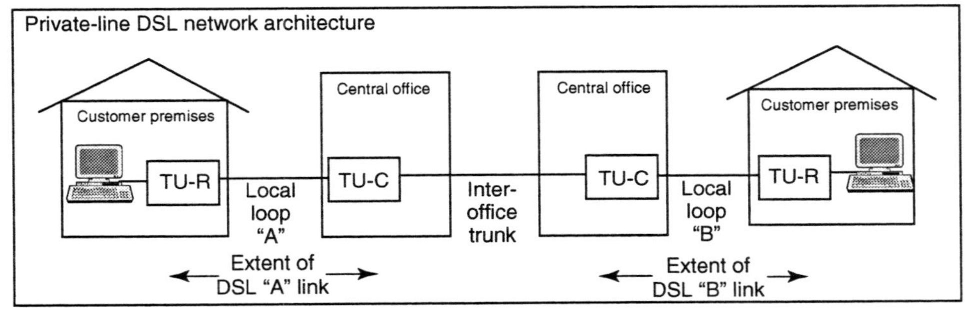

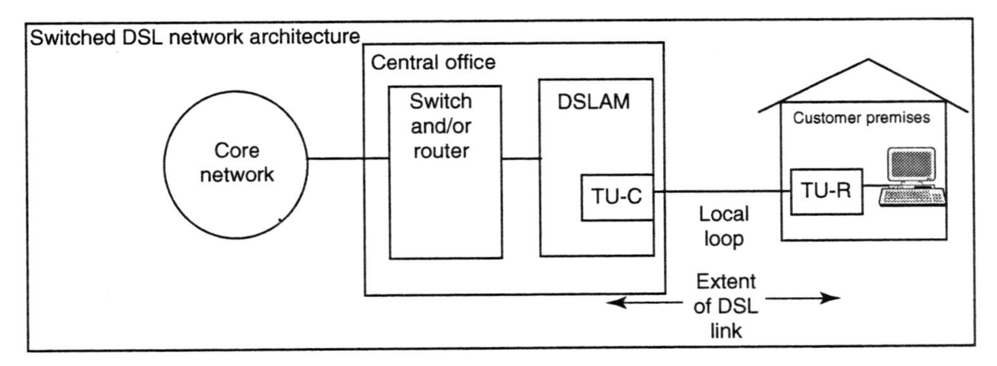

Figure 1.3 Comparative Network Architectures

wire, or a proportionate length of mixed wire gauges. HDSL2 is spectrally compatible with other services in the same cable within the CSA line lengths but may not be spectrally compatible if a midspan repeater is used to serve longer lines. HDSL4, using trellis-coded pulse amplitude modulation (TC-PAM) for two pairs of wires, achieves spectral compatibility for 1.5 Mb/s transport on longer loops. By reaching up to 11 kft on 26 AWG lines without repeaters, HDSL4 further reduces the need for repeaters. The complementary pair of technologies—HDSL2 (for CSA lines) and HDSL4 (for longer lines)—provide a lower cost and spectrally compatible means to provide symmetric 1.5 Mb/s for nearly all lines. Chapters 4 and 6 discuss HDSL2 and HDSL4, respectively. Chapter 6 also addresses the symmetric SHDSL technology.

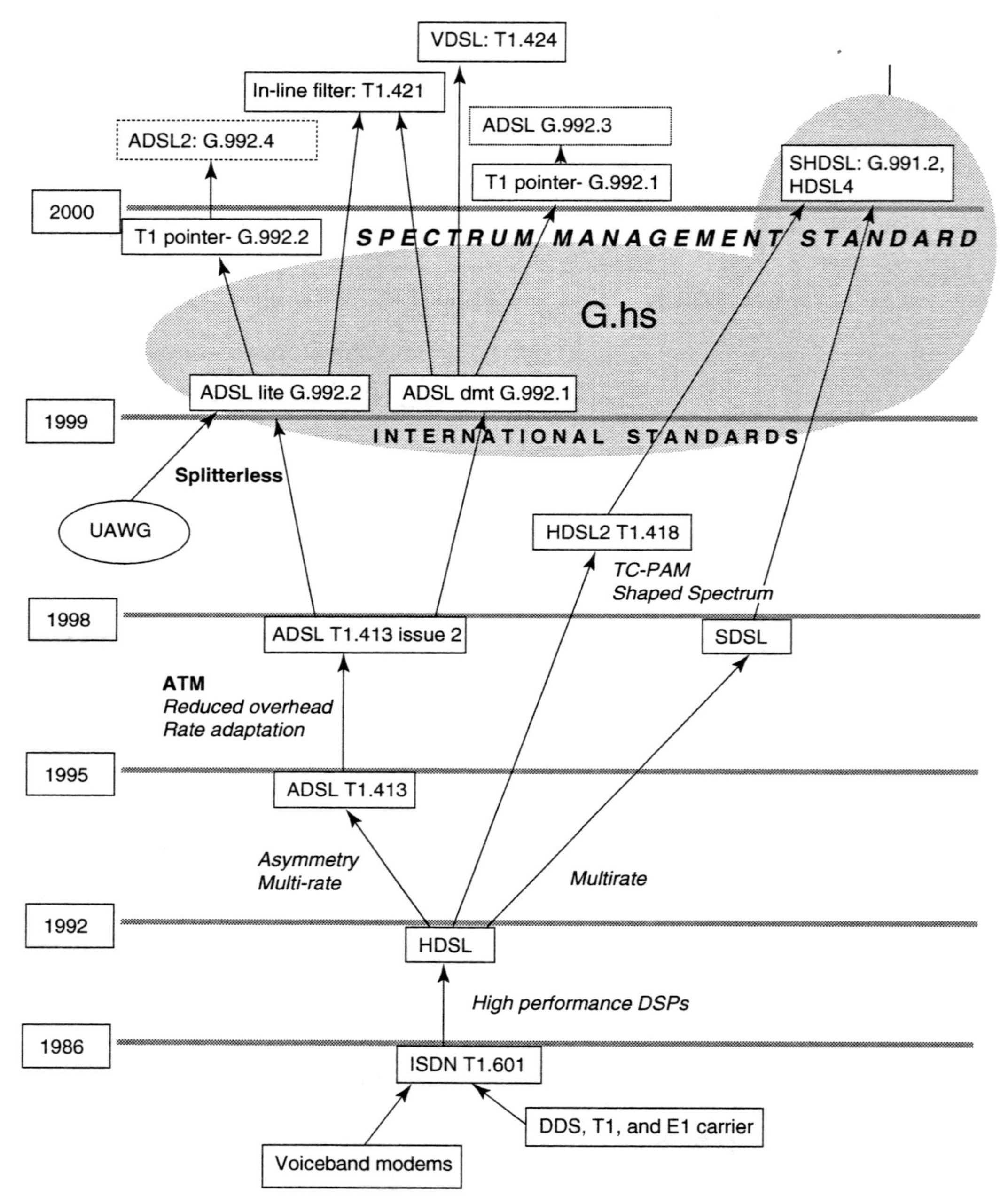

Figure 1.4 Evolution of DSL Technology (*Note:* Dates indicate publication of relevant standards.)

Asymmetric DSL (ADSL) service was introduced in 1995 and employed the following new technology aspects:

- Higher downstream bit rates are achieved via transmission asymmetry, using a wider bandwidth for downstream transmission and a narrower bandwidth for upstream transmission.
- Near-end crosstalk is reduced by partial or full separation of the upstream and downstream frequency bands.
- Simultaneous transport of POTS and data is achieved by transmitting data in a frequency band above voice telephony.[3]
- Use of advanced transmission techniques (trellis coding, Reed-Solomon codes with interleaving, and DMT modulation).
- Rate-adaptive transmission that adjusts to the highest bit rate allowed by the unique conditions for each line.

ADSL is widely used for applications benefiting from the bit-rate asymmetry, for example, high-speed Internet access and workstation access for small business offices and home work offices (SOHO). ADSL supports downstream bit-rates up to 8 Mb/s and upstream bit-rates up to 900 kb/s on short lines (less than 6 kft) with moderate line noise. However, to assure service to more lines with more noise, ADSL service is most often provided at bit-rates of 2 Mb/s or less downstream and 128 kb/s or less upstream. At mid-year 2002, there were 26 million ADSLs in service worldwide, with approximately 80 percent of the lines serving residential customers and 20 percent of the lines serving business customers. The early deployments of ADSL employed a splitter at both ends of the line to combine the 0–3.2 kHz analog voice signal with the ADSL signals in a higher frequency band. The development of the "G.lite" (ITU Rec. G.992.2) standard introduced the concept of enabling customer self-installation without a splitter at the customer end of the line. This reduced the labor cost to install the service. Field trials of G.lite demonstrated that an in-line filter must be inserted in series with most types of telephone sets to prevent problems for both the voice transmission as well as the digital transmission. Subsequently, it was determined that the in-line filters permitted effective operation of the full-rate ADSL (T1.413 and ITU Rec G.992.1), whereas G.lite is restricted to about 1.5 Mb/s downstream. As a result, the large majority of current ADSL installations use full-rate ADSL self-installed by the customer, placing an in-line filter by every telephone in their premises. Because ITU Recs. G.922.1 and G.992.2 were derived from the earlier T1.413 standard, all these ADSL standards are very similar, and most ADSL equipment supports all three standards. Chapter 3 discusses ADSL in more detail.

Single-pair high-bit-rate DSL (SHDSL) products were available by the end of 2000 based on the ITU Rec. G.991.2 standard. Like the nonstandard 2B1Q SDSL systems, SHDSL supports symmetric transmission at bit-rates from 192 kb/s to 2.32

---

[3]In some parts of Europe, ADSL operates in a frequency band above Basic Rate ISDN (4B3T).

Table 1.1 Comparison of DSL Technologies

XDSL Technology	ADSL Full rate	ADSL-Lite "g.Lite"	HDSL, HDSL2, HDSL4	SDSL	SHDSL	IDSL	VDSL
Data rates	192k–8 Mb/s DS 64–900 kb/s US	256k–1.5 Mb/s DS 64–400 kb/s US	1.5 Mb/s symmetric	256k– 2 Mb/s symmetric	192 k– 2.3 Mb/s symmetric	128k or 144kb/s symm.	12–52 Mb/s DS 6–26 Mb/s symmetric
Loop reach mixed gauge wire	15 kft reach at lower rates	15 kft reach at lower rates	HDSL & HDSL2—9 kft, HDSL4— 11 kft (×2 with repeater)	16 kft @ 256kb/s, 7 kft @ 1.5Mb/s	20 kft @ 256 kb/s, 9 kft @ 1.5 Mb/s	15 kft reach, repeater for ×2 reach	3 kft @ 26M/4M 1 kft @ 26 Mbs symmetric
Service types	Data & POTS Shared line	Data & POTS Shared line	DS1 private line	Data only	Data and optional digitized voice	Data only	Data and POTS
Principle applications	Internet access, data	Internet access, data	Data, voice trunking	Data	Data, voice	Data	Video, Internet access, data
Modulation	DMT	DMT	2B1Q-HDSL TC-PAM-HDSL2&4	2B1Q	TC-PAM	2B1Q	Multiple (CAP, QAM, DMT, FMT)
Common protocols	PPP over ATM	PPP over ATM	DS1	Frame Relay	ATM	Frame Relay	ATM
Standard	ITU G.992.1 T1.413	ITU G.992.2	ITU G.991.1-HDSL T1.418-HDSL2	None	ITU G.991.2	T1.601	T1 Trial use standard
Number of wire pairs	One pair	One pair	2 pairs: HDSL, HDSL4 1 pair: HDSL2	One pair	One pair (2 pair option doubles the bit rate)	One pair	One pair (1, 2 or 4 pairs in EFM area)

DS: Downstream—data rate towards the customer
US: Upstream—data rate from the customer
Symmetric: same data rate in both directions

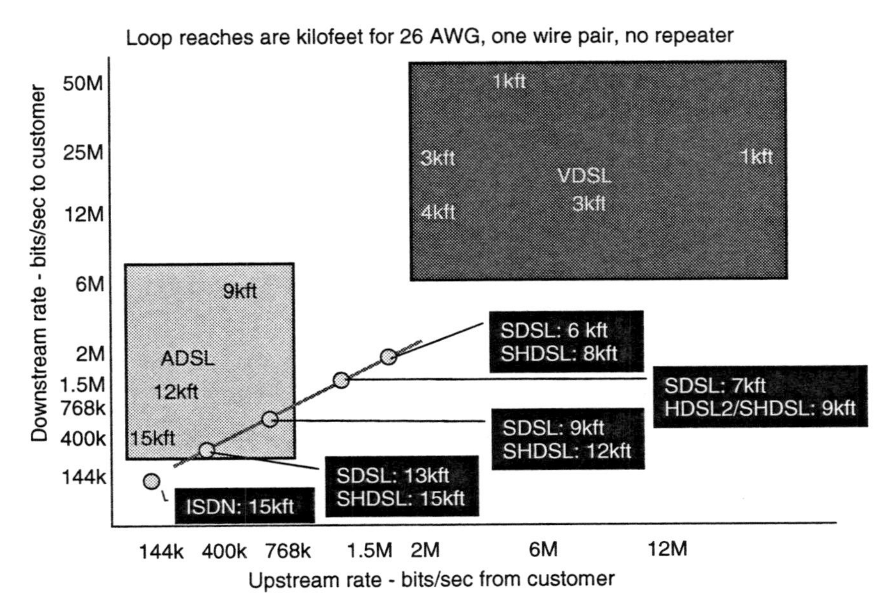

Figure 1.5 DSL Data Rates

Mb/s while providing at least 2,000 feet greater line reach than SDSL. Furthermore, the SHDSL specifications provided for the use of multiple pairs of wires and midspan repeaters to achieve greater bit-rates and line lengths. SHDSL uses trellis coded pulse amplitude modulation (TC-PAM), which is also used for HDSL2 and HDSL4. Chapter 6 discusses SHDSL in more detail.

Prestandard very high-bit-rate DSL (VDSL) systems were used in field trials in 2000. VDSL supports asymmetric bit-rates as high as 52 Mb/s downstream or symmetric bit-rates as high as 26 Mb/s. The key distinction of VDSL is its limitation to very short loops, as short as 1,000 feet for the highest bit-rates or up to about 4,000 feet for moderate data rates. The very short line length operation depends on shortening the copper line by placing an optical network unit (ONU) close to the customer site and then connecting one or more ONUs to the network with a fiber. Like ADSL, VDSL is a rate adaptive system that provides for simultaneous transmission of data and an analog voice signal. Chapter 7 discusses VDSL in more detail.

The term xDSL applies to most or all types of DSL technology. Chapter 5 discusses ITU G.994.1 (g.handshake), which DSL transceivers use to negotiate a common operating mode.

Figure 1.5 shows the upstream and downstream rates supported by the various DSL technologies with the symmetric technologies (ISDN, SHDSL, HDSL) residing along a line of symmetry, and the rate adaptive technologies (ADSL, VDSL) covering a broad range of bit-rates with the corresponding maximum line lengths indicated.

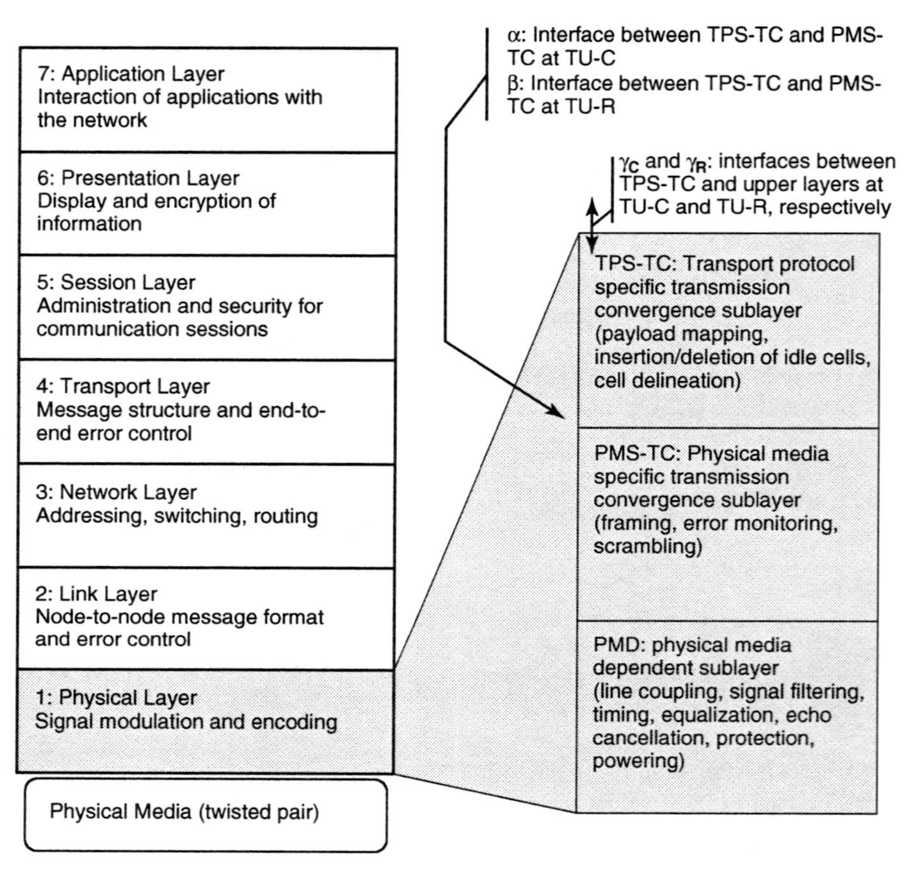

Figure 1.6 Protocol Reference Model

1.4 DSL PROTOCOL REFERENCE MODEL

Figure 1.6 shows the open systems interconnection (OSI) protocol reference model with additional detail shown for the physical layer. The structure within the physical layer is largely due to the contributions of Les Humphrey. The protocol reference model provides a structure to organize complex communications systems. The layered approach hides details of information from the subsystem, invoking the service of a particular layer. Thus, an application on a host requesting communication with its peer application does not need to know the details of its physical connection to a data network, communications used between itself and the network, or even the details of the protocols exchanged between the hosts supporting the application. The layered approach also simplifies the analysis of complex communication systems by segmenting the systems into several well-defined portions, facilitates reuse of portions of communications systems, and allows upgrades of portions of a communications system independently of each other.

CHAPTER 2

REVIEW OF TRANSMISSION FUNDAMENTALS FOR DSLs

2.1 INTRODUCTION

This chapter provides summary of useful concepts and formulae for the transmission-systems basics that have been applied to DSL systems. A more comprehensive treatment for DSL occurs in a predecessor to this book [1], and the reader can refer to that earlier book for more details and derivations. The intent is the successful use of these formulae and principles in DSL-system simulations and performance-feasibility studies.

Section 2.2 reviews the transmission methods used in the earliest DSL systems that were primarily baseband systems (i.e., systems with no analog POTS simultaneously preserved on the same line as the DSL), whereas Section 2.3 provides the same for the later DSL methods of QAM/CAP and multicarrier methods, particularly DMT. More details on all are found in [1], and the DMT method in particular will appear also in Chapters 3, 7, and 11 of this book. Section 2.4 concludes by summarizing some simple impairment models: background noise and NEXT and FEXT coupling models.

All transmission channels are fundamentally analog and thus may exhibit a variety of transmission effects. In particular, telephone lines are analog, and so DSLs all use some form of *modulation*. The basic purpose of modulation is to convert a stream of bits into equivalent analog signals that are suitable for the transmission line.

Figure 2.1 depicts a digital transmission system. The transmitter converts each successive group of b bits from a digital bit stream into one of 2^b **data symbols,** $\boldsymbol{x}_m$, via a one-to-one mapping known as an **encoder.** Each group of b bits constitutes a message m, with $M = 2^b$ possible values $m = 0, \ldots, M - 1$. The data symbols are N-dimensional (possibly complex) vectors, and the set of M vectors forms a **signal constellation. Modulation** is the process of converting each successive data symbol vector into a continuous-time analog **signal** $\{x_m(t)\}_{m=0,\ldots,M-1}$ that represents the message corresponding to each successive groups of b bits. The message may vary from use to use of the digital transmission system, and thus the message index m and the corresponding symbol $\boldsymbol{x}_m$ are considered to be random, taking one of M possible values each time a message is transmitted. This chapter assumes that each message is equally likely to occur with probability $1/M$. The encoder may be sequential, in which case the mapping from messages to data symbols can vary with time as indexed by an encoder state, corresponding to ν bits of past state information (function of previous input bit groups). There are 2^ν

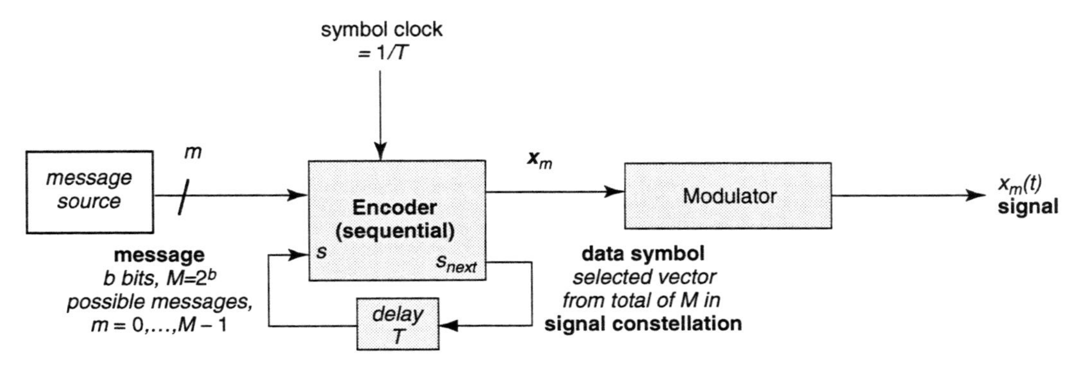

Figure 2.1 Transmitter of a digital transmission system.

possible states when the encoder is sequential. When $\nu = 0$, there is only one state and the encoder is **memoryless.**

Linear modulation also uses a set of N, orthogonal unit-energy basis functions, $\{\varphi_n(t)\}_{n=1:N}$, which are independent of the transmitted message, m. The basis functions thus satisfy the orthonormality condition:

$$\int_{-\infty}^{\infty} \varphi_n(t)\varphi_l^*(t)dt = \begin{cases} 1 & n = l \\ 0 & n \neq l \end{cases}.$$

The nth basis function corresponds to the signal waveform component produced by the nth element of the symbol $\boldsymbol{x}_m$.[1] Different **line codes** are determined by the choice of basis functions and by the choice of signal constellation symbol vectors, $\boldsymbol{x}_m$, $m = 0, \ldots, M-1$. Figure 2.2 depicts the function of linear modulation: For each **symbol period** of T seconds, the modulator accepts the corresponding data symbol vector elements, $x_{m1}, \ldots, x_{mN}$, and multiplies each by its corresponding basis function, $\varphi_1(t), \ldots, \varphi_N(t)$, respectively, before summing all to form the modulated waveform $x_m(t)$. This waveform is then input to the channel.

The **average energy,** E_x, of the transmitted signal can be computed as the average integrated squared value of $x(t)$ over all the possible signals,

$$E_x = \frac{1}{M} \cdot \sum_{m=0}^{M-1} \left\{ \int_{-\infty}^{\infty} |x_m(t)|^2 dt \right\},$$

or more easily by finding the average squared length of the data symbol vectors,

$$E_{\mathbf{x}} = \frac{1}{M} \cdot \sum_{m=0}^{M-1} ||x_m||^2.$$

---

[1]Complex basis functions occur only in the mathematically abstract case of baseband-equivalent channels, as in Section 2.3.5, and a superscript of * means complex conjugate (and also transpose when a vector).

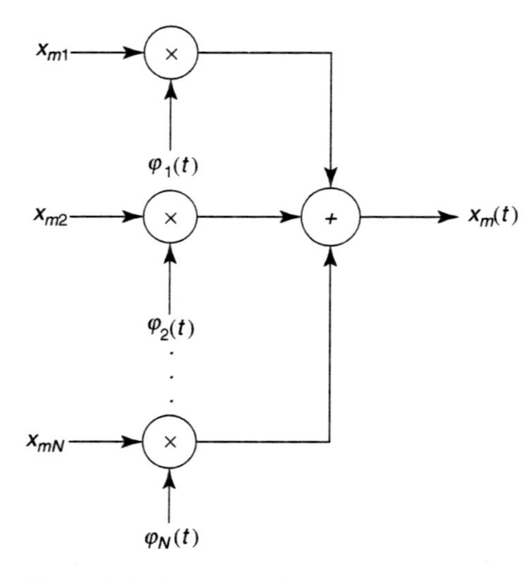

Figure 2.2 Linear modulator.

The **digital power** of the transmitted signals is then $S_x = E_x/T$. The **analog power,** P_x, is the digital power at the source driver output divided by the input impedance of the channel when the line and source impedances are real and matched. Generally, analog power is more difficult to calculate than digital power. Transmission analysts usually absorb the gain constants for a specific analog driver circuit into the definitions of the signal constellation points or symbol vector values $\boldsymbol{x}_m$ and the normalization of the basis functions. The digital power is then exactly equal to the analog power, effectively allowing the line and analog effects to be viewed as a 1-ohm resistor.

The channel in Figure 2.3 consists of two potential distortion sources: bandlimited filtering of the transmitted signals through the filter with Fourier transform $H(f)$ and additive Gaussian noise (unless specifically discussed otherwise) with zero mean and power spectral density $S_n(f)$. The designer should analyze the transmission system with an appropriately altered $H(f)$ $\left(H(f) \rightarrow \frac{H(f) \cdot \sigma}{\sqrt{S_n(f)}}\right)$ to include the effects of spectrally shaped noise, and then it is sufficient to investigate only the case of equivalent white noise where the noise power spectral density is a constant, σ^2.

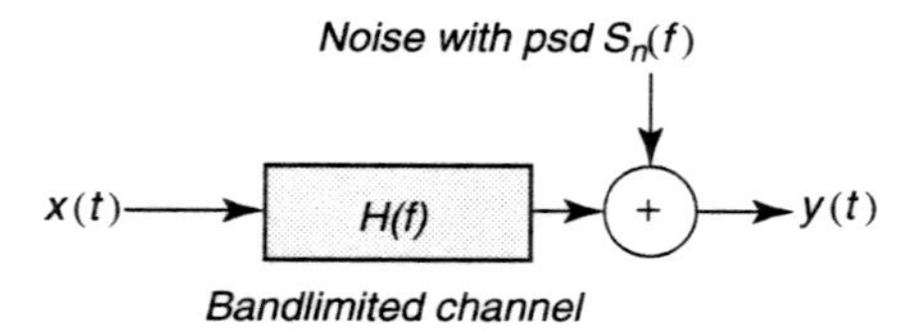

Figure 2.3 Bandlimited channel with Gaussian noise.

2.1.1 The Additive White Gaussian Noise (AWGN) channel

The additive white Gaussian noise (AWGN) channel is the most heavily studied in digital transmission. This channel simply models the transmitted signal as being disturbed by some additive noise. It has $|H(f)| = 1$, which means there is no bandlimited filtering in the channel (clearly an idealization). If the channel is distortionless, then $|H(f)| = 1$ and $\sigma^2 = 0$. On a distortionless channel, the receiver can recover the original data symbol by filtering the channel output $y(t) = x(t)$ with a bank (set) of N parallel matched filters with impulse responses $\varphi_n^*(-t)$ and by sampling these filters' outputs at time $t = T$, as shown in Figure 2.4. This recovery of the data symbol vector is called **demodulation.** A bidirectional digital transmission apparatus that implements the functions of "modulation" and "demodulation" is often more succinctly called a **modem**. The reversal of the one-to-one encoder mapping on the demodulator output vector is called **decoding.** With nonzero channel noise, the demodulator output vector $\boldsymbol{y}$ is not necessarily equal to the modulator input $\boldsymbol{x}$. The process of deciding which data symbol is closest to $\boldsymbol{y}$ is known as **detection.** When the noise is white Gaussian, the demodulator shown in Figure 2.4 is optimum. The optimum detector selects $\hat{\mathbf{x}}$ as the symbol vector value $\boldsymbol{x}_{\hat{m}}$ closest to $\boldsymbol{y}$ in terms of the vector distance/length,

$$\hat{m} = i \text{ if } \| y - x_i \| \leq \| y - x_j \| \text{ for all } j \neq i,\ j = 0,\ldots,M - 1.$$

Such a detector is known as a **maximum likelihood detector,** and the probability of an erroneous decision about $\boldsymbol{x}$ (and thus the corresponding group of b bits) is minimum. This type of detector is only optimum when the noise is white, Gaussian, and the channel has very little bandlimiting (essentially infinite bandwidth) and is known as a **symbol-by-symbol detector.** Each matched filter output has independent noise samples (of the other matched-filter output samples), and all have mean-square noise sample value σ^2. Thus the SNR is

$$SNR = \frac{E_x/N}{\sigma^2}.$$

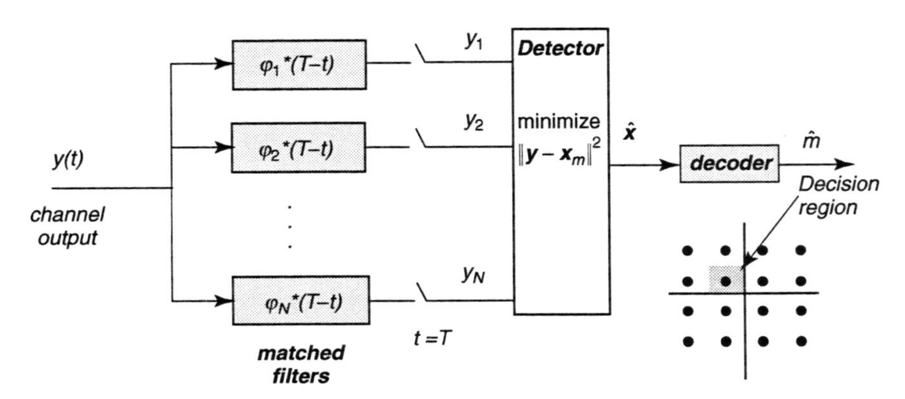

Figure 2.4 Demodulation, detection, and decoding.

Detector implementation usually defines regions of values for $\boldsymbol{y}$ that map through the ML detector into specific input values. These regions are often called **decision regions.**

An error occurs when $\hat{m} \neq m$, that is, $\boldsymbol{y}$ is closer to a different symbol vector than to the correct symbol vector. An error is thus caused by noise being so large that $\boldsymbol{y}$ lies in a decision region for a point $\boldsymbol{x}_j$, $j \neq m$ that is not equal to the transmitted symbol. The probability of such an error on the AWGN channel is less than or equal to the probability that the noise is greater than half the distance between the closest two signal-constellation points. This **minimum distance** between two constellation points, d_{min} , is easily computed as

$$d_{min} = \min_{i \neq j} \|x_i - x_j\|.$$

Symbol vectors in a constellation will each have a certain number of nearest neighbors at, or exceeding, this minimum distance, N_m. The average number of nearest neighbors is

$$N_e = \frac{1}{M} \cdot \sum_{m=0}^{M-1} N_m,$$

which essentially counts the number of most likely ways that an error can occur. So, the **probability of error** is often accurately approximated by

$$P_e \cong N_e Q\left[\frac{d_{min}}{2\sigma}\right],$$

where the **Q-function** is often used by DSL engineers. The quantity $Q(x)$ is the probability that a unit-variance zero-mean Gaussian random variable exceeds the value in the argument, x,

$$Q(x) = \int_x^{\infty} \frac{1}{\sqrt{2\pi}} e^{-u^2/2} du.$$

The Q-function must be evaluated by numerical integration methods, but Figure 2.5 plots the value of the Q-function versus its argument ($20log(x)$) in decibels. For the physicist at heart, $Q(x) = .5 \cdot erfc(x/\sqrt{2})$. For more error measures, see [1], but basically P_e is of the same magnitude of order as all the others.

2.1.2 Margin, Gap, and Capacity

It is desirable to characterize a transmission method and an associated transmission channel simply. Margin, gap, and capacity are related concepts that allow such a simple characterization. Many commonly used line codes are characterized by a **signal-to-noise ratio gap** or just **gap**. The gap, $\Gamma = \Gamma(P_e, C)$, is a function of a chosen probability of symbol error, P_e, and the line code, C. This gap measures efficiency of the transmission method with respect to best possible performance on an addi-

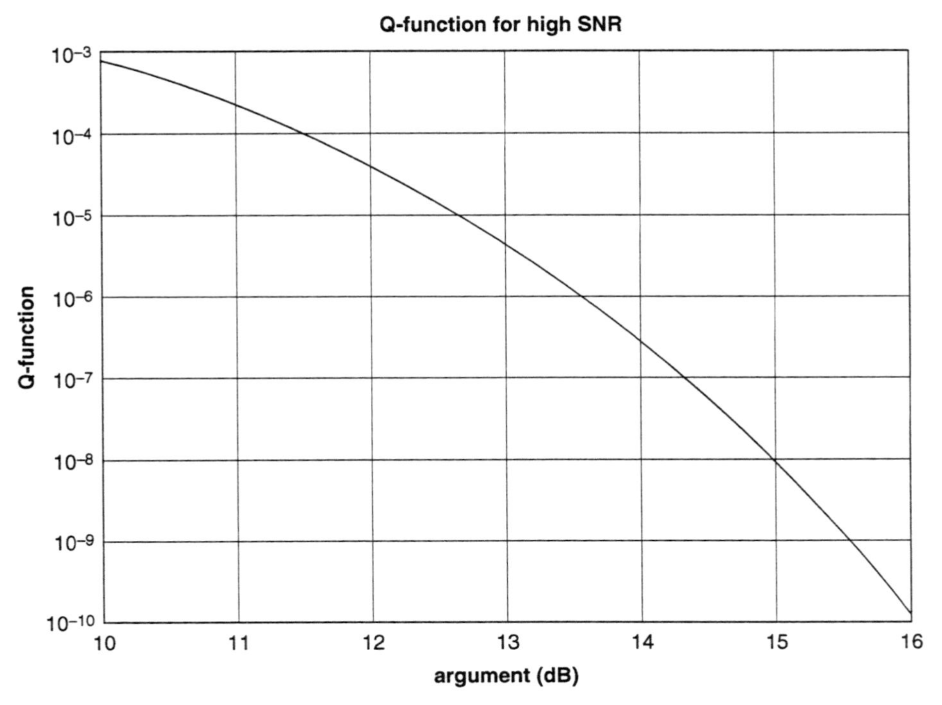

Figure 2.5 Q function versus SNR.

tive white Gaussian noise channel, and is often constant over a wide range of b (bits/symbol) that may be transmitted by the particular type of line code. Indeed, most line codes are quantified in terms of the achievable bit rate (at a given P_e) according to the following formula:

$$\bar{b} = \frac{1}{2} log_2\left(1 + \frac{SNR}{\Gamma}\right).$$

Thus, to compute data rate with a line code characterized by gap Γ, the designer need only know the gap and the SNR on the AWGN channel. An experienced DSL engineer usually knows the gaps for various line codes and can rapidly compute achievable data rates in their head. The SNR may be higher for better transmission modulation, but the gap is a function only of the encoder in Figure 2.1.

An optimum line code with a gap of $\Gamma = 1$ (0 dB) achieves a maximum data rate known as the **channel capacity.** Such an optimum code necessarily requires infinite complexity and infinite decoding/encoding delay. However, it has become practical at DSL speeds to design coding methods for which the gap is as low as 1–2 dB. It is also possible when combining such codes with DMT to achieve capacity in DSL systems.

Often, transmission systems are designed conservatively to ensure that a prescribed probability of error occurs. The **margin** of a design at a given performance level is the amount of additional signal to noise ratio of the design in excess of the minimum required for a given code with gap Γ. The margin can be computed according to

$$\gamma_m = \frac{SNR}{\Gamma \cdot (2^{2\bar{b}} - 1)}$$

2.2 BASEBAND TRANSMISSION

Baseband codes are distinguished from **passband codes** (See Section 2.3) in that baseband codes can transmit energy at DC ($f = 0$), while passband codes transmit at a frequency spectrum translated away from DC. Baseband line codes in DSLs have $N = 1$ and appear in the earliest DSLs: Examples are T1, ISDN, and HDSL.

2.2.1 The 2B1Q Line Code (ISDN and HDSL)

The 2B1Q baseband line code is heavily used in early DSLs. The 2B1Q line code *ideally* uses a basis function[2]

$$\varphi(t) = \frac{1}{\sqrt{T}} \text{sinc}\,(t/T),$$

which leads to the independence of successively transmitted data symbols when sampled at multiples of the symbol period. Thus, in the modulation of Section 2.1, $N = 1$, $b = 2$, $M = 4$, and the encoder is memoryless. On an ideal channel with no distortion, a corresponding matched filter in the receiver also maintains the independence of successive symbol transmissions. In practice, the sinc function cannot be exactly implemented, and transmission through the bandlimited channel distorts this basis function anyway. Thus, compromise pulse masks that attempt to predistort the transmit basis function to incur minimum distortion have been incorporated into the ISDN and HDSL standards documents [2,3] and are further detailed in [1].

The signal constellation for 2B1Q appears is basically $\pm$ 1, $\pm$ 3 (see [1]). The number of bits in a group is $b = 2$, and each group of bits maps into one of the four data symbol values. A one-dimensional value for x could be computed as $x = 2m - 3$, $m = 0,1,2,3$. The name "2B1Q" derives from the mneumonic "2 bits per 1 quartenary" symbol. Because the transformer/hybrid coupling to the transmission line does not pass DC, baseband line codes like 2B1Q undergo severe distortion, and a straightforward implementation of a symbol-by-symbol detector will likely not perform acceptably. The receiver must compensate for the transformer and line distortion using the DFE of Section 2.5.

For purposes of analysis, the following function characterizes the one-sided output power-spectral density of the transmit shaping filters where the use of a rec-

---

[2]The sinc function is $sinc\,(x) = {}^{sin(\pi x)}/_{\pi x}$.

tangular basis function (with additional gain $\sqrt{T} = 1/\sqrt{f_0}$) was assumed, along with a unity-DC-gain nth-order Butterworth filter,

$$PSD_{2B1Q}(f) = K_{2B1Q} \cdot \frac{2}{f_0} \cdot \frac{\left[sin\left({}^{\pi f}/_{f_0} \right) \right]}{\left({}^{\pi f}/_{f_0} \right)^2} \cdot \frac{1}{\left[1 + \left({}^{f}/_{f_{3dB}} \right)^{2n} \right]},$$

where $K_{2B1Q} = \frac{5}{9} \cdot \frac{V_p^2}{R}$ and $R = 135\ \Omega$. For ISDN: $n = 2$ $V_p = 2.5$ V, $1/T = f_0 = f_{3db} =$ 80 kHz, so the data rate is 160 kbps; and for HDSL: $n = 4$ $V_p = 2.7$ V, $1/T = f_0 =$ 392kHz, $f_{3db} = 196$ kHz, so the data rate is 784 kbps (per line).

The probability of error, given an AWGN channel, is

$$P_e = P_b = 1.5Q\left(\sqrt{\frac{SNR}{5}} \right),$$

with a straightforward encoder mapping. The gap for 2B1Q is 9.8 dB at $P_e = 10^{-7}$. ISDN and HDSL also mandate an additional 6 dB of margin, leaving ISDN and HDSL at least 16 dB below the best theoretical performance levels. Additional loss of several dB occurs with ISI (see Section 2.5) on most channels. Thus, uncoded 2B1Q is not a high-performance line code, but clearly the transmitter is fairly simple to implement. The receiver for improved performance is more complex than necessary, leading in part to 2B1Q's later reduced use in DSL.

2.2.2 Pulse Amplitude Modulation (PAM)

2B1Q generalizes into what is known as pulse amplitude modulation (PAM), but has the same basis function as 2B1Q modulation, and simply has $M = 2^b$ equally spaced levels symmetrically placed about zero, $b = 1, \ldots, \infty$. The number of dimensions remains $N = 1$, and PAM also uses a memoryless encoder, $x = 2m - (M - 1)$, $m = 0, \ldots, M - 1$ or a code like gray code in 2B1Q. Approximately a 6 dB increase in transmit power is necessary for the extra constellation points associated with each additional bit in a PAM constellation if the performance is to remain the same (i.e., P_e is constant). Eight-level PAM is sometimes called "3B1O" and is used in single-line HDSL or HDSL-2 standards[3] (see Chapters 5 and 6). In general for PAM,

$$P_x = \frac{E_x}{T \cdot R} = \frac{M + 1}{3(M - 1)} \cdot \frac{V_p^2}{T \cdot R}$$

where $T = 1/f_0$ is the symbol period, R is the line impedance (typically 100 to 135 Ohms), and $V_p = (M - 1)d_{min}/2$ is the peak voltage. The power spectral density of the transmitted signal similarly generalizes to

[3]The original HDSL makes use of 2 twisted pair (or 3 in Europe) while HDSL-2 uses only 1 twisted pair.

$$PSD_{PAM}(f) = K_{PAM} \cdot \frac{2}{f_0} \cdot \frac{\left[sin\left({}^{\pi f}\!/_{f_0} \right) \right]^2}{\left({}^{\pi f}\!/_{f_0} \right)} \cdot \frac{1}{\left[1 + \left({}^{f}\!/_{f_{3dB}} \right)^{2n} \right]}$$

with $K_{PAM} = \frac{M+1}{3(M-1)} \cdot \frac{V_P^2}{R}$ and f_{3dB} equal to the cut-off frequency of a nth order (Butterworth) transmit filter. The data rate is $\log_2 (M)/T$ bps. The probabilty of error is

$$P_e = 2(1 - 1/M) \cdot Q\left(\sqrt{\frac{3 \cdot SNR}{M^2 - 1}} \right)$$

$$\overline{P}_b = (1 - 1/M) \cdot Q\left(\sqrt{\frac{3 \cdot SNR}{M^2 - 1}} \right)$$

assuming again an input bit encoding that leads to only 1 bit error for nearest neighbor errors. For more details on different baseband transmission methods like Manchester, differential encoding, J-ISDN, BnZs, HDB, and 4B5B, see [1].

While relatively harmless at low speeds, 2B1Q, and more generally PAM, should be avoided in DSL at symbol rates exceeding 500 kHz. Wideband PAM unnecessarily creates unacceptably high and unyielding crosstalk into other DSLs.

2.2.3 Alternate Mark Inversion (AMI)

Alternate mark inversion (AMI) is less efficient than PAM and used in early T1 transmission. The modulation basis function is dependent on past data symbols in that a 0 bit is always transmitted as a zero level, but that the polarity of successive 1's alternate, hence the name. One recalls that early data communciations engineers called "1" a "mark." Thus, AMI is an example of sequential encoding. The basis function can contain DC because the alternating-polarity symbol sequence has no DC component, thus the sinc basis function can again be used. However, the distance between the signal level representing a zero, 0 Volts, and 1 (either plus or minus a nonzero level) is 3 dB less for the same transmit power as binary PAM, so

$$P_e = 1.5Q(SNR/2).$$

This 3 dB performance loss simplifies receiver processing. An AMI transmitter is shown in Figure 2.6. Differential encoding of binary PAM signals ("1" causes change, "0" causes no change) provides signal output levels of ½ and −½. The difference between successive outputs (+1, 0, or −1) is then input to the basis function shown to complete modulation. The receiver maps

$$\pm 1 \rightarrow 1$$

$$0 \rightarrow 0.$$

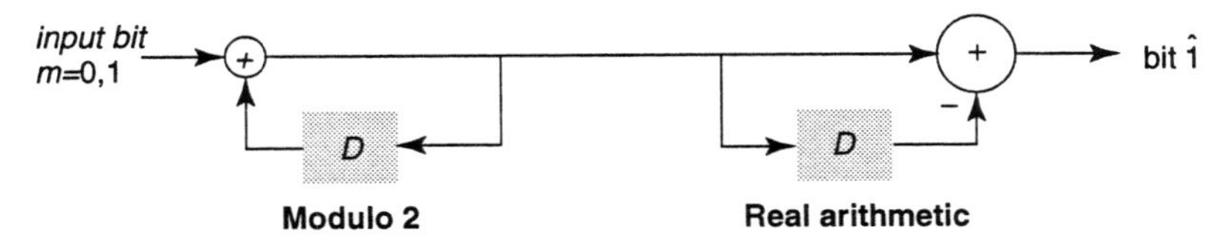

Figure 2.6 Encoder/modulator for AMI.

The transmit power spectral density of AMI is shaped as

$$S_{ami}(f) = \frac{V_p^2}{R} \cdot \frac{2}{f_0} \cdot \left[\frac{\sin(\frac{\pi f}{f_0})}{(\frac{\pi f}{f_0})} \right]^2 \cdot \sin^2(\tfrac{\pi f}{f_0}) \cdot \frac{1}{1 + (\frac{f}{f_{3dB}})^{12}},$$

assuming 6th-order Butterworth transmit filtering and no transformer high pass. A high pass filter that represents a transformer is sometimes modeled by an additional factor of $\left[1 + \left(f_t/f\right)^2\right]^{-2}$ where f_t is the 3dB cut-off frequency of the high pass).

ANSI T1 transmission uses AMI with the basis function shown in Figure 2.7.

2.2.4 Successive Transmission

Data transmission consists of successive transmission of independent messages over a transmission channel. The subscript of n denoting dimensionality is usually dropped for one dimensional transmission and a time-index subscript of k is instead used to denote the message transmitted at time $t = kT$, or equivalently the kth symbol. Thus, x_k corresponds to one of M possible symbols at time k. The corresponding modulated signal then becomes

$$x(t) = \sum_k x_k \varphi(t - kT),$$

a sum of translated single-message waveforms. The symbols x_k may be independent or may be generated by a sequential encoder. A receiver design could sample the output of matched filter $\varphi(-t)$ on an AWGN channel at time instants $t = kT$. The design of the modulation filter $\varphi(t)$ may not be such that successive translations by one symbol period are orthogonal to one another. However, this property is desirable and the function is called a **Nyquist pulse** when it satisfies the orthogonality constraint

$$\int_{-\infty}^{\infty} \varphi(t - kT)\, \varphi(t - lT)dt = \delta_{kl}.$$

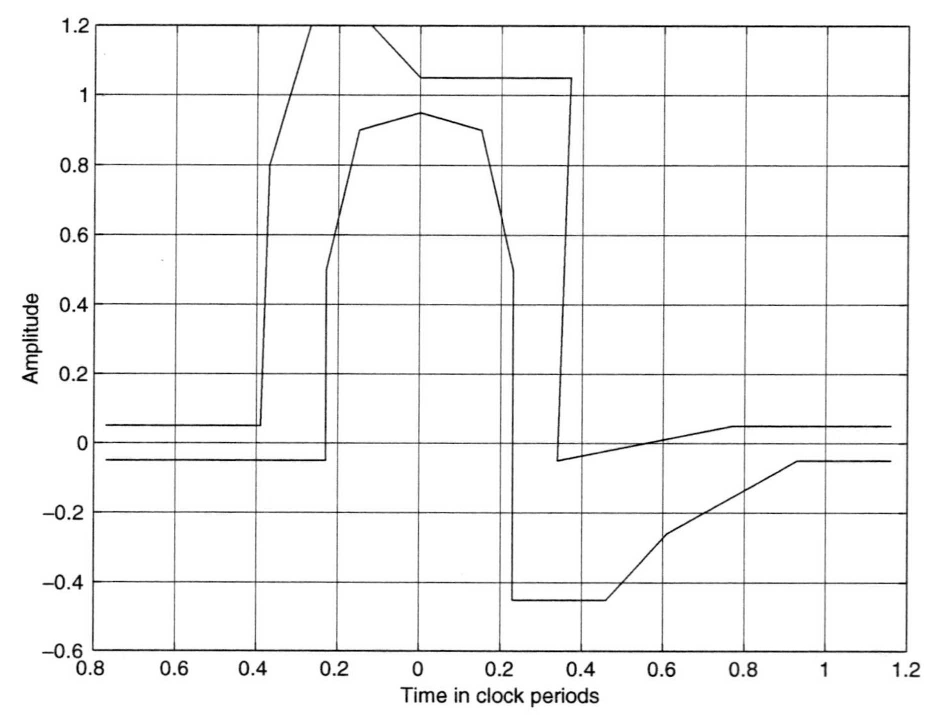

Figure 2.7 T1.403 "T1" pulse shape (basis function after 1-D operation in T1 alternate mark inversion (AMI), 1 = 648 ns).

Nyquist pulses exhibit no overlap or **intersymbol interference** between successive symbols at the matched-filter output in the receiver. The **Nyquist criterion** can be expressed in many equivalent forms, but one often encountered is that the combined transmit-filter/matched-filter shape $q(t) = \varphi(t) * \varphi(-t)$ should have samples $q(kT) = \delta_k$ or the "aliased" spectrum should be flat,

$$1 = \frac{1}{T} \cdot \sum_{n=\infty}^{\infty} Q\left(\omega + \frac{2\pi n}{T}\right).$$

The function $\varphi(t) = {}^{1}/\sqrt{T} \cdot sin\, c({}^{t}/_{T})$ is a Nyquist pulse, but also one that "rings" significantly with time, because the amplitude of the sinc function at non-sampling instants decays only linearly with time away from the maximum at time $t = 0$. This leaves a transmission system designed with sinc pulses highly susceptible to small timing-phase errors in the sampling clock of the receiver. There are many Nyquist pulses, but perhaps the best known are the **raised-cosine pulses,** which satisfy the Nyquist criterion and have their ringing decay with $1/t^3$ instead of $1/t$. These pulses are characterized by a parameter α that specifies the fraction of **excess bandwidth,** which is the bandwidth in excess of the minimum $1/T$ necessary to satisfy the Nyquist criterion with sinc pulses. The raised cosine pulses have response

$$q(t) = sin\,c\left(\frac{t}{T}\right) \cdot \left[\frac{cos\left({\alpha \pi t}/{T} \right)}{1 - \left({2\alpha t}/{T} \right)^2} \right] \quad or$$

$$Q(\omega) = \left\{ \frac{T}{2} \left[1 - sin \begin{pmatrix} T \\ \frac{T}{2\alpha}[|\omega| - \frac{\pi}{T}] \\ 0 \end{pmatrix} \right] \quad \begin{matrix} |\omega| \le \frac{\pi}{T}(1-\alpha) \\ \frac{\pi}{T}(1-\alpha) \le |\omega| \le \frac{\pi}{T}(1+\alpha) \\ \frac{\pi}{T}(1+\alpha) \le |\omega| \end{matrix} \right\}$$

The actual pulse shapes for transmit filter and receiver filter are **square-root raised-cosine pulses** and have transform $\sqrt{Q(\omega)}$ and time-domain impulse response:

$$\varphi(t) = \frac{4\alpha}{\pi\sqrt{T}} \cdot \frac{cos([1+\alpha]\pi / T) + T \cdot sin([1-\alpha]\pi t / T) / (4\alpha t)}{1 - [4\alpha t / T]^2}$$

This pulse convolved with itself (corresponding to the combination of a transmitter and receiver matched filter) is a raised-cosine pulse. The time and frequency response of a square-root raised cosine transmit filter appear in [1].

2.3 PASSBAND TRANSMISSION

Passband line codes have no energy at or near DC. The reason for their use in DSLs is because DSLs normally are transformer coupled (for purposes of isolation) from transceiver equipment, and thus the transformers pass no DC or low frequencies. This section reviews four of the most popular passband transmission line codes.

2.3.1 Quadrature Amplitude Modulation (QAM)

Quadrature amplitude modulation (QAM) is a $N = 2$-dimensional modulation method. The two basis functions are (for transmission at time 0):

$$\varphi_1(\mathrm{t}) = \sqrt{\frac{2}{T}} \cdot \varphi(t) \cdot cos(2\pi f_c t)$$

$$\varphi_1(\mathrm{t}) = -\sqrt{\frac{2}{T}} \cdot \varphi(t) \cdot sin(2\pi f_c t)$$

where $\varphi(t)$ is a baseband modulation function like sinc or a square-root raised cosine pulse shape. The multiplication of the pulse shape by sine and cosine moves energy away from baseband to avoid the DC notch of the twisted-pair transformer coupling. QAM pulses suffer severely from line attenuation in DSLs and compensation is expensive. Some proprietary DSL systems with QAM. QAM is often used in voiceband modem transmission where line-characteristics are considerably more mild over the small 3–4 kHz bandwidth, allowing DC to be avoided and a receiver to be implemented with tolerable complexity. (See Section 2.4 on equalization.)

For successive transmission, QAM is implemented according to

$$x(t) = \sqrt{2/T} \cdot \left[\sum_k x_{1,k}\, \varphi(t - kT)\, cos(2f_c\pi t) - x_{2,k}\, \varphi(t - kT) sin(2\pi f_c t) \right],$$

in which one notes the sinusoidal functions are not offset by kT on the kth symbol period. Because of the presence of the sinusoidal functions and the potential arbitrary choice of a carrier frequency with respect to the symbol rate, QAM functions do not appear the same within each symbol period. That is, QAM basis functions are not usually periodic at the symbol rate, $x_n(t) \neq x_n(t + kT)$, even if the same message is repeatedly transmitted. However, the baseband pulse is repeated every symbol period. This aperiodicity is not typically of concern, but the use of periodic functions can allow minor simplification in implementation in some cases with the so-called "CAP" methods of the next subsection.

2.3.2 Carrierless Amplitude/Phase Modulation (CAP)

Carrierless amplitude/phase modulation (CAP) was proposed by Werner, who notes that the carrier modulation in QAM is superfluous because the basic modulation is two-dimensional and a judicious choice of two DC-free basis functions can sometimes simplify the transmitter implementaton [10]. The potentially high receiver complexity of QAM is still apparent with CAP (see Section 8.4.2 of [1] on Equalization).

CAP basis functions appear the same within each symbol period:

$$\varphi_1(\mathrm{t}) = \sqrt{\frac{2}{T}} \cdot \varphi(t) \cdot cos(2\pi f_c t)$$

$$\varphi_2(\mathrm{t}) = -\sqrt{\frac{2}{T}} \cdot \varphi(t) \cdot sin(2\pi f_c t)$$

Successive transmission is implemented with

$$\begin{aligned} x(t) = \sqrt{2/T} \cdot \sum_k x_{1,k}\, \varphi(t - kT) cos(2\pi f_c[t - kT]) \\ - x_{2,k} \varphi(t - kT) sin(2\pi f_c[t - kT]) \end{aligned}$$

This form is a more natural extension of one-dimensional successive transmission and the basis-function concept. There is no carrier frequency, and f_c is simply a parameter that indicates the center of the passband used for transmission, whence the use of the term "carrierless," while nevertheless the amplitude and phase of $x(t)$ are modulated with the two-dimensional basis set. CAP transmission systems are not standardized for use in DSL, but have been used by a few vendors in proprietary implementations [9] and appear in HDSL reports (see Chapter 4).

CAP and QAM are fundamentally equivalent in performance on any given channel given the same receiver complexity—only the implementations differ, and only slightly. For other quadrature modulation schemes, see [1].

2.3.3 Constellations for QAM/CAP

The easiest constellations for QAM are the **QAM square constellations,** which are essentially the same as two PAM dimensions treated as a single two-dimensional quantity. Indeed, square QAM and PAM are essentially equivalent when appropriate normalization to the number of dimensions occurs, $P_e\ (PAM) = \bar{P}_e\ (SQ - QAM)$. Figure 2.8 shows SQ QAM constellations for even and odd numbers of bits per two-dimensional symbol. The expressions for probability of error are then

$$P_e(odd) < 4\left(1 - {}^{3}\!/_{2M}\right) \cdot Q\left(\sqrt{\frac{6 \cdot SNR}{2M - 1}}\right)$$

$$P_e(even) \leq 4\left(1 - {}^{1}\!/_{\sqrt{M}}\right) \cdot Q\left(\sqrt{\frac{3 \cdot SNR}{M - 1}}\right)$$

Another popular series of constellations for $b \geq 5$ are the cross constellations that use the general constellation shown in Figure 2.9 and have relationship to probability of symbol error

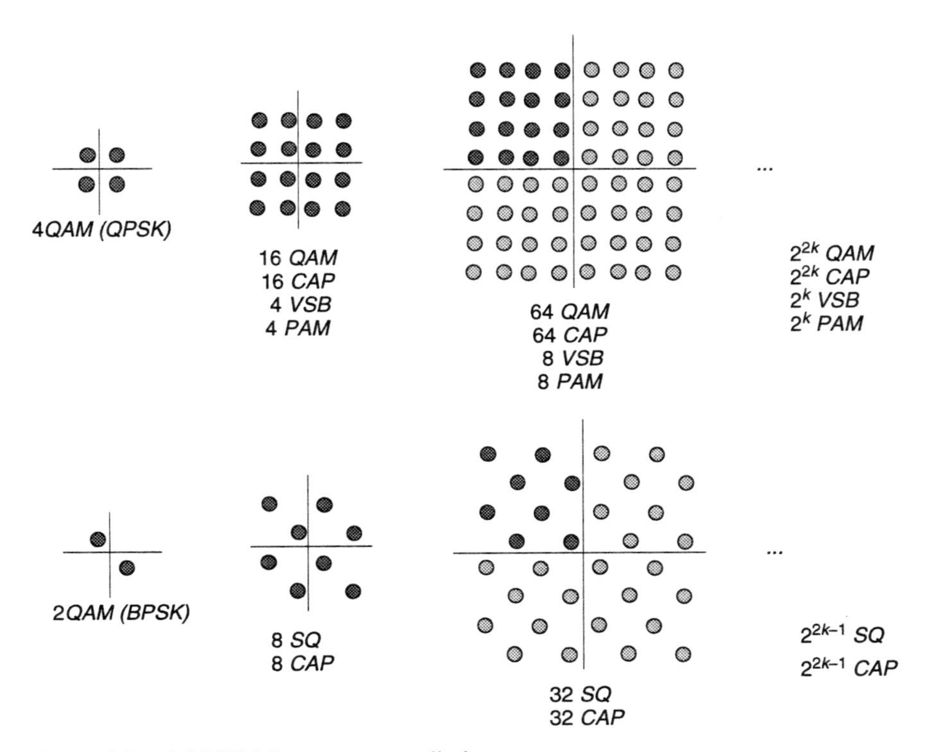

Figure 2.8 QAM/CAP square constellations.

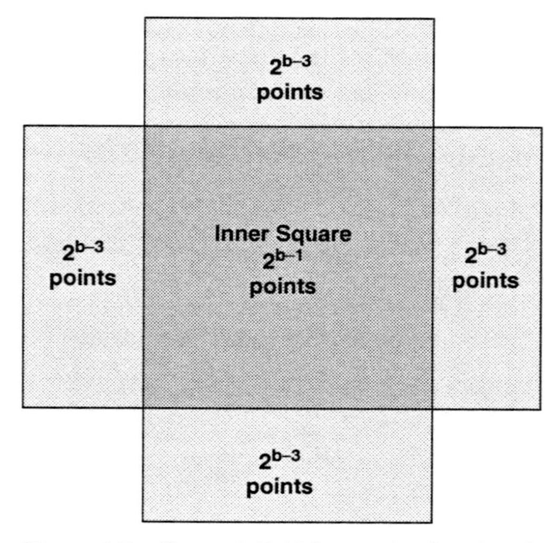

Figure 2.9 General QAM cross for $b > 4$ and odd.

$$P_e(cross) \leq 4\left(1 - \frac{1}{\sqrt{2M}}\right) \cdot Q\left(\sqrt{\frac{3 \cdot SNR}{\frac{31}{32} M - 1}}\right).$$

2.3.4 Complex Baseband Equivalents

Figure 2.10 shows a passband transmission system's energy location. The designer only cares about the region of bandwidth used and not the entire transmission band.

Typically, these systems are two-dimensional and many designers prefer to describe the transmission system with scalar complex (rather than real) functions and analysis. There are two types of complex representations in common use: **baseband equivalents** that are most useful in analyzing QAM systems and **analytic equivalents** that are most useful in analyzing CAP systems. We first define the complex data symbol $x_{b,k} = x_{1,k} + jx_{2,k}$.

A QAM waveform is determined by taking the real part of the complex waveform

$$x_\alpha(t) = \sum_k x_{b,k}\, \varphi(t - kT) \cdot e^{j2\pi f_c t}$$

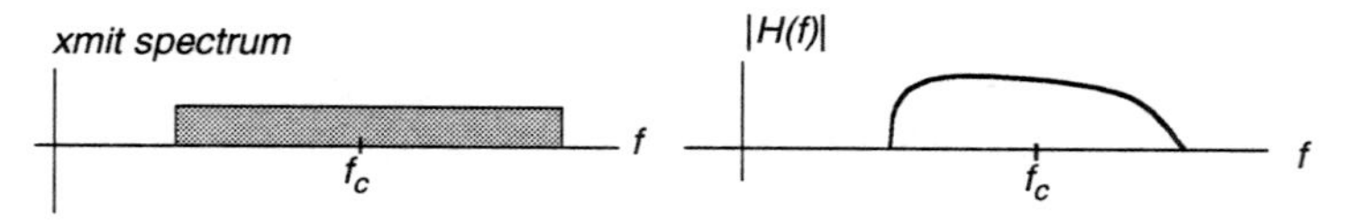

Figure 2.10 Passband channel input and output spectra.

where $x_a(t)$ is the **analytic equivalent** of the QAM signal. The **baseband equivalent** is just

$$x_b(t) = \sum_k x_{b,k}\, \varphi(t - kT) = x_\alpha(t) \cdot e^{-j2\pi f_c t}$$

The baseband equivalent does not explicitly appear to depend on the carrier frequency and essentially amounts to shifting the passband signal down to baseband. That is,

$$X_b(f) = 2 \cdot X(f - f_c)\, \forall f \geq -f_c.$$

The baseband equivalent output of a noiseless linear channel with transfer function $H(f)$ can be found as

$$Y_b(f) = H(f - f_c) \cdot X_b(f)\, \forall f \geq -f_c.$$

Defining $P(f) = H(f - f_c)\Phi(f)$ as the pulse response of the channel, and scaling the channel output (signal and noise) down by a factor of $\sqrt{2}$ (to eliminate the extra square-root 2 factors in the normalized basis functions of QAM and the artificial doubling of noise inherent in the complex representation) produces the convenient complex baseband channel model

$$y_b(t) = \sum_k x_{b,k} \cdot p(t - kT) + n_b(t),$$

where $x_{b,k}$ are complex symbols at the channel input and $n_b(t)$ is complex baseband equivalent white noise with power spectral density $2\sigma^2$ or equivalently σ^2 per real dimension. This baseband equivalent channel has exactly the same form as with successive transmission with real baseband signals such as PAM, except that all quantities are complex. Thus, PAM is a special case. This representation allows a consistent single theory of equalization/modulation as will be followed by subsequent sections of this text and generally throughout the DSL industry.

CAP systems can also be modeled by a complex equivalent system above except that the channel output is still passband and is the analytic signal

$$y_\alpha(t) = \sum_k x_{b,k} \cdot p_a(t - kT) + n_a(t),$$

where $P_a(f) = H(f) \cdot \Phi(f - f_c)\, \forall f > 0$ and the noise is statistically again WGN with power spectral density σ^2 per real dimension.

In either the QAM or CAP case, a complex equivalent channel has been generated and the effects of carrier and/or center frequencies can be subsequently ignored in the analysis, which can then concentrate on detecting $x_{b,k}$ from the complex channel output.

2.4 EQUALIZATION

Equalization is a term generally used to denote methods employed by DSL receivers to reduce the mean-square ISI. The most common forms of equalization for DSLs appear in this subsection.

2.4.1 Linear Equalization

Linear equalizers are very common and easy to understand, albeit they often can be replaced by nonlinear structures that are less complex to implement and that work better, but are harder to understand. Nonetheless, this study begins with linear equalization and then progresses to better equalization methods in later subsections. The basic idea of linear equalization is to invert the channel impulse response so that the channel exhibits no ISI. However, straightforward filtering would increase the noise substantially, so the design of an equalizer must simultaneously consider noise and ISI. Noise enhancement is depicted in Figure 2.11. The equalizer increases the gain of frequencies that have been relatively attenuated by the channel, but also simultaneously boosts the noise energy at those same frequencies. The boosting of noise is called **noise enhancement.**

The Minimum-Mean-Square Error Linear Equalizer (MMSE-LE) appears in Figure 2.12. A matched filter for the channel pulse response and sampler precede a linear filter whose transfer function is selected to minimize the mean-square difference between the channel input and the equalizer output. This filter's response has setting has D-transform

$$W(D) = \frac{1}{R(D) + {}^{1}\!/_{SNR}},$$

where SNR $= E_x/\sigma^2$ and $R(D)$ is the discrete-time transform of the samples of $r(t) = p(t) * p^*(-t)$.[4] The equalizer need only be symbol-spaced (implemented at the symbol sampling instants in discrete time), but some implementations often combine the matched filter and linear filter into a single filter called a **fractionally spaced equalizer.** This single combined filter must be implemented at a sampling rate at least twice the highest frequency of the channel pulse response. The MMSE value in either implementation is

$$MMSE = \sigma_{LE}^2 = \sigma^2 \cdot w_0 = \frac{\sigma^2 T}{2\pi} \int_{-\pi/T}^{\pi/T} \frac{1}{R(e^{j\omega T}) + 1/SNR} d\omega.$$

where w_0 is the center tap (time zero sample) in the discrete-time symbol-spaced response w_k. A symbol-by-symbol detector processes the equalizer output. The scaling just prior to the detector removes a small bias inherent in MMSE estima-

[4] E_x and σ^2 must consistently be for the same number of real dimensions.

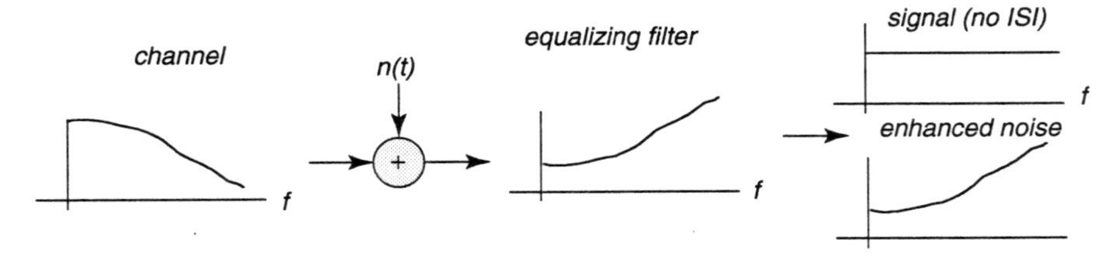

Figure 2.11 Illustration of noise enhancement with linear equalizer.

tion [12]. The performance of the MMSE-LE is characterized by an unbiased signal-to-noise ratio at the detector

$$SNR_{LE,U} = \frac{E_x}{\sigma_{LE}^2} - 1.$$

This receiver signal-to-noise ratio is equivalent to that characterizing an AWGN channel's ML receiver for independent decisions about successive transmitted symbols. Thus, a probability of error can be approximated by assuming the error signal is Gaussian and computing P_e according to well established formula for PAM and QAM. For instance, square QAM on a bandlimited channel using an MMSE-LE with $SNR_{LE,U}$ computed above has probability of error

$$\overline{P}_e \approx 2\left(1 - {}^{1}\!/\!\sqrt{M}\right)Q\left[\sqrt{\frac{3 \cdot SNR_{LE,U}}{M - 1}}\right].$$

Other formulae for other constellations with an MMSE-LE can similarly be used by simply substituting $SNR_{LE,U}$ for SNR in the usual AWGN channel expression.

2.4.2 Decision Feedback Equalization

Decision feedback equalization (DFE) avoids noise enhancement by assuming past decisions of the symbol-by-symbol detector are always correct and using these past decisions to cancel ISI. The DFE appears in Figure 2.13.

The linear filter characteristic changes so that the combination of the matched filter and linear **feedforward filter** adjust the phase of the intersymbol interference so that all ISI appears to have been caused by previously transmitted

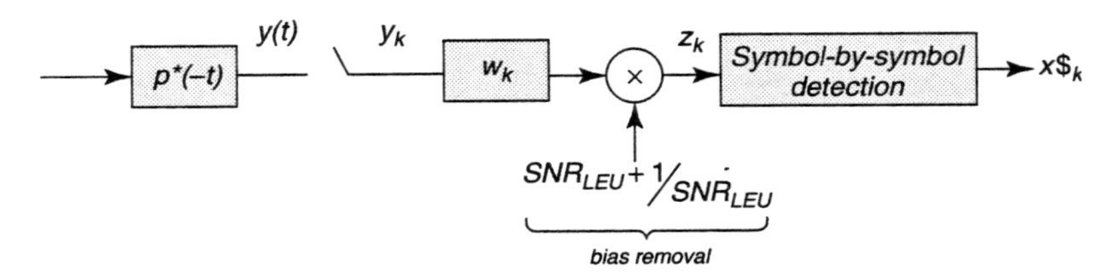

Figure 2.12 MMSE linear equalizer.

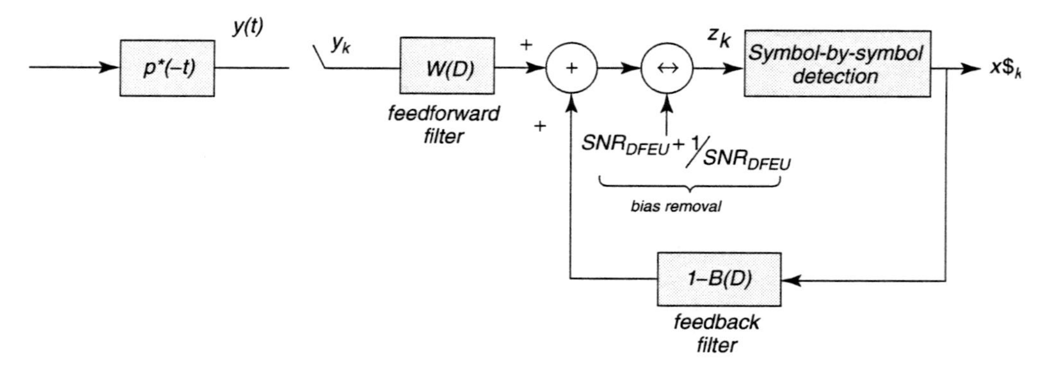

Figure 2.13 Decision-feedback equalizaton.

symbols. Because these previous symbols are available in the receiver, the **feedback filter** estimates the ISI, which is then subtracted from the feedforward filter output in the DFE. The DFE always performs at least as well as the LE and often performs considerably better on severe-ISI channels where the noise enhancement of the LE is unacceptable. Clearly, the LE is a special case of the DFE when there is no feedback section.

The best settings of the DFE filters can be obtained through spectral ("canonical") factorization of the channel autocorrelation function,

$$R(D) + 1/SNR = \gamma \cdot G(D) \cdot G^*(D^{-*})$$

where $\gamma > 0$ is a positive constant representing an inherent gain in the channel and $G(D) = 1 + g_1 \cdot D + g_2 \cdot D^2 + g_3 \cdot D^3 = ...$ is a monic ($g_0 = 1$), causal (exists only for non-negative time instants), and minimum-phase (all poles/zeros outside the unit circle) polynomial in D.[5] Also, $G^*(D^{-*}) = 1 + g_1^* \cdot D^{-1} + g_2^* \cdot D^{-1} + ...$ is monic, anti-causal, and maximum-phase. The feedforward filter has setting

$$W(D) = \frac{1}{\gamma \cdot G^*(D^{-*})},$$

and the combination of the matched-filter and feedforward filter is known as a mean-square whitening matched filter (MS-WMF). The output of the MS-WMF is the sequence $X(D)G(D)+E(D)$, a causally filtered version of the channel input plus an error sequence $E(D)$ that has minimum mean square amplitude among all possible filter settings for the DFE. Because this sequence is causal, trailing intersymbol interference may be eliminated without noise enhancement by set-

[5]The factor $G(D)$ can be obtained by finding roots of magnitude greater than one, or can be directly computed for finite lengths in practice through a well-conditioned matrix operation known as Cholesky factorization.

ting $B(D) = G(D)$. The sequence $E(D)$ is "white," with variance $\sigma^2_{DFE} = \sigma^2/\gamma$. Thus the signal to noise ratio at the DFE detector is

$$SR_{DFEU} = \gamma \frac{E_x}{\sigma^2} - 1.$$

The SNR of a DFE is necessarily no smaller than that of a linear equalizer in that the linear equalizer is a trivial special case of the DFE where the feedback filter is $B(D) = 1$. The DFE MS-WMF is approximately an all-pass filter (exactly all-pass as the SNR becomes infinite), which means the white input noise is not amplified nor spectrally shaped—thus, there is no noise enhancement. As such, the DFE is better suited to channels with severe ISI, in particular those with bridge taps or nulls. Also, recalling that the channel pulse response may be that of an equivalent white noise channel, narrowband noise appears as a notch in the equivalent channel. Thus the DFE can also mitigate narrowband noise much better than a linear equalizer. The DFE is intermediate in performance to a linear equalizer and the optimum maximum-likelihood detector.

The removal of bias is also evident again for the MMSE-DFE in Figure 2.13. Error propagation is a severe problem in decision-feedback equalization and occurs when mistakes are made in previous decisions fed into the feedback section. Reference [1] investigates the problem and solutions.

In practice, the filters of either a LE or a DFE are implemented as finite-length sums-of-products (or FIR filters). Such implementations are better understood and less prone to numerical inaccuracies, while simultaneously lending themselves to easy implementation with adaptive algorithms. See [1] for a development of finite-length equalizers.

2.5 DISCRETE MULTICHANNEL TRANSMISSION (DMT)

The concept of multichannel transmission is one often used against a formidable adversary, following the old adage: "divide and conquer," where the adversary in this case is the difficult transmission characteristic of the DSL twisted pair. Multichannel methods transform the DSL transmission line into hundreds to thousands of tiny transmission lines, each of which is easy to transmit on. The overall data rate is the sum of the data rates over all the easy channels. The most common approach to the "division" is transmission in non-overlapping narrow bands. The "conquer" part is that a simple line code on each such channel achieves best performance without concern for the formidable difficulty of intersymbol interference, which only occurs when wide bandwidth signals are transmitted.

Multichannel line codes have the highest performance and are fundamentally optimum for a channel with intersymbol interference. A key feature of multichannel transmission for DSLs is the adaption of the input signal to the individual characteristics of a particular phone line. This allows considerable improvement in range and reliability, two aspects of an overall system design that can dominate overall system cost. Thus, multichannel line codes have become increasingly used and popular for DSLs.

Multichannel transmission methods achieve the highest levels of performance and are used in ADSL and VDSL. The equalizers of Section 2.4 only partially mitigate intersymbol interference and are used in suboptimum detection schemes. As the ISI becomes severe, the equalizer complexity rises rapidly and then performance loss usually widens with respect to theoretical optimums. The solution, as originally posed by Shannon in his famous mathematical theory of communication (see references in [1]) is to partition the transmission channel into a large number of narrowband AWGN subchannels. Usually, these channels correspond to contiguous disjoint frequency bands, and the transmission is called **multicarrier** or **multitone** transmission. If such multitone subchannels have sufficiently narrow bandwidth, then each has little or no ISI, and each independently approximates an AWGN. The need for complicated equalization is replaced by the simpler need to multiplex and demultiplex the incoming bit stream to/from the subchannels. Multicarrier transmission is now standardized and used because the generation of the subchannels can be easily achieved with digital signal processing. Equalization with a single wideband carrier may then be replaced by no or little equalization with a set of carriers or "multicarrier," following Shannon's optimum transmission suggestion, and can be implemented and understood more easily. The capacity of a set of such parallel independent channels is the sum of the individual capacities, making computation of theoretical maximum data rates or use of gaps for practical rates easy.

The basic concept is illustrated in Figure 2.14. There, two DSL transmission line characteristics are posed, each of which would have severe ISI if a single wideband signal were transmitted. Instead, by partitioning the transmit spectrum into narrow bands, then those subchannels that pass through the channel can be loaded with the information to be transmitted. Note the receiver has a matched filter to each transmit bandpass filter, thus constituting an easily implemented maximum likelihood receiver (without need for Viterbi sequence detection, even on a channel with severe spectral filtering). Better subchannels get more information, whereas poor subchannels get little or no information. If the subchannels are sufficiently narrow, then no equalizer need be used.

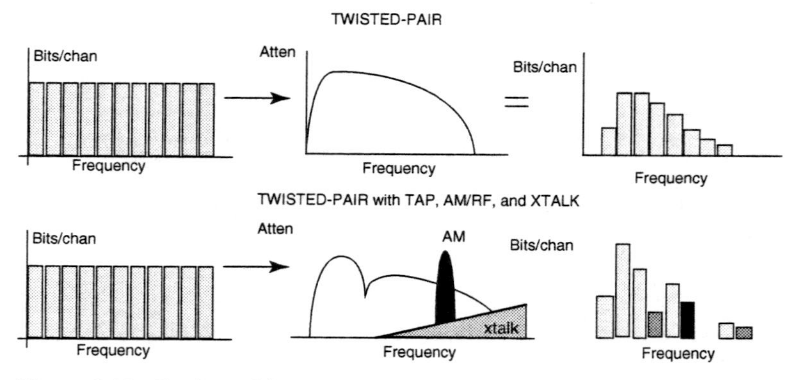

Figure 2.14 Basic multitone concept.

The set of signal-to-noise ratios that characterize each of the subchannels is important to performance calculation. It is assumed that there are N subchannels each carrying $\bar{b}_n = \frac{1}{2} log_2\left(1 + {SNR_n}/{\Gamma}\right)$ bits/dimension. The average number of bits is the sum of the numbers of bits carried on each channel divided by the number of dimensions (assumed here to be N) as

$$\bar{b} = 1/N \sum_{n=1}^{N} \frac{1}{2} log_2\left(1 + \frac{SNR_n}{\Gamma}\right) = \frac{1}{2} log_2\left[\prod_{n=1}^{N}\left(1 + \frac{SNR_n}{\Gamma}\right)\right]^{1/N}$$

$$= \frac{1}{2} log_2\left(1 + \frac{SNR_{geo}}{\Gamma}\right)$$

where SNR_{geo} is the **geometric signal-to-noise ratio,** or essentially the geometric mean of the $1 + SNR/\Gamma$ terms,

$$SNR_{geo} = \Gamma\left\{\left[\prod_{n=1}^{N}\left(1 + \frac{SNR_n}{\Gamma}\right)\right]^{1/N} - 1\right\}.$$

The entire set of parallel independent channels then behaves as one additive white Gaussian noise channel with SNR essentially equal to the geometric mean of the subchannel SNRs.[6] This geometric SNR can be compared against the unbiased SNR of equalized passband and baseband systems' SNR directly. This SNR can be improved considerably when the available energy is distributed nonuniformly over all or a subset of the parallel channels, allowing a higher performance in multichannel systems.[7] The process of optimizing the bit and energy distribution over a set of parallel channels is known as loading and is studied in the next subsection.

2.5.3 Loading Algorithms

The process of assigning information and energy to each of the subchannels is called **loading** in multichannel transmission. Reference [1] studies early loading algorithms in detail, and Chapter 3 introduces several improved methods for ADSL.

2.5.4 Channel Partitioning

Channel partitioning is the means of dividing the transmission channels into a set of parallel independent ISI-free subchannels. General channel partitioning is of strong interest, but as yet only DMT finds wide use. A general discussion of how to partition channels is found in Chapters 4 and 5 of [2].

---

[6]This geometric mean statement becomes very accurate for moderate to high SNR on all subchannels.

[7]Under certain severe restrictions, the DFE SNR can be made equal to the geometric SNR when MMSE-DFEs are used, the used subchannels are all next to each other in frequency (i.e., no gaps), and the same energy distribution is used by the DFE, see [12],[16].

2.5.5 Discrete Multitone Transmission (DMT)

In DMT, a packet of channel output samples, y, can be related to a packet of input samples, x, by $y = Hx + n$. n is a packet of noise samples, and H is a channel matrix. A guard period between blocks of N samples called DMT symbols contains a **cyclic prefix.** The samples in the prefix must repeat those at the end of the symbol, that is, $x_{-i} = x_{N-i}$ $i = 1, ..., \nu$. When a cyclic prefix is used, the matrix H becomes what is called a square "circulant" matrix (the last ν output samples of each transmitted packet are ideally ignored in DMT). Circulant matrices have the property they may be decomposed as

$$H = Q^* \Lambda Q,$$

where Q is a matrix corresponding to the discrete Fourier transform (DFT) and Λ is a diagonal matrix containing the N Fourier transform values for the sequence h_k that characterizes the channel. The klth element of Q starting from the bottom right at $k = 0$, $l = 0$ and counting up is

$$Q_{kl} = \frac{1}{\sqrt{N}} e^{-2\pi \frac{kl}{N}}.$$

The DFT is a heavily used and well-understood operation in digital signal processing and a variety of structures exist for its very efficient implementation in $N \log_2 (N)$ operations rather than the usual N^2 for most matrix multiplication. Thus, DMT is a very efficient multichannel partitioning method. Even when H is real, the matrices in the DMT decomposition are complex.

The input/output relation for the channel is $\boldsymbol{Y} = H\boldsymbol{x} + \boldsymbol{n}$. The $N \times N$ circulant matrix H is written for DMT as:

$$H = \begin{bmatrix} h_0 & h_1 & \cdots & h_v & 0 \\ 0 & h_0 & h_1 & \cdots & h_v \\ h_v & 0 & h_0 & \cdots & h_{v-1} \\ \vdots & \ddots & \ddots & \ddots & \ddots \\ 0 & 0 & h_v & \cdots & h_0 \end{bmatrix}$$

where the reader can verify implements circular convolution on the packet of inputs. The transmit symbol vectors are created by $\boldsymbol{x} = Q^*\boldsymbol{X}$, where $\boldsymbol{X}$ is a frequency-domain vector of channel inputs. Each element of $\boldsymbol{X}$ is complex and can be thought of as a QAM signal. When the channel is real and thus the symbol $\boldsymbol{x}$ must also be real, the frequency-domain input must have conjugate symmetry, which means that

$$X_n = X^*_{N-n},$$

meaning there are $N/2$ complex subchannels, not N. The energy of the cyclic prefix guard period is clearly wasted, thus effectively reducing the amount of power available for transmission by $N/(N + \nu)$ on the average. This power penalty is in addition to an excess bandwidth penalty of the cyclic prefix, which is an additional factor of S/N, but made for the extremely efficient implementation of DMT. If the guard-period length ν can be made small with respect to the packet length N, then this penalty is small. The receiver generates the outputs of the set of parallel channels by forming:

$$\boldsymbol{Y} = Q\boldsymbol{y} = Q(H\boldsymbol{x} + \boldsymbol{n}) = QHQ^{*}\boldsymbol{X} + \boldsymbol{N} = \Lambda\boldsymbol{X} + \boldsymbol{N},$$

again a set of parallel independent channels when the input noise is white. When the input noise is not white, the output noise vector tends to white as long as the block size is long enough, so noise prewhitening need not be used in reasonable DMT implementations.

As block length goes to infinity, DMT becomes multitone and optimum. DMT for ADSL is described in more detail in Chapter 3. For more generally on the DMT system, see [1].

2.6 IMPAIRMENT MODELING

General impairment modeling is treated in great depth in [1]. Chapters 3 and 7 also study some specifics pertinent to ADSL and VDSL respectively.

2.6.1 Background Noise

Background noise in DSL systems today is often presumed to be –140 dBm/Hz. There is no justification for this level other than early DSL designers agreed that this would represent a noise floor caused by a reasonable ADCs quantization level in 1990.[8] The number stuck. Today, noise floors can be considerably lower as conversion technology has advanced. DSL modems with lower quantization noise will often perform better. The actual thermal limit of line noise at room temperature is –173 dBm/Hz. Usually line noise is slightly higher than this limit and between –160 dBm/Hz and –170 dBm/Hz.

2.6.2 Other Noises

The most common and largest noise in DSL is crosstalk, which is studied in detail in [1]. Crosstalk noise in DSLs arises because the individual wires in a cable of twisted pairs radiate electromagnetically. The electric and magnetic fields thus created induce currents in other neighboring twisted pairs, leading to an undesired crosstalk signal on those other pairs. Figure 2.15 illustrates two types of crosstalk commonly encountered in DSLs. **Near-end crosstalk (NEXT)** is the type of crosstalk that occurs from signals traveling in opposite directions on two twisted

[8]Some claim that exhaustive studies were performed one place or another to get this number, but that is simply not correct. It occurred in a telephone conversation about what was reasonable as no one had studied it in depth just days before the release of a Bellcore Technical Advisory on DSLs in 1990. Subsequent studies sometimes reinforce the number and sometimes show it is too high or too low

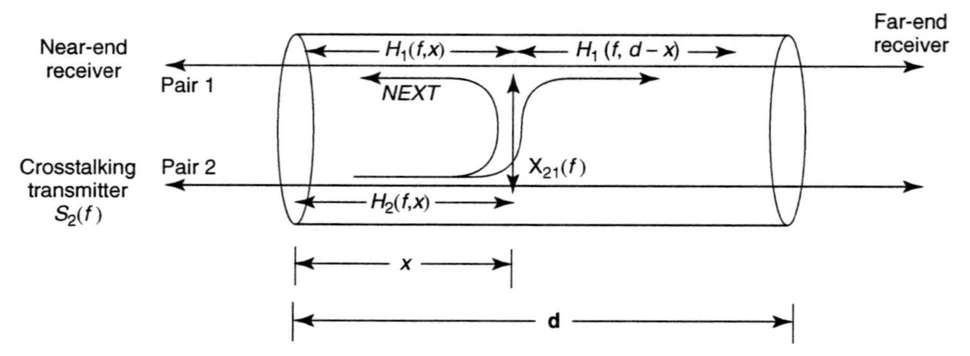

Figure 2.15 Illustration of crosstalk.

pairs (or from a transmitter into a "near-end" receiver). **Far-end crosstalk (FEXT)** occurs from signals traveling in the same direction on two-twisted pairs (or from a transmitter into a "far-end" receiver).

Crosstalk can be the largest noise impairment in a twisted pair and often substantially reduces DSL performance when it cannot be eliminated.

NEXT Modeling

DSL standards and theoretical studies have modeled crosstalk in a 50-pair binder with the coupling function

$$S_n(f) = k_{next} \cdot f^{1.5} \cdot S_{transmit}(f)$$

where K_{next} has been determined by ANSI studies to be

$$k_{next} = 10^{-13} \cdot \left({}^{N}\!/_{49} \right)^{0.6}$$

and N is the number of pairs in the binder expected to be carrying similar DSL service. This value is believed to be worse than 99% of the twisted-pair situations.

Thus, for example, to find the crosstalk noise from an ISDN circuit into another twisted pair for a binder containing 24 ISDN circuits, the power spectral density on any line in the binder is modeled by

$$S_n(f) = \left({}^{24}\!/_{49} \right)^{0.6} \cdot 10^{-13} \cdot f^{1.5} \cdot S_{ISDN}(f).$$

For more on NEXT models, see [1].

FEXT Modeling

FEXT modeling parallels NEXT modeling with formula

$$S_f(f) = k_{fext} \cdot f^2 \cdot d \cdot |H(f,d)|^2 \cdot S_{transmit}(f),$$

where d is the length in feet, $|H(f,d)|^2$ is the transfer function from the line input (insertion loss) for the length of transmission line being investigated, $S_{transmit}(f)$ is again the power spectral density input to the line (and not at the source), and finally

$$k^{fext} = \left(N/_{49}\right)^{0.6} \cdot 9 \times 10^{-20}.$$

FEXT is important in ADSL and VDSL and is considered more in Chapters 3 and 7.

2.6.3 Radio Noise

Radio noise is the remnant of wireless transmission signals on phone lines, particularly AM radio broadcasts and amateur (HAM) operator transmissions.

Radio-frequency signals impinge on twisted-pair phone lines, especially aerial lines. Phone lines, being pieces of copper, make relatively good antennae with electromagnetic waves incident on them leading to an induced charge flux with respect to earth ground. The common-mode voltage for a twisted pair is for either of the two wires with respect to ground—usually these two voltages are about the same because of the similarity of the two wires in a twisted pair. Well-balanced phone lines thus should see a significant reduction in differential RF signals on the pair with respect to common-mode signals. However, balance decreases with increasing frequency and so at frequencies of DSLs from 560 kHz to 30 MHz, DSL systems can overlap radio bands and will receive some level of RF noise along with the differential DSL signals on the same phone lines. This type of DSL noise is known as **RF ingress.** Chapter 3 and [1] both have more on RF ingress.

Amateur Radio Interference Ingress

Amateur radio transmissions occur in the bands in Table 2.1. These bands overlap the transmission band of VDSL, but avert the lower transmission bands of other DSLs. Thus, HAM radio interference is largely a problem only for VDSL.

Table 2.1 Amateur Radio Bands

HAM operator bands (MHz)	
Lowest Freq.	***Highest Freq.***
1.81	2.0
3.5	4.0
7.0	7.1
10.1	10.15
14.0	14.35
18.068	18.168
21	21.45
24.89	24.99
28.0	29.7

A HAM operator may use as much as 1.5 KW of power, although such large use of power is rare and may not be in residential neighborhoods or areas with many phone lines. A 400W transmitter at a distance of 10 meters (about 30 feet) leads to an induced common-mode (longitudinal) voltage of approximately 11 Volts on a telephone line. With balance of 33 dB, the corresponding differential (metallic) voltage is about 300 mV, which is 0 dBm of power on a $Z_0 = 100\ \Omega$ transmission line. HAM operators use a frequency band of 2.5 kHz intermittently with either audio (voice) or digital (Morse code, FSK) signals, leading to a noise PSD of approximately –34 dBm/Hz. More typically, HAM operators transmit at lower levels or may be spaced more than 10 meters when transmitting at higher levels, so that more typical HAM RF ingress levels may be 20–25 dB less powerful. Nonetheless, this still leads to PSDs for noise in the range from –35 dBm/Hz to –60 dBm/Hz, well above the levels of crosstalk in this section. Furthermore, such high voltage levels may saturate analog front-end electronics.

HAM operators tend to switch carrier frequency every few minutes and the transmitted signal is zero (SSB modulation) when there is no signal. Thus, a receiver may not be able to predict the presence of HAM ingress.

Fortunately, HAM radio signals are narrowband and so transmission methods attempt to notch the relatively few and narrow bands occupied by this noise, essentially avoiding the noise (for some transmission methods, see Chapter 8) rather than try to transmit through it. Some degree of receiver notch filtering is also necessary to eliminate the effect completely (see Chapter 8).

Emissions of signals in radio bands is also of importance. Some studies have found annihilation of radio receivers by VDSL at distances as far as 30 meters from a phone line. This problem is resolved by zeroing tones in the known amateur radio bands in DMT. QAM has no solution to the problem other than to use multiple QAM bands in between the known reserved radio bands. (Notch filters do not solve the problem in QAM, even if decision-feedback equalization is used.)

AM ingress is investigated in Chapter 3 for ADSL.

2.6.4 Combining Different Crosstalk Signals and Noise

Recently, the modeling of several types of different crosstalk signals in a single signal has been investigated by a group of phone companies known as the "full-service access network" group, and has been standardized in [10]. That model is now used in many standards and allows combination of n_i crosstalkers of type i by raising each term to the power 1/0.6 before summing. Then, after the summation, the resultant expression is raised to the power 0.6. This can be expressed as:

$$Xtalk\left(f, n = \sum_{i=1}^{N} n_i\right) = \left(\sum_{i=1}^{N} Xtalk(f, n_i)^{1/0.6}\right)^{0.6},$$

where X_{talk} is either NEXT or FEXT, n is the total number of crosstalk disturbers, N is the number of types of unlike disturbers, and n_j is the number of each type of disturber. Example uses of this equation are given in the following subsections.

Take the case of two sources of NEXT at a given receiver. The NEXT coupling length is the length of line segment where crosstalkers are in the same binder as the affected DSL. In this case there are n_1 disturber systems of spectrum S_1 (f) and NEXT coupling length L_1 and n_2 disturber systems of spectrum S_2 (f) and NEXT coupling length L_2. The combined NEXT is expressed as:

$$NEXT[f,n] = \left(\left(s_1(f)X_N f^{3/2} n_1^{0.6} \cdot \left(1 - |H_1(f, L_1)|^4\right)\right)^{1/0.6} + \left(S_2(f)X_N f^{3/2} n_2{}^{0.6} \cdot \left(1 - |H_2(f, L_2)|^4\right)\right)^{1/0.6}$$

The dependency of NEXT on $H_1(f, L_1)$ and $H_2(f, L_2)$ is a slight refinement to the model in Section 2.6.2. The $|1-H|^4$ factor is usually close to unity, making the two formulae nearly identical in most situations.

For three sources of FEXT at a given receiver, there are n_1 disturber systems of spectrum S_1 (f) at range l_1, a further n_2 disturber systems of spectrum S_2 (f) at range l_2 and yet another n_3 disturber systems of spectrum S_3 (f) at range l_3. The expected crosstalk is built in exactly the same way as before, taking the base model for each source, raising it to power 1/0.6, adding these expressions, and raising the sum to power 0.6:

$$FEXT[f,n,l] = \begin{pmatrix} (S_1(f)|H_1(f)|^2 X_F f^2 l_1 n_1{}^{0.6})^{1/0.6} \\ + (S_2(f)|H_2(f)|^2 X_F f^2 l_2 n_2{}^{0.6})^{1/0.6} \\ + (S_3(f)|H_3(f)|^2 X_F f^2 l_3 n_3{}^{0.6})^{1/0.6} \end{pmatrix}$$

2.6.5 Total Noise Power Spectral Density

The NEXT term and the FEXT term are computed to arrive at separate NEXT and FEXT disturbance power spectra. These power spectra should then be summed with the background noise and radio noise to determine the total disturbance seen by the victim receiver. The total noise seen by the victim receiver is given by:

$$N(f) = NEXT[f,n] + FEXT[f,n,l] + 10^{-14} + S_{radio}(f)\ mW/Hz.$$

REFERENCES

These references are provided for further (and more detailed) reading on the data transmission and data communications principles.

[1] T. Starr, J. Cioffi, and P. Silverman, *Understanding Digital Subscriber Line Technology.* Upper Saddle River, NJ: Prentice Hall, 1999.

[2] J. Cioffi, *Digital Transmission Theory*. Available at: *http://www.stanford.edu/class/ee379a/* and *http://www.stanford.edu/class/ee379c/* 2002.
[3] E. Lee and D. Messerschmitt, *Digital Communications,* Boston: Kluwer, 1993.
[4] J. Proakis, *Digital Communications* (4th ed.). New York: McGraw-Hill, 2000.
[5] R. Gitlin, S. Weinstein, & J. Hayes, *Data Communications Principles (Applications of Communications Theory).* New York: Plenum, 1992.
[6] C. Schlegal, *Trellis Code.* New York: IEEE Press, 1997.
[7] S. Wicker, *Error Control Codes.* Upper Saddle River, NJ: Prentice Hall, 1995.
[8] M. Sorbara, J. J. Werner, N. A. Zervos, "Carrierless AM/PM: Tutorial," T1E1.4/90-154, September, 1990.
[9] J. J. Werner, "Tutorial on Carrierless AM/PM—Performance of Bandwidth-Efficient Line Codes," T1E1.4/93-058A, March 1993.
[10] ANSI Spectrum Management Standard, T1.417-2002.

CHAPTER 3

ADSL

Asymmetric digital subscriber lines (ADSL) service is by far the most popular and deployed DSL service with nearly 30 million subscribers active worldwide by the end of 2002 and the number keeps growing. ADSL is *asymmetric,* with a downstream-to-upstream data-rate ratio of about 4:1 in early deployments (500 kbps downstream and 128 kbps upstream) at the longest ranges. As the ADSL connection speed increases to a maximum of 8 Mbps downstream, the data-rate asymmetry ratio may increase to as much as 10:1. Higher speed ADSL connections (> 1.5 Mbps), while feasible and implemented in available systems and standards, are used less frequently than lower speed connections for two basic reasons: (1) lower speed can be achieved on a greater percentage of loops, and phone companies thus transmit intentionally at lower speeds to try to ensure successful connection even on the longest of loops, and (2) Internet connections/gateways to DSL often cannot presently accommodate higher speeds per user. However, some cities in Asia (e.g., Tokyo) have mainly short lines; 8 Mb/s ASDL service is the norm in these cities.

As with any relatively new service, a number of improvements and issues have become known. ADSL was first standardized in 1995 by the ANSI T1E1.4 group. The early ADSL standard had its own internal protocol (ATM not being in use at the time and Ethernet not having the required functionality at the time) that allowed multi-line voice, data, and video traffic. This early standard was used largely in trials that evaluated all these data types, and many issues and improvements were learned. The ADSL Forum (now DSL Forum) was formed in 1995 to focus on the data/fast-Internet application of ADSL and successfully ignited worldwide interest in ADSL for this application. A second issue of the American standard appeared in 1998 [2] and the ITU's international versions G.992.1 and G.992.2 appeared a year later [3] in 1999. Following G.992.1 and G.991.2 in 1999, the ITU developed a second generation set of ADSL Recommendations in 2002 that are known as G.992.3 (previously G.dmt.bis) and G.992.4 (previously G.lite.bis). The new aspects found in the second generation ADSL Recommendations are described in this chapter. ADSL's internal protocol was reduced in later standards, and external channelization of the services began when widespread deployment commenced. The basic DMT engine, first tested in 1993, remains essentially unchanged in all the standards and is today implemented very cost effectively in silicon that is interoperable and widely available from multiple suppliers. The DSL Forum now has also successfully promoted the multi-line voice application through both IP and circuit-switched ATM protocols. Video has yet to become widely deployed with ADSL in 2002.

This chapter is not meant to be a comprehensive review of ADSL. The authors have previously published considerable material in a predecessor book [1] on DSL. This chapter is intended to update that earlier material and to focus on some of the enhancements and issues. Section 3.1 begins with some basic trends in the performance improvement of ADSL, and Section 3.2 discusses the potential use of more advanced coding and framing methods in conjunction with the DMT physical layer. Channel time variation and the robustness of ADSL are addressed in Section 3.3, whereas the important area of mutual compatibility with AM radio appears in Section 3.4. Sections 3.5 to 3.9 complete this chapter with a number of topics/issues that enhance deployment and applications.

3.1 BASIC PERFORMANCE ENHANCEMENT

A key issue that has limited deployment speeds, especially in the United States where labor is very expensive, is the need in the earliest ADSL systems to send trained field personnel to the customer premises to eliminate a problem that prevents or degrades service. Problems with service may often not be the physical layer and can be related to computer software incompatibilities, external-to-ADSL equipment-interface problems, and so forth. However, as those nonphysical-layer problems have diminished, the issue of unusual noise disturbances or line effects that limit the range become more important. Replacement of the line, or "pair-selection" as it is often called, requires labor and thus increases the cost of ADSL deployment. Thus, most improvement in ADSL has focused simply on the robust delivery of service, that is, to eliminate the need for service technicians to go to the customer. More than one visit to a customer can cause the service to be financially unviable at monthly prices that ADSL consumers can be expected to pay. Splitterless ADSL [1] also attempts to eliminate even the initial visit and is described by customers/providers as a "self-install."

Perhaps the most important characteristic of ADSL service is reliability. Customers want DSL service to work well when they need it. Typically, ADSL service is characterized as being "always on" in that the modem physical layer is continuously energized as long as the equipment itself is powered. Thus, no initialization-period time delay annoys the customer. Telephone companies want to achieve this "always-on" connectivity reliably without need for service visits to the customer.

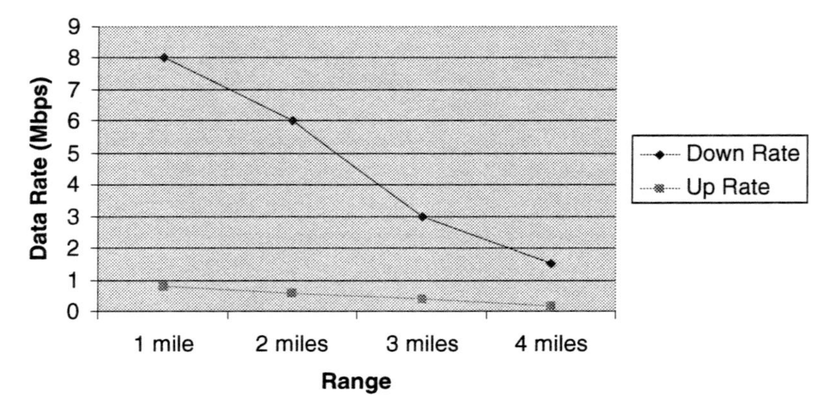

Figure 3.1 ADSL data rates.

To achieve this reliable continuous service, the data rate of ADSL service may be sacrificed.

The trade-off then for a given level of good reliability becomes speed of service versus range from a telephone company central office. Figure 3.1 illustrates the classic trade-off between these two parameters. Within the parameter of speed, the level of asymmetry of service becomes important also. Different levels of asymmetry correspond to different ranges for a given downstream data rate. Thus, the ADSL service provider has a difficult compromise to evaluate in attempting to achieve a highly reliable level of speed/service to all customers. Today, the practice is to design for absolute worst-case situations and to guarantee service with little need for visits to the customer. Chapter 11 suggests that such worst-case design is overwhelmingly pessimistic and could be improved to more adaptive automated service provisioning in the future, while still further reducing the frequency of customer service visits.

3.1.1 Increasing Range

Figure 3.2 illustrates the basic range problem simply.[1] For a given radius around a telephone central office in Figure 3.2, the number of customers served is proportional roughly to the area of the circle. Although it is true that most COs are not at the center of a circle, the basic conclusions inferred from Figure 3.2 will not change substantially. The radius of the circle can be thought of as the length of the longest ADSL loop. As the radius of coverage increases, the ADSL speed decreases because the ADSL signals are attenuated by a longer length of copper wire and thus are more harmed by various noises. One of the noises for an ADSL transmission line is the ADSL upstream crosstalk of other users (or similarly from upstream signals of other symmetric DSLs deployed within the circle). The greater the upstream data rate, the more noise it creates, thus "robbing" data rate from the downstream direction of ADSL. Typically an asymmetry ratio of 8:1 or more dou-

[1]Reference [1] has a more complete discussion of the topology of typical DSL loop plants, and here we choose instead to illustrate the basic problem conceptually.

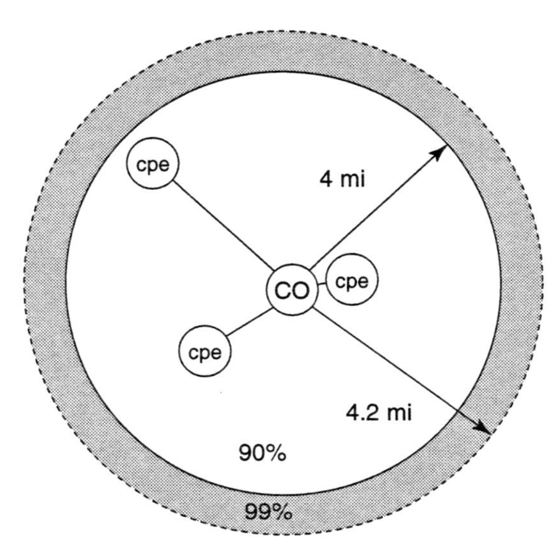

Figure 3.2 Illustration of area around central office (CO) with regard to line length.

bles the radius that can be achieved (thereby quadrupling the number of customers served without need for visits to those customers) compared with symmetric transmission. However, the actual speed is also important.

As an example, most (over 90%) U.S. phone lines are believed to conform to what are called RRD rules [1], which basically means they are less than 4 miles in length.[2] Thus, an ADSL system planner might state that at Area = $\pi(4 \text{ mi})^2$, 90 percent of the customers are served and the other 10 percent will need extra customer service visits to operate properly. Thus, $k\pi(4 \text{ mi})^2 = .90$, meaning at this 4-mile radius, the area of the circle covers 90 percent of the customers (k is a proportionality constant, $k = \frac{.9}{16\pi^2}$.) Thus, of 1,000,000 customers, then 900,000 would have their premises located within a set of circles that have radius 4 miles from the various central offices servicing all the 1,000,000 customers. Thus also, the remaining 100,000 would be outside the circle and require service visits in this example. If these 1,000,000 customers are to be served at 1.5 Mbps/160 kbps and each customer beyond the 4-mile range requires on the average $1,000 to correct service, then it costs $100M extra in labor to visit/service those million customers, who at $30/month will generate a cumulative revenue of a little over a billion dollars in 3 years. The cost of equipment for DSL service is otherwise about $200/line, so that total equipment cost is $200M. One can see that the extra $100M in labor is significant. If the number of "problem" customer loops is to be reduced from 10 percent in this example to 1 percent, then only 10,000 customers need service visits and this costs $10M. Thus $90M is saved if the radius of no-problem service increases to just

[2]RRD (revised resistance design) rules basically limit loops to 17 kft or less.

4.2 miles ($k\pi(4.2 \text{ mi})^2 = .99$). This extra .2 miles increases line attenuation by 6 dB, or effectively reduces ADSL data rate by 1 Mbps. Obviously, this is a crucial trade-off between revenue and service speed. Today's customers are unlikely to pay the extra $90M for a higher data rate. Thus, service providers are ultraconservative on the data rates that they attempt with ADSL to try to avoid the very costly service visits that dominate the deployment costs of ADSL. Clearly a better ADSL modem that could reliably cover 4.2 miles (or in general increase the range just a little) can offer a dramatic savings to service providers even if it costs slightly more. The cheapest modem is not always the one that makes ADSL service the most economically attractive.

ADSL systems can achieve 1.5 Mbps down and 160 kbps up at 4-mile range in typical situations with splitters and about 500 kbps down and 128 kbps up in worst-case situations without splitters at 4.2 miles. Splitterless operation avoids an initial customer visit for the phone company, but reduces the achievable data rate and reliability, perhaps causing an alternative need for the visit because of service reliability. Thus, the phone company is faced with a classic trade-off. The service provider's choice to date has been to lower the data rate and to use splitterless service to try to ensure a minimum of required service visits, allowing the lowest possible (profitable) per-monthly charges per customer for ADSL service. Although the numbers change in each situation, the reader can appreciate the degree of the problem: increasing 4-mile range just a little bit has a big effect on ADSL service profit. Thus, methods to ensure reliable transmission of higher data rates at the longest ranges have been an initial focus of ADSL system vendors. Some of these methods are described later in this chapter.

3.1.2 Increasing Speed

The history of data communications and the computer industry has been a steady progression to higher speeds and shows no signs of abatement. Thus, ADSL customers will eventually demand higher speeds. Although 500 kbps service may be initially excellent as a 10× increase over 56 kbps modems, 500 kbps will not remain sufficient in the future. Phone companies anticipate this demand and have invested in programs to shorten phone-line lengths. In the early 1980s, the old Bell System introduced what are called CSA (carrier serving area) design rules, which basically limited new installed phone lines to be less than 2 miles in length. Thirty years of progress with CSA installations has left about 70 percent of the American network within 2 miles of a telephone-company central office or "remote terminal."[3] Southwest Bell Corporation's (SBC) "project Pronto" is probably the best known example of a program designed to bring nearly all telephone lines within 12 kft to increase DSL speeds. This program promises to spend $6 billion over a short period of time (few years) to bring the last approximately 10 million SBC customers (of about 45 million total) within 2 miles of a CO.

---

[3]An RT (remote terminal) is a small environmentally controlled enclosure in which telephone company equipment is placed and connected to the CO by either fiber or some other transmission mechanism of wide bandwidth.

A 2-mile radius considerably increases ADSL speeds. In most situations, 6 Mbps down and 600 kbps up is possible. At the very least, with splitterless operation, CSA range means reliable delivery of 1.5 Mbps down and 384 kbps up in telephone-company conservative terms. More generally, yet shorter loops mean even greater speeds, such as discussed in Chapter 7 on VDSL. At least one VDSL draft standard is interoperable with ADSL, so that VDSL offers a mechanism for speeds to increase as phone companies see the opportunity to pay for more fiber and shorter loops. Fiber installation on average is very expensive, and thus its costs need to be shared over a number of subscribers. As line lengths get shorter (i.e., fiber gets longer), fewer customers share the fiber, and so the fiber cost case becomes more difficult. Thus, methods for squeezing more bandwidth on shorter lines are and will continue to be of great interest as customers grow to know and love their DSL service, and thereby start demanding greater bandwidth. It is however, unlikely in the foreseeable future that fiber will be economically deployed directly to the customer on any wide scale.

3.1.3 Improving Reliability

Both at the longest ranges and also as speeds increase on shorter loops, reliable service is still most important for telephone companies. Customers expect the service to be available when they need it. Given splitterless installations will increase, which means that ADSL (and any following VDSL in the future) will need to work well in a plethora of different situations so that issues with system performance will eventually return to the reliability of the physical layer transmission line. The remainder of this chapter addresses a variety of effects/issues in maintaining or improving good reliability. Among these are coding methods intended to improve resistance to nonstationary impulse effects and other time-varying line disturbances. Also discussed are analog effects and radio interference issues.

3.2 CODING

Codes can be concatenated with the basic DMT system in ADSL to improve performance, nominally by as much as 6.5 to 7.5 dB with random background crosstalk and line noise, and by a larger amount with impulsive disturbances. At the time of first ADSL standardization, a concatenated coding system (trellis code with outer interleaved Reed Solomon) was mandated. That system has considerable flexibility and thus, while implemented without full ability to understand codes of the future, fortunately can be used to provide more transmission robustness than originally anticipated. The designer needs to know only how to best decode that earliest specified system. Present-day knowledge allows high coding gain of 6–8 dB with a more complicated decoder than originally planned. However, the gain versus complexity improvement of a better decoder may not merit the effort. Even the most modern of codes and decoder methods can be implemented as a fraction of the digital-signal processing on ADSL digital chips that today cost less than $5/line on average to make.

An aspect of coding not addressed in previous work [1] is the proper loading of the ADSL DMT modem and the proper swapping mechanisms when codes are used. This is addressed in Section 3.3.

Also of importance in coded DSL is calculation of performance somewhat more precisely than simple approximations, which can be inaccurate, as well as improvement of this performance level in the presence of AWGN. Section 3.2.1 addresses performance calculation and advanced codes. Also of importance is impulse noise. Section 3.2.2 adds to previous models and understandings of impulse noise in [1] and also provides a method for analysis of any impulse in terms of performance loss. Section 3.2.3 discusses interleaving and its use.

3.2.1 High-Level Analysis and Advanced Codes (and Pointers for Decoding)

Trellis codes and Reed-Solomon (RS) codes for ADSL were detailed in [1]. Section 3.3 details exact coding gains of the RS code, while the trellis code adds approximately 1.5 dB to the gain of the RS code, which makes the total coding gain about 5 to 5.5 dB for white Gaussian noise with hard decoding of the RS code and Viterbi decoding of the trellis code. Soft or iterative decoding of the RS and trellis together could lead to 7–8 dB of coding gain. Impulse noise is discussed in the next section. Since early standardization, more easily implemented powerful coding systems have come into greater acceptance and understanding. These *turbo* and *low-density parity check* (*LDPC*) codes have longer block lengths (or more exactly greater memory span) than the trellis code, and have been proposed as optional replacements for the earlier trellis codes.[4] The advanced codes follow the basic outline of information theory that suggests more random structure to the encoder is usually good if the block length is sufficiently long. Turbo codes generally achieve the randomness through interleaving two simpler convolutional (or in effect in ADSL, trellis) codes. LDPC codes are based upon random construction of a parity matrix. The main improvement over trellis codes is achieved by implementation of the decoder through suboptimal iterative calculation/approximation of the log-likelihood of the probability distribution for each bit transmitted. Such algorithms almost always perform very close to optimum with a complexity far less than traditional exhaustively searching detectors or soft decoding of the RS code, and are often easier to understand than optimum decoders, but do require numbers of decoder operations well in excess of the trellis Viterbi decoder alone. The February 2001 issue of the *IEEE Transactions on Information Theory* [19] and the May 2001 issue of *IEEE JSAC* [20] are dedicated to these newer codes and contain much information. This chapter outlines performance gains of two specific proposals for ADSL by AT&T [4] and IBM [5], turbo and LDPC, respectively. The reader is also referred to the Web site [21] for worked examples of encoders and decoders based on these two codes in a forthcoming academic textbook.

Turbo and LDPC codes are binary and thus do not uniformly map bits to an ADSL DMT symbol, which uses multilevel constellations on each tone that may vary the number of bits/tone from 0 to 15. Lauer [22] found the nonuniformity to

[4]While still maintaining hard decoding of the RS code.

be of little consequence and that straightforward mapping of the encoder output bits through the constellation mapper for the DMT modulator tones with proper construction of the a priori likelihoods from the corresponding receiver FFT outputs allows achievement of the nominal gain of the code. For construction of the likelihoods necessary to initiate iterative decoding for such systems, see the examples at the Web site [21]. Such normal gain is typically about 7 dB, but is a little larger on tones with smaller numbers of bits and a little smaller on tones with a larger number of bits. However, such an approach has three drawbacks for ADSL:

1. The decoder has higher complexity because all bits are encoded.
2. The number of parity bits punctured to achieve a given rate changes the Coding gain and creates a problem similar to that in Section 3.3 for DMT loading with RS, creating a huge search space for best concatenated advanced-code and RS.
3. The interleaver depth for the advanced code is unnecessarily long, increasing delay and complexity.

Several researchers at AT&T [4] studied these issues carefully and proposed a solution that involves use of a binary rate-one-third turbo code, use of a special interleaver designed to match the ADSL frame structure and to eliminate turbo-code error floor effects, and variable puncturing of parity bits according to the number of bits per symbol to handle any constellation size. The code can thus be characterized as two-dimensional in the same sense as the existing ADSL trellis code is four-dimensional. The code has a gain at 2 bits per symbol of nearly 9 dB, which drops to about 6.2 dB for the largest constellations, making the gain about 7 dB overall. The code is based on two interleaved uses of an eight-state convolutional code of rate one-half with generator $G(D) = \left[1 \frac{1 + D + D^2 + D^3}{1 + D^2 + D^3} \right]$.

Several IBM researchers [5] have instead proposed a class of LDPC codes based on array codes that only encode the lowest 1 or 2 bits per dimension, thus simplifying the decoder. The output bits of the coder selects cosets in the same way that trellis code encoders do (see [1]), that is, in one dimension with exclusively square constellations. Special cases for small constellations are enumerated in [5]. The intrapartition distance is thus 12 to 18 dB, well above that of the code. The structure thus allows a lower complexity decoder and the same gain against AWGN, which is again 6–7 dB for the LDPC code. Impulses that exceed the intrapartition distance will cause errors and thus force the use of external FEC, but that is nominally implemented in any case.

The advanced encoding architecture for any of the cases is illustrated in Figure 3.3. The dual latency structure of ADSL is preserved with only short-block FEC applied in the "fast" path and both advanced coding (LDPC or turbo) and interleaved FEC applied in the "interleave" (or "slow") path. Although no advanced codes have yet been agreed for standardization, this area is likely to be one of advance in the G.dmt.ter standard of the future in the ITU. Decoding is more computationally complex than hard RS decoding combined with Viterbi-trellis-code

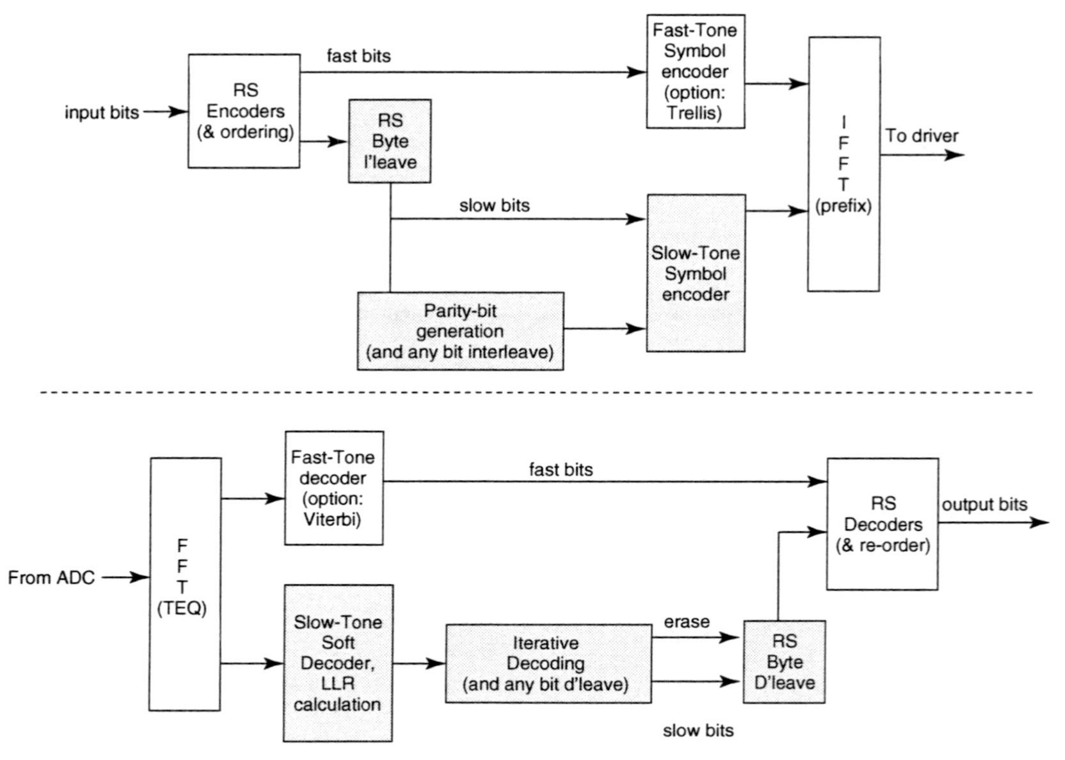

Figure 3.3 Advanced coding architecture.

decoding, and the reader is referred to [21] and [23] for detailed information on decoding and exact analysis.

3.2.2 Impulse Noise Characterization

Impulse noise has become one of the dominant issues in ADSL deployment, affecting range and reliability for all data rates. With the increase in ADSL deployments, several phone companies have studied the modeling of impulses and have derived models to help in qualification of service. These models are also useful for testing and design of better ADSL modems. This section summarizes models and improves upon the art in terms of providing models that could be used in testing as well as in design of better ADSL modems.

Figure 3.4 illustrates an impulse generation mechanism. There are three key elements in the generator:

1. Impulse waveform storage
2. Interimpulse-delay generation
3. Impulse amplitude scaling

The waveform storage may have measured impulses or mathematically modeled impulses stored, perhaps several of them. (Alternatively, the waveforms may

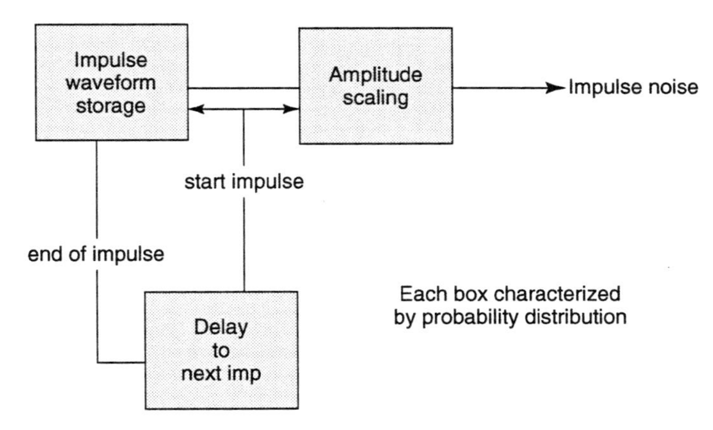

Figure 3.4 Impulse generation mechanism.

be statistically generated.) Each stored waveform may have a certain probability of being selected when the generator extracts the next impulse from storage. The time of extraction is measured from the termination of the last impulse generated and is determined by the interimpulse-delay generator or "delay to next impulse" in Figure 3.4. The delay generator is reset each time the last impulse is finished transmitting. The impulse waveform is scaled by the amplitude generator. The amplitude is selected randomly from a distribution of amplitude or possibly fixed at some constant, depending on the specifics of the impulses that the telephone company desires to test (or that the ADSL designer desires to analyze). Sometimes a random amplitude is selected for each stored sample of the impulse in addition to a gross gain applied to the entire impulse.

Impulse Amplitude Distribution

The amplitude box in Figure 3.4 simply scales the impulse waveform by some (usually randomly chosen) constant. Of interest is the distribution for the constant, which is often randomly selected each time a sample of an impulse is generated.

France Telecom [6] provides histograms of various impulse-noise measurements. In particular, Figure 3.5 (repeated from [6]) illustrates a histogram of impulse energy. Small-energy impulses have the highest probability, whereas very large impulses tend to be less probable. One can see that about 90 percent of impulses have a duration of less than 250 μs (see also Figure 3.7) and have amplitudes of less than about 10 mV. However, there are a small fraction of impulses that may also have very high amplitudes. Thus a theoretical model that perhaps somewhat conservatively overestimates the probability of large peak impulses may be useful in making theoretical projections based on the energy histogram in Figure 3.5. Impulses with amplitudes greater than 200 mV are very unlikely. For impulse sizes below 10 mV, the probability grows roughly linearly with the log of the peak voltage (or energy, assuming the logarithm of the two are linearly related).

The author has empirically guessed a distribution for the energy from the histogram in dBm/Hz.

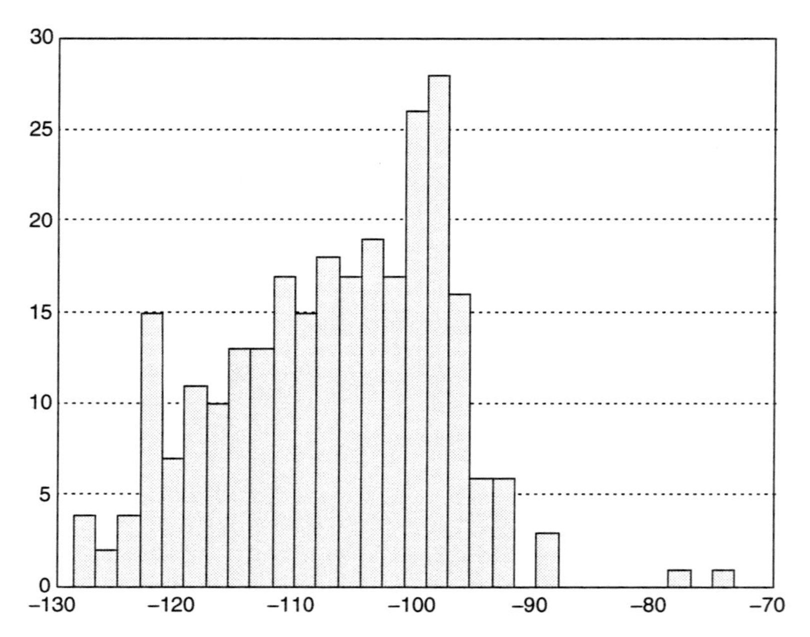

Figure 3.5 Impulse energy histogram (horizontal axis is dB in Watts/Hz). Courtesy France Telecom [6].

$$P(E) = \begin{cases} .288\left(1 + {}^{E}\!/_{100}\right) & E < -75 \text{ dBm/Hz} \\ \dfrac{1.1}{(\text{E} + 83.6)^2} & E > -75 \text{ dBm/Hz} \end{cases}$$

which has a discontinuity at –75 dBm/Hz but projects 90 percent of impulses below this energy level and 10 percent above. A distribution such as the one above, even given it is not perfect, allows projection of ADSL error statistics theoretically. This particular model does not predict a large number of tiny impulses. It is useful only for a gross gain per generated impulse and not for each sample of an impulse. This model would be used when a set of a few impulses is stored and generated as in Figure 3.4.

A more elegant approach to impulse energy characterization has been provided by the work of Henkel and colleagues [24],[25]. Two distributions are provided that are shown [25] to model well measurements by British Telecom and Deutsche Telekom. The first model is a simple exponential distribution[5] given by

[5]One can verify the coefficient 240 by using the substitution $\frac{x}{\sqrt{2}} = \left(\frac{a}{A}\right)^{.1}$ in the integral and then using the unit-variance one-sided odd-moment Gaussian distribution formula $\int_0^\infty x^{2k+1} \cdot \frac{e^{-x^2/2}}{\sqrt{2\pi}} dx = \frac{(2k)(2k-2)\Lambda\, 2}{\sqrt{2\pi}}$.

$$p(a) = \frac{1}{240A} \cdot e^{-|a/A|^{0.2}},$$

which has mean zero and variance $\sigma_a^2 = \frac{5(14!)}{2^{13}} A^2$. Typical values for the parameter "A" range between 10 nV and 10 μV with the lower part of the range being more typical. The tail of this distribution above 10 nanovolts is plotted in Figure 3.6 for $A =$ 30 nV. One can see an enormous concentration of probability at very low impulse amplitudes. Impulses that exceed 1 mV in amplitude have a probability of occurrence less than 0.1 percent. This is somewhat of an artifact of the impulse model, but experts [24], [25] have found that the tails of this distribution tend to model well the probabilities of large impulse occurrences. They also show that this distribution is dominated most of the time by typical Gaussian noise so the overweighting of small impulses is inconsequential. The difficulty in generating this random variable from other more available distributions like uniform or Gaussian motivated Mann et al. [25] to find another distribution with more mathematically pleasant properties, most notably the Weibull distribution that is related to the gamma distribution family and can be generated with specified autocorrelation properties according to a recent method of Tough and Ward [8]. However, the authors [25] did find greater error with the Weibull distribution. The author of this chapter notes instead that generation of any random variable with density $p(a)$ from a uniform random variable essentially involves solving the differential equation $g'(a) = p(a)$ and then inverting the function to obtain $a = g^{-1}(u)$ where u is a uniform random variable. The differential equation corresponds to a family for the exponential distribution $p(a)$ that must be evaluated numerically to form a table look up for $g(a) = \int_{-\infty}^{a} p(x)dx$. The impulse amplitude scaling apparatus inverts that table (which is just another look-up table) to get $a = g^{-1}(u)$. The latter alternative may be preferred for many implementation reasons in generating impulses for actual laboratory measurement or simulation. Clearly this same method of table look up can be used for the empirical impulse-energy distribution above also. These impulse distributions apply (unlike $P(E)$ above) to each sample of a transmitted impulse.

The alternative Weibull distribution offered by Mann et al. [25] is

$$p_w(a) = \frac{\alpha b}{2} \cdot |a|^{\alpha-1} \cdot e^{-b|a|^{\alpha}}.$$

This distribution also overemphasizes small values excessively, but well approximates the tail of the impulse distribution because of the same exponential dependence. Generation of this distribution is described in [8] and can be tailored to provide a given autocorrelation function also. However, the generator in Figure 3.5 (unlike the structures in [25] and considered by ETSI) does not need specification of the autocorrelation function of the impulse because several impulses can simply be stored and then selected each with some individual probability, following more closely the impulse classification in [7] into a handful of representative shapes.

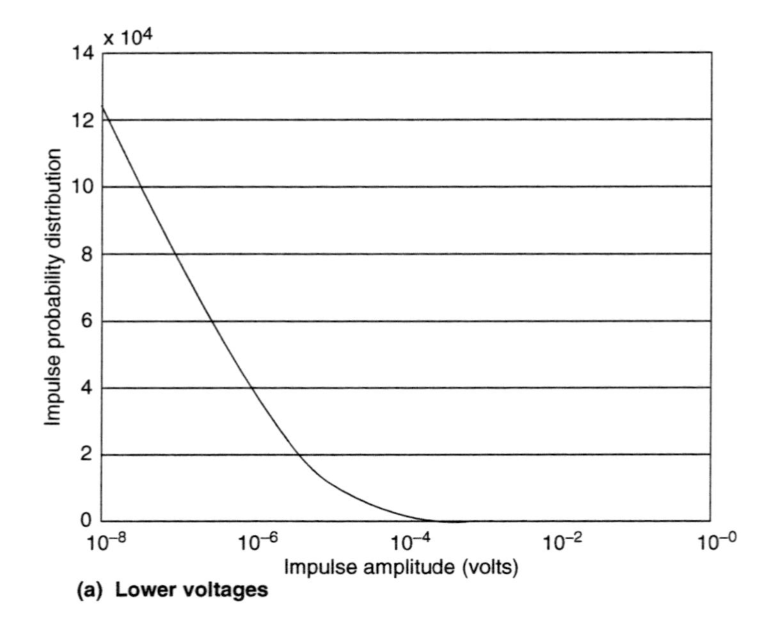

(a) Lower voltages

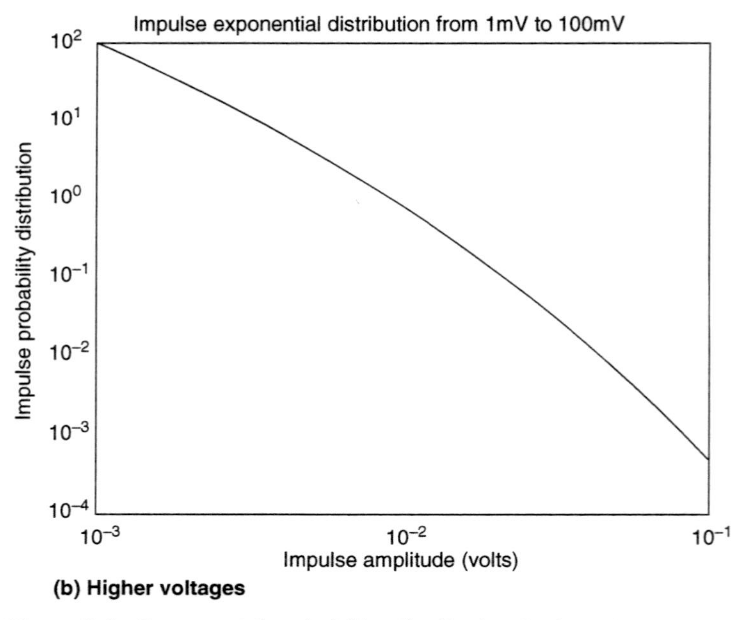

(b) Higher voltages

Figure 3.6 Exponential probability distribution for impulses.

Various phone companies have measured the values for A in the exponential distribution and α,b in the Weibull distribution. We list a few here, partially repeated from [7]:

α in Volts for Parameters in This Table	Weibull α	b	Exponential A
BT (CPE)	.263	4.77	9.12μV
DT (CPE)	.486	44.40	23.3nV
DT (CO)	.216	12.47	30.67nV

North American phone company impulse measurements were not available at time of writing.

Impulse Duration Distribution

Figure 3.7 illustrates a histogram of measured impulse durations. Clearly some impulses are very long, but have low probability of occurrence. Indeed, many impulses are very short, less than 100 microseconds in duration. These short impulses will most often have small amplitudes, and indeed it is possible for many small impulses to follow each other closely in time without disruption of service. The approach of standards groups so far has been to define models that are accurate in the tails of probability distributions, that is, that big impulse samples occur infrequently and last a long time whereas small impulse samples of little consequence can occur

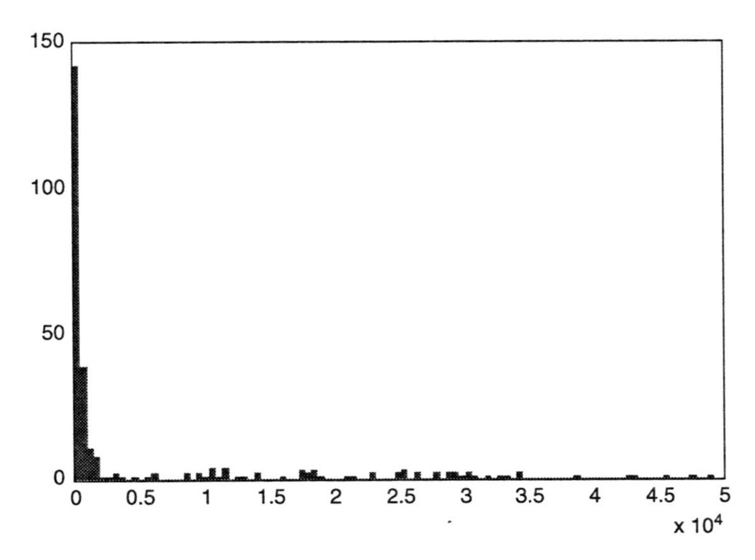

Figure 3.7 Impulse duration histogram (horizontal axis is μs, to 50 ms spanned). Courtesy France Telecom [6].

often. If the tails are modeled correctly by testing or analysis, then the frequent small impulses may not significantly alter performance projections (and indeed match measurements anyway as evident from Figure 3.7). Extremely long impulses typically have energy concentrated in a small number of frequency bands.

Specifically in the model of Figure 3.4, the impulses are separated by a delay. Figure 3.8 illustrates a model of the generation of the inter-impulse delay. Basically, the generator contains a 2-state machine (Markov model) with one of the states being that the next impulse will occur in less than 1 ms and the other state being that the next impulse will occur after more than 1 ms of delay. Each time an impulse is sent, the state machine can make a transition, either to stay in the same state or to change states.

The probability that a short delay is followed by another short delay is written as $p_{s/s}$ whereas the probability of a long delay following a short delay is $p_{l/s}$. The probabilities can be summarized in the matrix

$$P = \begin{bmatrix} P_{s/s} & P_{l/s} \\ P_{s/l} & P_{l/l} \end{bmatrix}$$

and the so-called Markov distribution for the probability of short delays and long delays is determined by

$$P_s = \frac{P_{s/l}}{P_{l/s} + P_{s/l}} \; ; \; p_l = 1 - p_s.$$

There are two distributions for the interimpulse delay τ:

$$p_s(\tau) = \lambda e^{-\lambda\tau} \quad \tau < 1 \text{ ms}$$

with typical $\lambda = .16$ and

$$p_l(t) = \alpha\tau^{-(\alpha+1)} \quad \tau > 1 \text{ ms}$$

with typical $\alpha = 1.5$ and τ in ms. The first distribution produces with small probability delays that can exceed 1 ms, and these delays are discarded and the random

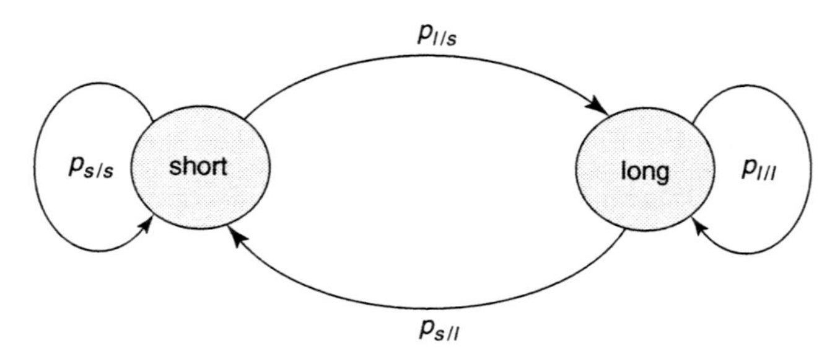

Figure 3.8 State transition diagram for interimpulse delay.

selection procedure cycled again until a delay less than 1 ms occurs. The second distribution exists from 1 ms to infinity. Both can be easily generated from a uniform random variable u on the unit interval [0,1] by the transformations

$$\tau_s = -\lambda \ln(u)$$

and

$$\tau_l = \frac{1}{\alpha\sqrt{1-u}},$$

respectively.

A 2×2 matrix is then provided with upper left entry corresponding to the probability that a short interarrival time is followed by a second short interarrival, the lower right corresponding to long followed by long. For instance, ETSI uses a model based on British measurements:

$$P_{\text{default}} = \begin{bmatrix} .8 & .2 \\ .4 & .6 \end{bmatrix},$$

$$P_{\text{alternate}} = \begin{bmatrix} .5 & .5 \\ .2 & .8 \end{bmatrix}$$

which (e.g., for the default) means that 80 percent of the time a short interarrival occurs, the next interarrival is also short, and the other 20 percent of the time the next impulse is more than 1 ms away and has the probability of interarrival given by the long distribution. Similarly, 60 percent of long-interarrival impulses are followed by long interarrival impulses. The default matrix provides that one-third of the impulses are separated by at least 1 ms, whereas the other two-thirds are very close together. Most samples of any impulse are again small, so even very many samples of those impulses with long interarrival times between them are small. The possibility of two large impulses in a row is thus very small, but not zero. France Telecom has found yet another model that is more characteristic of their network and is

$$P_{France} = \begin{bmatrix} .92 & .08 \\ .86 & .14 \end{bmatrix}.$$

Impulse Waveforms

One particularly in-depth study by C. Collobert and colleagues at France Telecom [7] has found after looking at enormous volumes of ADSL impulse data that impulses can be classified through the use of neural networks into categories that have similar profiles in terms of a cumulative spectral density. FT was able to find five classes that well characterized all their impulses. Each ADSL class is then characterized by a single impulse for test purposes, but the amplitude of that impulse may be nominal (0 dB) or increased to 6 dB larger, and decreased –6, −12, −18, and –24 dB lower in amplitude. An impulse used in one of the ADSL classes derived by FT appears in Figure 3.9. Here the amplitude of the impulse is very large, corresponding to a scaling with rare occurrence in the model of Figure 3.4. The associated frequency spectrum is also plotted in Figure 3.9. Note the impulse is large and long and could be expected to cause difficulty for transmission. With only a few classes, one could evaluate the probability of the different classes empirically (or perhaps adaptively in real time) and then complete the model of Figure 3.4 by storing representative impulse waveforms in the impulse-waveform-storage box. By contrast, in statistical generation using the exponential or Weibull distributions, the autocorrelation of classes of impulses may be stored, and the impulses are generated using these autocorrelations [25].

Others have tried to create mathematical models for the impulse. For instance, Reference [1] describes impulses and a characterization published by BT known as the Cook Pulse. The Cook Pulse has a discrete-time model when sampled at interval T, or equivalently at times kT, that is, zero at time $k = 0$, and otherwise is

$$v(kT) = V_{peak} \cdot |k|^{-.75} \cdot \text{sgn}(k) \text{ mV}.$$

Approximately 45 percent of the impulse energy is in the peak sample and the rest decays with time. Longer impulses tend to have larger peak voltages, and the peak voltage also increases with increased bandwidth use (sampling rate, $1/T$) of the DSL system, roughly as

$$V_{peak} \propto \frac{1}{T^{.75}}.$$

In recent years, many phone companies have further studied impulse noise and found the Cook pulse to be insufficient in characterization. Thus other models have been attempted. One such model was proposed in [25] for ETSI specification. This model essentially models the impulse as an exponentially windowed sinusoid with the exponential decay of the window and the frequency of the sinusoid each determined by random selection from a Guassian distribution. The duration of the window is also random and determined by selection from a convex sum of two log-normal distributions.

The model for the autocorrelation function of the impulse is thus

$$r(t) = e^{-|\beta t|} \cdot \cos(2\pi g t) \qquad -\frac{T_{imp}}{2} \leq t \leq \frac{T_{imp}}{2},$$

where g and β are the Gaussian parameters. The parameter g is selected to be one of 3 Gaussian random variables as shown in Table 3.1, whereas the window parameter β is selected from one of four Gaussians characterized by the randomly selected length of the impulse as shown in Table 3.2.

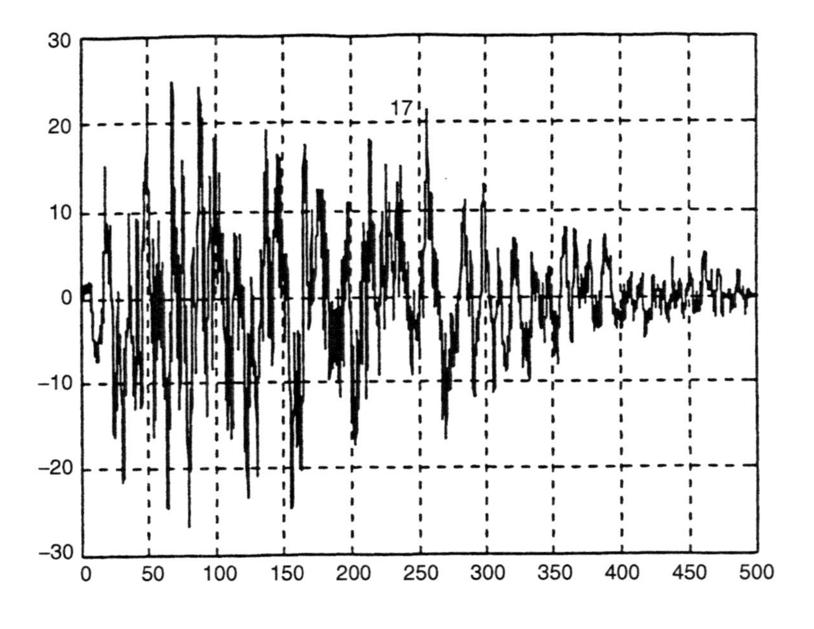

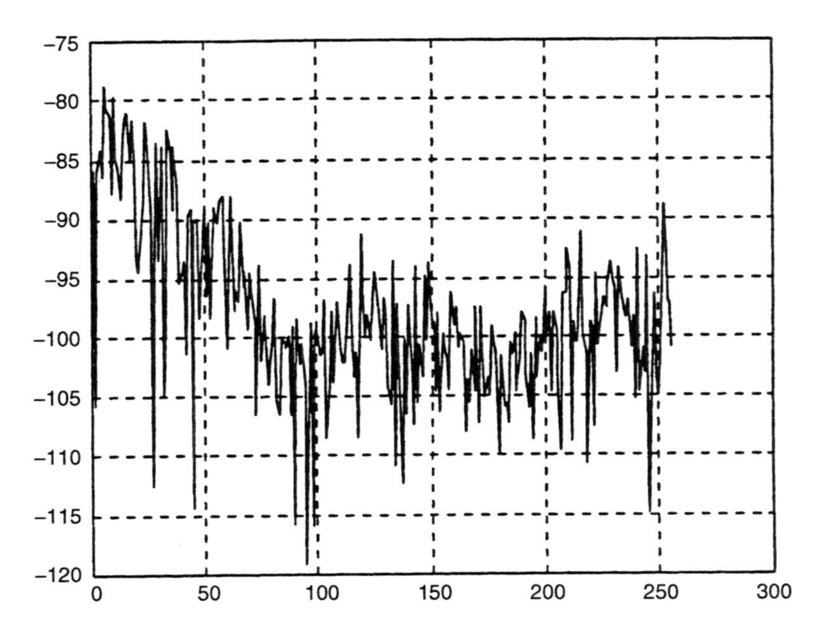

Figure 3.9 France Telecom's "Imp" class impulse (horizontal axes in μs and ADSL DMT tone index, respectively, vertical axes in mV and dBm/Hz).

Table 3.1 g Values

Probability of Gaussian	Mean (ns)	Standard Deviation (ns)
.24	115	25.4
.56	486	178.7
.20	640	9.3

The distribution for the length of the impulse (equivalently the length of the exponential window) is

$$p_{Timp}(t) = \frac{B}{\sqrt{2\pi s_1 t}} \cdot \left(\frac{t_1}{t}\right)^{\frac{1}{2s_1}} + \frac{1 - B}{\sqrt{2\pi s_2 t}} \cdot \left(\frac{t_2}{t}\right)^{\frac{1}{2s_2}}$$

where the customer premises parameters are $B = 1$, $s_1 = 1.15$, $t_1 = 18\ \mu s$, and thus the second term above is zero, and the CO parameters are $B = .25$, $s_1 = .75$, $t_1 = 8\ \mu s$, $s_2 = 1.0$, $t_2 = 125\ \mu s$. This model is likely less representative, and certainly much more complex to implement and understand, than Collobert's representative waveforms in [7]. ETSI is now evaluating the France Telecom impulse-class model.

Probability of Error Analysis with Impulse Noise

The authors would especially like to thank Dr. Wei Yu and Mr. Daniel Gardan for inputs to this section.

ADSL uses Reed-Solomon codes to provide coding gain and also to mitigate impulse noise disturbance. The RS code is a block code of block size up to 255 based on GF(2^8) (or bytes). An RS code word can therefore be up to 255 bytes long (in ADSL systems). The RS code word boundary is aligned with the DMT symbol boundary. Because of the rate adaptive nature of a DMT system, the number of bits in each DMT symbol varies depending on the data rate. ADSL [3] allows a variable number (1, 2, 4, 8, or 16) of DMT symbols for each RS code word. ADSL also allows an even number of parity bytes (up to 16) to be included in an RS code word. The decoder is able to correct up to half as many error bytes.

In order to take full advantage of the error-protecting ability from RS code, it is important to correctly choose the number of DMT symbols per RS code word and the number of parity bytes included in each code word. From a pure error-correcting point of view, for the same amount of parity overhead, it is always the best to have as many DMT symbols in an RS code word as possible. For example, a system that includes 8 DMT symbols and 16 bytes of parity for each RS code word performs better than a system that includes 4 DMT symbols and 8 bytes of parity

Table 3.2 β Values

Length of Impulse (μs)	Mean (ns)	Std. Dev. (ns)
$0 \le T_{imp} < 1$	38.3	6.4
$1 \le T_{imp} < 3$	26.7	7.9
$3 \le T_{imp} < 10$	10.7	3.7
$10 \le T_{imp} < \infty$	30.7	1.6

for each RS code word. This is because the former system can correct at least as many errors as the latter system. A longer RS code word means longer delay, and possibly more decoder complexity. The analysis here concentrates on reducing the impact of impulses as best as is possible, and will therefore include as many DMT symbols per RS code word as possible for impulse protection, necessarily enduring delay. Then it only remains to decide how many parity bytes to include in each RS code word.

In most cases, the maximum error-correcting protection against Gaussian noise when the system operates near capacity is obtained when the parity overhead is approximately 6 to 10 percent [18], which is about 16 parity bytes in 128–256 bytes. The RS code word length depends on the number of DMT symbols per RS code word; thus, it depends on the system data rate. (A framing overhead of 128 kbps is assumed here.) For example, at 608 kbps, each DMT symbol is about $(608 + 128)/4 = 184$ bits = 23 bytes long. So, up to 8 DMT symbols can be grouped into an RS code word. Assuming 16 bytes of parity are inserted, the resulting RS code word is $23 \cdot 8 + 16 = 200$ bytes long. At 1.216 Mbps, each DMT symbol is about $(1216 + 128)/4 = 336$ bits = 42 bytes long. So, up to 4 DMT symbols can be grouped into an RS code word. Again assuming 16 bytes of parity, the code word length is $42 \cdot 4 + 16 = 184$ bytes long. At 2.048 Mbps, each DMT symbol is $(2048 + 128)/4 = 68$ bytes long. So, 2 DMT symbols are grouped into an RS code word, resulting in a code word length of $68 \cdot 2 + 16 = 152$ bytes.

Assuming 16 bytes of parity, the RS code is able to correct up to 8 bytes of error. When more than 8 errors occur, the Reed-Solomon decoder will either decode to a false code word, in which case nearly all the bytes (or half of the bits) are incorrect; or the decoder will declare decoding failure, in which case the original code word is unchanged so the output has the same number of error bytes as before. The following calculation shows that decoder failure is a much more likely possibility:

The RS code operates on GF(256). The total number of length -256 valid code words is 256^{240}. (This is because the 16-byte parity is to be added to any 240-byte message.) For a correctable error to occur, at most 8 bytes can be in error. These errors can be located in $\binom{256}{8}$ possible byte locations. In each error location, the number of ways that an error can occur is 255. So, for each code word, there are $255^8\binom{256}{8}$ "neighbors" that a decoder can perfectly correct. But, this is a small fraction of the total number of errors, as $255^8\binom{256}{8}256^{240} \Big/ 256^{256} \approx 10^{-5}$, where 256^{256} is the total number of possible 256-byte strings. So, when an impulse hits and an uncorrectable error occurs, the probability that the uncorrectable error will fall in the neighborhood of some other code words is small. So, the RS decode is likely to declare decoding failure, thus keeping the number of error bytes same as before.

Many ADSL receivers use erasures for impulses. A simple erasure mechanism that works very well is to compare the sum of squared differences between tone decoder slicers' outputs and inputs. Nominally this sum is nearly zero if there is no impulse. When this sum exceeds a threshold, all bytes in the DMT symbol are "erased" (marked). The RS decoder can then correct erased bytes (twice as much). Even in this case, failed RS decoding is easily detected.

The following analyzes the coding gain of an RS code in an AWGN channel. In an AWGN channel (or a channel with a well-designed equalizer), the probability of error for each byte is approximately the same. Let *Pe* denote the probability of a byte error. When *Pe* is small, the probability of code word error is closely approximated by the probability that 9 byte errors occur. When 9 byte errors occur, the Reed-Solomon decoder will declare decoding failure, so the total number of error bytes will be 9. For each byte error, the most probable channel defect is that a single bit is likely to be wrong. So the probability of bit error in the code word is 9/200/8 = $4.4 \cdot 10^{-3}$, when a decoding error occurs. Because the required probability of bit error is 10^{-7}, the required probability of code word error is $10^{-7}/4.4 \cdot 10^{-3}$. In this case, a 608 kbps ADSL system with 200 bytes per RS code word needs to satisfy:

$$10^{-7}/4.4 \cdot 10^{-3} = \binom{200}{9} Pe^{9}(1 - Pe)^{191}.$$

Thus the uncoded system needs to have $Pe = 0.0065$ for each byte to attain an overall probability of bit error 10^{-7}. Since the probability of byte error is 0.0065, the uncoded probability of bit error is then[6] $0.0065/8 = 8.1 \cdot 10^{-4}$. The gap for QAM at $P_b = 8.1 \cdot 10^{-4}$ is found by noticing $2Q(10.5\text{dB}) = 8.1 \cdot 10^{-4}$. So instead of requiring the argument of the Q-function to be 14.5dB, with coding, only 10.5dB is needed. In other words, coding provided a gain of 14.5dB − 10.5dB = 4.0dB. This is the coding gain for a 608 kbps system with 8 DMT symbols per RS code word and 16 parity bytes. This (raw) coding gain does not take the extra parity overhead into account. In reality, having 16 bytes of parity for every 8 DMT symbols translates to an extra 64kbps coding overhead. The loss in dB for the overhead is $64/(256 \cdot 4)$ = 1/16 or .4 dB so the actual coding gain is 3.6 dB. This method of calculating the RS coding gain is more accurate than assuming 3dB coding gain for all systems. One needs to exercise caution in impulse analysis as opposed to AWGN-analysis in Section 3.3.5—here the probability of bit error is in the range of 10^{-4} whereas in Section 3.3.5, the same probability is computed under an assumption of 6 dB margin at 10^{-7} and can thus be expected to be higher (coding gain is constant only as probability of error gets small, or SNR is high, which is not true in this section). Use of erasures does not improve performance with Gaussian noise.

Probability of Error Calculation for Impulses

Uncoded System. The filtered impulse samples are treated as a deterministic signal. Its interpolated and resampled version is passed to an FFT demodulator. The resampling rate is 2.208MHz, and the FFT size is 512, representing 512 real dimensions in a DMT symbol. The noise samples contained in the cyclic prefix are

---

[6] Taking the first term of the binomial expansion: $P_{byte} = 8 \cdot P_{bit} (1 - P_{bit})^8 = 8 \cdot P_{bit}$.

discarded. The output of the 512-FFT is then combined into 256 complex samples, representing the real and imaginary parts of the impulse noise in each tone. The complex disturbance is the perturbation to the constellation caused by impulses. The perturbation decreases the minimum distance between constellation points, and hence increases the probability of bit error. Strictly speaking, the increase in probability of error caused by impulses depends on the constellation size (because the boundary points are less susceptible to impulse disturbance than points in the middle of the constellation.) A simplification assumes all QAM points have exactly four nearest neighbors, and four next-nearest neighbors, and computes the probability of error based on these neighbors. Figure 3.10 illustrates the computation.

In Figure 3.10, suppose that the constellation point X is being sent. Without impulse, the probability of error in the presence of AWGN is closely approximated by $Pe = 4Q(d/2\sigma)$, where d is the minimum distance between constellation points, and σ^2 is the AWGN noise variance. The coefficient 4 represents the four nearest neighbors of each constellation point.

When the system is disrupted by an impulse, the impulse is regarded as a deterministic signal, shown in the diagram as an arrow from the original constellation point X to the new location Y. If Y is outside of the decision region, an error is almost certain to occur. In this case, $P_e = 1$. If Y is inside of the decision boundary, the probability of error is computed by summing over the probability that the point Y may be confused for each of the neighbors:

$$P_e \leq \sum_i Q\left(\frac{d_i}{2\sigma}\right),$$

where d_i are the distances between Y and its neighbors A, B, C, D, and E. Strictly speaking, summing the probability is only valid when the error events are mutually exclusive. The above formula is thus an upper bound.

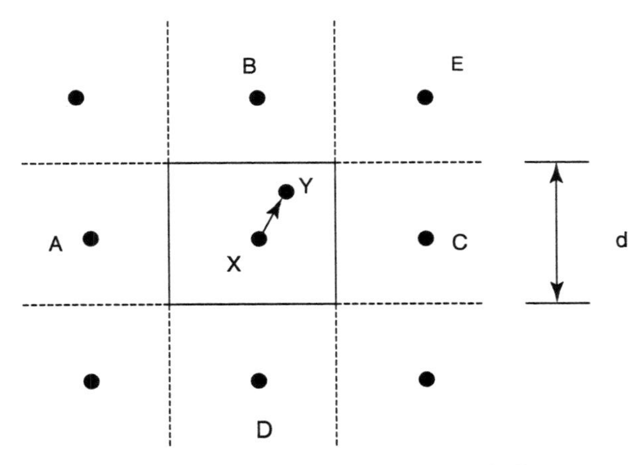

Figure 3.10 Illustration of impulse analysis *Pe* calculation.

Following the preceding calculation, the probability of symbol error for each tone may be calculated separately. Because each tone carries different number of bits, the average probability of bit error is computed as

$$P_{ave} = \sum_i b_i P_i \Big/ \sum_i b_i,$$

where P_i is the probability of symbol error on each tone, and b_i is the number of bits carried on ith tone.

Coded System. In a coded system, the minimum distance between constellation points is smaller than an uncoded system because a coded system partially relies on the error correcting code to compensate for the smaller *dmin*. The goal is to compute the expected number of error bits when an impulse occurs. Evaluation first requires translation of the probability of bit error into the probability of byte error, because RS code works on bytes. Every 8 bits are grouped into a byte, and the probability of byte error $P_{byte} = 1 - \prod_{i=1}^{8}(1 - P_{bit}(i))$, where $P_{bit}(i)$ is the probability of error for ith bit. When an impulse occurs, the probabilities of bit error are different from tone to tone, and the byte boundary does not necessarily coincide with the tone boundary. So, in an ADSL RS code word, the probabilities of byte error are different for each byte.

The performance measure of interest is the expected number of byte errors. When there is no coding, the expected number of byte errors is just the sum of $P_{byte}(k)$, where $P_{byte}(k)$ denotes the probability of byte error in kth byte. When an FEC is implemented, the probability events with fewer than nine errors are successfully corrected, so an expression for the average number of byte error is:[7]

$$\begin{aligned} \mathrm{E}[ErrorBytes] &= \sum_{m>8} m \Pr(m\_error\_bytes) \\ &= \left(\sum_k P_{byte}(k)\right) - \sum_{m\leq 8} m \Pr(m\_error\_bytes) \end{aligned}$$

The above expression is relatively easy to compute. Although enumerating all 8-byte errors in a 200-byte code word involves summing over $\binom{200}{8}$ probabilities, only a handful of these probabilities are likely to contribute significantly to the sum. Simplified calculation can more efficiently select only the large probability terms. For example, if most $P_b(k)$ are small except a few, it is safe to assume that all errors will occur in those handful number of bytes with large $P_b(k)$.

---

[7]This expression recognizes that $\sum_{m=1}^{\infty} m \Pr(m\_error\_bytes) = \sum_k P_{byte}(k)$ = expected number of byte errors in a symbol without FEC.

Table 3.3 Interleaving Parameters on an ADSL Modem

		Medium Delay			Long Delay		
	FAST mode	**608 k**	**1.2 M**	**2.0 M**	**608 k**	**1.2 M**	**2.0 M**
No. of DMT per RS	1	8	4	2	8	4	2
Interleaver depth	1	2	4	8	8	16	32
Delay	2ms		6–9ms			~20ms	

Interleaving. It is interesting to note that major service providers have indicated that approximately 99 percent of ADSL lines are operating without interleaving (fast mode). This is motivated by the desire to provide low latency transport. Many of the 1 percent of lines with interleaving use the interleaved mode to increase line reach or attempt to resolve service troubles. ADSL allows RS code word to be interleaved with an interleaver depth of 1, 2, 4, 8, 16, 32, or 64. If the interleaver size is D, byte i in an RS code word is delayed by $(D - 1)i$ byte transmission times. The interleaved version is sent over the channel and is corrupted by impulses. The deinterleaved version is decoded, so, in effect, the number of corrupted bytes decreases by a factor of D. To calculate the effect of interleaving on isolated impulse, analysis can assume that impulses are far enough apart so that after deinterleaving, no two impulses occur in the same code word. Table 3.3 summarizes typical interleaver parameters for three data rates.

Impulse Alignment. In real systems, impulse may occur at any time instant. So an impulse typically spans more than one DMT symbol, and could also span more than one RS code word. This effect needs to be taken into consideration in any simulation. For results of simulations using methods like those described here, see [6]. Often detailed uses of this type of analysis for specific impulses is held secret by the service providers who endeavor to compute them.

3.2.3 Interleaving and Decoding Improvements

Interleaving and deinterleaving are described in [1] for ADSL and allow an interleave depth, D, of up to 64 code words. Up to sixteen DMT symbols per code word are also allowed at very low data rates, so that a delay can be up to 256 ms. Fast path delay is limited to 2 ms, and the number of symbols per code word is forced to one as is the interleave depth. Minimum delay is desirable for interactive applications like voice or games, although realistically a user does not notice (on an echo canceled network link) delays of less than 100 ms (no matter what standards and experts say).[8] Thus it is often advantageous to use the maximum delay and corresponding interleave depth of sixty-four whenever possible for extra impulse protection. At higher speeds where the number of symbols per code word is four or less, the delay is inconsequential to most applications. More important though is

[8]For instance, wireless portable phones insert 100 ms of delay into the link without notice of the user, even though some feel voice-over ADSL must have no more than 2 ms delay. Because modern DSLAMs will undoubtedly be connected to modern echo-canceled switches, the longer delay is tolerable for voice, in practice. However, some customers are acutely sensitive to more than about 2 ms delay for data and interactive gaming applications.

that increased delay leads to increased buffering of data throughout the network, which then can impose a more stringent delay constraint on the ADSL link. This delay has nothing to do with applications, but instead is caused by insufficient buffer-size decisions in other parts of the network or protocol used. Network designers of the future will consider the impact that better ADSL end service will have on their sales and revenues and understand that excessively tight delay constraints on the DSL modem is quite counterproductive for their overall objectives, and otherwise arbitrarily imposed by poor network choices. Also, reference [29] has recently made significant progress in mitigating all impulses with very low latency.

This area still remains one of active interest and study today as to how to obtain the best possible impulse-noise rejection with the least delay. Fortunately, DMT offers a theoretical advantage of 24 dB (and a measured advantage of about 10 dB) in impulse-noise rejection with respect to other non-DMT-based DSLs, making the trade-off somewhat more palatable for ADSL than for other non-DMT DSLs.

3.2.4 Impulse-Cognizant Loading and Erasure Methods

A strong feature of RS codes is their capability to correct double the number of errors if the location of the error bytes is known. This is called *erasure decoding,* and is very useful for impulse noise. Clearly, such a capability also allows halving the interleave depth (or halving the delay and the amount of memory required for implementation) for a given level of impulse protection. Several ADSL modem manufacturers exploit this capability. Impulses must be detected for erasures, typically of all bytes in any DMT symbol known to be simultaneous with an impulse. As mentioned earlier, an easy impulse-defect metric is simply the sum of the squared tone-slicer errors.

Reference [29] is a very recent study of 11,000 France Telecom impulses that uses an erasure RS decoder to remove all errors from all impulses with less than 5 ms latency. It is the first public report of successful use of erasures. Cioffi suggests that recent soft-decoding algorithms with trellis coding can be applied; for instance, the well-known SOVA method [9], soft-output Viterbi algorithm. These methods are more complex than the usual Viterbi detector, but retain a reliability indicator for the selected bits/symbol strings that can be simply processed to provide a reliability indicator for each byte input to the following FEC decoder. This may allow more than impulse-type erasure. In normal operation with 6 dB margin, the reliability indicator will be excellent and will suggest no errors are being made. When impulse or other non-Gaussian noise is present, the reliability indicator will be poor, suggesting that many errors are being made. There is an independent reliability indication for every DMT tone symbol (or at least every two symbols in the four-dimensional code). The reliability indicator for a byte is essentially the sum of all symbol reliability indicators for the symbols that contribute bits to that byte. This is an unusual use of SOVA, but consistent with iterative decoding methods that heavily and productively use SOVA for AWGN disturbances. To further this concept, recent turbo and LDPC code proposals for ADSL (see Section 3.2.1) may find acceptance. In these proposals, the trellis code is replaced by a more powerful turbo/LDPC, and the reliability indicator is inherent in the decoding methods used for those codes. While already having 6.5 dB or more of coding gain, and better performance against impulses themselves, the reliability indicator of these codes can also be passed to the FEC decoder and used as an erasure indicator. The probability of false indication of an erasure is far less with these advanced decoding algorithms (such as SOVA or iterative log-likelihood construction for LDPC), and thus they offer an enormous possible improvement for impulse noise rejection. This area is still under study by several groups and likely to be one of the improvement areas of ADSL in the future, but verifiable results have yet to be published.

Another method for improving impulse reliability that is simpler, but reduces data rate, is **impulse-cognizant loading.** In impulse-cognizant loading, the distribution of energy of the impulse is either known or estimated on-line. Most impulses while generally perceived to be wideband, do not cover the entire frequency spectra equally (indeed the reason why the normal 24 dB [= 10 $\log_{10}$ number of tones] advantage of DMT is more like 10 dB in practice). However, often lower frequencies are affected with larger amounts of energy. Thus, the margin for the lower frequencies can be increased (while upper frequencies, or more generally those known to be relatively unaffected by impulse energy, can have their margins in turn decreased). This procedure either reduces data rate or overall margin against Gaussian noise, but can improve margin against impulse noise. Such a procedure can dramatically reduce the amount of interleaving necessary for the same-level impulse protection (perhaps by as much as a factor of three or four) for perhaps the cost of several hundred kilobits in aggregate maximum-achievable data rate or throughput, which may be an acceptable trade in many cases (especially if the implemented data rate is already well below the maximum achievable). The difficulty is knowing what frequencies/tones are disturbed most on any given line, which may have many sources of impulses. Again, the soft-information/reliability-indicator of more advanced codes may create opportunity in the future to do greater on-line characterization of impulses on any given line in the receiver, thus creating the opportunity for frequency-selective loading.

The authors expect impulse rejection methods to improve significantly in the future, and likely the data rates of ADSL in deployments will be increased in many regions from 500–800 kbps downstream rates (which are very conservative largely because of impulse noise and a perceived need for low latency) to 1.5 Mbps on 4-mile lines. Theoretically, impulses have very little energy over all time, even over short intervals in time, and thus should not dramatically reduce performance. To induce huge margins on DSL service to protect against impulse problems is therefore unnecessary and inhibits the long-term capabilities and revenues of DSL.

3.2.5 Decoupling of ADSL Frame and Code

Decoupling of the ADSL framing from the DMT symbols allows a greater continuum of FEC parameters and has been incorporated into the lastest ADSL standards drafts.

New Framing

The ITU second-generation ADSL recommendations G.992.3 and G.992.4 [3] include a new reduced-overhead framing mode for ADSL. Earlier ADSL modems supported four framing modes having different amounts of minimum overhead with one or two latency paths. These paths can be configured to have different overheads and thus have different framing efficiencies. With an RS code word enabled, the minimum overhead is 64 kbps or higher, corresponding to 2 bytes of parity per DMT symbol. This causes substantial framing inefficiencies at low-line rates such as 224 kbps or lower, quite typical in the upstream directions on long loops. The same framing inefficiency also existed for downstream link on long loops. The earlier standardized ADSL framing overhead used at least one overhead byte per frame and only an integer-ratio choice of R (RS code word parity bytes) and S (Number of DMT symbols over which the RS code word spans) parameter values for RS code word. The new ADSL standards make appropriate adjustments in these parameters and can minimize these inefficiencies. In addition, the early standardized framing was byte aligned and required the DMT symbols to carry an integer number of bytes, causing additional inefficiencies up to 28 kbps. Together, up to 92 kbps minus the required minimum overhead is now available to improve user payload in the new framing. The new standards achieve this reduction by removing the integer constraint on the ratio of R/S and allowing the overhead to be configurable to smaller values than 32 kbps. This, in effect, also removes the constraint of the DMT symbols to carry an integer number of bytes or the line rate to be an integer multiple of 32 kbps. These modifications can provide very significant improvements toward the net data rate of the modem. For example, by configuring the current framing to have a 4 kbps overhead and 4 kbps RS overhead, ADSL can avail 88 kbps toward the net data rate just because of the improved framing.

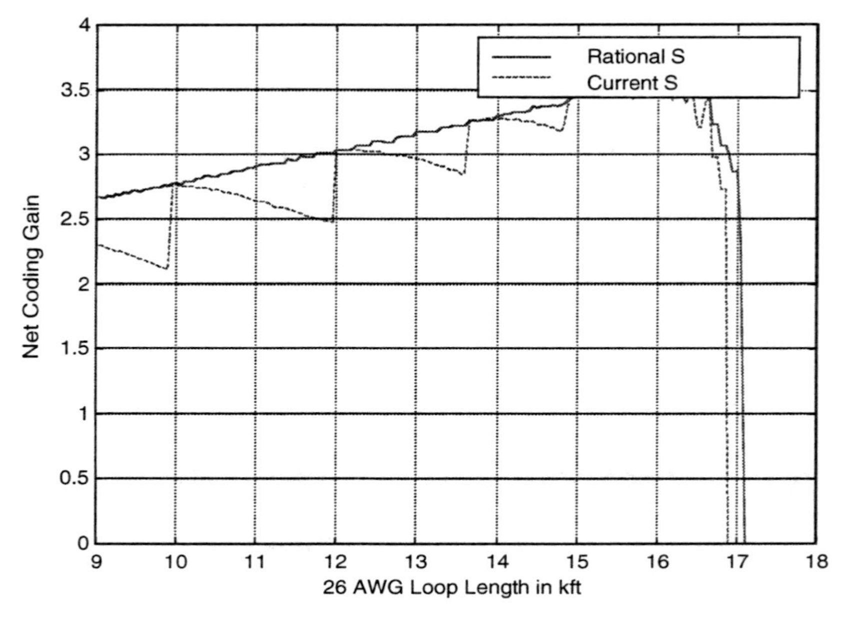

Figure 3.11 Net coding gain produced by new ADSL framing proposal (see [6]).

Allowing S to be a rational number further helps optimize the coding gain on all loops. The granularity of the S does not permit maximized coding gain on all loop lengths. The dotted curve in Figure 3.11 shows the effect of current S values as a function of 26 AWG loop length with link latency less than or equal to 16 ms. Because of coarse granularity in current S values, there is a reduction in net coding gain of the modem as a function of loop length and then a jump back to optimum coding gain as the next optimum S value becomes available. On the other hand, rational S values optimize the coding gain on all loop lengths. The maximum net coding gain advantage is about 0.64 dB in this example. This advantage varies on different loops but rational S always performs equally or better.

There is a direct relation between the line rate and the coding gain. For example, an increase in coding gain corresponds to a certain amount of line rate increase on a given loop. After taking into consideration the overhead associated with the coding scheme, this transforms into a net data rate increase. Thus, increase in net data rate due to increased framing efficiency would also correspond to a certain amount of coding gain. In other words, increased framing efficiency corresponds to increased coding gain.

1-Bit Constellation

ITU Recommendations G.991.3 and G.992.4 also include provisions for 1-bit signal constellations. Earlier ADSL modems do not support the use of 1-bit constellations on subcarriers. The 1-bit constellation allows the use of those subcarriers that do not have sufficient SNR to support the 4 QAM constellation but sufficient to support 2 QAM. This helps increase the total capacity of the ADSL modems without a significant increase for implementation complexity. One-bit carriers also simplify swapping, as in Section 3.3. In a theoretical simulation of the ADSL modems on ISDN # 7 loop with 24 HDSL crosstalk, the capacity without the use of 1-bit constellation was 468 kbps downstream and 168 kbps upstream, respectively. The same improved to 596 kbps downstream and 196 kbps upstream, respectively, when the use of 1-bit constellation was enabled. This advantage would vary among different loop and crosstalk scenarios, but, in general, there would be some increase in capacity due to the use of 1-bit constellation. The implementation of transceivers with 1-bit constellation must also address the bit assignment for trellis coding.

Rate Adaptation

The transmission environment encountered by ADSL modems is not long-term stationary. Crosstalk noise changes as the other users in a binder group connect and disconnect modems. Other noise changes are due to radio transmitters transmitting more or less power. The signal attenuation changes with time of the day due to changes in cable temperature. A service margin of 6 dB in the current ADSL modems should provide protection against changes due to these factors for all but the most extreme cases. However, this results in many lines operating with far more margin than necessary, and some lines operating at some times with virtually no margin. Rate adaptation in the improved ADSL modems would allow for dynamic (in-service) changes in the line rate without losing payload data. This is possible because of the new framing that allows transparent reconfiguration and a change in the modem line rate through a message-based line reconfiguration protocol, which is run over the ADSL overhead channel. This line reconfiguration protocol controls the reconfiguration of the line rates, bit swapping among different subcarriers to maintain margin, and reconfiguration of the framing parameters to support new line rates or partitioning of the bandwidth among different latency paths.

The ability to change transmission bit rate without causing errors in the payload is known as seamless rate adaptation (SRA). Because most existing implementations cause a loss of data for a few seconds when changing transmission bit rate, many ADSL services will change to a lower bit rate only when service is severely impaired. Current ADSL systems generally do not increase the transmission bit rate if the channel conditions improve, unless the ADSL modem is disconnected and then reconnected. Thus, existing systems tend to ratchet the bit rate down to meet the worst condition over the long term. Seamless rate adaptation could enable the ADSL line rate to continuously track the best possible bit rate at the time. Higher operating bit rates could be achieved in some cases because SRA would allow the line to adjust the bit rate to maintain an SNR margin that is enough but not too much to assure high quality service. SRA would provide no benefit for lines operating near their performance limit because these lines would have no excess margin. Also, SRA would provide little benefit for lines that are operating at the maximum bit rate specified for the customer's service class. Thus, SRA provides its benefits for lines that would have excess margin while operating at less than the maximum service bit rate.

3.3 SWAPPING AND ADVANCED SWAPPING/LOADING METHODS

One of ADSL's most important and essential features is the bit-swapping described in [1]. Bit swapping accomplishes several features either crucial or desirable for DSL operation by:

1. Allowing continuous response to line changes caused by environmental effects.
2. Responding to changing crosstalk situations as other DSLs energize or deenergize.
3. Allowing the data rate to be changed when/if desirable in some applications without restart.
4. Allowing multiple DSL lines to be mutually compatible within power-spectral-density constraints.
5. Enabling features such as the SRA described earlier.

The first two features are essential to proper operation of the DSL modem. DMT modems use minimal equalization, which is why they require far less digital signal processing operations for high-performance operation than wideband ("single carrier" or "baseband") modulation methods that alternatively require long equalizers (DFEs, see [1] or Chapter 2) for best operation. The price paid for such low-cost high performance is the need for continuous handshaking on a control channel between the receiver and transmitter that continuously optimizes the allocation of information, and possibly also energy, to the different subchannels in a DMT modem. This handshaking not only implements the function of the adaptive equalizer in wideband modems, but it also continuously ensures that the optimum transmission band is maintained, or equivalently that the best possible reliability is maintained. Such continuous optimization occurs through the bit-swapping control channel (also known as the AOC—see Section 3.3.1). The swapping channel is highly robust as discussed in Section 3.3.1, but may be undesirably slow to react in some situations, so Section 3.3.2 discusses accelerated swapping protocols (sometimes known as "express swapping" that are allowed in advanced ADSL and DMT-based VDSL modems). Sections 3.3.3 and 3.3.4 discuss some advanced uses of swapping that have come to be more common as ADSL systems are increasingly improved by manufacturers. Section 3.3.5 discusses advanced loading algorithms that can be used for swapping or loading, basically due to H. Levin [10]–[13] (this area was not covered in [1]). In particular the often-encountered question of "how to load" with various types of coding and associated redundancy present is discussed in Section 3.3.5.

3.3.1 Explanation of Swapping

Figure 3.12 illustrates the basic concept of bit swapping. There is a DMT controller in both transmitting and receiving DMT modems. This controller sets the number of bits and transmit power level of each subchannel. A protocol exists on the AOC[9] between the two modems. There are essentially three types of commands on the AOC: swap request, swap acknowledge, and extended swap request. Swapping commands are identified on the overall control channel by a leading byte of all ones. A swap request from the receiver to the transmitter on the AOC has an additional 8 bytes of information, that is, four 2-byte fields that specify a tone idex (0 to 255) in the first byte and an action (increase or decrease the number of bits allocated to that tone by 1, or increase or decrease the power of that tone by 1, 2, or 3 dB, or do nothing to that tone) in the second byte. Usually the same tone index is transmitted twice with first an increase/decrease in number of bits, and subsequently a corresponding energy adjustment. However, it is possible to increment a single tone four times with no energy change and many other combinations as well. A swap acknowledgment from the transmitter to the receiver on the AOC is an extra 2 bytes that repeats the all ones pattern in the first of these two additional bytes, and then specifies a superframe number (from 0 to 256) on which the swap will be implemented by the transmitter. Superframes start with count number equal to 0 on the very first transmission of customer data, and subsequently, increment modulo 256 every 68 DMT symbols (or 69, counting any synch symbol). Loading algorithms for incrementing and decrementing bits and energy are discussed in

[9]AOC = auxiliary control channel, or sometimes A = ADSL or Amati. (Amati was a company originally very active in the development of ADSL and original proposer of the AOC.)

Section 3.3.5, most specifically Campello de Sousa's method in [14]—now also known as having been proposed earlier by Levin of Motorola [10]. An extended swap has 12 additional bytes, allowing up to 6 tones to be specified for alterations of bits and/or energy. Typically, extended swap is used to swap from 2 bits to 0 or vice versa on a subchannel by sending an increment/decrement command twice, along with a corresponding energy adjustment for each of donor and recipient tones.

The bit swap protocol is resilient to loss of handshake commands, both bit swap requests and/or bit swap acknowledges. Although loss of commands is extremely rare because of an error-corrected-plus-three-of-five command protocol, it is conceivable that a gross channel disturbance might cause a command to be missed: A situation sometimes suggested as a "flaw" in bit swap suggests that such a miss causes modem disfunction. Actually, this statement is incorrect, and there are many correct implementations that easily and simply recover without error from a command loss. The following section offers the conceptual implementation of Figure 3.12 as one of many that suffice.

In the implementation of Figure 3.12, the receiver typically initiates a swap of a bit from one tone to another tone through the AOC DMT-control channel shown (this channel is actually embedded in the data channel, but shown separately for illustration in Figure 3.12). A bit swap "request" command is sent from the receiver to the transmitter through a heavily protected reverse "AOC" channel.[10] The transmitting modem then "acknowledges" the request through the forward "AOC" channel and specifies a time for the new bit table to be implemented. If either command is not received for any reason, the receiver can simply monitor the incoming signal using two bit tables (this can be implemented simply in several fashions), OLD and NEW. The FEC corrected/detected error flag (syndrome) can be monitored for which of the tables, OLD or NEW, is correct. The receiver then knows whether the swap was implemented by the transmitter or not, even if one or both of the AOC commands could not properly traverse the channel. *Good bit swap implementations can thus be fully resistant to AOC channel errors.* This also means that the channel itself need not be burdened with excessive use of commands for double acknowledgments, and so on, which leads to an increase in overhead bandwidth. Another method for detection of an implemented bit swap is to monitor the tone to which a bit is added. This extra bit leads to a larger constellation, with the probability of a new outer point corresponding to the extra bit, tending toward one in just a very short time. This method, while extending the error burst by 1 percent or so, is extremely simple to implement, corresponding to a complexity increase of less than 0.1 percent in the receiver signal processing.

Bit swap is an essential feature of a DMT DSL modem. A simple situation is illustrated in Figures 3.13 and 3.14 for a 4.25 km 26-gauge twisted pair. One bit distribution corresponds to the DSL being the first one deployed/used in a cable. The second distribution corresponds to the new bit distribution that is best when one

---

[10]The AOC channel uses a triple match (out of five tries) protocol with 16-bit header, which even when the channel bit rate is a coin/flip probability of one-half leads to a probability of false command receipt of once in one billion years. (False commands do not occur in practice—the only possibility is a missed command.)

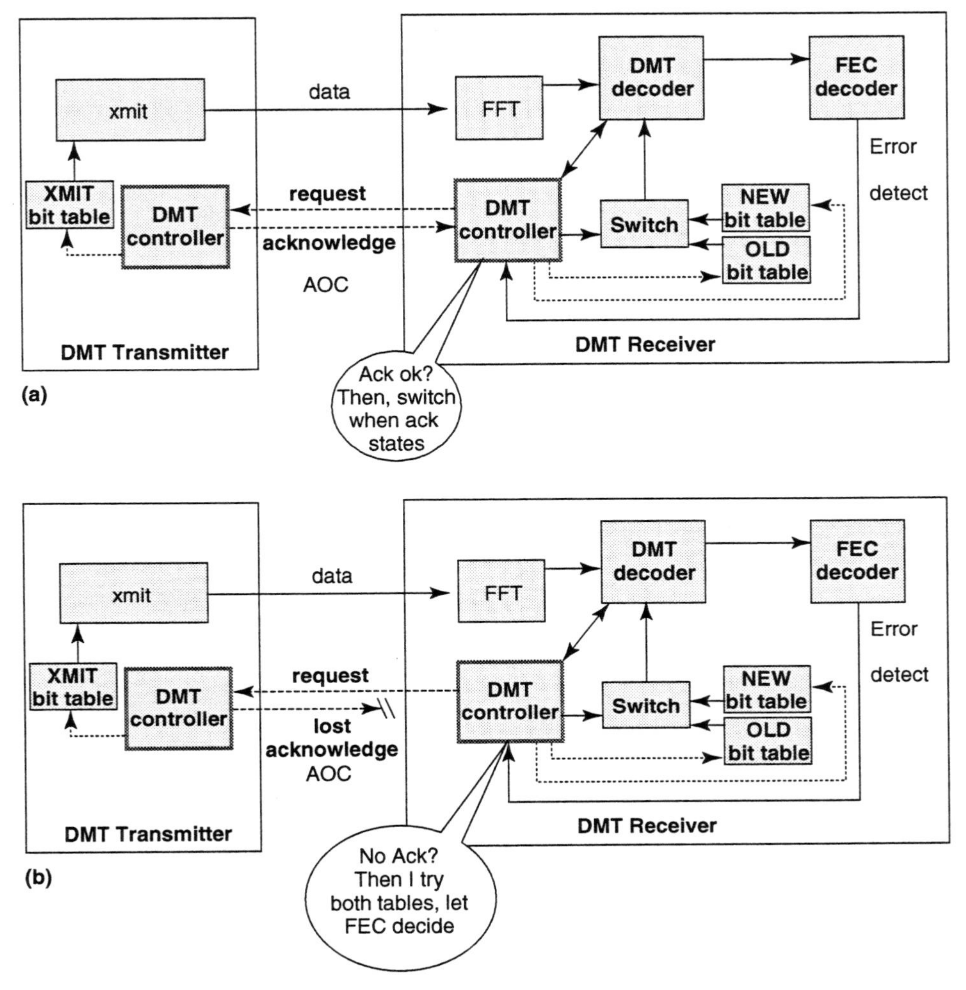

Figure 3.12 Illustration of functional and dysfunctional AOC. (a) Bit swap with AOC functioning properly. Receiver DMT controller switches to new bit table at time specified by AOC acknowledgment. (b) Bit swap with lost AOC acknowledgment. Receiver DMT controller tries both switch settings using the FEC decoder correct/errors detect to arbitrate.

crosstalking DSL appears. This crosstalker is an HDSL, but one could readily produce similar results for other types of crosstalkers and/or noises. The 1.5 Mbps data rate is readily achieved with the HDSL crosstalker both when the bit distribution is properly optimized for 0 crosstalkers (34 dB margin) or for 1 crosstalker (19.3 dB margin). These margins do not include FEC coding gain.

However, if the original crosstalk-free bit distribution were to be maintained after the single HDSL crosstalker turned on, then the margin is *negative* 6dB (−6 dB) on some of the tones. These tones will make errors with a probability of about 1/10, leading to a few tones in error on each symbol. Forward error correction can correct these errors temporarily, but there will no longer be any

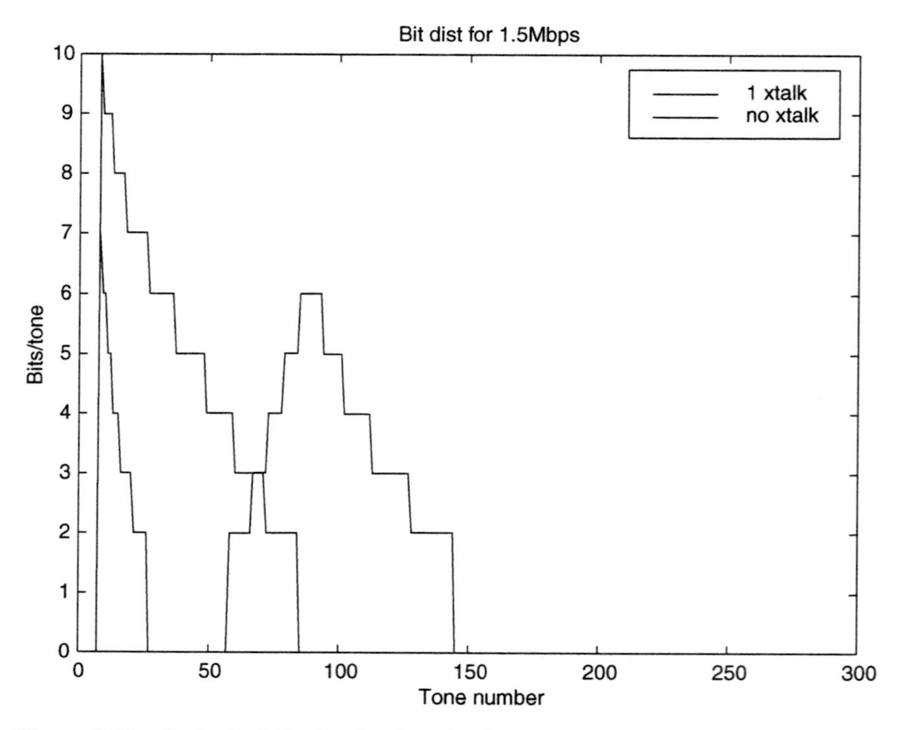

Figure 3.13 Optimized bit distributions for 0 crosstalkers and 1 crosstalker on 4.25 km 26-gauge twisted pair (xtalk is HDSL, background noise is –140 dBm/Hz).

immunity to impulse noise nor to further changes in the channel noise like additional crosstalkers. Any further slight disturbance results in unnecessary retraining of the modem and loss of service when bit swapping is not used. The performance loss is over 25 dB if bit swap is not used (19.3 dB – (−6 dB) = 25.3 dB). Bit swapping by contrast allows the first modem to vary its bit distribution to the new optimum bit distribution, allowing a full 19.3 dB of margin to be achieved. The swapping also protects the HDSL system as the ADSL actually creates less crosstalk for HDSL after swapping to the new bit distribution. Optimum bit swapping methods for discrete distributions (see [1]) will actually correct the worst tone (the one with lowest margin after the crosstalker ignites) first, then each successively better tone until the margin is restored. The very first few swaps will eliminate the most negative margins and begin to restore the link's ability to deal with additional noises. Eventually, depending on the speed of swapping (usually 20–100 swaps per second is a good number), the optimum 19.3 dB will be restored, although standards only mandate an ability to implement 1.25 swaps/second. However, a vendor's modem can request swaps more frequently and if they are implemented by the other modem, profit from the increased speed of swapping. For 20–100 swaps/second, the new bit distribution in our example would be implemented in a couple of seconds without any bit errors or loss of service or additional errors (beyond those that occur for other reasons). This bit swapping process is seamless without errors, and without service inter-

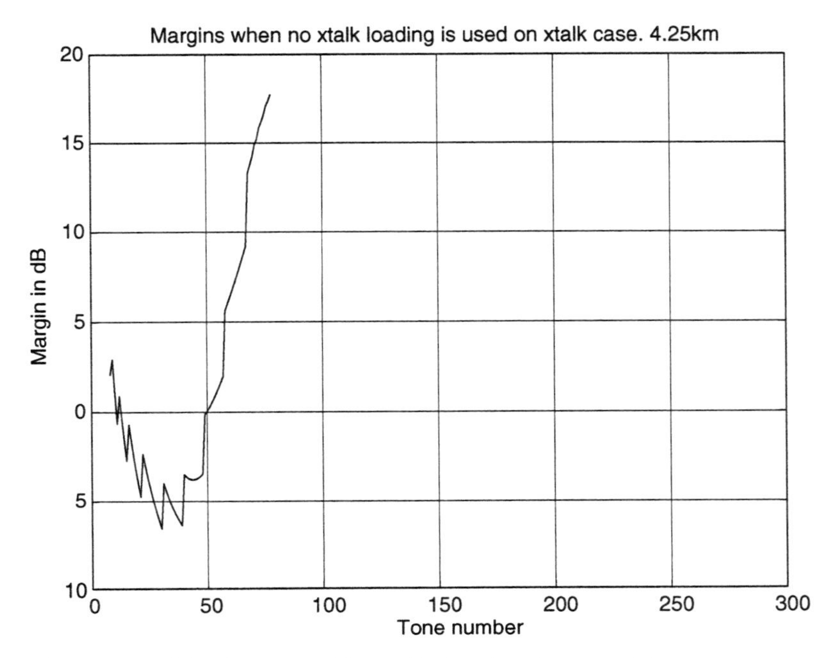

Figure 3.14 Margin for NO SWAPPING when the crosstalker is turned on.

ruption or need for retraining. There are many other types of crosstalk and other noise examples that lead to bit swapped correction of the DMT distribution. This simple and realistic example shows one case in which bit swap is necessary for restoration of margin and continued flawless operation of the DMT modem. There are many other similar cases that could be produced. Retraining or fast retraining need only be used for gross channel changes that essentially would otherwise have permanently disabled the modem.

3.3.2 Express Swapping Methods

A concern in ADSL modems is the time that it takes for swapping to implement a request. Data could be lost during this time if a large change occurred and the FEC were temporarily overwhelmed during the time that the swaps are correcting the situation. The solution to this is known as "express swapping," which allows an entire bit table, or equivalently any number of tones/subchannels, to be simultaneously specified in one command. This command is standardized for DMT VDSL, which has as a subordinate mode of operation an advanced ADSL modem [15]. The designer needs to be careful in the use of express swapping commands in that crosstalk into other channels can be dramatically altered in a short period of time, so while the present channel is corrected, the modem should not violate spectrum constraints into other modems. One can show that in all but pathological situations, a series of express swaps (often called distributed iterative water-filling; see Chapter 10) does actually converge to a stable energy distribution for all the lines involved. Fortunately, swapping tends to move energy away from crosstalkers rather than into the common spectra and thus is a stable process even when implemented swiftly in an uncoordinated fashion on multiple lines in the same binder.

The express swapping AOC command can dramatically increase the speed of swapping (by a factor of more than 100). Express swapping adds one additional AOC command and an associated express swap time-out mechanism. The command format is:

AOC Message Header	AOC Message Field Total Length Including Message Header (Bytes)	Interpretation
11110011	$2n+4$	Express swap request for n tones

An express swap command is sent only one time and has an internal 2 byte CRC protection for error detection. The first byte is a marker for the express command format, and the last two bytes are the CRC for error detection in the command receipt by the transmitter.

The express bit-swap request message has the following format:

Message Header	ES Control	1st Tone Index	1st Tone Total Bits/ Gain	.....	nth Tone Index	nth Tone Total Bits/ Gain	CRC
11110011 (1 byte)	Sup number & Tone count (1 byte)	Tone number (1 byte)	# of bits/gain (1 byte)		Tone number (1 byte)	# of bits/gain (1 byte)	2 bytes

The express swap (ES) control byte has its most significant bit set to 0 if the transmitter should implement the ES on the next superframe. This bit is set to 1 if the transmitter should implement the ES on the next-to-next superframe. The remaining 7 bits enumerate the number of tones, n, that are changed by the next $2n$ bytes in the command. Each tone has two bytes—the first indicates which tone should be changed and the second byte quantifies the change. In the second byte for each tone, the new absolute number of bits is a number between 0 and 15 and is encoded in the upper nibble (4 bits) according to 0000 for no bits, 0010 for two bits, ..., 1111 for 15 bits.[11] The relative gain is a 2's complement 4-bit quantity between –4 and +3.5 dB (with .5 dB increments) with most significant (sign) bit. 0 bits implies 0 energy.

There is no ES acknowledge command. The receiver that initiates an ES shall be responsible for monitoring the returned DMT signal to determine if the command has been implemented by the transmitter. If the swap has not been detected on the correct superframe, the receiver shall assume the transmitter did not imple-

[11]One might note that 0001 for 1 bit constellations currently is not used in ADSL standards, but should that change, then this setting could also then be included.

ment the command. The ES initiating DMT receiver may then elect to again initiate a second ES command, another AOC command, or a retrain. This command is not sent five times, but instead sent only once to improve speed. The CRC at the end of the command follows the same byte CRC protocol as used in initialization for confirmation of correct receipt of message fields. The polynomial used is $g(Z) = Z^{16} + Z^{12} + Z^5 + 1$, where Z is an advance of one bit period—equivalently numbering the bits starting with the first bit of the message header as m_0 through m_{16n+32} and forming $m(Z) = m_0 \cdot Z^{16n+32} + m_1 \cdot Z^{16n+31} + \ldots + m_{16n+32}$, the check bits are $c(Z) = m(Z)\text{modulo } g(Z)$.

The maximum number of bytes in a command (which likely is rare in occurrence) would be 260. A transmitter unable to accommodate this length of command would simply ignore it, thus forcing the receiver to perhaps attempt different corrective mechanisms, which could include breaking this large-length command into smaller commands.

Dynamic Rate Adaption

ES also provides a mechanism by which data rate could be adapted if allowed without disruption of data flow across link, which is also possible in present versions of bit-swap commands. However, use of such a facility requires higher-level control of the DSL link to allow such rate change. Thus, express swap does not change this particular aspect of ADSL operation, it just provides a faster mechanism for executing it if desirable. Otherwise, the receiver should ask for bit distribution changes that maintain the same data rate.

3.3.3 Intentional Tone Zeroing and Q-Mode

Tones may be zeroed in a DMT modem, which amounts to setting the corresponding entry in the bit table to 0 bits, which implies also that zero energy should be sent. Such zeroing has come into use for many reasons:

1. Conformance to spectrum masks associated with ADSL downstream and upstream transmissions, which may vary considerably on a worldwide basis.
2. Insertion of noncustomer-data carrying facility on the tone that is known to transmitter and receiver (and perhaps proprietary), such as mechanisms to reduce peak-to-average power ratio [1].
3. Power reduction when the customer is not transmitting data, but the modem remains on.
4. Aversion of egress in frequency bands that may be known or detected to be in use by various radio systems in the vicinity of the ADSL line.

Additionally, G.992.3 and G.992.4 ADSL modems have considered the use of a **Q-mode** of operation when idle or useless data occupy the line. The basic idea is to transmit known idle symbols that have reduced PAR ratios during this time period. If the PAR is reduced (by perhaps as much as 10 dB for some known and/or repeated DMT line signal transmitted during Q-mode), then the current delivered to the line driver can be reduced, saving power on each DSL line. In the DSLAM, this can save 1–2 watts/line (of a figure that today without statistical multiplexing factors is about three watts/line). In the upstream direction, this may be a smaller savings for a DSL modem in a battery-powered lap top, but total power consumption of the CPE modem can be then a few hundred milliwatts. The area is best summarized at time of writing by Carlson in [16], and no proposal had been agreed for standardization.

Basic Q-mode proposals share the following constraints:

1. They maintain the same average power spectral density as during normal operation to prevent time-variation of crosstalk spectra.
2. They transmit points from circular constellations (QPSK or 4 QAM) on all used tones.
3. At a detailed level, maintenance channels such as AOC and EOC are kept partially operational.

Areas of dispute seem to be how to communicate the "idle" condition from higher levels of protocol to the modem physical layer and whether to use very low PAR signals (basically a "chirp" DMT signal has the lowest PAR—see [17]—of 3 dB and various proposals try to approximate this) but also perhaps desire to maintain the signal constellation points used during normal transmission to simplify implementation. Statistical multiplexing and dynamic spectra management may be able to exploit the type of Q-mode signal used (see Chapter 10). The work in ITU Q4/15 has suggested that Q-mode (using reduced PAR transmission) would provide little benefit for transmitters using newer types of line drivers. Q-mode was not included in G.992.3/4, instead L2-mode was provided that is a reduced power mode used for periods while no payload data is being sent. Signal energy is maintained on the line during L2-mode to minimize changes in crosstalk energy to other lines (non-stationary behavior), and to enable rapid resumption of normal transmission.

3.3.4 Time-Varying Crosstalk

Usually with always-on DSL systems, crosstalk is relatively stationary in spectra (it may appear slightly time variant simply because of clock differences between DSL systems). A single excitation of another DSL in the binder to an always-on state can usually be accommodated by the swapping mechanisms described earlier. However, some crosstalkers periodically vary with time intentionally. Still others make gross adjustments in their spectra using nonstandard methods. These may create a problem for some DSLs, depending on the time-variation pattern of the non-stationary crosstalk.

The greatest single offender in terms of time-varying crosstalk are nonstandard ping-pong DSL systems that have been installed by some competitive local exchange carriers in unbundled DSL COs.[12] Such systems, while proprietary, can switch up-

[12]Some proprietary symmetric "ping-pong" DSL systems were introduced early in the history of DSL, and have introduced unique type of crosstalk that subsequent systems must attempt to deal with. Later, a large group of DMT ADSL/VDSL suppliers, instead dropped a "ping-pong" system and proposal for which they had a leading position in silicon-availability simply because it appeared that such systems might harm existing DSL systems. Spectrum management standards (see Chapters 9, 10, and 11) attempt to equalize the rewards and punishments for adhering to good-neighbor based crosstalk restrictions.

stream and downstream transmission at a rate of, for instance, 8000 times per second, or every 125 ms. However, any "ping-pong" rate is possible, and some systems dynamically vary the duty-cycle as a result of the payload data rate. The DMT receiver thus needs to know if crosstalk is time-varying. A classic error in the DMT receiver would be to measure noise on each tone as an average over time without regard to whether individual noise samples over a reasonable period of time were exhibiting statistically aberrant variation. For instance, with a ping-pong modem crosstalker one might expect to see roughly periodic variation in the noise variance. An ADSL modem with such advanced noise monitoring can bit load for the worst-case noise expected rather than the average, thus improving reliability (at the expense of data rate or range). Such a modem can also code differently or even try to vary the bit distribution with time to counter the crosstalk variation with time-varying bit distributions.

Annex C of the worldwide ITU G.992.1 and G.992.2 standards [3] is another example of a system designed for time-varying crosstalk in Japan. ISDN transmission in Japan is "ping-pong" or TDMA with a periodic variation of 400 Hz. Such a system introduces time-varying crosstalk into ADSL at least over the common ADSL band. Thus the ADSL system needs to correspondingly adjust bit tables according to a 345-DMT-symbol "hyperframe" that aligns with five successive 17 ms-long DMT-ADSL superframes (of 68 data and 1 synch symbol per superframe). This hyperframe is 85 ms long and corresponds also to exactly thirty-four "ping-pong" cycles of the Japanese ISDN signal. There are consequently 2 bit tables in each direction, one for each direction of transmission of the ISDN crosstalker. Swapping occurs automatically at regular intervals between the two tables—their use is standardized with respect to a 400 Hz network clock supplied to the modems in Japan. Additional bit-swapping is implemented within each of the tables with a slightly modified bit swap command that allows specification of which table the command applies as in Annex C of [3]. This is sometimes called *dual-bit loading.* Such noise would disable conventional PAM transmission, so SHDSL methods are dual-bit loaded symmetric DMT systems that use 1.104 MHz (256 tones) in both directions.

3.3.5 Loading with Codes

Reference [1] reports on several loading algorithms, including the basic greedy concepts suggested by Hughes-Hartog and developed by Campello. Levin and colleagues [10]–[13] have extensively studied this area and deserve perhaps equal or greater credit for the methods previously known as Campello algorithms, and henceforth now called "Levin-Campello" or just LC.

Energy Functions and Table

Although reference [1] simplified the explanation of loading algorithms using the "gap" approximation, both Levin and Campello independently noted that the gap is not exactly constant and that with integer numbers of bits per tone as a restriction, this constant-gap presumption can unduly reduce performance (in a worst-case situation by a dB of margin). The solution is to construct a table of next energies to transmit the next bit (for 0 bits to 15 bits) for each and every tone. This incremental energy table essentially scales with the inverse of the subchannel SNR

$g_n = |H_n|^2 / \sigma_n^2$ (where $|H_n|^2$ is the measured channel gain and σ_n^2 is the measured noise variance for tone n). Thus, the receiver needs only to store one table with sixteen entries (the last is "infinity" [large] as more than 15 bits is not allowed) and then scale for each tone. In some cases when a PSD constraint may be violated, a separate sixteen-entry table need be kept for each tone, or at least the maximum number of bits possible on that tone needs to be stored. The concept can be viewed as in Figure 3.15. Basically, the LC algorithms search the tone tables for the tone and associated index n_{next}, at which increasing the constellation size by one bit incurs least energy cost and also for the tone and associated index n_{last} at which decreasing the constellation by one bit saves the most energy. This concept is equivalent to the incremental energy concept used in [1], which simplified expressions using the gap approximation (and that simplification need not occur and causes performance loss).

A certain number of bits, including any parity or overhead bits, needs to be transmitted. The LC algorithm simply allocates each successive bit (up to the total) to the place of least incremental energy. Swapping is executed in the algorithm whenever the amount of energy to be saved on tone n_{last} exceeds the amount of energy to be added on tone n_{next} by a threshold amount. The threshold is set sufficiently high to prevent continuous swapping unnecessarily and may be on the order of 1 dB (a 1 dB savings on one tone does not correspond to a 1 dB overall savings, and indeed is far less overall with many tones used in DMT). In older ADSL standards, 1-bit constellations were not allowed, but that restriction has now been removed so that swapping is simplified and the algorithm here applies directly. When

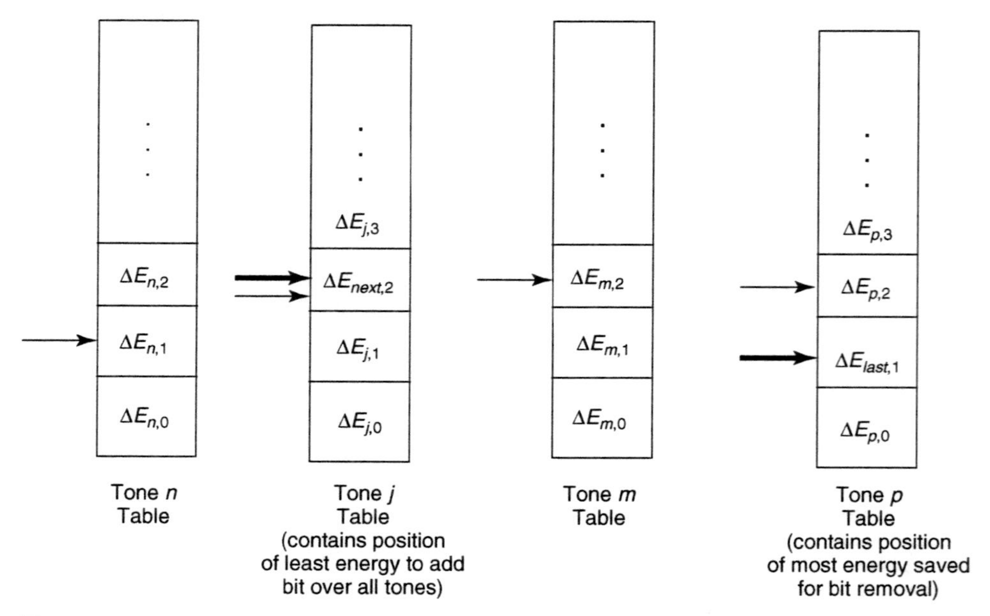

Figure 3.15 Bit-swapping table illustration. Light arrows indicate position to which energy for adding a bit to constellation on current tone is listed; heavy arrows indicate the positions of least cost and most saved energy over all tones.

swapping must occur with a 2-bit minimum, Levin has created and patented a series of exception procedures that address most of the contingencies in [11].

Two-dimensional trellis coding (which is not used in ADSL) or two-dimensional turbo or LDPC codes (which may be used in future ADSL) require only a simple modification of the energy table if parity and puncturing are well defined.

Four-Dimensional Trellis Coding Redundant Bits

When four-dimensional trellis coding is used with a system such as in Figure 3.15, the number of extra or redundant bits varies with the number of tones. Basically, there is one redundant bit for every two tones. When the number of tones is odd, one of the tones is paired with a tone carrying zero bits to create the appearance of an even number of tones. The total number of bits, including the trellis-code extra redundancy bits, is loaded on the tones as if the system were not coded. For encoding and decoding purposes, each successive pair of tones is presumed then to have exactly one redundant trellis coding bit with respect to the input of the encoder and the output of the decoder for that pair of tones.

When bits are moved into previously unused tones by the LC algorithm, or all bits have been removed from previously used tones by the LC algorithm, the number of used tones can vary. Thus, the number of redundant bits can vary. Levin's modification of the LC procedure in this case is to retain a counter for N^* the number of "used" tones (i.e., the number of tones carrying 1–15 bits of information). When N^* is even and increases to the next higher odd number, the LC algorithm adds one extra bit at tone n_{next}. When N^* is odd and decreases to the next lower even number, the LC algorithm deletes one extra bit at tone n_{last}. Thus, AOC swapping commands may carry two added bits and one deleted bit, or possibly one added bit and two deleted bits when trellis coding is used, but the actual data rate remains constant.

Gain-Swapping or Gain Adjustment

Variation of gain on a tone in ADSL is allowed up to an increase of 2.5 dB and any level of decrease (with zero gain meaning the tone is not used). Gain swapping can be used to move energy from one tone that may have an optimal solution to exceed a power spectral density mask to another that would not, or just to improve performance slightly. The gains are multiplicative factors that are applied to each tone in the ADSL DMT transmitter, and thus essentially can be viewed as part of the channel and absorbed into the table for the LC algorithm. Essentially, when the transmitted energy on a tone will exceed the maximum allowed by a PSD, that tone has extra energy. That energy can be reallocated to a tone that would not exceed the mask. The table for LC is appropriately scaled so that n_{next} will be forced to occur on tones that have will not exceed the maximum power spectral density.

Generally PSD masks in ADSL are not a good idea even though imposed by most transmission standards, including ADSL. They are imposed under the assumption that limiting the mask level will limit crosstalk. However, they may limit total power transmitted on a long loop at low frequencies below the 20 dBm limit, which reduces range substantially. A higher PSD at such lower frequencies will not cause damage to other services statistically because the crosstalk coupling is relatively low and ADSL modems will actually bit swap away from one another automatically to a globally better use of spectra between the modems. Although this still works with the PSD limit, it works better without it. Additionally, actual crosstalk coupling functions are highly variable so the likelihood of a problem occurring as opposed to just letting the modems find a better mutual optimum is actually favorable to the modems by a large margin. Imposition of spectrum masks (like –40 dBm/Hz in ADSL) in standards does allow for some level of variation, like 3 dB, which helps a bit. Nonetheless, engineers have for the longest time argued about and felt the need to impose spectral masks as perhaps a learned legacy of questionable need and habit than of technical necessity. More advanced systems such as those in Chapter 10 will further mitigate the need for spectral masks.

A classic use of gain adjustment is the situation where the channel SNR characteristic is fairly smooth when ordered. As the LC algorithm progresses, there is usually a jump in transmitted energy from the last tone, using b_n bits to the first tone allocated $b_n + 1$ bits. This produces a saw tooth character to the transmitted energy characteristic. Because the transmitter accepts during initialization a recommendation from the receiver to adjust its transmit energy levels according to the gain factors g_n, these factors may need to be adjusted when swaps occur so that the correct amount of transmit energy is maintained on each subchannel. For instance, if the LC algorithm suggests that a bit be moved from n_{last} to n_{next}, then there is a corresponding transmit energy adjustment on each tone, which is

$$E_{n_{next}}\left(b_{n_{next}} + 1\right) = E_{n_{next}}\left(b_{n_{next}}\right) + \Delta E_{n_{next}}\left(b_{n_{next}} \rightarrow b_{n_{next}} + 1\right)$$

and

$$E_{n_{last}}\left(b_{n_{last}} - 1\right) = E_{n_{last}}\left(b_{n_{last}}\right) - \Delta E_{n_{last}}\left(b_{n_{last}} \rightarrow b_{n_{last}} - 1\right)$$

and thus the gain adjustments are then

$$g_{n_{next}} = g_{n_{next}}\left(b_{n_{next}}\right) \cdot \frac{E_{n_{next}}\left(b_{n_{next}} + 1\right)}{E_{n_{next}}\left(b_{n_{next}}\right)}$$

and

$$g_{n_{last}} = g_{n_{last}}\left(b_{n_{last}}\right) \cdot \frac{E_{n_{last}}\left(b_{n_{last}} - 1\right)}{E_{n_{last}}\left(b_{n_{last}}\right)}.$$

For more details on loading and swapping, see [21].

Forward Error Correction Redundancy Percentage

Early work [18] on forward-error correction overhead with RS codes for ADSL suggested a rough 4 percent for 5–8 Mbps transmission, 6 percent for 2.5–5 Mbps, and 8 percent for less than 2.5 Mbps. These simulations were conducted for a few lines. FEC percentages are restricted to use even numbers of parity bytes and so these numbers can only be approximated in practice. Furthermore, the exact best level of FEC to use on any line depends on the line, the noise, any impulse noise, and finally the delay requirement. Thus, many have noted that the above rule of thumb can reduce performance in situations different from what were initially derived.

Ideally, the designer knows in advance exactly how many parity bytes (see [1]) will be added in the RS code in ADSL beforehand, perhaps based on impulse-noise/delay constraints. The total number of bits, data plus parity for RS, are then loaded (with or without trellis coding according to the LC above) to the tones. The energy tables for incremental addition of a bit on any tone could be reduced by the presumed gain of the code. This presumed gain of the code would take into account the redundancy. However, the problem is that the amount of coding gain is not constant because greater redundancy increases the gain at the expense of data rate, so there is a trade-off. Thus, as Levin notes [13], it is possible to compute a "gross coding gain" for any FEC system, which is simply the amount of coding gain (as determined by graphing probability of error with and without coding at the same clock rate) without regard to the data rate loss. This number is relatively fixed for any given block length and number of parity bits at probability of error 10^{-7}, which is the standardized design point in ADSL. Levin provides the following useful table of such coding gain:

Gross Coding Gain at 1e-7 probability of error for GF(256) RS codes (hard decoding) – in dB

Block/Parity	2	4	6	8	10	12	14	16
20	2.76	4.54	5.85	6.88	7.73	8.46	9.09	9.65
40	2.62	4.35	5.64	6.65	7.49	8.21	8.83	9.38
70	2.51	4.20	5.46	6.47	7.30	8.01	8.63	9.18
110	2.42	4.08	5.32	6.32	7.14	7.85	8.46	9.01
180	2.32	3.94	5.17	6.15	6.97	7.67	8.28	8.82
255	2.25	3.84	5.05	6.03	6.84	7.54	8.15	8.69

Each entry in the other table of incremental energies used in the LC algorithm can be scaled down by the amount in this table corresponding to the code parameters (block length 20, 40, ..., 255) and parity (2, 4, ..., 16). The LC algorithm is then performed for each possible combination of block and parity, and the system with the highest margin (or highest data rate for rate adaptive) over the RS code-parameter choices can be maintained. Interleave depth for impulse noise protection may then be determined. If impulsive noise is present, other factors may enter the choice of code parameters outside of the loading algorithm described here. The gain can be 1 dB with respect to using the above rule of thumb.

3.4 RF ISSUES

Radio frequency issues were discussed in [1]. However, some new information has emerged with respect to AM radio egress that merits further comment for ADSL. This section also addresses some issues not fully appreciated with early ADSL.

3.4.1 Egress into AM Radio

The egress problem is related to the fringe coverage area of an AM radio station. Some regulators in various countries (for instance, the United Kingdom) have suggested that an AM radio receiver may be as close as 1 m to a telephone line. A radio that close on the fringe area of coverage of an AM radio station can easily be affected by egress of average telephone lines. A radio station could then argue that they should pay less for their radio license because fewer customers are being reached.

A compromise that seems to work is to regulate that a DSL receiver must lower its spectra by 20 dB in the band of any AM radio station that is fringe in the area of the line. Such a radio signal may not be sensed by the DSL line as a noise disturbance. The programmable mask of DMT must be used in such cases to zero tones in the band of the known, fringe AM station. This with the use of Bingham's canceller in [1], for instance, can reduce the spectra by 20 dB and satisfy the objective. It is particularly fortuitous that DMT allows this reduction. An interesting observation is that the highest speeds of what are called SHDSL and also single-carrier-based VDSL do not have any mechanism to implement this reduction in the AM radio band from 560 kHz to 1600 kHz (DMT-based VDSL does have this capability, and does allow symmetric transmission of the same speeds as SHDSL also) or in other bands for other radio signals. In any case, ADSL solves this problem but the operator deploying ADSL must know to preset the modem to zero such bands in a CO neighborhood on the fringe of an AM geographic coverage area.

The performance loss of zeroing one or a few tones is minimal. This particular egress problem was not observed until ADSL was deployed on a wide scale. Fortunately, DMT-based ADSL standards anticipated the general problem and have a provision for it.

3.4.2 Ingress Issue

AM radio ingress usually does not occur on the fringe, but rather close to the radio tower where AM signals are strong enough to couple into phone line at sufficiently high voltages. As in [1], DMT loading will sense the problem and zero the tones in the band of the AM radio signal.

3.4.3 Spectrum above 1.1 MHz

Downstream tones 128 through 256 have images in the band from 1.104 MHz to 1.656 MHz, and thus also overlap the AM transmission band. ADSL line balance at these higher frequencies may be less and thus filtering above 1.104 MHz in the ADSL modem may need to consider AM radio egress. However, the solution afforded by DMT is that again in the fringe area of a radio station, the image frequencies may be zeroed to reduce energy in the band of an AM radio signal anywhere between 1.104 MHz and 1600 kHz. Also, amateur radios in the 1.8 to 2.0 MHz band can alias to frequencies from 200 kHz to 400 kHz, so that often DMT ADSL modems exhibit notches in these low bands.

3.5 THE ANALOG FRONT END (AFE)

ADSL transmission systems benefit greatly from well-designed analog interface to the telephone line. Transmit filter designs are covered in [1] in detail. This section instead discusses precision requirements and sampling methods for both transmit DAC and receiver ADC on an ADSL line. Basically the requirements are stringent. Fortunately, although perceived as difficult at the time of standardization of first ADSLs, today, analog front ends (or AFEs) for ADSL abound in low-cost high-performance packages.

3.5.1 Linearity and Noise Requirements

ADSL at time of conception, and still today, is one of the more challenging analog designs known to engineers. The basic problem can almost seem insurmountable in that a signal with up to 20 dBm of transmit power, with peaks to 32 dBm, is present on the same line with a severely attenuated signal from the other direction, possibly by as much as 60 dB with (even with that attenuation) an SNR of 30 dB. That means the dynamic range of at least a few components, presuming hybrid reduction of 12 dB, may need to be as high as 110 dB. Prior to ADSL, no one had attempted a linearity for phone-line coupling that exceeded 70 dB. Fortunately, not all analog components need exhibit this huge dynamic range. Additionally, the noise floor of an ADSL modem is sometimes limited by the modem's internal components. Although ADSL standards suggest –140 dBm/Hz as the lowest noise PSD of concern, thermal noise on a telephone line can actually be as low as the fundamental physical limit of –173 dBm/Hz (kT at room temperature). Designers who desire maximum range on DSL lines actually try to set internal noise floors between –155 and –160 dBm/Hz. An early DSL company called Amati was able to prototype a "Prelude ADSL" modem with in excess of 110 dB of linearity and –156 dBm/Hz noise floor (of which about 200 were constructed and delivered to various ADSL-interested companies around the world who were subsequently able to evaluate and then devise methods for reproducing such performance in mass). Others then proceeded to capture the concepts in mass-produced cost-effective packages.

Figure 3.16 depicts the essential components of the *analog front end* (AFE), which is often 1–2 integrated circuits and associated discrete components in a DSL modem of any type. The asymmetry of ADSL allows some cost reduction and opportunity for performance with respect to symmetric transmission, and the AFE component at each end may be different. At the ATU-C side, the ADSL modems are colocated for many lines within a DSLAM, and there may be several then integrated on a single chip to save space and power at the CO. At the ATU-R side, there is only one line typically so that integration of multiple AFE's is of limited value if any unless multiple lines are coordinated (see Sections 7.5 and 10.5).

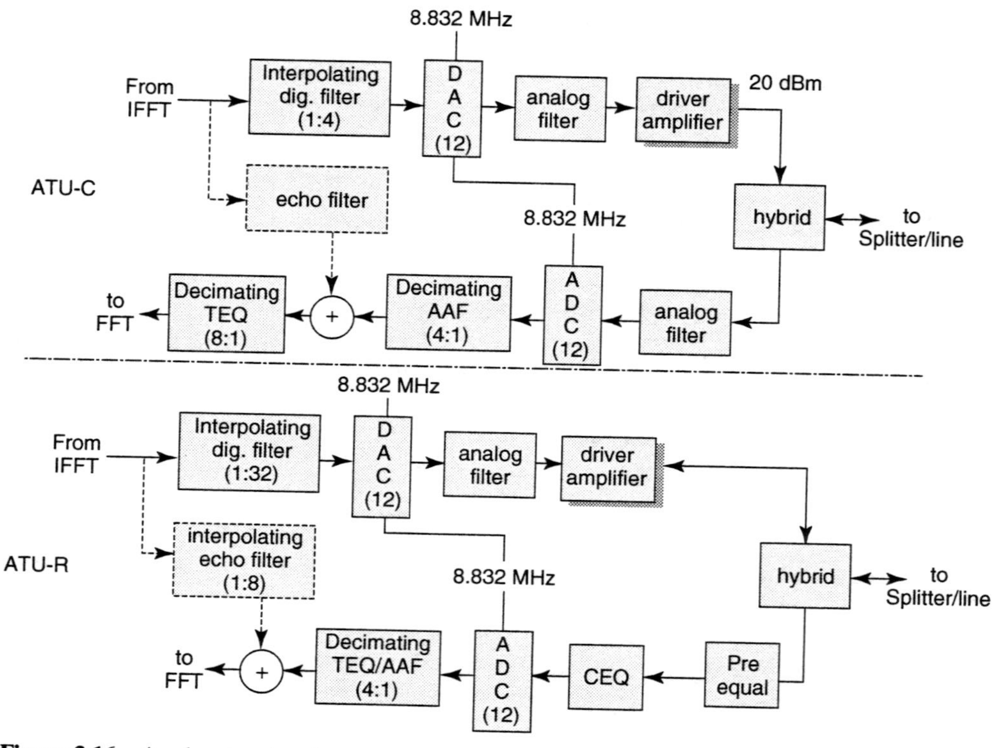

Figure 3.16 Analog front ends.

Typically, AFE chips are isolated from other ADSL-DSP chips for cost/performance reasons. The point of separation is usually the interface to/from the DAC, and ADC so Figure 3.16 contains more than what is typically on the AFE chip.

3.5.2 Central Office Side

The transformer that couples to the line is one device that needs to exhibit high linearity.[13] Fortunately, linearity is a function of magnetic core choices and winding techniques. With sufficient volume in production, these factors have not yet been significant. However, as the rest of the integrated ADSL modem continues to drop in price, the transformer could again become a significant cost item. It is also possible to introduce magnetic feedback in designs to achieve in excess of –120 dB linearity. Some of the most creative concepts involve the use of magneto-electro-mechanical modules that may ultimately allow integration of the transformer into the AFE chip. If the splitter circuit is implemented in analog, it also needs to main-

[13]Many DSL suppliers today may sacrifice linearity and try to deal with the consequent distortion in digital signal processing, which may become more common in the future to squeeze the last costs from mass-produced products.

tain the high linearity, although today increasing interest has shifted to ADSL systems that do not require a central office (nor customer premises) splitter and instead deliver voice service digitally (see Section 3.7).

The line driver is ADSL's largest consumer of power. Power consumption is determined by peak line-drive power (not average), which is 32 dBm in a well-designed DMT downstream transmitter. Even the best designs today consume 1 watt in the driver circuit, which is about 10 percent efficiency relative to the 100 mW transmit power. This is the dominant part of transmit power and the entire modem may require otherwise less than another watt. The solution that has started to emerge is the exploitation of quiet periods of the always-on DSL modem to transmit a signal with only about 6 dB peak-to-average ratio. This is the L2 mode of G.992.3/4 mentioned earlier, possibly augmented by a Q-mode in the future.

The driver power consumption may be reduced by over a factor of 2 statistically when the ADSL modem traffic is predominantly data. This requires a line driver that scales its consumed power according to a supplied control signal, which is now common in ADSL designs. Furthermore, DSLAMs with many modems draining power from a common source may reduce transmit power where it is not necessary on shorter lines, thus on average reducing power consumption per line. (See also the iterative-waterfilling concept in Chapter 11.) The driver and associated filter also need to maintain linearity at high levels. Frequency-division multiplexed (FDM) ADSL modems, also known as Annex A and B (G.992 standards [3]) divide upstream and downstream transmission by frequency. Such a design relaxes the linearity requirement for the line driver because the large downstream signals do not overlap the frequency band of the small received upstream signals because the transmit high-pass and receive low-pass filters prevent such overlap. The DAC and linearity requirements in this case are determined by the downstream path (presuming some analog high-frequency boost on long lines in the downstream receiver) at 14 bits, or equivalently 90 dB for the filter and driver analog components. See [1] for filter and driver designs.

With increasing stress on higher upstream and downstream data rates, FDM designs have become increasingly replaced by echo-canceled designs. Echo cancellation allows overlap of upstream and downstream signals over at least the lower 138 kHz (and sometimes lower 276 kHz when ISDN is also on line). Echo canceled modems achieve higher data rates. Echo cancelers can be used in FDM modems to simplify the analog filter design in the overlapping transition band. Full overlap can force the use of 16 bit DACs and up to 18 bit ADCs to maintain the 110 dB linearity requirement. However, full overlap is rare. Because CO upstream bandwidth is 138 kHz, oversampling at 8 MHz or higher is used with clever digital filters and a smaller number of bits in the ADC, to meet its requirements and typically also the transmit DAC. Often in practice, manufacturers do not completely echo cancel nor completely implement FDM, but instead find an acceptable trade-off point in terms of cost and performance.

Transformer and driver linearity may also be reduced by hybrid circuits. Typically a fixed hybrid circuit provides at least 12 dB of signal reduction, which was used in the above calculations. A reduction of 24–30 dB is possible with adaptive hybrids that match their internal balancing impedance to the line. Such a design reduces all linearity requirements and ADC/DAC requirements by 12–18 dB or equivalently 2–3 bits, allowing for super low-cost and high-performance ADCs. An automatic gain control (AGC) circuit precedes the ADC and is discussed in [1]; its use ensures the maximum number of bits are always best used by the ADC. It can also be used to reduce the number of bits in the ADC, although effectively the ADC and AGC can be viewed at a higher level as an ADC with the same degree of linearity.

3.5.3 Customer Premises Side

The CPE side AFE is also shown in Figure 3.16. The essential differences are that the line driver need only transmit 14 dBm of power, saving a bit everywhere nominally, but the receiver sampling rate is eight times higher so that oversampling of the ADC provides less increase in resolution than at the CO side. A splitter circuit may or may not be used, and linearity requirements are essentially about 1 bit less than the CO side. Low noise is particularly important at the CPE end where it can extend the range of usable frequencies and thereby range at any data rate downstream. Echo cancellation and adaptive hybrids can also be used to reduce requirements.

The prefilter before the ADC is usually a set of filters with increasing amounts of high-frequency boost for increasing long lines. This has the effect of altering the channel signal and noise levels with respect to the ADC noise floor at higher frequencies, effectively making the AFE look as if its ADC has more effective bits. For a complete analysis of this type of filter, see Chapter 7 in VDSL, where it is also used.

The receiver analog filter (Figure 3.16) needs to have very low noise. Often, this can happen only with the use of very special design of gain control circuits. On a very long line, the signal needs to be amplified after the prefiltering to significant signal levels. However, on a short line or null loop, such amplification is unnecessary and would lead to saturation of the ADC circuit. The low noise floor is necessary only on the long loop. Thus gain control circuits and design need exhibit the low noise floor (−140 dBm/Hz or lower) only on long loops.

Perhaps the greatest single noise source often overlooked by new ADSL designers is the noise of the other ADSL components themselves. Layout of components in an ADSL receiver is very important. Digital electronics needs to be isolated well from analog components, because the noise associated with digital power supplies and digital logic is often well above the levels needed for ADSL. This is why AFE components are most often separate from digital-signal-processing components.[14] Optical isolation of the analog and digital sections is extremely effective but expensive, although some groups have been able to achieve the same type of effect with clever proprietary designs. Analog sections and ground planes should be isolated as best as possible from digital ground planes. Additionally, the inside of a PC is a very harsh environment and requires careful attention to the radiation of energy from other parts of the computer into the ADSL board. Special metal (or mu metal) enclosure of critical analog components (analog transformer,

---

[14]Despite warnings and understanding of the problem, one major early modem supplier attempted a single-chip ADSL modem and was late-to-market due to noise problems, and consequently dropped from the ADSL market.

and early filter stage components) may be necessary to ensure that noise levels are not artificially high. After engineers have labored for years to design effective ADSL standards, it is self-defeating for the final component layout and enclosure to cause more noise and distortion than the telephone line itself.

Although this area is somewhat of a "black art" (i.e., one dominated by those who know special secrets and tricks that remain proprietary to the employing company), good design practice and careful consideration of the AFE may lead to enormous advantage of one vendor's ADSL product over another. With range being particularly important to phone companies because of the labor costs associated with special service visits, it may not be the cheapest component that actually is the lowest cost for the ADSL system. This can be a lesson hard learned for some, and ADSL designers are well advised not to underestimate this area.

3.6 OTHER WIRING ISSUES

3.6.1 Customer Premises Wiring Issues

Customer premises wiring has become important for ADSL with the advent of splitterless service. Figure 3.17 shows two internal wiring configurations for a residence. In the first, a splitter appears near the entry to the home, separating the internal POTS network from the ADSL network. The ADSL modem is on a separate wire that is often well maintained and free of further serious transmission degradation. Often this modem is close to the splitter and may be a part of a home gateway that serves as a bridge to CAT-5 quad internal wiring, coaxial cable, wireless LAN, or possibly combinations thereof, premises distribution. An alternative that is actually the dominant installation of ADSL today is the splitterless configuration also in Figure 3.17. In this case, the ADSL modem is in or near the PC and is simply used as an upgrade to earlier voiceband modem internet connection (indeed the modem may have v.90 or v.92 backward compatibility). At or near the personal computer (PC) can be a router (Ethernet, USB, or possibly other) that passes signals to the PC and possibly to other PCs or application devices (i.e., Internet telephones, televisions, etc.) within the home. Alternatively, the modem may be internal to the PC. In any case, the second configuration is simpler for installation by the customer, but allows the ADSL signals to be subject to internal wiring that may have bridged taps, flat-pair with increased noise power spectral densities, phone-on/off-hook transients, and greater attenuation. Thus, the range of the ADSL system from CO to CPE may be reduced because of the internal-wiring at the CPE end that reduces performance margins. Redistribution of signals to other application devices within the CPE may occur on the same twisted pair at frequencies above the ADSL band, which is sometimes known as HomePNA (home phone network alliance) or G.pnt in the second configuration.

The advantage of splitterless is the ease of customer installation, which cannot be underestimated in terms of impact on cost and desirability of service. The disadvantage is the harder transmission path, resulting in a lower data rate. The design of splitters was previously treated in [1]. Since that time, so-called in-line

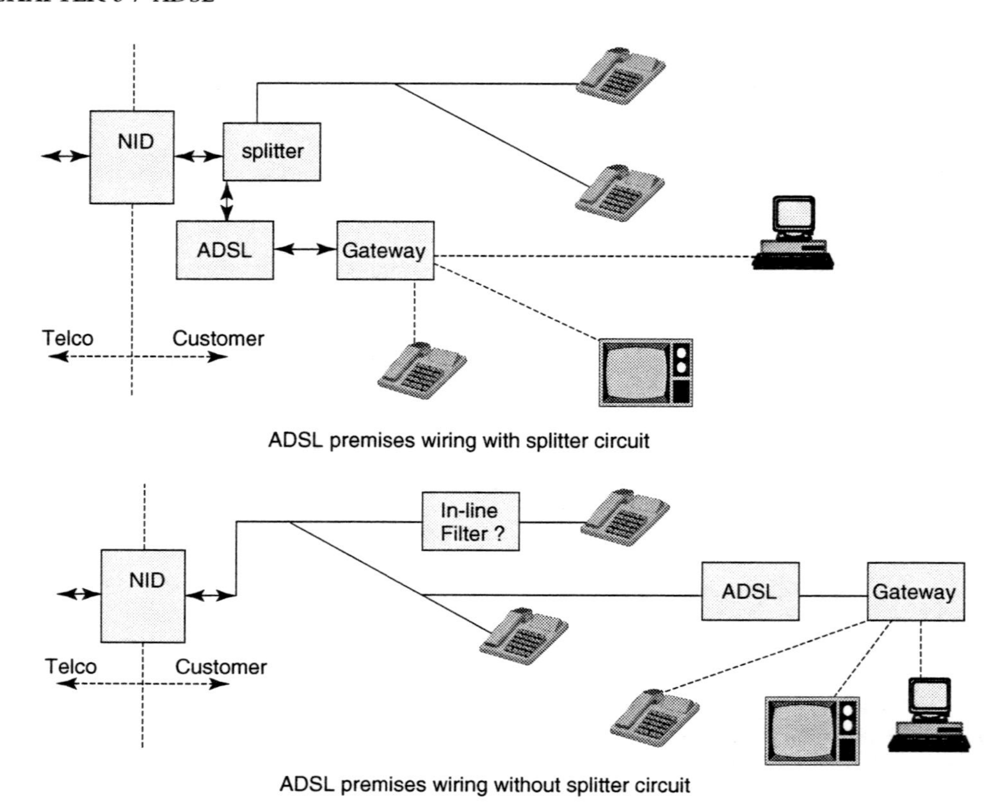

Figure 3.17 Customer premises wiring options.

filters have been standardized [26]. An in-line filter is a two-way low-pass filter that is installed at or near each (and hopefully every) telephone that shares the same line as ADSL. The in-line filter protects the phone from the upper frequencies used by ADSL and also prevents the higher-frequency impulse like noise from ring and on/off hook transients from entering the ADSL modem. The in-line filter also often can be used at the customer's point of access to "hide" internal bridge-taps, noises, and attenuation if the modem is also placed there. If the modem is still placed deep within the premises, the in-line filter does not hide these effects.

This subsection proceeds by investigating some of the more pronounced issues with internal-premises wiring, starting with bridged taps, then premises noises, and signal propagation loss.

Bridged Taps

Bridged taps were modeled in [1] and are discussed at length in [1] and Chapter 7. There are two basic propagation issues with the bridged tap: signal loss and reflections.

Signal loss is a simple concept: the energy traversing the phone line from the central office is divided at the bridged tap position. Half the energy remains on the line and half goes into the bridged-tap section of the line. If the bridged-tap is terminated in the 100 Ohm characteristic impedance of the line, then half the energy is lost each time a tap occurs. So, several terminated taps before the ADSL modem could result in severe signal loss, for instance, 3 terminated taps before the modem corresponding to three extension phones would mean a factor of 8, or 9 dB, of signal loss. However, usually taps are not terminated and instead are an open circuit at higher frequencies, meaning that the signal energy is reflected. The reflected energy has a delay at any frequency with respect to the energy at the same frequency on the main line. Thus, although some energy does return to the main line (which is good generally for DMT modems as they can collect and use this energy, something in-line filter designers may forget when trying to design the low-pass filter to match line impedance, to the detriment of the ADSL system), the energy may add destructively at some frequencies and constructively at other frequencies, creating a "rippled" frequency characteristic. The characteristic can never exceed in magnitude at any frequency the level that would have been present had there been no bridged tap. Thus, even with in-line filters, bridged taps cause performance loss if they are not "behind" the filter. DMT modems can recover some of the loss (whereas other modems like QAM or PAM often cannot, even with infinite-complexity decision feedback equalizers); there is nonetheless some level of signal loss. This is a fundamental trade-off for splitterless ADSL. Some bridged taps of course occur before customer premises and thus cause signal loss even when splitters are used. Phone companies may elect to cut or remove these bridged-taps when they can find them (see Chapter 8). The longer the bridged tap, the closer the loss will be to the full 3 dB that characterizes the perfect termination case, and also the lower the frequency at which destructive interference occurs (and thus perhaps within the ADSL band).

The reflection of a bridged tap is in some sense good as signal energy returns to the line and some will pass to the ADSL receiver with delay. This energy is recoverable, but requires an integrating mechanism in the receiver to recover both first and delayed energy. In DMT, this occurs because of the long FFT block length, that is, the FFT has enough delay in it. However, the TEQ of [1] can also be useful in ensuring that energy delay is appropriately phased so that it is effectively partially contained within a single symbol boundary of the FFT. In QAM and PAM modems, this integration occurs in the feed-forward filter of the DFE (when it is long enough, which can be 10 times longer than is in use in practice for cost reasons). However, fundamentally, depending on the length of the tap and the group, a very deep notch can occur from reflected energy on short bridged taps or those of less propagation attenuation (fatter gauge wires for instance, as in flat wiring). A DMT transmission system will sense the problem and vacate that and adjacent frequencies—this is optimum and significantly enhances performance. DFEs *cannot* implement the same effect and will suffer performance loss.

Overall, DMT ADSL has some degree of capability to work with reasonable numbers of bridged taps on most lines, thus enabling splitterless DSL operation. However, bridged taps do reduce performance even in the best of situations, and so fewer taps means higher data rates and reliability of the DSL connection.

Radio Noise Ingress

Many homes use flat-pair internal wiring on telephones. Without twisting, radio noise is more easily coupled into phone lines within customer premises. The effect is similar to the bridged tap, except there is no signal loss (other than the propagation loss of the flat pair itself), in that the signal-to-noise ratio reduces at the radio frequencies of AM broadcast (or other less well-known) signals in the vicinity of the phone line. Some signals will couple much more strongly than others depending on the geometry of the radio antenna and the relative orientation of the phone line. The proper solution in this case, theoretically and implemented in DMT ADSL, is to turn off the bands that correspond to the radio signal. This happens automatically in the DMT modem.[15] Such silencing thus reciprocally protects radios in the vicinity of the phone line from egress/emissions of DSL signals. It is usual for ADSL lines to see 2–3 AM radio signals that are sufficiently strong to cause silencing of a few tones each of the DMT ADSL signal.

Impulse Noise on CPE

Impulse noise couples better into flat pairs than into twisted pairs, and thus can be increased in amplitude within customer premises. Also sources of impulse noise are often within the customer premises and close to internal wiring (e.g., phone lines in an elevator shaft close to magnetic or electric motors that generate noise, refrigerator motors, dimmer lights, etc.). Impulse noise has been previously discussed, but splitterless operation increases such noise and its performance reducing effects.

Propagation Loss

Customer premises wiring is simply an extension of the phone line and thus additional loss occurs from attenuation, with highest frequencies attenuated most on average. Customer premises wiring can sometimes be up to 1,000 feet, which may lead to as much as 6 dB of additional signal attenuation. This occurs even in systems with splitters. Many in-home telephone lines use PVC insulation that has very high attenuation at high frequencies. This causes additional loss.

Repeatered ADSL

Some companies have investigated ADSL repeaters within CPE—with the cost of ADSL modems presently being very low, it is possible to simply cascade two of them at the premises entry. The signals are effectively "repeatered" at that point. Other companies simply try to amplify electronically the ADSL signal as it enters CPE before the signal undergoes addition of any additional noises (impulse, radio) or sees the signal loss of bridged taps or long internal lines.

Other companies have designed ADSL repeatered remote terminals that effectively concentrate ADSL modems as well for remote regions where ADSL lines

---

[15]Incidentally, any other transmission system necessarily suffers performance loss if it tries to use these frequencies that theoretically should have been silenced.

may be much longer than normal. These repeatered modems are sold to telephone companies rather than to the customer directly.

3.6.2 Wired and Wireless Home Gateways and Distribution

Many companies today have focused on what is called the home or CPE gateway. A gateway is simply another name for a router where one network, namely, the ADSL network, is connected to one or more networks within the CPE. These other networks may be Ethernet on CAT-5 (or CAT-7) wiring, coaxial cable, wireless, or may even reuse the existing copper. Some have even considered using power lines. The gateway contains the physical-layer modems as well as higher-layer routing of the application signals and conversion into a format that the DSLAM on the CO side can decode and appropriately send to the proper service interface (class 5 switch, Internet service provider, video service provider, etc.).

Two more dominant means of distribution of signals within CPE have emerged. Ethernet today runs at 10 Mbps (10BT), 100 Mbps (100BT) and 1000 Mbps (Gig Ethernet) on cat-5 wiring. Typically, the cable is well designed, has little interference or emissions, and relatively low crosstalk. Simple transmission is used at the physical layer and distances are limited to 100 meters. Several locations may branch from the same gateway in a "home-run" (or "star") configuration where each separate device (i.e., computer) has a dedicated physical connection to the router. The customer pays to install (or previously paid to install) the category 5 (or 7) wiring. Typically, this is far less expensive than running any kind of cable (typically fiber) to the CPE. Wiring within the CPE is usually easier because there are existing passage ways for signals under floors, behind walls, or perhaps designed into the structure. Cost of installation may be a few hundred dollars by a professional installer or possibly free if installed by the customer themselves. Ethernet connections and routing mechanisms are well accepted and used by the customer already and somewhat ubiquitous. Thus ADSL-Ethernet bridges/routers or gateways are common. This tends to assume TCP/IP protocol is used on the Ethernet section.

Ethernet signals (100 Mbps and 10 Mbps, but not 1000 Mbps) may be sent on coaxial cables also (although that should not be the same cable as for TV systems).

Wireless interfaces make use today of the 2.4 GHz-band IEEE 802.11(b) 11 Mbps standardized interface [27] for distribution within premises. This transmission is less reliable and slower than Ethernet and can be affected by the position of the receiver within CPE (much like portable phones work in some locations but not others). The IEEE 802.11(a) standard [28] is increasingly of interest in the less-crowded 5 GHz transmission band, and can transmit to 55 Mbps and possibly beyond. Reliability issues in 802.11(a) and low-speed in 802.11(b) have led to the formation of an 802.11wng group that will standardize multiple-antennae wireless LANs with speeds to 500 Mbps.

The gateway allows a number of application devices to share DSL, thereby amortizing the cost of the ADSL service deployment and modems over the various application service fees. For example, a voice-over DSL service could be shared by several neighbors in the vicinity of a phone line with ADSL service, with wireline or wireless distribution of the POTS signals. Then the cost of the single ADSL service provides a mechanism to have significantly higher revenue than might be provided by a single customer. Clearly the same is true of the high-speed Internet service, or possibly even a video service. Such sharing may be particularly accelerative of DSL demand and particularly effective in crowded population areas and hotels.

3.6.3 Central Office Wiring Issues

CO wiring is simpler than the customer premises, but still complicated by ADSL's splitter. Figure 3.18 shows the main distribution frame in the central office (or remote CO terminal) where lines are essentially connected and reconnected to DSLAM (or POTS switch) connections. Where one connection on the MDF was necessary for POTS, now three occur for ADSL and POTS. Additionally, space for the splitters has to be provided. Some companies have investigated DSLAMs that also accepted digital POTS inputs and convert on the ADSL modem card the mu-law or A-law digital POTS signals to analog and then combine them with the ADSL signals (perhaps in a single DAC) for transport. Thus, the MDF needs only one connection, although clearly the POTS switch now needs to be connected to the DSLAM somehow, presumably through high-speed fiber interface that aggregates all the digital POTS signals for the lines shared with the ADSL service by the DSLAM.

Digital POTS service within ADSL has become increasingly of interest and eliminates some of the wiring issues at both CPE and CO end and is discussed in the next section. There are also new MDF solutions that integrate splitters and thereby reduce CO wiring.

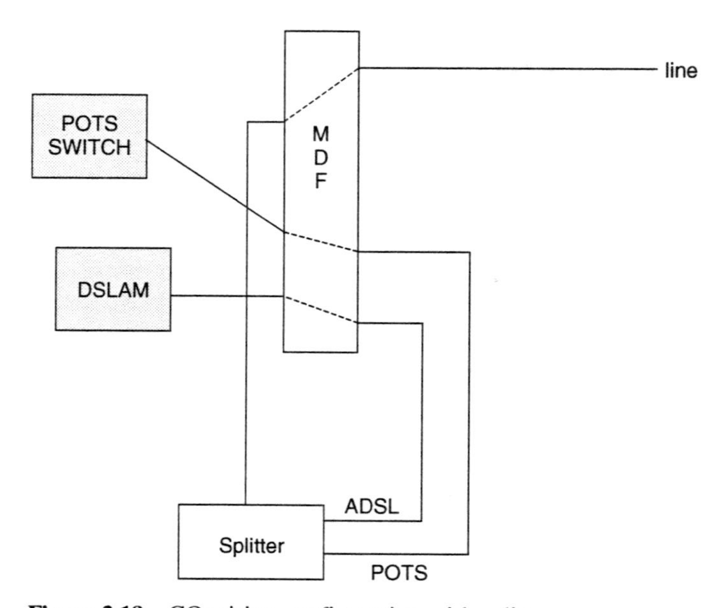

Figure 3.18 CO wiring configuration with splitters.

3.7 ALL-DIGITAL LOOP

The ITU Q4/15 update to the Recommendation G.992 includes an all-digital mode providing for the use of the 0 to 25 kHz band for upstream and/or downstream digital transmission. In essence, DMT tones fill this band in place of the traditional analog voice signal. The all-digital loop provides approximately 256 kb/s of additional capacity in each direction. The benefits of all-digital loop are most pronounced on long lines (over 14 kft) where there is little capacity at higher frequencies. The additional upstream capacity is particularly useful for voice-over DSL systems that require more nearly symmetric data rates.

3.8 ADSL2 SUMMARY

The following summary of the new aspects of the second generation ITU ADSL Recommendations G.992.3 and G.992.4 (2002) is adapted from information provided by Frank Van der Putten, the ITU Associate Rapporteur for G.992.3.

3.8.1 Improved Performance

ADSL2 increases the bit-rate and line-reach performance of ADSL, enabling up to 256 kb/s of additional performance on typical lines. A further increase in performance is possible when using the all-digital mode described in Section 3.8.7 in combination with the enhancements described here. The specifications now address bit rates up to 8 Mbit/s downstream and 800 kb/s upstream. An emerging ADSL standard can be scaled up to 15 Mbit/s downstream and 1.5 Mb/s upstream (on short lines). Trellis coding is now mandatory, whereas it was previously optional. 1-bit constellations enable additional capacity, using tones that would previously not have been used. FEC overhead support of 16 parity bytes is also mandated. Data can now be modulated on the pilot tone. Framing overhead has been reduced. The easily met performance requirements in G.992.1 have been replaced by far more complete and strict performance specifications in DSL Forum TR-048 for North America (annex A), and ETSI TS 101 388 V1.3.1 for Europe (annex A & B).

3.8.2 Loop Diagnostic Tools

A special diagnostic mode initialization has been added to help start-up on troublesome lines. Additional double-ended line testing has been specified for trouble resolution; this defines the following measurements to be performed by both the ATU-C and ATU-R: line attenuation, quiet line noise, and signal-to-noise ratio over the ADSL band. Messages are defined to convey these measurements to the other end of the line.

3.8.3 Improved Initialization

To improve robustness against bridged taps and RFI, the receiver can allocate the location of the pilot tone. RFI cancellation techniques are enabled by turning off transmission in selected tones when necessary, and a tone interleaving or shuffling option is included to disperse RFI distortion. Equalization is improved with spectrum shaped initialization signals. Rate negotiation is improved by replacing G.992.1's method of offering four rate options with the receiver choosing the bit-rate. To improve robustness, the receiver now determines the duration of initialization signals.

3.8.4 On-Line Reconfiguration (OLR)

Seamless Rate Adaptation (SRA) provides for in-service bit-rate changes to track application and BER requirements. The speed of the bit swap protocol has been improved. Dynamic rate repartitioning has been added for applications such as channelized VoDSL.

3.8.5 Power Management

The L2 low power mode has been added to enable statistical power savings based on user activity. Service is kept alive and full-rate operation is restored within 0.5 ms by maintaining low-bit-rate transmission during the L2 mode. Since signal energy is maintained during the L2 mode, non-stationary crosstalk behavior is minimized.

3.8.6 Framing

A new reduced overhead framing mode supports up to 4 frame bearers and up to 4 latency paths with delay and BER configurable per frame bearer. Framing now scales to support high data rates without using S = 1/2. The RS coding now scales better for long loop performance.

3.8.7 All-Digital Mode (No Underlying Analog Voice or ISDN Service)

The new all-digital mode extends ADSL transmission through the 0 to 25 kHz band to provide a total of 32 (Annex I) or 64 (Annex J) upstream tones. This enables an additional 256 kbit/s upstream data rate in addition to the other performance improvements described above, and is particularly important for improving performance on long lines.

3.8.8 Higher Layer Adaptation

Bonding of multi-pair operation via IMA for ATM-based ADSL is specified. Packet-based ADSL (e.g., Ethernet) support is described.

3.8.9 Home Installation

The specified architecture now includes in-line filters (splitterless), and a high impedance state is specified to permit multiple ATU-R to be connected to the line provided that only one ATU-R is active at a time.

3.8.10 Fast Start-Up (3 Seconds)

Fast start-up is provided from stand-by mode, sleep mode, and as an error recovery during Showtime mode.

3.8.11 Backwards Compatibility with First-Generation ADSL

G.handshake (G.994.1) will indicate support of multiple ITU-T ADSL (and other) Recommendations: G.992.1, G.992.2, G.992.3, G.992.4. ADSL modem implementations are expected (but not mandated) to support G.992.1/2 and G.992.3/4 for interoperability with existing deployments. G.992.1/2 equipment practice (e.g., DSLAM port density and power consumption) is expected not to be impacted by multimode support for G.992.3/4. PSD Masks are identical to G.992.1 for operation over POTS and ISDN, and spectrum management and deployment considerations are the same as for G.992.1.

REFERENCES

[1] T. Starr, J. M. Cioffi, and P. Silverman, *Understanding Digital Subscriber Lines.* Upper Saddle River, NJ: Prentice Hall, 1999.

[2] *Asymmetric Digital Subscriber Lines Metallic Interface.* American National Standards Institute, Alliance for Telecommunications Insititute Standards (ANSI/ATIS) North American Standard T1.413-1998 New York: ANSI, 1998.

[3] *Digital Transmission System for ADSL on Metallic Local Lines, with Provisions for Operation in Conjunction with Other Services,* International Telecommunications Union Standards G.992.1 and G.992.2 (G.lite), 1999, Geneva, Switzerland.

[4] H. Sadjadpour, "Encoder Structure of Multitone Turbo-Trellis Coded Modulation," *ITU SG15/Q4 Contribution RN-027,* Red Bank, NJ, May 21, 2001.

[5] S. Olcer, "LDPC Coding Proposal for G.dmt.bis and G.lite.bis," *ITU SG15/Q4 Contribution CF-061,* Clearwater, FL, January 8, 2001.

[6] W. Yu, J. Cioffi, D. Gardan, and F. Gauthier, "The Impact of Impulsive Noise on DMT ADSL," submitted to *IEEE JSAC,* April 2001.

[7] B. Rolland, D. Bardouil, F. Clerot, and D. Collobert, "Impulse Noise Classification by a Non-Supervised Training Method," France Telecom Research and Development Report, December 21, 2000.

[8] R. J. A. Tough and K. D. Ward, "The Correlation Properties of Gamma and Other Non-Gaussian Processes Generated by Memoryless Nonlinear Transformation," *Journal of Applied Physics* 32 (1999): 3075–3084.

[9] J. Hagenauer and P. Hoeher, "A Viterbi Algorithm with Soft-Decision Outputs and Its Applications," *Proceedings 1989 Global Telecommunications Conference (Globecom)* (November 1989): 1680–1686, Dallas, TX.

[10] H. Levin and N. Teitler, "Method for Reallocating Data in a DMT Communication System," U.S. Patent # 6,122,247, issued: September 19, 2000, filed November 24, 1997.

[11] H. Levin, "A Complete and Optimal Data Allocation Method for Practical DMT Systems," submitted *IEEE JSAC,* April 2001.

[12] H. Levin, J. Djordjevic, and J. Kosmach, "Real-Time Selection of Error-Control Codes for DMT Systems," submitted for publication, *IEEE JSAC,* April 2001.

[13] H. Levin and J. Kosmach, "Method of Identifying an Improved Configuration for a Communication System Using Coding Gain and an Apparatus Therefore," European Patent 1,094,629 A2, issued April 25, 2001, filed September 29, 2000.

[14] J. Campello de Sousa, *Discrete Bit Loading for Multicarrier Modulation Systems,* Ph.D. dissertation, Stanford University, 1999.

[15] S. Schelestrate, ed., "Very-High-Speed Digital Subscriber Line (VDSL) Metallic Interface, Part 3: Multicarrier Modulation (MCM) Specification," *ANSI Draft Standard, T1E1.4/99-013R4,* February 19, 2001, Costa Mesa, CA.

[16] A. Carlson, "A Comparison of Q-Mode Proposals," *ITU SG15/Q4 Contribution RN-024*, Red Bank, NJ, May 21, 2001.

[17] J. M. Cioffi and J. A. C. Bingham, "A Data-Driven Multitone Echo Canceller," *IEEE Transactions on Communications* 42, no. 10 (October 1994): 2853–2869.

[18] T. N. Zogakis, P. T. Tong, and J. M. Cioffi, "Performance Comparison of FEC/Interleave Choices with DMT for ADSL," *ANSI Contribution T1E1.4/93-091,* April 14, 1993, Chicago, IL.

[19] B. J. Frey, R. Koetter, G. D. Forney, F. R. Kschischang, R. J. McEliece, and D. A. Spielman, Special issue on Codes on Graphs and Iterative Algorithms, *IEEE Transactions on Information Theory* 47 no. 2 (February 2001).

[20] P. H. Siegel, D. Divsalar, E. Eleftheriou, J. Hagenauer, D. Rowitch, and W. H. Tranter, Special issue on Codes on the Turbo Principle: From Theory to Practice, *IEEE Journal on Selected Areas in Communication* 19 no. 5 (May 2001).

[21] J. M. Cioffi, Class Textbook, EE379C—Advanced Digital Communications, http://www.stanford.edu/class/ee379c/.

[22] J. Lauer, "Turbo Coding for Discrete Multitone Transmission," Ph.D. dissertation, Stanford University, May 2000.

[23] B. Vucetic and J. Yuan, *Turbo Codes: Principles and Applications.* Boston: Kluwer, 2000.

[24] W. Henkel, T. Kessler, and H. Y. Chung, "Coded 64-CAP ADSL in an Impulse Noise Environment—Modelling of Impulse Noise and First Simulation Results," *IEEE Journal on Selected Areas in Communication* 13, no. 9 (December 1995): 1611–1621.

[25] I. Mann, S. McLaughlin, W. Henkel, R. Kirkby, and T. Kessler, "Impulse Generation with appropriate Length, Inter-arrival, and Spectral Characteristics," *IEEE Journal on Selected Areas in Communications,* May 2002, to appear. See also ETSI TM6 contribution TD20, Sophia Antipolis, France, February 2001, "Realistic Impulse Noise Model," by R. Kirkby of BTexact.

[26] ANSI/ATIS Draft In-Line Filter Standard, *ANSI Contribution T1E1.4/2001-007R2,* D. Brooks, ed., May 2001, Tampa, FL.

[27] IEEE 802.11(a) *Standard, Wireless LAN Medium Access Control (MAC) and Physical Layer (PHY) Specifications: High-Speed Physical Layer in the 5 GHz Band,* September 16, 1999, IEEE, New York, NY.

[28] IEEE 802.11(b) *Standard, Wireless LAN Medium Access Control (MAC) and Physical Layer (PHY) Specifications: High-Speed Physical Layer in the 2.4 GHz Band,* September 16, 1999, IEEE, New York, NY.

[29] D. Toumpakaris, J. Cioffi, and D. Gardan, ANSI Contribution TIE1.4/2002-128, April 8, 2002, Atlanta, GA.

CHAPTER 4

HDSL AND SECOND-GENERATION HDSL (HDSL2)

The origin of the high-rate digital subscriber line (HDSL) began in the late 1980s after the successful demonstration of the feasibility and manufacturability of ISDN basic rate (160 kb/s) transceivers. The driving application of

HDSL was the provisioning of T1 (bit synchronous 1.544 Mb/s) service to end customers via the local loop plant. The T1 signal itself is also referred to as DS1, which is the first-level digital signal in the North American network hierarchy. The objective of HDSL is to transport a bit synchronous DS1 signal from the CO to the customer premises (CP) without repeaters on loops that met the carrier serving area (CSA) requirements [4]. This same concept was also adopted in Europe for the transport of G.703/704 [1],[2] 2.048 Mb/s bit synchronous payload. The 2.048 Mb/s signal, often referred to as E1, is the first level digital signal in the European network hierarchy. The E1 signal in the Europe is analogous to the T1 signal in North America. The elimination of repeaters on most loops greatly reduced the cost to provide DS1 services, and also enabled more rapid turn-up of service.

HDSL's benefits are largely due to the elimination of midspan repeaters. Each repeater site must be custom engineered to assure that each section of the line remains within the limits for signal loss. The repeated signals can cause severe crosstalk to other systems; thus special care must be taken in the design of repeatered facilities to avoid excessive crosstalk to other transmission systems. The repeater is placed in an environmentally hardened apparatus case in a manhole or on a pole. The apparatus case must be spliced into the cable. The apparatus case costs far more than the repeaters it holds. A repeater failure results in a field service visit. Repeaters are usually line powered; this requires a special line feed power supply at the CO. Most of the power fed by the CO power supply is wasted due to loop resistance and power supply inefficiencies.

HDSL also is preferred over a traditional T1 carrier because HDSL provides more extensive diagnostic features (including SNR measurement) and HDSL causes less crosstalk to other transmission systems because HDSL's transmit signal is confined to a narrower bandwidth than traditional T1 carrier.

The first generation of HDSL was an extension of the 2B1Q technology used in ISDN [3], operating at a higher bit rate to support the transport of a DS1 payload. Other technologies considered for HDSL include carrierless amplitude and phase (CAP) modulation and discrete multitone (DMT) modulation. A standard for first-generation HDSL was never developed in North America; however Committee T1 published a technical report describing the three different line codes for HDSL. In 1994, Bellcore (now Telcordia) published a technical advisory (TA) on HDSL that described the preferred 2B1Q approach for HDSL. Although it was never fully completed, this TA was the closest to defining an interface specification for HDSL. From 1994 through 1996, Committee T1 working group T1E1.4 was developing a more complete specification for HDSL [6], which included definitions of 2B1Q and CAP line codes. This draft specification, although very thorough in scope and completion, was never balloted for publication by Committee T1 (i.e., neither as a technical report nor a standard).

Contrary to the activity in North America, in 1993 ETSI started the development of an HDSL technical report, which was later converted to an ETSI technical specification TS 101 135 [7], with the driving application of transporting an E1 payload on subscriber lines. The ETSI specification defines two normative approaches to the implementation HDSL. The first approach defines HDSL using the 2B1Q line code. The TS contains definitions of the 2B1Q approach for E1 signal transport on three wire pairs, two wire pairs, and a single wire pair. The second approach for HDSL uses the CAP line code; however, the CAP specification is defined for operation on two wire pairs and a single wire pair.

In 1998, the ITU-T published an HDSL Recommendation based on the contents of ETSI TS 101 135 [7] and Committee T1 TR-28 [4]. The ITU-T Recommendation on HDSL is G.991.1 [9].

Also in 1996, T1E1.4 began work on studying the feasibility of transporting a DS1 payload on a single wire pair for deployment on CSA loops. In other words, the goal was to investigate a modulation approach that would transport a DS1 payload on a single wire with similar performance to that of a 2B1Q HDSL transceiver operating on a single wire pair. The result is the generation of the second generation HDSL (HDSL2) standard T1.418 in the year 2000.

This chapter provides a description of second-generation HDSL and shows the improvements achieved over the first generation. We also provide a brief description of first generation HDSL as well as the background of HDSL.

4.1 REVIEW OF FIRST-GENERATION HDSL

The driving application of HDSL is to transport a DS1 payload without repeaters on loops that meet carrier serving area (CSA) requirements. CSA loops basically define a range of 9 kft on 26-gauge wire and 12 kft on 24-gauge wire. Meeting this goal would greatly simplify and cost reduce the provisioning of "high-capacity" DS1 Service to end customers. Hence, HDSL would be an alternative choice for the network operator in provisioning a DS1 access line.

Figure 4.1 shows a diagram of a DS1 access loop using repeatered T1 alternate mark inversion (AMI) technology. Conventional T1 AMI is deployed on a specially engineered repeatered line. The loop uses two wire pairs. One loop carries the full 1.544 Mb/s bit rate downstream (i.e., from the CO to the customer premises), and the other pair carries the upstream signal. Repeaters are used on the specially engineered T1 lines to reach the customer and provide the required quality of service.

Provisioning of the repeatered lines is very labor intensive and time consuming, especially when apparatus cases for housing repeaters need to be installed. The time for provisioning the repeatered link may take up to two or three months, depending on the amount of labor needed to install all the necessary equipment.

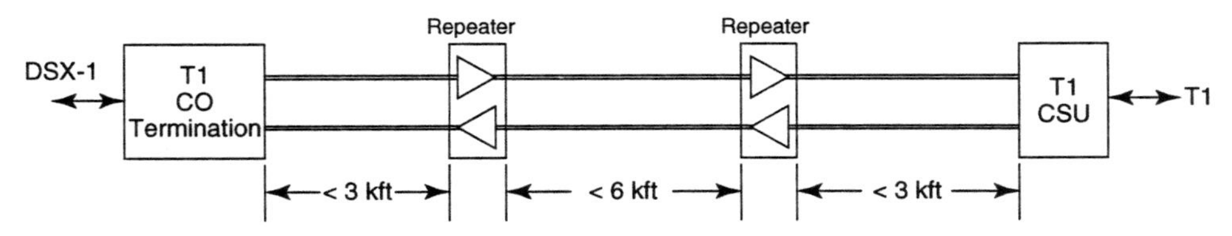

Figure 4.1 Conventional repeatered T1 AMI service provisioning.

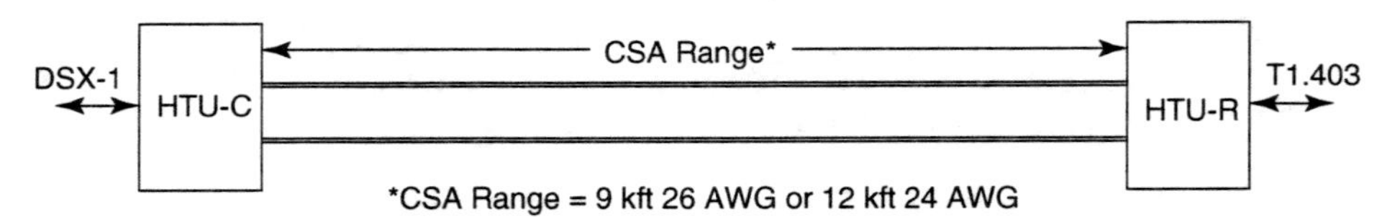

Figure 4.2 HDSL dual-duplex provisioning of T1 service.

The goal for HDSL is to deploy a circuit that transports a DS1 payload without repeaters on loops that are within CSA range. By simply identifying loops that meet CSA requirements, the provisioning time of a DS1 circuit with HDSL could be as low as two or three days. This is a significant savings and labor costs. Also note that in addition to prequalifying a loop for CSA requirements, it must be certain that the loop does not contain load coils.

The feasibility studies show that 2B1Q cannot meet the performance objective of transporting the DS1 payload to CSA range on a single wire pair in the presence of crosstalk; however, 2B1Q was observed to be robust enough to transport half the DS1 rate at CSA range in presence of self near-end crosstalk. With this observation in mind, it was agreed to define HDSL using two wire pairs to transport the DS1 payload where each wire pair transports one-half the DS1 payload in both directions (duplex). The result is the dual-duplex architecture shown in Figure 4.2.

In the dual duplex architecture, each wire pair carries one-half of the DS1 payload along with some additional overhead for operations and maintenance reasons. The transmission on each wire pair is full-duplex transmission, where echo cancellation is the mechanism used to separate the two directions of signal transmission.

4.1.1 Dual Duplex Architecture

Figure 4.3 shows the dual-duplex architecture incorporated in first-generation HDSL. The HDSL modem is called the HDSL transceiver unit in the CO, or simply HTU-C. In the CO, the HTU-C connects to the DS1 signal via a DSX-1 interface.[1] The T1 AMI block terminates the AMI signal and converts it into a bit stream together with the recovered signal clock. The DS1 payload is then split evenly between loops #1 and #2 and the contents inserted to HDSL frames and then transmitted on their respective loops by transceiver #1 and #2. Both transceivers support duplex transmission using echo cancellation. Also, the transceivers in the HDSL pass the DS1 timing information end to end via the use of stuff and delete operations.

4.1.2 DS1 Frame Structure

Figure 4.4 shows the frame structure of a DS1 payload. The basic frame structure is 125 μsec long, which is synchronized to the 8 kHz PCM sampling clock (32 ppm) in the central office. The frame is broken up into 24 time slots of 8 bits each; because each bit in the frame represents a capacity of 8 kb/s, each time slot has a capacity of

[1]The DSX-1 interface is a DS1 passive cross-connect defined in AT&T compatibility bulletin CB-119.

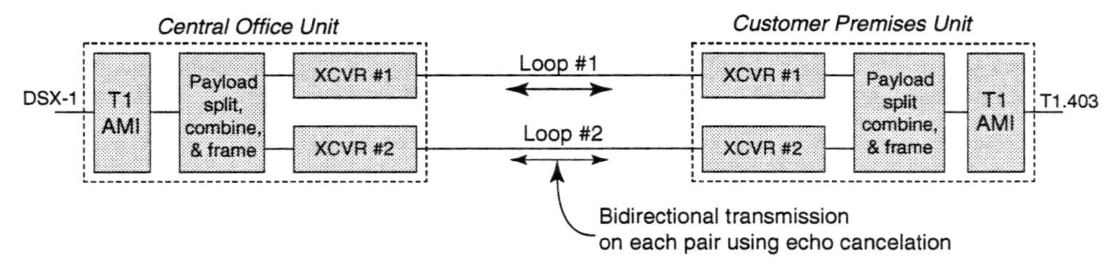

Figure 4.3 Dual-duplex architecture of HDSL.

64 kb/s. The F-bit is the frame indicator bit that defines the boundary of the frame. In the 125 μsec frame period, there are 192 bits from the 24 time slots plus 1 F-bit for a total of 193 bits. This corresponds to the total bit rate of 1.544 Mb/s.

In addition to the core frame, the DS1 circuit is provisioned with a superframe structure that groups multiple core frames for the purpose facilitating frame boundary identification, performance monitoring, and operations and maintenance. There are two types of superframe structures defined for DS1 signals. The first is the original 12-frame superframe (SF) structure defined in AT&T Tech Pub 62411 [10], and the second is a 24-frame extended superframe (ESF) structure defined in Committee T1 Standard T1.403 [11]. The T1 AMI circuit in the HDSL modems of Figure 4.3 must first know which superframe structure the DS1 circuit is provisioned with, and it identifies the DS1 frame boundary by searching for a unique sequence of F-bits.

Once the HDSL modem synchronizes to the DS1 frame, the modem splits DS1 frame contents into two groups for transmission on the two loops. The splitting can be done in numerous ways, but the two most frequently used approaches are:

1. Alternate the times to each pair, for example, assign the odd numbered time slots to loop #1 and the even numbered time slots to loop #2.
2. Assign the first twelve contiguous time slots to one loop and the remainder to the other loop, for example, time slots 1–12 are assigned to loop #1 and time slots 13–24 are assigned to loop #2.

Having the HDSL modems know something about the DS1 circuit provisioning could enable certain service features such as passing half the DS1 payload in the unlikely event that operation on one of the HDSL loops fails. To enable such a capability, there must be consistent allocation of customer data in the DS1 frame with the splitting mechanism used in the HDSL modem.

Because HDSL transmitter locks on to the DS1 frame and splits the payload evenly between the two loops, the receiver must combine the two recovered bit

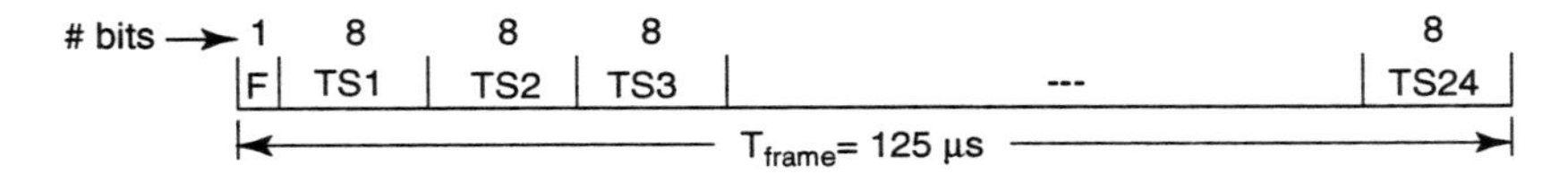

Figure 4.4 DS1 frame structure.

streams to reconstruct the original DS1 signal. To improve the reliability in the reconstruction of the DS1 payload at the receiver, the F-bit is transmitted on both loop #1 and #2. In the event that one loop fails and the other loop remains active, the receiver must still interface to a DS1 circuit with one half of the payload. Having the F-bit transported on both loops facilitates this capability. Furthermore, an HDSL embedded operations channel (EOC) and indicator bits are also replicated on both loops.

4.2.3 HDSL Transmitter Structure

Figure 4.5 shows the transmitter structure of an HDSL system. As mentioned earlier, the T1 AMI interface circuit terminates the AMI signal and converts the AMI pulses to a bit stream and its associated clock. The splitter block divides the DS1 payload into two equal bit streams of approximately one-half the bit rate for transmission on two wire pairs in the local loop. The two bit streams are then fed to the core 2B1Q transmitters, which consist of a scrambler, bit-to-symbol mapper, and a spectral shaper. The HDSL framer and core transceiver blocks are described in the following subsections.

4.1.4 HDSL Frame Structure

In order to reliably split the DS1 payload for transmission on two wire pairs and then reconstruct the original DS1 payload, the HDSL transceiver encapsulates the DS1 bits in an HDSL frame prior to transmission on the two wire pairs. Figure 4.6 shows a block diagram of the HDSL frame structure. In addition to identifying the specific bits and/or time slots of the DS1 payload, the HDSL frame format is constructed to support the following capabilities:

1. Frame synchronization
2. Performance monitoring using a 6 bit cyclic redundancy check (CRC-6)
3. Indicator bits
4. Embedded operations channel (EOC)
5. Passing of DS1 timing information (rate adaptation)

These capabilities are supported by incorporating 8 kb/s of overhead in the frame. Both transceivers in the HDSL modem implement this frame structure. At the receiver, the frame overhead is removed prior to reconstructing the DS1 bit stream.

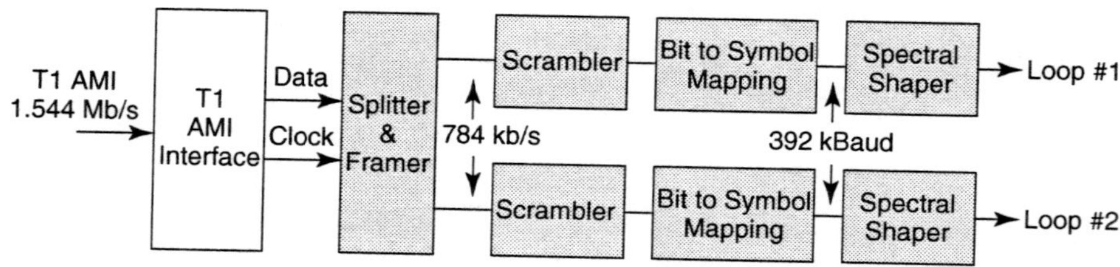

Figure 4.5 HDSL transmitter structure for lines #1 and #2.

As seen in Figure 4.6, the HDSL frame structure contains a superframe that is 6 msec. This period corresponds to 48 DS1 core frame periods. For every DS1 core frame included in the HDSL superframe, we include one additional bit of overhead, which corresponds to including 8 kb/s or over in the HDSL line signal. Hence, each HDSL superframe contains 48 overhead bits and the appropriate data from 48 DS1 core frames.

The frame is broken up into the following components of overhead and payload:

1. Four payload blocks, labeled PB1–PB4, in which is placed the DS1 data assigned to the specific line
2. Four overhead blocks labeled OH1–OH4
3. Stuff bits (2 stuff bits constitute a nominal frame) labeled SB

Each payload block contains 12 sub-blocks of DS1 data. As seen in the figure, each sub-block contains one DS1 F-bit and 12 time slots from the DS1 payload. In this example, we assign the odd numbered time slots to the HDSL frame of loop #1 and the even numbered slots to the frame in loop #2. The same DS1 F-bit is transported on both lines. There are $12 \cdot \left(\frac{192}{2} + 1\right) = 1{,}164$ in each payload block, so the entire superframe contains 4,656 payload bits. If we add the 48 overhead bits, we get 4,704 total bits in each 6 msec superframe, which corresponds to a bit rate of 784 kb/s that is transmitted on each wire pair.

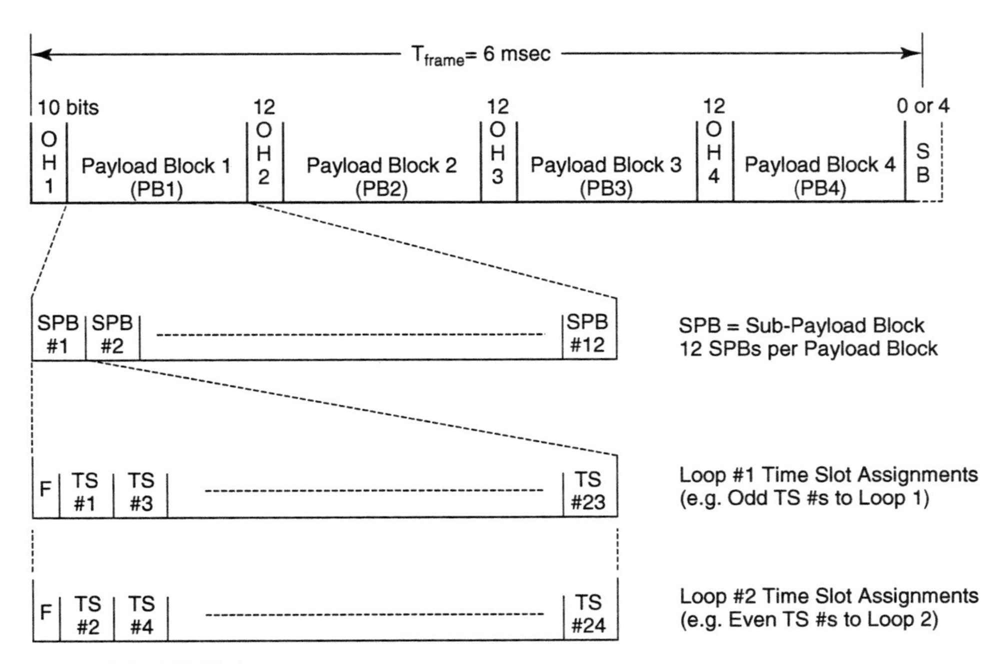

Figure 4.6 HDSL frame structure.

The first overhead block contains 14 bits that are dedicated for frame identification and synchronization. The frame sync word (FSW) is a double (or interleaved) Barker code defined as 11111100001100. Using the 2B1Q line code, this FSW sequence is transmitted with the outermost 2B1Q symbols, namely, sequence +3 +3 +3 −3 −3 +3 −3. Barker codes have superior autocorrelation properties and are effective in timely synchronization of the HDSL frame. Note that dibit 11 is mapped to 2B1Q symbol +3 and dibit 00 is mapped to 2B1Q symbol −3. The downstream direction of transmission uses the above-mentioned FSW; the upstream direction uses the reverse FSW sequence, namely, −3 +3 −3 −3 +3 +3 +3.

Overhead blocks OH2–OH4 each have 10 bits. Six bits are allocated for transport of a CRC output for the purposes of performance monitoring. The remaining bits are allocated for indicator bits and an EOC. The assignment of 24 bits to an EOC provides a capacity of 4 kb/s for data communication. There have been different assignments of these in various HDSL systems; G.991.1 provides definitions for HDSL systems in Europe and in North America.

The stuff bits at the end of the superframe provide the mechanism for transferring DS1 signal timing information end to end. If the HDSL frame buffer is approaching overflow because the HDSL line clock is not reading the buffer data fast enough, then the framer would eliminate transmission of the stuff bits at the end of the current frame. Alternatively, if the HDSL frame buffer is approaching empty because the HDSL line clock is reading the data too fast, then the framer would insert four stuff bits into the current frame to allow more payload bits to be loaded into the buffer.

The stuff and delete operations for the 2B1Q systems defined in G.991.1 transmit either 0 or 4 stuff bits; the nominal frame of 6 msec is never sent, it is only seen on the average transmission of many superframes. In the CAP-based systems defined in G.991.1, the framer may include 0, 2, or 4 stuff bits; in these systems, the nominal superframe of 6 msec is often sent. Having the finer stuffing resolution allows for better timing jitter performance.

4.1.5 Scrambler

The HDSL transceivers use a self-synchronizing scrambler to randomize the data for transmission on the line. All of the bits except for the 14-bit frame sync word and the stuff bits are scrambled. In all cases the scramblers use a 23rd order polynomial. In the downstream direction, the scrambler polynomial is $p(x) = 1 + x^{-5} + x^{-23}$ and that for the upstream channel is $p(x) = 1 + x^{-5} + x^{-23}$, where the + sign indicates modulo-2 addition. Figure 4.7 shows the block diagrams of the downstream and upstream channel scramblers and descramblers. Note that these are the same scramblers and descramblers used in basic rate ISDN [3].

4.1.6 Bit-to-Symbol Mapping

The bit-to-symbol map block converts the serial bit stream into a sequence of multilevel pulses. For 2B1Q, the mapping block takes two bits at the input and produces a sequence of pulses that contain four possible levels. Figure 4.8 shows an example 2B1Q bit mapping taken from ETSI TS 101 135 [7]. In this example, the first bit defines the sign of the output pulse where a one maps to a positive ampli-

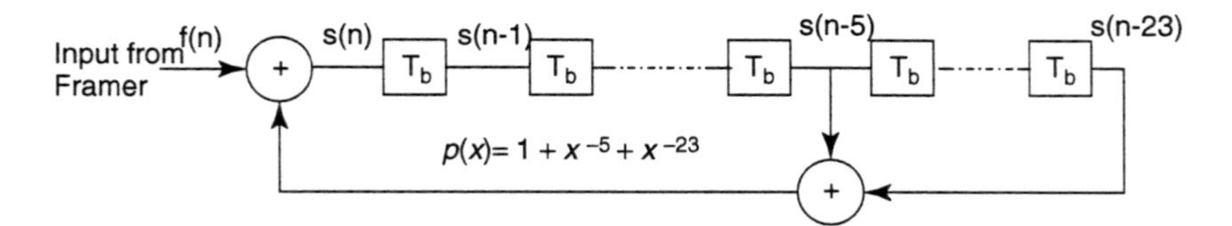

(a) Downstream Channel Scrambler

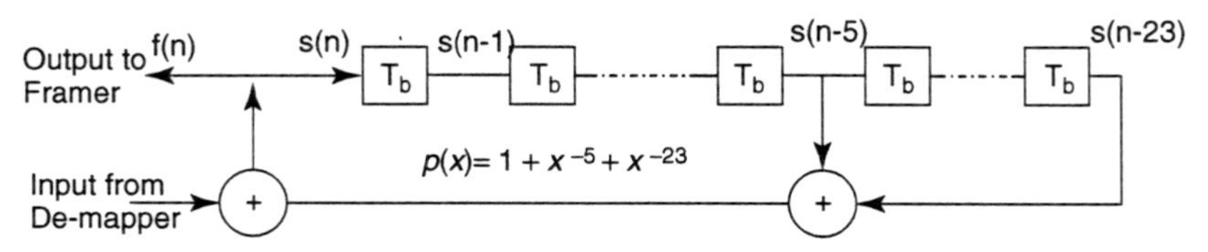

(b) Downstream Channel Descrambler

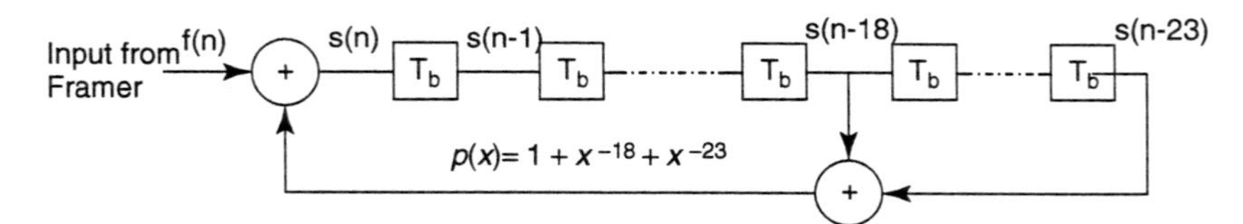

(c) Upstream Channel Scrambler

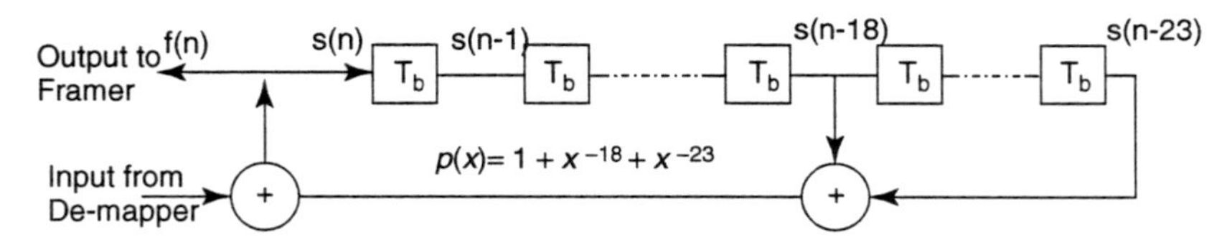

(d) Upstream Channel Descrambler

Figure 4.7 Downstream and upstream channel scramblers and descramblers.

tude pulse and a zero maps into a negative amplitude pulse. The second bit defines the magnitude of the output pulse where a one maps to the lower pulse amplitude and the zero maps to the larger pulse amplitude.

The input bit sequence has a bit rate of R_b bits/sec; the corresponding bit interval is $T_b = 1/R_b$ seconds. Since the mapping block collects two bits for each multi-level symbol, the output symbol interval is twice the bit interval; hence the symbol rate is $R_S = 1/2T_b = R_b/2$. For 2B1Q, the output symbol rate is one half that of the input bit rate.

For two-pair HDSL in North America [4],[9] and three-pair HDSL in Europe [7],[9], the bit rate on each wire pair is 784 kb/s, and the corresponding symbol rate is 392 kBaud (kSymbols/s). The 3-dB signal bandwidth is 196 kHz.

For two-pair HDSL in Europe transporting a 2.048 Mb/s E1 payload, the line bit rate is 1,168 kb/s on each wire pair, and the corresponding symbol rate is 584 kBaud. The 3-dB signal bandwidth is 292 kHz.

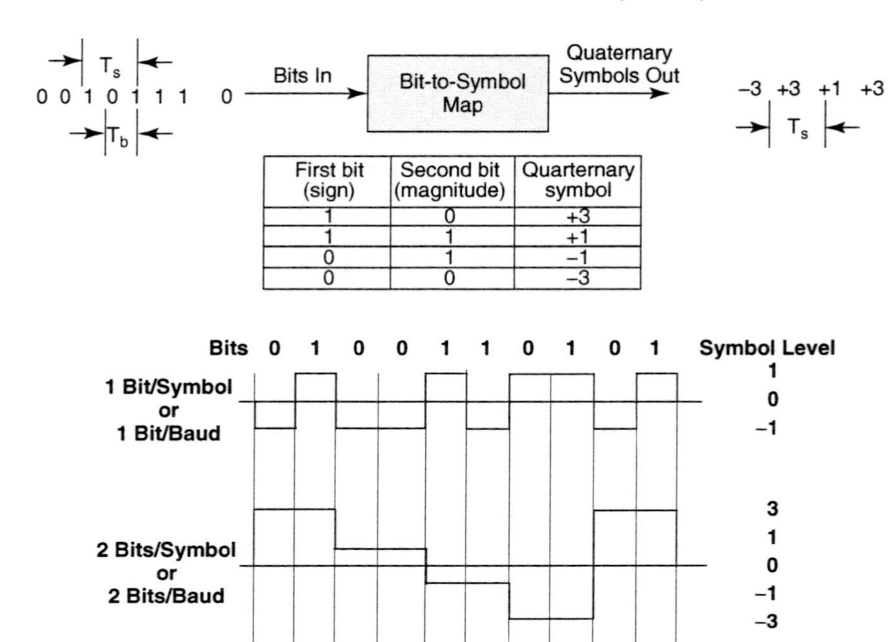

First bit (sign)	Second bit (magnitude)	Quarternary symbol
1	0	+3
1	1	+1
0	1	−1
0	0	−3

Figure 4.8 Bit-to-symbol mapping in 2B1Q HDSL.

4.1.7 2B1Q Spectral Shaper

The spectral shaper is a low pass filter that shapes the transmit signal spectrum to a form suitable for transmission in loop plant. The 2B1Q transceiver uses a simple spectral shaper, that is, one that provides a fourth order roll-off. The fourth order roll-off is defined via a PSD mask in Section 5.8.4 of the ETSI technical specification on HDSL [7]. Typical implementation of HDSL systems pass a 2B1Q NRZ (non-return to zero) spectrum through a fourth-order Butterworth filter and scale that signal such that the total transmit power is 13.5 dBm.

The HDSL transmit signal power spectral density (PSD) can be mathematically modeled by the following equation:

$$PSD_{HDSL}(f) = K_{HDSL} \cdot \frac{2}{f_0} \cdot \frac{\left[sin\left(\frac{\pi f}{f_0}\right)\right]^2}{\left(\frac{\pi f}{f_0}\right)^2} \cdot \frac{1}{1 + \left(\frac{f}{f_{3dB}}\right)^8}$$

where $f_0 = 392$ kHz, $f_{3dB} = 196$ kHz, $K_{HDSL} = \frac{5}{9} \cdot \frac{V_p^2}{R}$, $V_p = 270$ Volts, and $R = 135\ \Omega$.

The above PSD expression models the passing of a rectangular pulse through a fourth order Butterworth filter. The magnitude of the expression is scaled to pro-

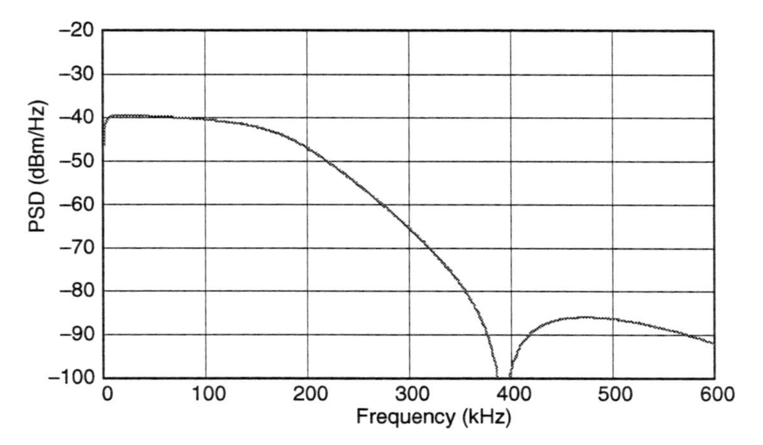

Figure 4.9 Nominal 2B1Q HDSL transmit PSD (784 kb/s) with 4th-order roll-off.

duce a transmit signal power of 13.5 dBm. Figure 4.9 shows a plot of a nominal HDSL transmit PSD scaled to have a 13.5 dBm transmit power in the interval from 0 to 392 kHz.

Note that for frequencies above the first null at f_0 = 392 kHz, there is still significant energy in the image lobes. This out-of-band energy is important when considering the crosstalk into other systems. Note that crosstalk coupling to other wire pairs in a cable increases with increasing frequency. A shaping filter with a higher roll-off could further reduce this out-of-band energy, which would reduce the amount of crosstalk into other signals deployed in the cable.

4.1.8 2B1Q HDSL Transceiver Structure

There are numerous ways to build a 2B1Q transceiver. Figure 4.10 shows a functional block diagram of an HDSL transceiver supporting the 2B1Q line code. The blocks shown are those that implement the core modem. The top row of blocks implement the transmitter, and the bottom row of blocks implement the receiver function. Because the upstream and downstream channels share the same frequency band, the duplexing function is provided with echo cancellation.

As discussed earlier, the transmitter of the core modem consists of the scrambler[2], bit-to-symbol mapping, spectral shaping filter, and an analog front end that includes a line driver. The scrambler feeds the bit-to-symbol mapping block that converts the serial bit stream to a sequence of four level pulses as shown in the example of Figure 4.8. The spectral shaper filters the multilevel pulse stream for the bit-to-symbol map block to a spectral shape suitable for transmission on the line. The filter may be implemented with either analog or digital processing. If analog

[2]The scrambler and descrambler blocks are not shown in the figure; however, the scrambler connects to transmitter input and the receiver output (symbol-to-bit mapper) connects to the descrambler input.

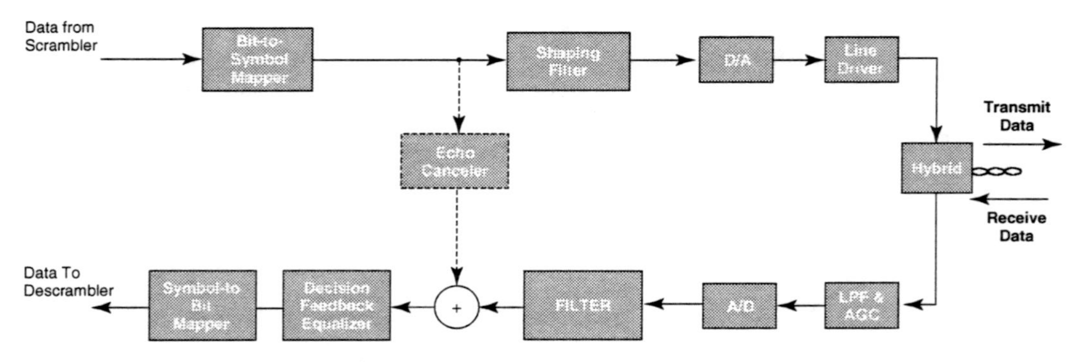

Figure 4.10 Functional block diagram of a 2B1Q HDSL transceiver.

processing is used, then the filtered signal may be fed directly to the line driver. If the filtering is done with digital processing, then the digital-to-analog (D/A) converter converts the digital samples into analog samples, which are then fed to the line driver for transmission on the subscriber line.

The hybrid circuit provides the four-to-two wire coupling of the transceiver to the subscriber line. Both the transmitter output and receiver input have two-wire connections to the hybrid circuit. The subscriber line is a single wire pair that transports signals in both directions. A balancing circuit (not shown in the figure) minimizes the leakage of the transmitter output signal into the local receiver input. An ideal hybrid would have no leakage of the local transmit signal into the local receiver, that is, infinite trans-hybrid loss; however, practical hybrids would have around 12 dB of trans-hybrid loss. This amount of trans-hybrid loss allows a significant amount of energy from the local transmitter that must be compensated for in the receiver.

Given that the hybrid circuit is far from an ideal device, the level of the signal leaking from the local transmitter into the local receiver, hereby referred to as the local echo, is typically higher than the level of the desired signal from the far-end transmitter. It is the job of the echo canceler to remove this local echo seen at the receiver input so that information in the desired received signal may be recovered with a high degree of confidence.

For example, let's assume that the subscriber loop attenuates the far-end transmit signal by about 40 dB and the local hybrid circuit has 12 dB of trans-hybrid. Assuming that the transmit signals at both ends of the line each have the same transmit power, then the level of the local echo will be 28 dB higher than the level of the local echo. If for reliable decoding of the desired received signal we desire the level local echo to be 30 dB below the desired received signal, then the echo canceler will need to provide at least 58 dB (28 dB + 30 dB) of echo cancellation. Good echo canceler designs typically provide at least 60 dB of echo cancellation.

As mentioned earlier, the input to the receiver includes the desired receive signal sent from the far-end transmitter across the subscriber line and a local echo resulting from leakage through the hybrid. The receiver first filters the received signal to remove any unwanted out-of-band noise and automatically adjusts the level of the total received signal, using either an automatic gain control (AGC) circuit or a programmable gain amplifier (PGA), for optimum use of the analog to digital (A/D) converter's dynamic range. Additional digital filtering may then be done on the digitized samples of the received signal.

Prior to recovering the desired signal, the local echo must be removed from the total received filtered signal. At initialization, the echo canceller circuit automatically learns the echo path between the near-end transmitter and near-end receiver. The echo path is that through the shaping filter, transmitter's analog front end, hybrid circuit, the receiver's filters and A/D converter. The reconstructed echo signal is subtracted from the filtered total receive signal. The receiver then uses a conventional decision feedback equalizer to compensate for the channel impairments, namely loop attenuation and phase distortion, crosstalk, residual echo, and background noise. Finally, the symbol-to-bit mapping block converts the 2B1Q samples to bits, which are then sent to the descrambler.

4.1.9 CAP-Based HDSL

As shown in Chapter 2, for any given baseband system it is always possible to design an equivalent passband system. By equivalent systems we mean that the two systems have the same bit rate, the same transmit power, and they both utilize the same bandwidth. If we consider a reference four-level PAM (pulse amplitude modulation) system as an example, a corresponding equivalent passband system may be implemented using 16-level CAP (carrierless amplitude and phase modulation) or QAM (quadrature amplitude modulation) as shown in Figure 4.11.

A baseband PAM has a one-dimensional constellation. The minimum required bandwidth is one-half the symbol rate, which is shown in the spectrum dia-

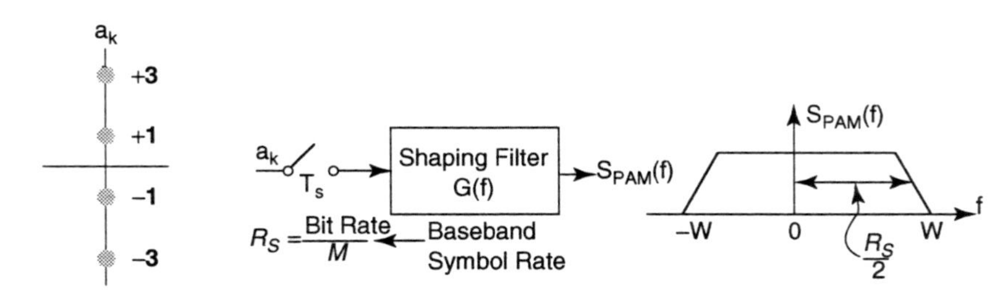

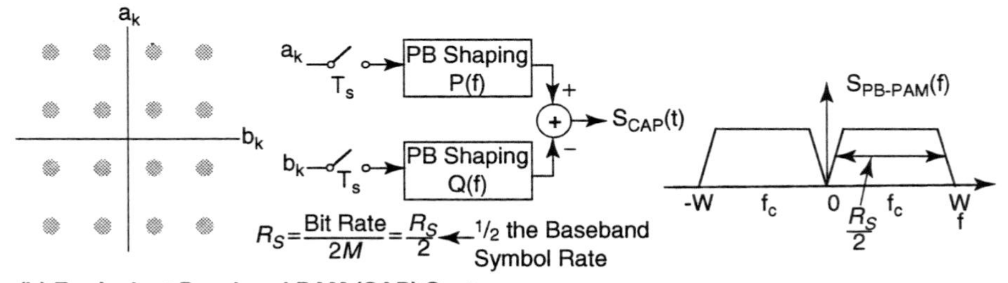

Figure 4.11 Example equivalent baseband and passband systems.

gram Figure 4.11(a) as $R_S/2$, where R_S is the symbol rate of the PAM system. If a single carrier of frequency f_c modulated the baseband spectrum, then the bandwidth of the resulting passband spectrum would be doubled. To obtain an equivalent bandwidth as the baseband system, the passband symbol rate would need to be halved. This is achieved by taking the original bit stream and converting it into two parallel symbol sequences, each running at one-half the baseband symbol rate; the two parallel symbol sequences may be modulated with quadrature carriers and added together to form a passband transmit signal having the same equivalent bandwidth as the corresponding baseband system. The differences are that the passband system has a two-dimensional constellation that is square the size and twice as many bits per symbol as the reference baseband system and the symbol rate of the passband system is half that of the reference baseband system. The bit rate and bandwidth are equivalent and both spectra may be scaled to produce the same transmit power.

In general, a baseband PAM system will have the following parameters:

- Bit rate R_B
- Symbol rate: $R_S = R_B/b$
- One-dimensional constellation with $N = 2^b$ levels
- b bits per baseband symbol
- Minimum bandwidth of $R_S/2$

The corresponding equivalent passband PAM (i.e., using CAP or QAM) system will have the following parameters

- Bit rate R_B
- Symbol rate: $R_S' = R_S/2 = R_B/2b$
- Two-dimensional constellation with $N' = N^2 = 2^{2b}$ levels
- $2b$ bit per passband symbol
- Minimum bandwidth of $R_S' = R_S/2$.

The functional block diagram of a CAP based transceiver is shown in Figure 4.12. There are two fundamental differences in this structure compared with the 2B1Q functional block diagram of Figure 4.10. Because CAP is a passband system, the symbol processing is done in two dimensions; in the figure, double lines connecting the functional blocks represent the passing of two-dimensional symbols. Second, the CAP transceiver uses a trellis code to provide added immunity to crosstalk. A channel precoder is used together with the trellis encoder so that the transceiver can achieve the joint benefits of the trellis code and a decision feedback equalizer.

The trellis code chosen for the CAP system is based on the two-dimensional eight-state (2D8S) code [40] that is used in V.32 modems [15]. This provides an asymptotic code gain of approximately 4 dB. This additional coding gain adds roughly a 1-kft reach improvement on 26-gauge wire over 2B1Q systems.

The rate of the trellis code is n/(n+1), where for each n input information bits, the trellis code outputs one additional bit for redundancy. Hence, the CAP trans-

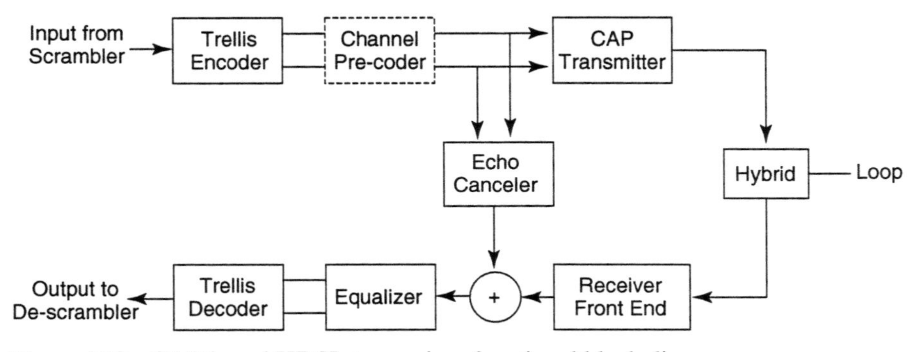

Figure 4.12 CAP-based HDSL transceiver functional block diagram.

mitter will transmit a constellation base on n + 1 bits. When compared with the 2B1Q transceiver, the equivalent uncoded CAP system has sixteen points in the constellation. When we add the 2D8S trellis code, the equivalent CAP system contains thirty-two levels.

The actual CAP system defined in Annex B of the ETSI HDSL technical specification [7] uses a sixty-four-point coded constellation. The spectral shaping is that of a square-root raised-cosine spectrum with at least 30 dB out of band rejection. Although the coded thirty-two-point constellation gives optimum performance against the worst-case self-NEXT crosstalk, the operation of coded 64-point CAP has less than a 0.5 dB degradation. However, the narrower bandwidth and efficient spectral shaping provides significantly less crosstalk into other systems in the cable. Of particular interest is the spectral compatibility with ADSL. The narrower bandwidth of coded 64-CAP and the low out-of-band energy provides significantly better spectral compatibility with ADSL than does the 2B1Q HDSL.

4.2 SECOND-GENERATION HDSL (HDSL2)

The primary goal of second-generation HDSL, hereby referred to as HDSL2, is to utilize the advancements in integrated circuit design to develop a digital subscriber line transmission system that transports a DS1 payload on a single wire pair on loops that meet carrier serving area (CSA) requirements. This objective has the same deployment coverage as conventional HDSL that uses two wire pairs. Figure 4.13 shows a provisioning diagram of a T1 access circuit provisioned using second-generation HDSL technology. Although the core technology for the transceivers is more complex than that for first-generation HDSL, the total provisioning and equipment costs will be lower than for conventional HDSL when deploying service on a single wire pair as opposed to two pairs.

Given that core second-generation transceivers can support the transmission of twice the bit rate at the same distance as the first-generation transceivers, an extension of this new technology should allow transmission of the original bit rate at a significantly greater distance than the first generation technology. Hence, we

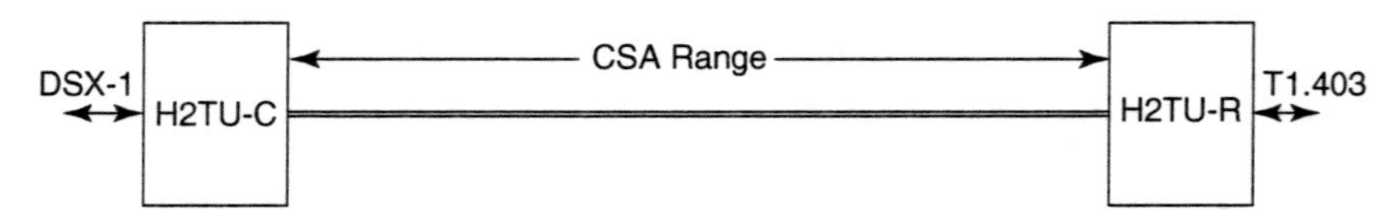

Figure 4.13 Provisioning of T1 service using HDSL2.

expect to be able to deploy dual-duplex HDSL using the new technology at distances significantly greater than CSA.

The following sections describe the HDSL2 technology in detail. We also provide some background in the development of the new approach for HDSL transmission. The complete HDSL2 standard is defined in T1.418 [8].

4.2.1 Performance Objectives for HDSL2

Although the HDSL2 project began in 1995 [16], industry interest did not take affect until 1996. At that time, the operators provided a list of requirements and objectives for DS1 transport on a single wire pair [17], [18]. Based on feedback from feasibility studies, the requirements were revised [19], [20], [21]. A summary of the performance requirements and objectives are as follows:

- Loop Reach: CSA Requirements (i.e., 9 kft 26 AWG and 12 kft 24 AWG)
- Performance against crosstalk impairments: minimum of 5 dB margin with one percent worst case crosstalk from the following interfering signals
  - 49 HDSL
  - 39 HDSL2
  - 39 Echo canceled (EC) ADSL
  - 49 Frequency-division multiplexed (FDM) ADSL
  - 25 T1 AMI disturbers
  - 24 T1 + 24 HDSL2
  - 24 FDM-ADSL + 24 HDSL
- Spectral Compatibility: In general, the goal for spectral compatibility of HDSL2 is to not disturb all existing DSL services more than what is tolerated from other services at the time of the development of the technology. However, based on feasibility study and market need, the following exceptions were agreed:
  - HDSL2 shall not degrade HDSL [4] by more than 2 dB [22], [23]
  - HDSL2 shall not degrade ADSL [12] by more than 1 dB [24]
- Latency: The maximum latency for HDSL2 is 500 μsec.

HDSL2 was the first DSL standards development to address crosstalk from mixed sources (e.g., the last two interferer models in the list above). Because the mixed noise model is more severe than the traditional consideration of only homogeneous noise, it was considered appropriate to design HDSL2 for a 5 dB margin instead of the traditional 6 dB margin.

4.2.2 Evolution of the Modulation Method for HDSL2

Numerous modulation methods (or line codes) were considered as possible candidates for HDSL2 prior to the selection of sixteen-level trellis code pulse amplitude modulation (TC-PAM). In order to gain significant performance over the 2B1Q line code for HDSL, the general approach taken for performance improvement in the modem signal processing is to use a high-state trellis code with channel precoding. In addition, special spectral shaping would be required to find the proper balance between performance and spectral compatibility with other signals in the cable. The upstream and downstream channels would share common frequencies so the duplexing method for HDSL2 would be echo cancellation. Figure 4.14 shows the generalized functional block diagram of the transceiver structure for HDSL2. Note that this block is applicable for baseband systems such as PAM and single-carrier passband systems such as QAM or CAP.

Early on in the development of HDSL2, the line codes considered for HDSL2 were multilevel PAM and multilevel CAP (or QAM). At first, all of the systems considered had used symmetric upstream and downstream spectra with the bandwidth set to one-half the symbol rate for the PAM-based systems and equal to the symbol rate for the CAP/QAM systems. The performance of all these systems was limited by self near-end crosstalk (SNEXT), and the best configuration of each system fell about 2 to 3 dB short of the requirements [38], [39], [41], [42]. The SNEXT limitation needed to be overcome by some method in order to meet or get close to meeting the objective requirements.

An alternative to using fully overlapped symmetric spectra is to use frequency-division multiplexing (FDM) of the upstream and downstream signal spectra. If the upstream and downstream spectra were totally nonoverlapping, then there would be no SNEXT. The performance limitation would be NEXT from other systems or self far-end crosstalk (SFEXT) from other similar systems. Although the latter case is certainly preferable, in a multiservice cable environment there will be a mixture of overlapped and nonoverlapped signal spectra in the cable. Also, the higher frequency band in an FDM system would be susceptible to greater loop attenuation because of its placement at higher frequencies. Crosstalk coupling is also greater at the higher frequencies, so crosstalk impact to other services needs to be considered. Hence, the spectrum design of HDSL2 must be such that it operates in numerous NEXT and FEXT crosstalk environments. An early proposal that

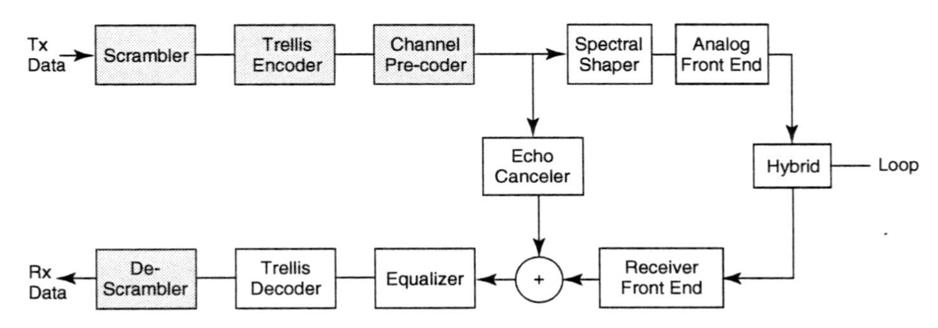

Figure 4.14 General transceiver functional block diagram for HDSL2.

considered these issues was that of a "staggered FDM" scheme described in T1E1.4/96-340 [25].

In T1E1.4/97-073 [26], partially overlapped echo-canceled transmission (POET) was proposed, which involved overlapping nonidentical upstream and downstream spectra. The spectra were carefully shaped so as to maximize performance in the presence of crosstalk and to minimize the spectral compatibility impact into other services in the cable. Other approaches working off of the same core principle included POET-PAM [26], OverCAPed (oversampled CAP/QAM) [27], OPTIS [28], [29], and MONET [30], [31].

Most of the modulation schemes later proposed had PSDs with boosted energies at higher frequencies that were above their nominal low-frequency values. Testing of HDSL systems in the presence of crosstalk from systems using the OPTIS shaping showed significant impact on the performance of the HDSL. Based on this testing, the OPTIS spectra were modified such that their spectral compatibility impact on HDSL would be less than 2-dB of degradation. The network operators in North America unanimously agreed upon this degree of degradation. The resulting PSDs adopted for HDSL2 are shown in Figure 4.23 for the downstream channel and Figure 4.24 for the upstream channel.

One element introduced in the evaluation of the above proposals is that of decoupling of the transmit signal bandwidth from the symbol rate, so that the bandwidth of the PSD could have significant energy in bandwidth greater than that determined directly by the symbol rate. An implication of such an approach would require the use of oversampling in the transceiver design, particularly in the receiver where at least a 2× or 3× sampling would be required, depending on the modulation approach chosen. The benefits of this decoupling were found in T1E1.4/97-237 [28] when measuring the performance of proposed HDSL2 spectra in the presence of mixed crosstalk for PAM-based systems. This extension of the bandwidth showed particular benefit to the PAM approach under certain conditions of mixed crosstalk in the downstream channel. With a conventional DFE, the alias terms of the PAM spectrum reflect back at half the PAM symbol rate. With the excess bandwidth of the downstream OPTIS spectrum (i.e., the portion of the downstream spectrum above the overlapped upstream-downstream region), the folding is such that this region fills the area of poor SNR due to NEXT from the upstream region.

The CAP/QAM system is processed as an analytic signal (i.e., a complex signal having only positive frequency components), and it folds in a nonreflective manner at the symbol rate. With the OPTIS spectrum, the folding does not benefit the passband system as it does the baseband PAM system because the frequencies that fold into the SNR-poor high end of the passband are substantially higher in frequency. The result is that with the OPTIS spectrum, the downstream channel administers 3 to 4.5 dB better performance in mixed crosstalk containing T1 AMI disturbers for the PAM system than for the CAP/QAM-based system. It was for this region that the PAM line code was chosen over CAP/QAM for HDSL2.

In summary, the core of the HDSL2 transceiver uses the OPTIS PSD with the PAM line code. The shape of the transmit spectrum is decoupled from the symbol rate to allow for flexible use of the excess bandwidth. The excess bandwidth provides improved performance in the downstream channel against mixed crosstalk that contains T1 AMI. The selected PAM configuration uses 3 information bits per symbol, which corresponds to an eight-level signal. The unique spectral shaping of OPTIS is the best compromise found in optimizing the performance against crosstalk in the loop plant and maintaining spectral compatibility with other systems. The nominal transmit power for the upstream channel is 16.5 dBm and that for the downstream channel is 16.8 dBm. This modulation technique was shown (using ideal DFE calculations) to have a minimum of 1.0 dB of margin on the worst-case test loop. In order to achieve the highly desired objective of 6-dB margin against worst-case crosstalk, an advanced coding mechanism with a minimum coding gain of 5 dB is required. It was agreed to use a high order trellis code to help achieve the objective margin. The following sections provide an overview of the configuration defined in HDSL2 standard T1.418 [8].

4.2.3 System Reference Model

As with first-generation HDSL, the driving application is to provision a T1 access circuit on loops that meet the carrier serving area requirements. Figure 4.15 shows the system reference model for an HDSL2 transceiver unit. This is very similar to that of HDSL except that core transceiver provides transmission on a single wire pair.

The payload interface for HDSL is that of a T1 AMI signal. In the CO, the interface is a DSX-1 interface that is defined in AT&T compatibility bulletin CB-119. At the customer premises, Committee T1 Standard T1.403 [11] defines the T1 AMI interface.

The T1 AMI interface circuit terminates the AMI signal and converts the signal into a 1.544 Mb/s bit stream together with the recovered signal clock. The HDSL2 transceiver unit will transport the 1.544 Mb/s payload end to end along with the bit synchronous timing information. Transfer of bit synchronous timing is referred to as *pass through timing*. This is required for the transparent support of T1 AMI service.

The HDSL2 framer is required for supporting operations, administration, and maintenance functions for the HDSL2 access circuit. The framer adds 8 kb/s of overhead, which provides frame synchronization, performance monitoring, performance indication, and an embedded operations channel (EOC) for communication between the CO and remote HDSL2 units.

The HDSL2 transceiver is the core HDSL2 modem that provides the key modulation and demodulation functions. The following sections describe each of the key blocks in detail as defined in the HDSL2 standard in T1.418 [8].

4.2.4 HDSL2 Framing

The structure of the HDSL2 core frame is based on the super frame structure of first-generation HDSL, except that there is no alignment defined in the HDSL2

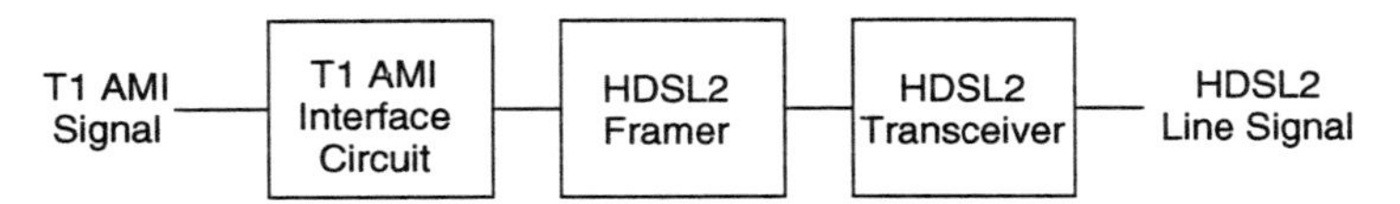

Figure 4.15 System reference model for and HDSL2 transceiver unit.

standard [8] to the DS1 time slots, nor the DS1 frame for that matter, in the HDSL2 frame. It is generally assumed that synchronization with the DS1 frame is provided separately from the HDSL2 frame.

Figure 4.16 shows the core HDSL2 physical layer frame structure. The nominal frame period (T_{frame}) is 6 msec. Because the line bit rate is 1552 kb/s (1544 kb/s payload and 8 kb/s overhead), there are nominally 9,312 bits in an HDSL superframe: 9,264 bits are for the DS1 payload and nominally 48 bits are for the overhead functions.

As shown in the figure, the HDSL2 superframe is broken up into four payload blocks and five overhead blocks. The four payload blocks are labeled PB1–PB4, and the overhead blocks are labeled OH1–OH4 and SB. Each payload block contains 2,316 bits. Note that there is no alignment of the DS1 frame or time slots with the HDSL2 payload blocks.

The overhead bits are distributed throughout five overhead blocks. Unlike the payload blocks, the overhead bits are not distributed as evenly across the blocks. Overhead block OH1 has 10 bits, blocks OH2–OH4 each have 12 bits, and the last block (the stuff bits) may have either 0 or 4 bits. The following is a summary of the overhead bit definitions.

- OH1 (10 bits): This block contains a 10-bit frame synchronization word (FSW). The purpose of the definition of the FSW is vendor specific. During preactivation, the transmitter forwards its preferred FSW to the far-end receiver via the specified exchange protocol.
- OH2 (12 bits):
  - Bits 1 and 2: crc-1 and crc-2
  - Bit 3: Stuff bit id (first copy)
  - Bit 4: Far-end DS1 loss of signal defect (losd) indicator bit. This bit identifies any anomaly or loss of signal associated with the DS1 signal sent from the far end. Under normal conditions, this bit is set to a 1; when an anomaly occurs, the losd bit is set to a 0 in the next superframe.
  - Bits 5–12: first 8 bits of the embedded operations channel (eoc01–eoc08)
- OH2 (12 bits):
  - Bits 1 and 2: crc-3 and crc-4
  - Bit 3: Unspecified indicator bit. This bit is undefined and reserved for future use.

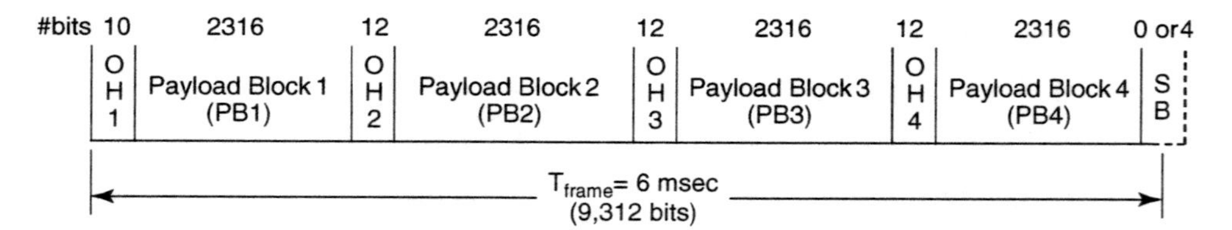

Figure 4.16 HDSL2 core frame structure.

  - Bit 4: Segment Anomaly (sega) indicator bit. This bit is used to identify any crc-6 errors encountered in the reception of a superframe. The error indication is that associated with the previous superframe. The purpose for this bit is performance monitoring.
  - Bits 5–12: second 8 bits of embedded operations channel (eoc09–eoc16)
- OH3 (12 bits):
  - Bits 1 and 2: crc-5 and crc-6
  - Bit 3: Stuff bit id (second copy)
  - Bit 4: Segment defect (segd) indicator bit. A repeater sets this bit to a zero to indicate a loss of sync word on an incoming frame. Normally, the end units set this bit to a 1.
  - Bits 5–12: third 8 bits of embedded operations channel (eoc17–eoc24)
- SB (0 or 4 bits):
  - Bits 1–4: Stuff bits
  - Either 0 or 4 stuff bits are inserted in an HDSL2 superframe. The nominal frame is never transmitted; the nominal frame length only appears on average.

The three indicator bits—losd, sega, and segd—are transmitted every superframe.

The six CRC bits are used for performance monitoring of the subscriber line. The six CRC check bits, crc-1–crc-6 are those associated with the contents of the previous HDSL2 superframe. The CRC is computed over all of the bits in the superframe except for the 10 sync word bits, the 6 CRC bits, and the nominally 2 stuff bits; hence, the data message contains 9,294 bits. The message polynomial is constructed such that first CRC computable bit is the coefficient of the term x^{9293}, and the last bit is the coefficient of the term x^0. The polynomial is then multiplied by a factor of x^6 and the result is divided (modulo 2) by the generator polynomial $x^6 + x + 1$. The coefficients of the remainder polynomial are used in the order of occurrence as the ordered set of check bits crc-1–crc-6. In the remainder polynomial, the coefficient of the term x^5 is crc-1 and that for the term x^0 is crc-6.

The embedded operations channel has 24 bits allocated in a 6 msec frame. This corresponds to a 4 kb/s clear data communications channel. This channel is used to pass operations and maintenance information between the CO and customer premises HDSL2 transceiver units.

4.2.5 Core Transceiver Structure

Figure 4.17 shows the structure of the HDSL2 core modem transmitter. Accurate definition of these blocks is required in the standard to help assure interoperability of transceivers from different manufacturers. The core transmitter defines the signal processing of the bits coming from output of the HDSL2 framer and the modulated signal output on the subscriber line.

The scrambler receives the bits at the output of the framer and randomizes the bit sequence so that key adaptive signal processing functions, such as the echo canceler and equalizer, will optimally perform.

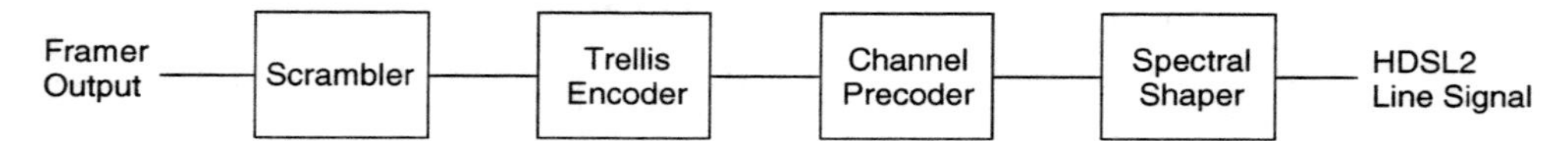

Figure 4.17 HDSL2 transceiver core modem transmitter.

The trellis encoder takes the randomized bit stream at the scrambler output and outputs an encoded symbol sequence containing redundancy per a predefined encoding procedure that is determined at initialization. For each transmitted symbol, the trellis encoder collects 3 information bits from the scrambler, adds one additional bit to accommodate the redundancy added by the encoding algorithm, and outputs a 4-bit symbol for transmission on the line. The resulting constellation for the transmit signal has $2^4 = 16$ levels; hence the line code name is trellis coded 16-level PAM (16 TC-PAM). When we compare the HDSL2 core transmitter blocks with that of the original HDSL 2B1Q system in Figure 4.5, the trellis encoder takes the place of the bit-to-symbol map function. The trellis encoder together with a Viterbi decoder (or other signal processing efficient decoder) provides the additional coding gain to help meet the system performance objectives.

In an uncoded system, a decision feedback equalizer (DFE) has been shown to provide near optimum performance in the detection of signals. Depending on the channel and noise characteristics, the DFE will provide at least 2 dB of output SNR improvement over linear equalizer. However, when a DFE is used directly with a trellis code, any error propagation through the DFE would significantly degrade any coding gain that is achievable with the trellis code. To achieve the performance gain of both the trellis code and the DFE, the error propagation must be removed. The channel precoder implements the equivalent of a decision feedback equalizer in the transmitter to effectively precode the transmit data prior to transmission on the line. In the transmitter, no decision errors will be made because there is no external injected noise to corrupt the signal. The equalization in the receiver, as shown in generalized transceiver structure of Figure 4.14, is effectively a linear equalizer so there is no error propagation to degrade performance. Details of the channel precoder are provided in a subsection below.

Finally, the spectral shaper is a filter with an impulse response that shapes the transmit signal per the OPTIS spectral shapes shown in Figure 4.23 for the downstream channel and Figure 4.24 for the upstream channel. A typical way of implementing the spectral shaping is to use digital transversal filters with oversampling.

The following subsections discuss the blocks of the core HDSL2 transmitter in more detail.

4.2.6 Scrambler

Figure 4.18 shows the scramble structure of the HDSL2 downstream and upstream transmitters. These scramblers are the same as those defined for HDSL and ISDN.

4.2.7 Trellis Encoder

In order to meet the demanding performance objectives of HDSL2, coded modulation is required to increase the immunity to crosstalk encountered in the loop

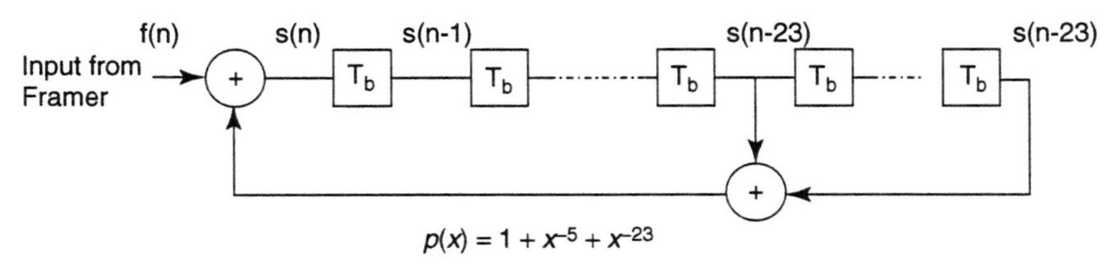

(a) Downstream Channel Scrambler

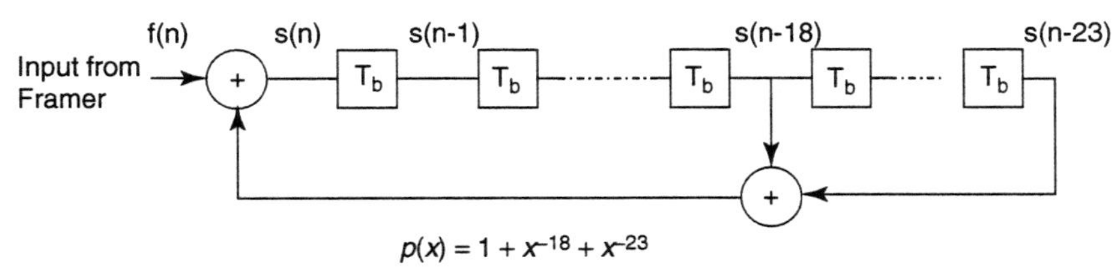

(b) Upstream Channel Scrambler

Figure 4.18 HDSL2 upstream and downstream channel scrambler

plant. Given the latency requirement of 500 μs, any types of block code or concatenated codes that require interleaving were ruled out because the latency introduced was too large. Interleaved block codes (such as Reed-Solomon codes), concatenated codes with interleaving, and turbo coding techniques proved to be difficult to use because they require latencies significantly greater than 500 μs to meet the performance objective. What remained was conventional trellis coding with channel precoding (such as the Tomlinson-Harashima precoding). Although multidimensional trellis codes were examined, it was determined that the simple one-dimensional Ungerboeck codes [33] were suitable meeting up to 5 dB of coding gain (ideal asymptotic coding gain) within a latency of 500 μs.

For the one-dimensional Ungerboeck codes [33], thirty-two states were sufficient to achieve a code gain of 4.0 dB. To achieve 5 dB of coding gain would require implementation of a 512 state code; the challenge here is the design of a decoder such that the implementation loss is minimized and the latency requirement is still met. Two proposals were provided for 512-state trellis codes: one from Pairgain in [36] and the other from Adtran in [37]. Both codes were linear codes claiming coding gains about 5 dB. The commonality is that the two proposed codes used rate one-half convolutional codes, where the convolutional coding was performed on one information bit while the other information bits were passed uncoded. The general structure of the trellis is shown in Figure 4.19, where the value of k is 3 (i.e., for every three information bits, there are four coded output bits).

To address the numerous codes possible, the agreed trellis code structure includes a programmable nonsystematic feed-forward convolutional encoder that codes the least significant bit of the three-bit information symbol [32]. The structure of the programmable convolutional encoder is shown in Figure 4.20. The convolutional code is a nonsystematic rate one-half code, where for each input data bit

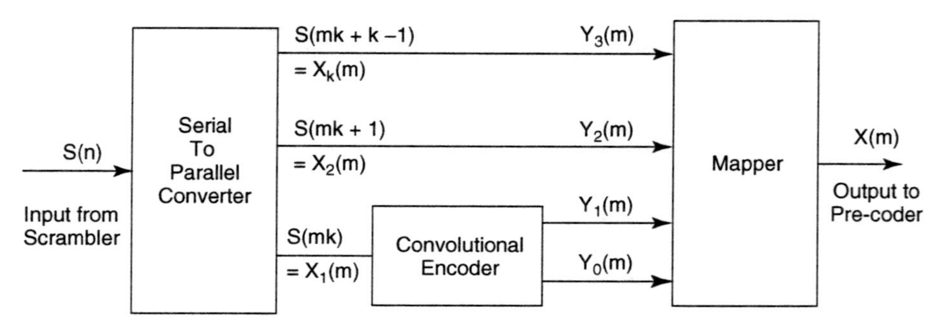

Figure 4.19 HDSL2 trellis encoder structure.

there are two output bits. The generator polynomials of the two output paths of the convolutional encoder are

$$P_a(x) = a_0 + a_1x^{-1} + a_2x^{-2} + \Lambda + a_{20}x^{-20}$$

and

$$P_b(x) = b_0 + b_1x^{-1} + b_2x^{-2} + \Lambda + b_{20}x^{-20}$$

for outputs Y_0 and Y_1, respectively. The coefficients $a_0, a_1, \ldots, a_{20}$ and $b_0, b_1, \ldots, b_{20}$ are binary coefficients and the operator "+" represents modulo 2 addition. A reference code having a coding gain of approximately 5 dB has coefficients $\{a_9, a_8, \ldots, a_0\}$ = {0101101110} and $\{b_9, b_8, \ldots, b_0\}$ = {1100110001}; the same coefficients are represented in octal as A = 556 and B = 1461, where all of the digits are octal digits. Another code with approximately the same coding gain has coefficients A = 732 and B = 1063.

The entire trellis code in Figure 4.19 is a rate 3/4 code, where the encoder accepts 3 information bits and outputs 4 coded bits. Because the convolutional code that operates on the least significant bit of the input code word is nonsystematic, the trellis code is also nonsystematic. The remaining two information bits are

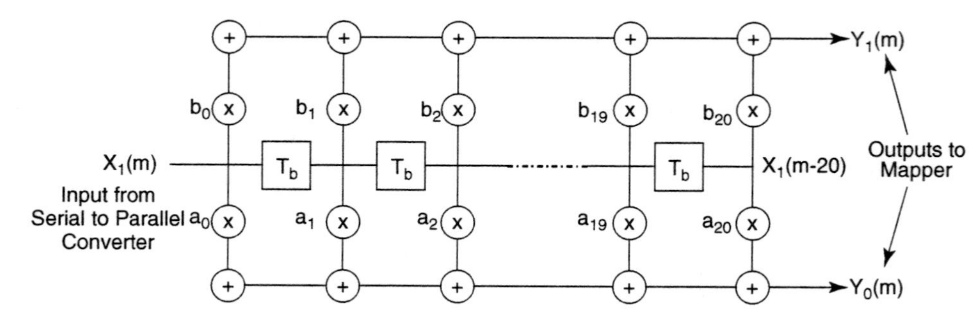

Figure 4.20 Convolutional encoder structure.

passed to the symbol mapper uncoded. The coefficients of the convolutional encoder, defined by two 20th order polynomials, are provided by the manufacturer's equipment that contains the receiver. During initialization, the coefficients are passed from receiver to encoder. An advantage of this programmable approach is that manufacturers could provide codes that are suitable to the type of decoder that they have implemented (e.g., a Viterbi decoder or sequential decoder). Also note that with this configuration, it is possible that the upstream and downstream channel could have different trellis codes.

The parallel bits at the output of the trellis encode must be mapped in symbols suitable for transmission on the line. The four bits at the trellis-encoder output are mapped into sixteen possible levels. The bit mapping of each of the levels are shown in Figure 4.21.

The input bits to the trellis encoder are $X_1(m)$, $X_2(m)$, and $X_3(m)$; the output bits are $Y_1(m)$, $Y_2(m)$, $Y_3(m)$, and $Y_4(m)$. $X_1(m)$ is the least significant bit of the trellis coder input bits, and $Y_1(m)$ is the least significant bit of the output symbol. The table in Figure 4.21 shows the mapping of the trellis encoder output bits to the output symbol. The mapping is also shown pictorially in the constellation diagram in the bottom of the figure.

4.2.8 Channel Precoder

The functional block diagram of the channel precoder is shown in Figure 4.22. The receiver computes the feedback filter coefficients $\{C_1, C_2, \ldots, C_N)$ during the training phase in the initialization process and then transfers the coefficients to transmitter during the parameter exchange phase. The input sequence $x(mT_s)$ is the

Mapping Table

Trellis Encoder Output, $Y_3(m)\ Y_2(m)\ Y_1(m)\ Y_0(m)$	Level, x(m)
0000	–15/16
0000	–13/16
0010	–11/16
0011	–9/16
0100	–7/16
0101	–5/16
0111	–1/16
1100	1/16
1101	3/16
1110	7/16
1111	9/16
1000	11/16
1010	13/16
1011	15/16

Mapper Output Level

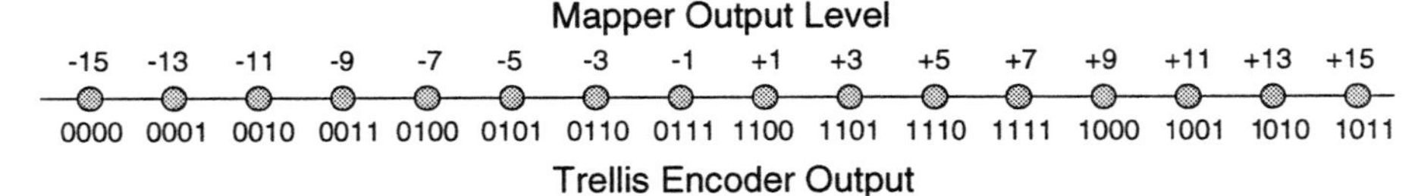

Figure 4.21 HDSL2 bit-to-symbol mapping.

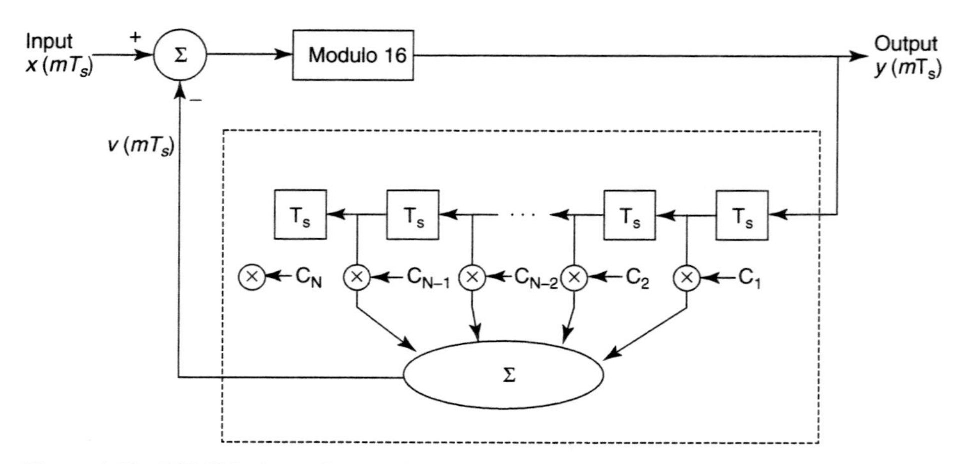

Figure 4.22 HDSL2 channel precoder.

output of the bit-to-symbol mapping in the trellis encoder. The output of the feedback filter, $v(mT_s)$, is computed by

$$v(mT_s) = \sum_{k=1}^{N} C_k \cdot y(mT_s - kT_s)$$

where T_s is the symbol interval, $y(mT_s)$ is the precoded output sample, m is the sample time index, and N is number of coefficients in the feedback filter. A modulo-16 operator then operates on the difference between the input sample and feedback filter output sample $u(mT_s)$. Operation of the modulo-16 block is to find an integer $d(mT_s)$ such that the sum $u(mT_s) + 2\,d(mT_s)$ falls between the values of -1 and $+1$. The resulting value of $y(mT_s)$ is then $u(mT_s) + 2\,d(mT_s)$.

In the HDSL2, the receiver determines the value of N and the value is passed on to the transmitter at initialization during the parameter exchange phase. The value of N has minimum value of 128 and a maximum of 180 samples; any value in between is valid.

4.2.9 Spectral Shaper

As described earlier, the spectral shape chosen for HDSL2 is that of OPTIS as proposed in [29]. A plot of the downstream mask is shown in Figure 4.23 and that of the upstream spectrum is given in Figure 4.24.

4.3 INITIALIZATION

When an HDSL2 based service is installed, the HDSL2 transceivers need to be configured automatically. The initialization process for an HDSL2 connection has the following phases: "Wake-up," preactivation, core activation, and data mode. A

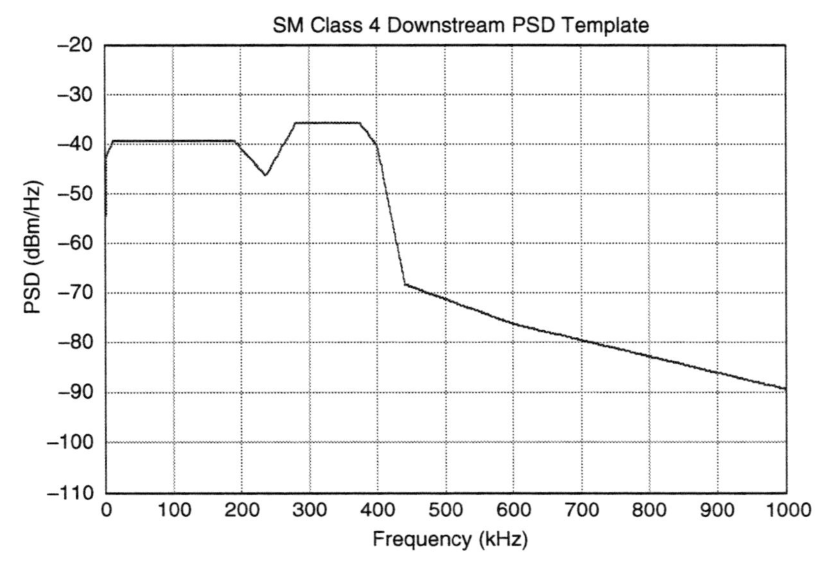

Figure 4.23 Downstream channel spectrum.

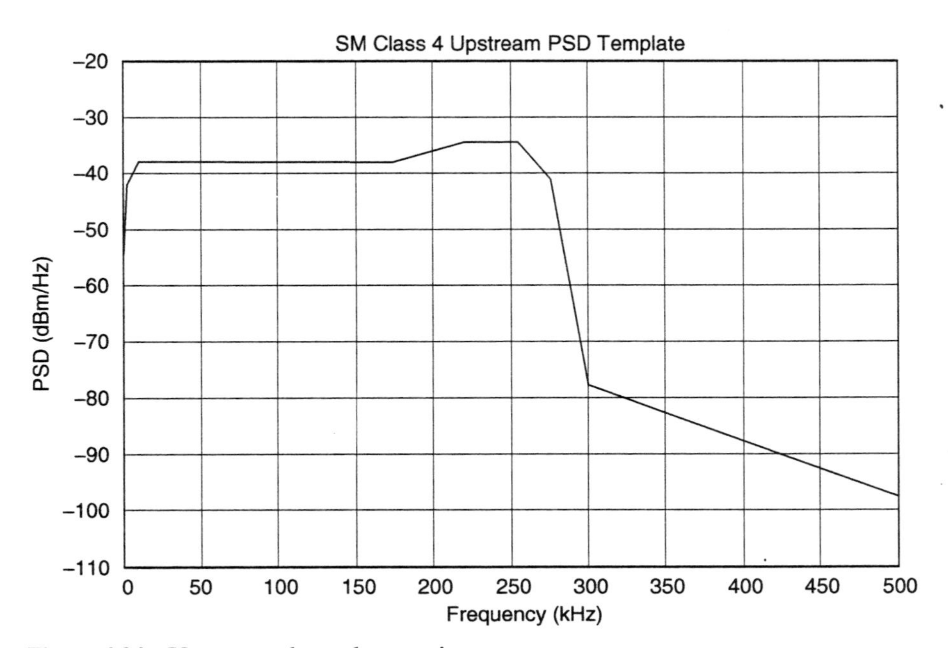

Figure 4.24 Upstream channel transmit spectrum.

typical startup time for HDSL2 initialization may range from 15 to 30 seconds. This section provides a brief overview of the HDSL2 initialization procedure.

The main objectives of the initialization process are to

- Determine the operation settings, for example, the transmit signal power level.
- Train the core transceiver block components, for example, the echo canceler, equalizer, and so on.
- Exchange key system parameters, for example, channel precoder coefficients, coefficients to the rate one-half convolutional encoder in the trellis encoder, and so on.
- Transition seamlessly to data mode.

To maximize robustness in the initialization process, the HDSL2 transceiver operates using two-level PAM, because it can operate with the lowest SNR of the multilevel PAM configurations. The transmitter structure at initialization is at its simplest, that is, it basically consists of a scrambler, bit-to-symbol map, and transmit shaping filter. Figure 4.25 shows the transmitter reference model of the HDSL2 transmitter during initialization. The input sequence is either the "all ones" sequence or the "activation frame" that passes transceiver parameters.

Figure 4.26 shows the timing diagram of the initialization sequence. The diagram shows sequences in the preactivation and core activation phases; at the completion of the core activation phase, the modems transition into data mode.

Sequences A_r and B_c are generated by sending the "all ones" pattern to the scrambler in the reference startup transmitter model of Figure 4.25. The resulting line signal is 2-PAM. These signals serve primarily as alert (wake-up) and response signal. The receiver can measure receive signal level for determining the level of transmit signal power. The duration of these sequences are 200 ms each.

Sequences A_r' and B_c' offer a slight extension to those of A_r and B_c. Although the two-level PAM sequences for A_r' and B_c' are generated in the same manner as for A_r and B_c by inputting the all ones pattern into the scrambler of the reference startup transmitter, the resulting PAM sequence is amplitude modulated with a binary sequence to transmit a preactivation frame that identifies the transmit signal power level. The modulation format of the two-level PAM signal is as follows:

- Each bit has a 10 msec period.
- A logic 1 is the two-level PAM sequence sent at maximum transmit power (i.e., 0 dB power cutback).
- A logic 0 is the two-level PAM sequence sent at 7 dB reduced transmit power (i.e., 7 dB power cutback).

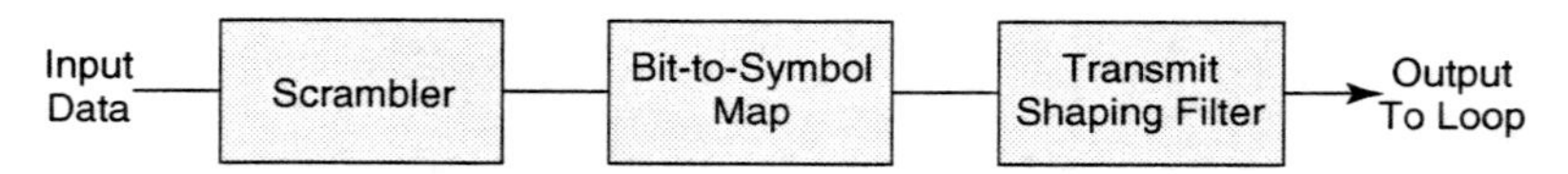

Figure 4.25 HDSL2 reference transmitter during startup.

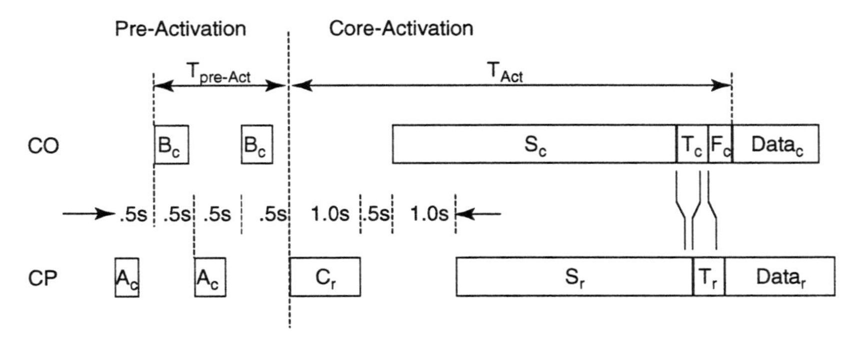

Figure 4.26 Initialization sequence timing diagram.

Within this sequence, the CO and CP (customer premises) units transmit a preactivation frame that defines the transmit signal power for the corresponding far end transmitters. The power levels are determined based on the level of the received signals during A_r and B_c sequence exchange. The 20 preactivation frame bits are allocated as follows:

- 2 bits to identify the frame start
- 1 bit for CRC error indication
- 4 bits for mask indication (i.e., power cutback level)
- 4 bits for CRC-4
- 9 bits reserved (undefined)

The HDSL2 standard defines sixteen levels of power cutback, which range from 0 to 15 dB of cutback in increments of 1 dB steps. Each level of power cutback can be viewed as an array of PSD masks, msk0–msk15, where the integer represents the level of power cutback. Once the sequences of A_r' and B_c' have completed and the power cutback levels have been set, the core transceiver training begins. All core activation sequences that follow are sent using the corresponding power cutback levels transmitted during the A_r' and B_c' exchange.

Core initialization begins with the CP unit transmitting sequence C_r for a period of 1 second. The CO unit may use the received sequence to perform timing recovery and train its equalizer. The CP unit may use this sequence to train an echo canceller. The duration of C_r is 300 ms nominal.

Upon the completion of sequence C_r, the CO unit begins sending sequence S_c one-half second later. Sequence S_c is the two-level PAM signal formed by inputting ones into the scrambler of the reference startup transmitter. Recall that the transmit power is set to the value specified in the preactivation frame sent during the A_r' and B_c' exchange. For the first second, S_c is the only signal on the line. The CP unit may train its equalizer and timing recovery circuit from the received signal; during this same time, the CO unit may train its echo canceler. After one second later, the CP unit begins transmitting Sequence S_r, and now there is simultaneous transmission of both upstream and downstream data. The sequence S_c is transmitted for a minimum of 5 seconds, which is the time required (timer T_{PLL}) for the CP unit to synchronize its phase locked loop.

Sequence S_r is also a two-level PAM signal resulting from an input of ones into the scramble of the reference startup transmitter, and the transmit power is that specified in the pre-activation frame sent during the A_r' and B_c' exchange. The CP unit begins transmitting about 1.5 seconds after it concludes transmitting C_r. During the period of simultaneous S_c and S_r transmission, the transceivers continue training their equalizer, echo canceler, and other necessary functions. If the transceiver functions have not converged by conclusion of S_c and S_r, then the transceiver enters an exception state. At that point the initialization process would need to be restarted.

After the CO unit transceiver has converged and it has been sending the S_c signal for at least 5 seconds (i.e., the value of the T_{PLL} timer), it transitions to sending signal T_c. During the transmission of T_c, the channel precoder coder coefficients and other system information is sent to the CP unit from the CO unit. Once the CP unit has converged and has begun detecting the T_c signal, it begins transmitting the T_r signal to the CO unit. As with T_c, the T_r signal passes the channel precoder coefficients and other signal parameters to the CO transceiver. The information transferred in the T_c and T_r signals is contained in a core activation frame.

Once the CO unit has detected the T_r signal and has completed transmission of the core activation frame, it begins sending signal F_c. Signal F_c sends the core-activation frame of T_c except that the frame sync word is reversed and all of the remaining information bits are set to arbitrary values. Two of these frames are sent during F_c, and this can serve as an acknowledgment that the CO unit received T_r.

Upon conclusion of S_c transmission, the CO unit begins sending data, and the CP unit begins sending data upon completion of T_r.

4.4 SHDSL AND HDSL4

See Chapter 6 for a discussion of SHDSL (Single-pair High-speed Digital Subscriber Line) and HDSL4 (extended reach DSL1 transmission).

REFERENCES

[1] G.703, ITU-T Recommendation "Physical/Electrical Characteristics of Hierarchical Digital Interfaces," November 2001.

[2] G.704, ITU-T Recommendation "Synchronous Frame Structures Used at 1544, 6312, 2048, 8448 and 44 736 kbit/s Hierarchical Levels," October 1998.

[3] Committee T1, "ISDN Basic Access Interface for Use on Metallic Loops for Application at the Network Side of NT, Layer 1 Specification," Document T1.601-1999.

[4] Committee T1, "High-Bit-Rate Digital Subscriber Line (HDSL)," Technical Report TR-28, February 1994.

[5] Bellcore TA-1210 on HDSL.
[6] Committee T1, Working Group T1E1.4, "A Draft Technical Report on High-Bit-Rate Digital Subscriber Line (HDSL)," T1E1.4/96-006, April 22, 1996.
[7] ETSI TS 101 135 V1.5.3 (2000-09), "Transmission and Multiplexing (TM); High Bit-Rate Digital Subscriber Line (HDSL) Transmission Systems on Metallic Local Lines; HDSL Core Specification and Applications for Combined ISDN-BA and 2,048 kbit/s transmission," September 2000.
[8] Committee T1 for Telecommunications—High Bit Rate Digital Subscriber Line—2nd Generation (HDSL2), T1.418-2000.
[9] ITU-T Recommendation G.991.1, "High Bit Rate Digital Subscriber Line (HDSL) Transceivers," October 1998.
[10] AT&T Tech Pub 62411.
[11] ANSI T1.403-1999, "Network and Customer Installation Interfaces—DS1 Electrical Interface."
[12] ANSI T1.413-1998, "Network to Customer Installation Interfaces—Asymmetric Digital Subscriber Line (ADSL) Metallic Interface."
[13] ITU-T Recommendation G.991.2, "Single-Pair High-Speed Digital Subscriber Line (SHDSL) Transceivers," February 2001.
[14] L. F. Wei, "Trellis-Coded Modulation with Multidimensional Constellations," *IEEE Transactions on Information Theory* IT-33 (1987): 483.
[15] ITU-T Recommendation V.32, "A Family of 2-Wire, Duplex Modems Operating at Data Signalling Rates of Up to 9600 bit/s for Use on the General Switched Telephone Network and on Leased Telephone-Type Circuits," March 1993.
[16] Ameritech, "Third Generation HDSL," T1E1.4/95-044, June 5, 1995.
[17] Ameritech, "Second Generation HDSL," T1E1.4/96-094, April 22, 1996.
[18] Ameritech, "HDSL2 crosstalk Interferer Model," T1E1.4/96-095, April 22, 1996.
[19] Pairgain Technologies, "Normative Text for Spectral Compatibility Evaluations," T1E1.4/97-180, June 30, 1997.
[20] Pairgain Technologies, "On the Importance of Crosstalk from Mixed Sources," T1E1.4/97-181, May 12, 1997.
[21] ADC, Adtran, Level One, Pairgain, "Performance Requirements for HDSL2 Systems," T1E1.4/97-469, December 8, 1997.
[22] ADC Telecommunications, "Measured Spectral Compatibility of HDSL2 with Deployed HDSL," T1E1.4/97-434, December 8, 1997.
[23] Adtran, "Test Results of HDSL Units with HDSL2 Noise Generator," T1E1.4/97-440R1, December 8, 1997.
[24] Adtran, "Simulated Performance of HDSL2 Transceivers," T1E1.4/97-444, December 8, 1997.
[25] Pairgain Technologies, "A Modulation Strategy for HDSL2," T1E1.4/96-340, November 11, 1996.
[26] Adtran, "A Modulation Technique for CSA Range HDSL2," T1E1.4/97-073, February 3, 1997.
[27] Pairgain, "Performance and Spectral Compatibility Comparison of POET PAM and OverCAPped Transmission for HDSL2," T1E1.4/97-179, May 15, 1997.

[28] Pairgain, "Performance and Spectral Compatibility of OPTIS HDSL2," T1E1.4/97-237, June 30, 1997.
[29] Level One, ADC, Pairgain, "OPTIS PSD Mask and Power Specification for HDSL2," T1E1.4/97-320, September 22, 1997.
[30] Cicada Semiconductor, "Performance and Spectral Compatibility of MONET-PAM HDSL2," T1E1.4/97-307, September 22, 1997.
[31] Cicada Semiconductor, "Performance and Spectral Compatibiity of MONET (R1) HDSL2," T1E1.4/97-412R1, December 8, 1997.
[32] Adtran, Cicada, Siemens, Tellabs, Westell, "Proposal to Break the FEC Logjam for HDSL2," T1E1.4/97-443, December 8, 1997.
[33] G. Ungerboeck, "Channel Coding with Multilevel/Phase Signals," *IEEE Transactions on Information Theory* IT-28, No.1, January 1982.
[34] G. Ungerboeck, "Trellis-Coded Modulation with Redundant Signal Sets Part I: Introduction," *IEEE Communications* 25 no. 2 (February 1987): 5–11.
[35] G. Ungerboeck, "Trellis-Coded Modulation with Redundant Signal Sets Part II: State of the Art," *IEEE Communications* 25 no. 2 (February 1987): 12–26.
[36] Pairgain Technologies, "A 512-State PAM TCM Code for HDSL2," T1E1.4/97-300, September 22, 1997.
[37] Adtran, "Performance and Characteristics of One-Dimensional Codes for HDSL2," T1E1.4/97-337 September 25, 1997.
[38] Adtran, "Single-Loop HDSL CAP/PAM Comparison," T1E1.4/95-107, August 21, 1995.
[39] AT&T Network Systems, "Performance of CAP and PAM for Single Pair HDSL in Presence of SNEXT," T1E1.4/95-106, August 21, 1995.
[40] AT&T Network Systems, "Text for HDSL Technical Report Issue 2: CAP Interoperability," June 5, 1995.
[41] Adtran, "Proposal for Single-Loop HDSL Using Simple Coded PAM," T1E1.4/96-037, January 26, 1996.
[42] AT&T Network Systems, "Single Pair T1 HDSL: Spectral Compatibility with ADSL & Implementation Issues," T1E1.4/96-038, January 26, 1996.

CHAPTER 5

HANDSHAKE FOR THE ITU-T SUITE OF DSL SYSTEMS

The ITU-T has defined a suite of digital subscriber line (DSL) Recommendations that include symmetric and asymmetric DSL transceivers. Each of the DSL Recommendations contains numerous options and capabilities, and before proper operation can begin, the transceivers need to negotiate and agree on a common set of parameters. ITU-T Recommendation G.994.1 defines handshake (HS) procedures for DSL transceivers to exchange transceiver capabilities and to negotiate and select a common mode of operation.

Handshake Recommendation G.994.1 (also referred to as G.hs) defines the modulation method, protocol, and messages for exchanging capabilities information and negotiating session configuration. The initial issues of G.hs apply to ADSL and SHDSL systems. We expect that later revisions of G.hs will apply to VDSL when the transmission approach is defined in the ITU-T.

This chapter provides an overview of the handshake mechanism defined in G.994.1 [1]. The goal is to provide assistance toward the understanding of the handshake mechanism for DSL systems. We describe the handshake signals and modulation method, initialization and clear-down, the set of commands, message structure, message coding format, and example transaction sequences.

5.1 HANDSHAKE MODULATION METHOD

All ITU-T compliant DSL modems provide support of the handshake mechanism defined in ITU-T Recommendation G.994.1 prior to the core initialization of the specific DSL. The purpose of the handshake is to identify the type of DSL modem at each end of the loop and to negotiate and agree on a common set of parameters for the communication session. As shown in Figure 5.1, handshake sits "in front" of the core DSL modem to operate in preactivation sessions. During the handshake session, the two modems exchange messages negotiating configuration parameters for the communication session. Once the common set of parameters is identified and agreed upon, the handshake session clears down and the core initialization of the DSL modem begins.

Because handshake is common to all ITU-T compliant modems, it needs to operate under the most extreme conditions, namely, on long loops and under relatively noisy conditions. At a minimum, it must operate at a range that is longer than that which can be supported by the core DSL modem. The modulation method used during handshake needs to be very robust as a top priority, while transmission speed on the link is not as important. Another key requirement is that

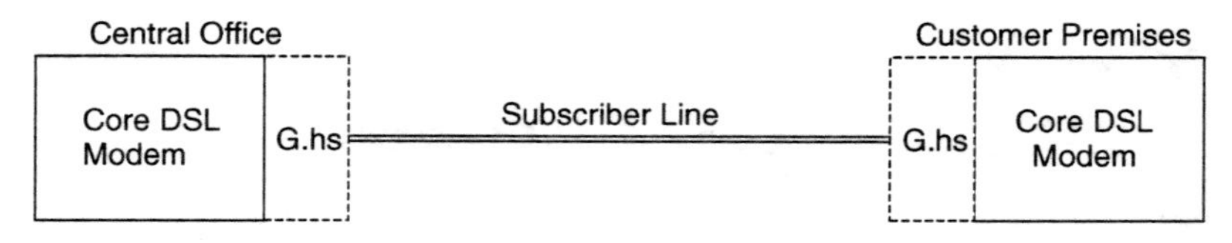

Figure 5.1 System reference model for handshake.

the modulation method needs to be simple, preferably based on or related to the transmission method of the core DSL modem.

The modulation method chosen for handshake is binary differential phase shift keying (DPSK). The core modulator is modeled as a one-dimension double sideband suppressed carrier amplitude modulator as shown in Figure 5.2. The differential encoding takes place in the transmit sequence A_n, where A_n takes on the values of +1 or –1. Transmit symbol A_n is rotated 180 degrees from the previous symbol if the transmit bit is a logic 1 and A_n is the same phase as the previous symbol if the transmit bit is a logic 0. Mathematically, the differential encoding rule is specified as $A_n = A_{n-1}$ if $b_n = 0$, and $A_n = -A_{n-1}$ if $b_n = 1$, where b_n is the transmit data bit at time sample index n.

The transmit signal $s(t)$ per the DPSK modulator in Figure 5.2 is represented by

$$s(t) = \sum_n A_n \cdot g(t - nT) \cdot \cos(2\pi f_i t + \phi_i),$$

where A_n is the differentially encoded bit sequence per the rules of

$$A_n = A_{n-1} \text{ if } b_n = 0$$
$$A_n = -A_{n-1} \text{ if } b_n = 1$$

where b_n is the transmit data bit sequence, $g(t)$ is the impulse response of the transmit shaping filter, T is the data symbol interval, f_i is the carrier frequency of index i, and ϕ_i is an arbitrary carrier phase for carrier index i.

G.994.1 defines different carrier sets, and a carrier set may contain more than one carrier that transports the same information in each carrier. In the case where more than one carrier is used to transmit the same data sequence in each carrier, the transmit signal may be represented as the sum of all the modulated carriers, namely

$$s(t) = \sum_i \left(\sum_n A_n \cdot g(t - nT) \cdot \cos(2\pi f_i t + \phi_i) \right).$$

Because handshake operates "in front" of the DSL modem, it makes sense to define a set of carriers consistent with the frequency spectrum used by the specific

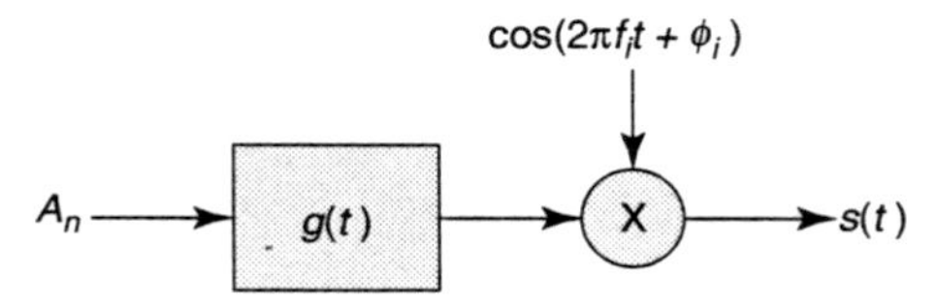

Figure 5.2 DPSK modulator block diagram.

Table 5.1 Carrier Sets for the 4.3125 kHz Signaling Family

Carrier Set	Upstream Carriers						Downstream Carriers					
	N_{u1}	F_{u1} *(kHz)*	N_{u2}	F_{u2} *(kHz)*	N_{u3}	F_{u3} *(kHz)*	N_{d1}	F_{d1} *(kHz)*	N_{d2}	F_{d2} *(kHz)*	N_{d3}	F_{d3} *(kHz)*
A43	9	38.8125	17	73.3125	25	107.8125	40	172.5000	56	241.5000	64	276.0000
B43	37	159.5625	45	194.0625	53	228.5625	72	310.5000	88	379.5000	96	414.0000
C43	-	-	7	30.1875	9	38.8125	12	51.7500	14	60.3750	64	267.0000

DSL and to utilize the corresponding core signal processing element(s) for generation of the handshake tones. The ITU-T suite of DSL modems contains a family of ADSL modems that uses multicarrier modulation with carrier frequency spacing of 4.3125 kHz; the suite also contains the symmetric high-rate DSL (SHDSL), whose signaling is derived from a fundamental 8 kHz reference clock. Because various DSL modems in the ITU-T suite are based on different fundamental frequencies, G.994.1 defines two signaling families: one family is based on 4.3125 kHz signaling (i.e., the ADSL modems), and the other is based on 4 kHz signaling (i.e., the SHDSL modems).

5.1.1 The 4.3125 kHz Signaling Family

The 4.3125 kHz signaling family has three carrier sets in support of ADSL. Table 5.1 shows the carrier sets for the 4.3125 signaling family; each carrier is identified by an index (multiplier) value *N* and corresponding frequency *F*. The subscripts *u* and *d* identify upstream and downstream, respectively.

The sets are labeled A43, B43, and C43, and they are designed in support of Annexes A, B, and C, respectively, in G.992.1 (full-rate ADSL) and G.992.2 (splitterless ADSL). Each carrier set contains three tones transmitting in the upstream direction and three tones transmitting in the downstream direction. The maximum transmit power for each upstream carrier set is –1.65 dBm and that for each downstream carrier set is –3.65 dBm.

5.1.2 The 4 kHz Signaling Family

Table 5.2 shows the 4 kHz signaling family carrier set. This set consists of one upstream tone at 12 kHz and one downstream tone at 20 kHz. The maximum transmit power for each tone is +5 dBm. G.991.2 (SHDSL) is the digital subscriber line technology that uses this carrier set during handshake preactivation.

Table 5.2 Carrier Set for the 4 kHz Signaling Family

Carrier Set	Upstream Carriers		Downstream Carriers	
	N_{u1}	F_{u1} *(kHz)*	N_{d1}	F_{d1} *(kHz)*
A4	3	12	5	20

5.2 MESSAGES AND COMMANDS

During a handshake session, the DSL modems at each end of the line exchange messages to negotiate the configuration of the DSL communication session. There are fourteen message types defined in G.994.1. The list below contains a summary of ten basic messages; note that some of the messages have different types associated with them, hence the total of fourteen messages overall.

- CL—Capabilities List: The central office unit (xTU-C) sends this message, and it contains a list of possible modes of operation that it can support.
- CLR—Capabilities List and Request: The remote or customer premises unit (XTU-R) sends this message to the central office unit. This message contains a list of the possible modes of operation that the customer premises unit can support and also request that the central office unit sends a CL message in response.
- MR—Mode Request: The customer premises unit sends this message to the central office unit requesting that (XTU-C) send a Mode Select message. Basically the CP unit requests the CO unit to select a particular mode of operation.
- MS—Mode Select: Either the central office or customer premises unit may initiate this command to select a particular mode of operation.
- MP—Mode Proposal: The customer premises unit sends this command to the CO unit proposing a particular mode of operation and requests that the CO unit send a mode select (MS) message in response.
- ACK—Acknowledgment of messages [2 types: ACK(1) and ACK(2)]. Either the central office unit or customer premises unit may send this message.
  - ACK(1) acknowledges one of the following:
    - Receipt of a complete CL or an intermediate frame of a segmented CL message and ends a handshake transaction; or
    - Receipt of a complete MS message of intermediate frame of a segmented MS message and initiates the handshake message clear-down procedure.
  - ACK(2) acknowledges receipt of an intermediate frame that contains a segmented CL, CLR, MP, or MS message. It also requests transmission of the next frame of the segmented message.
- NAK—Negative acknowledgment of messages (4 types)
  - NAK-EF: This is a negative acknowledgment due to reception of an *error frame.*
  - NAK-NR: This message acknowledges complete reception of a MS message or intermediate frame of a segmented MS message, but the negative acknowledgment indicates that the receiving unit is temporarily unable (or *not ready*) to invoke the mode requested in the MS message.
  - NAK-NS: This message acknowledges complete reception of a MP or MS message (or intermediate frames of the segmented forms of each), but the negative acknowledgment indicates that the receiving unit either doesn't support or has disabled the requested mode sent by the sending unit.

- NAK-CD: This message is a negative acknowledgment indicating that the received message was not understood, and it initiates the *clear-down* procedure in response.
- REQ-MS—Request MS message: The central office unit sends this message to the customer premises unit in response to an MR message, in that it requests the customer premises unit to send an MS command. In this case, the CO unit does not want to decide on the mode of operation, and it leaves the choice to the CP unit.
- REQ-MR—Request MR message: The central office unit sends this message to the customer premises unit in response to an MS message, in that it requests the customer premises unit to send an MR command. In this case, the CO unit tells the CP unit that it wants to make the choice of operating mode.
- REQ-CLR—Request CLR message: The central office unit sends this message to the customer premises unit in response to reception of either an MR message, an MP message, or an MS message (or intermediate frames of the segmented forms of each) in that it requests the customer premises unit to send a CLR command. In this case, the CO unit wishes to perform a capabilities exchange with the CP unit.

5.3 G.hs FRAME STRUCTURE

Now that the set of messages and commands are defined, the next step is to describe the format in which the messages and commands are transmitted.

A message is defined by a sequence of octets, where each message may contain one or more segments. The packets transmitted during a handshake session contain message segments. Figure 5.3 shows the frame structure of the message segment to be transmitted. The beginning of a frame is identified using the HDLC

8 7 6 5 4 3 2 1	Octet
Flag #1	1
Flag #2	2
Flag #3	3
Flag #4 (Optional)	
Flag #5 (Optional)	
Message Segment	
FCS (1st Octet)	
FCS (2nd Octet)	
Flag #1 end	N–2
Flag #2 end	N–1
Flag #3 end (Optional)	N

Figure 5.3 Message segment frame structure.

Flag octet ($01111110_2 = 7E_{16}$) using a minimum of three flag octets or up to a maximum of five flag octets. The end of a frame is identified with a minimum of two flag octets or a maximum of three flag octets. Following the beginning flags is the message segment. The frame check sequence (FCS) field occupies two octets; this field provides a check on the received data to see if any errors were received. If any errors were detected in the received signal, then a retransmission is required.

The octet convention is as follows. The bits are numbered horizontally 1 to 8 as shown in Figure 5.3, where bit 1 is the *least significant bit* (*LSB*) and bit 8 is the *most significant bit* (*MSB*). The octets are numbered vertically from 1 to N, which is also shown in Figure 5.3. Octets are transmitted in ascending order with the LSB sent first.

In general, bit 8 in any octet is the MSB, and bit 1 is the LSB. If the field spans more than one octet, then the order of the bit values increases with decreasing octet numbers per the octet numbering convention in Figure 5.3. This is shown in the following table:

8	7	6	5	4	3	2	1	Octet
2^{15}	2^{14}	2^{13}	2^{12}	2^{11}	2^{10}	2^9	2^8	**k**
2^7	2^6	2^5	2^4	2^3	2^2	2^1	2^0	**k+1**

In Figure 5.3 the FCS field is shown to occupy two octets. For the FCS field, the MSB resides in bit 1 of the first octet, and the LSB resides in bit 8 of the second octet. The convention for the FCS field is opposite to the general convention for octets as described earlier. The FCS field is 16 bits, and uses the CRC-16 as defined in ISO/IEC 3309.

Once the FCS field is computed, the entire frame is examined for occurrence of the following special characters: the HDLC flag sequence character $7E_{16}$ and the control escape character $7D_{16}$. G.994.1 uses the HDLC byte stuffing mechanism to obtain octet transparency, which is also defined in ISO/IEC 3309. Any octet equal to flag sequence character or control escape character is replaced by a sequence of two octets as follows:

- The data octet of 7E16 is encoded as the two octet sequence of $\{7D_{16}, 5E_{16}\}$
- The data octet of 7D16 is encoded as the two octet sequence of $\{7D_{16}, 5D_{16}\}$

When receiving the message segment frame, the control escape octets ($7D_{16}$) must be removed, and the original sequence restored prior to computation of the received FCS. The FCS of the received signal is compared with that transmitted in the message segment. If the two match, then no errors were detected; if the two do not match, then there were errors received, and the message segment must be retransmitted.

5.4 INFORMATION FIELD STRUCTURE

The next step is to define the data structure of the message field. The message field, shown in Figure 5.4, consists of three components in the following sequence:

Identification (I) Field	Standard Information (S) Field	Non-standard Information (NS) Field

Figure 5.4 Message field structure.

- The identification (I) field
- The standard information (S) field
- The nonstandard information (NS) field

The identification field identifies the message type, the revision number of the G.994.1 Recommendation that the equipment is following, identification of the equipment vendor or manufacturer, and identification of parameters independent of the mode of operation. The standard information field contains parameters that represent the modes of operation or capabilities of the DSL transceivers. The non-standard information field is used to pass non-standard information. Prior knowledge of equipment is required to read and understand the information transmitted via the nonstandard facilities, and this is identified via the vendor ID codes in the information field. A data structure for the non-standard field is also defined in G.994.1.

Both the identification (I) and standard information (SI) fields contain an array of parameters relating to particular modes of operation, features, and/or capabilities associated with the two end modems. G.994.1 defines a data structure that allows future extension of the parameter list in such a way as to permit present and future G.994.1 implementations to correctly parse the information fields. The following summarizes the rules for encoding data to be transmitted in the I and S information fields.

The general data structure for defining parameters in the I and S fields that is adopted in G.994.1 is a three-level tree structure. This three-level tree structure is illustrated in Figure 5.5.

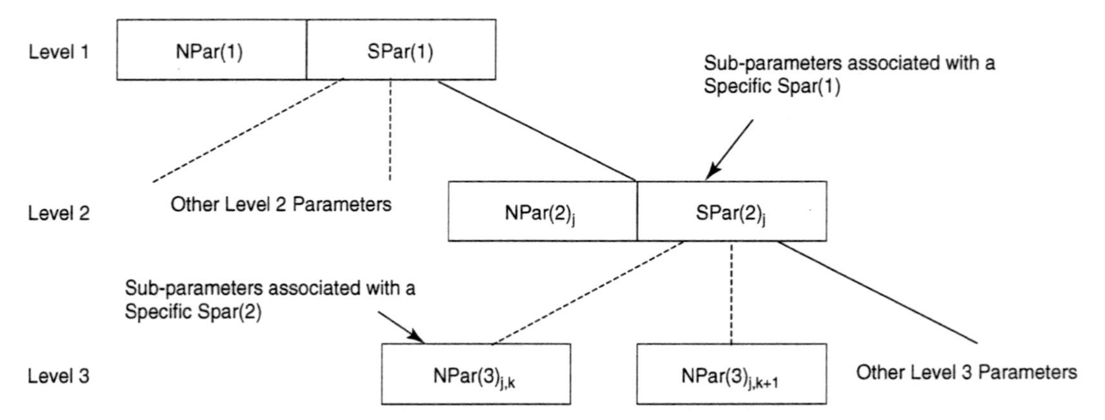

Figure 5.5 Tree data structure of the I and S parameter fields.

5.5 PARAMETER DATA STRUCTURE

The parameters, as defined in G.994.1, are generic and the data structure is used in all messages where parameters are communicated. The parameters have two types of classifications: NPars and SPars.

- NPars are parameters that have *no* subparameters associated with them.
- SPars are parameters that have subparameters with them.

Level 1 is the highest level in the tree, which contains an array of NPars and SPars. The general notation for parameters defined at Level 1 is NPar(1) and SPar(1). Each of the SPar(1) parameters have a set of NPars and SPars at Level 2, identified as NPar(2) and SPar(2) parameters, respectively. Because Level 3 is the lowest level in the three-level tree structure, there can be no SPar values at Level 3.

In Figure 5.5, the general notation is as follows:

- $NPar(2)_j$ and $SPar(2)_j$ indicate the set of Level 2 parameters associated with the *j*th Level 1 Spar.
- $NPar(3)_{j,k}$ indicates the set of Level 3 parameters associated with the *k*th Level 2 Spar value(s), which is associated with the *j*th Level 1 SPar.

5.6 TRANSMISSION ORDER OF THE PARAMETERS

The parameters are transmitted sequentially as a block consisting of an integer number of octets. The order of transmission of the parameters from the tree structure is shown in Figure 5.6. First the NPar(1) and SPar(1) parameters are passed,

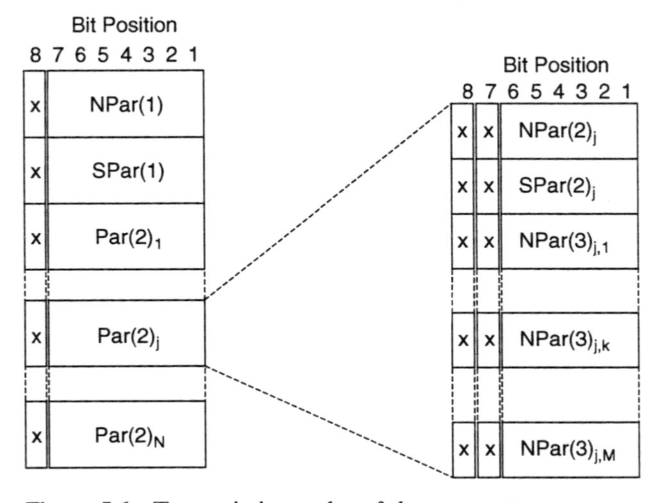

Figure 5.6 Transmission order of the parameters.

followed by the NPar(2) and SPar(2) subparameters. For each SPar(2), all corresponding NPar(3) subparameters must be passed.

In Figure 5.6 we show there to be N level 2 parameters, which are the subparameters to first level SPar set. $Par(2)_j$ corresponds to the set (block) of Level 2 parameters associated with the *j*th Level 1 SPar, which may contain NPars and SPars. For each second level SPar, there are a set of third-level subparameter blocks labeled $NPar(3)_{j,k}$, where k would range from 1 to M and the value of M would be different for each second-level SPar set (i.e., for each value of j). Transmission begins with the first octet of the NPar(1) set and ends with the last octet of the $Par(2)_N$ set. For each Par(2) set [i.e., $Par(2)_j$], transmission begins with the first octet of $NPar(2)_j$ and ends with the last octet in $NPar(3)_{j,M}$.

In order to determine the beginning and end of a parameter block, we need to define a set of delimiting rules. Following are the delimiting rules adopted in G.994.1.

In each octet of a parameter block, at least one bit is defined as a delimiter bit, which defines the last bit to be transmitted in the block. A logic 0 in the delimiter bit position indicates that there is at least one additional octet in the block to be transmitted; a logic 1 in the delimiter bit position indicates the last octet in the block to be transmitted.

For the NPar(1), SPar(1), and all of the Par(2) blocks, bit 8 is used as a delimiter bit, and the remaining bits are used to encode the parameter data. For the above parsing rule to work properly, the identification (I) and standard information (S) fields each need to include at least one NPar(1) octet and one SPar(1) octet. As shown in Figure 5.6, there are N Par(2) blocks that correspond to the SPar(1) block. The delimiting bit is shown as an x in the bit #8 position.

For each Par(2) block, bit 7 serves as the delimiter bit in the corresponding NPar(2) block, Spar(2) block, and each of the associated NPar(3) blocks. The illustration in Figure 5.6 shows there to be M of the NPar(3) blocks for each SPar(2) block, where the value of M may be different for each SPar(2) block. The delimiting bits are shown with an x in the bit #8 and #7 positions.

Note that each Par(2) block may transmit both NPar(2) and SPar(2) blocks or only NPar(2) blocks. If only NPar(2) block is transmitted, the final octet in the block will contain a logic 1 in each delimiter bit position.

In order to be compatible with future versions of G.994.1, a receiver needs to parse all parameter blocks and ignore information that is not understood.

5.7 IDENTIFICATION (I) FIELD

The identification field (or simply the I field) identified in Figure 5.4 contains four components as shown in Figure 5.7. The components are

- Message Type Field (1 octet)
- Revision Number Field (1 octet)
- Vendor ID Field (8 octets)
- Bit-encoded parameter field

Message Type Field	Revision Number Field	Vendor ID Field	Bit-encoded Parameter Field

Figure 5.7 Identification field structure.

The message type is one of the message or command types defined earlier. The message type list and the corresponding bit encoding are provided in Table 5.3.

The revision number field is a single octet that identifies the revision number of the G.994.1 Recommendation that the equipment conforms to. Note that the revision number is updated only when a structural change is made to the handshake protocol, for example, the addition of new command or message types. The addition of code points will not constitute a change in revision number.

The vendor ID field is an 8-octet field that identifies the manufacturer of the handshake functionality, which is usually the same as the manufacturer of the integrated circuit implementing the physical layer DSL. The octets in this 8-octet field are defined as follows:

- The first two octets contain the country code of the manufacturer. The codes are defined in Recommendation T.35.
- The next four octets define the "provider code," which is the vendor identification. Regional standards bodies, such as Committee T1 and ETSI, provide the vendor identification numbers.
- The final two octets are reserved for vendor specific information.

The parameter field contains octets with bit-encoded parameters. The parameters defined in the identification field are independent of the operation mode to be selected. Instead, these parameters are typically service or application related.

Inside the parameter field are Par(1) and Par(2) parameters. In the identification parameters field, there are no Par(3) parameters defined. In this chapter, we

Table 5.3 Message Type Bit Encoding

Message	Encoded Bits							
Type	8	7	6	5	4	3	2	1
MS	0	0	0	0	0	0	0	0
MR	0	0	0	0	0	0	0	1
CL	0	0	0	0	0	0	1	0
CLR	0	0	0	0	0	0	1	1
MP	0	0	0	0	0	1	0	0
ACK(1)	0	0	0	1	0	0	0	0
ACK(2)	0	0	0	1	0	0	0	1
NAK-EF	0	0	1	0	0	0	0	0
NAK-NR	0	0	1	0	0	0	0	1
NAK-NS	0	0	1	0	0	0	1	0
NAK-CD	0	0	1	0	0	0	1	1
REQ-MS	0	0	1	1	0	1	0	0
REQ-MR	0	0	1	1	0	1	1	1
REQ-CLR	0	0	1	1	0	1	1	1

describe the general structure of the parameters used in G.994.1. We do not define the complete set of parameter code points; all of the code points are defined in ITU-T Recommendation G.994.1 [1].

The Par(1) parameters in the standard information field contain one NPar(1) octet, three SPar(1) octets, and 20 NPar(2) octets defining the subparameter values in support of the parameters identified in the three SPar(1) octets. The only parameter defined in the NPar(1) octet is to identify use of the nonstandard information field. The SPar(1) octets identify parameters such as net upstream and downstream data rates, upstream and downstream data flow characteristics, and central office and customer premises splitter information. The 20 NPar(2) octets in support of the SPar(1) parameters are distributed as follows:

- Net upstream data rate—3 octets
- Net downstream data rate—3 octets
- Upstream data flow characteristics—2 octets
- Downstream data flow characteristics—2 octets
- Customer premises splitter information—1 octet
- Central office splitter information—1 octet
- Relative power level/carrier for upstream carrier set A43—1 octet
- Relative power level/carrier for downstream carrier set A43—1 octet
- Relative power level/carrier for upstream carrier set B43—1 octet
- Relative power level/carrier for downstream carrier set B43—1 octet
- Relative power level/carrier for upstream carrier set C43—1 octet
- Relative power level/carrier for downstream carrier set C43—1 octet
- Relative power level/carrier for upstream carrier set B4—1 octet
- Relative power level/carrier for downstream carrier set B4—1 octet

5.8 STANDARD INFORMATION FIELD

The parameters defined in the standard information field represent modes of operation or capabilities that relate to the DSL modems. The standard information field is not used in all of the messages. The messages that use the standard information field include CL, CLR, MP, and MS; those that do not use the standard information field are MR, ACK, NAK, and REQ.

The Par(1) parameters in the standard information field contain one NPar(1) octet and two SPar(1) octets. The NPar(1) octet identify availability or use of V.8 or V.8bis in the voice-band, and it identifies enabling of the Silent Period in the G.994.1 clear-down procedure. Together, the two SPar(1) octets identify availability or use of G.992.1 Annexes A, B, C, and/or H; G.992.2 Annexes A, B, and/or C; and G.991.2 Annexes A and/or B. Also included are code points for the multicarrier and single-carrier VDSL systems defined in the ETSI VDSL standard [5], [6] and the Committee T1 VDSL Trial Use Standard [7]. Each of the SPar(1) parameters have an array of Par(2) and Par(3) subparameters; in total there are 191 octets defined, that is, total Par(2) and Par(3) octets, in support of the SPar(1) standard information parameters. The breakdown on the distribution of octets is as follows:

- G.992.1 Annex A
  - NPar(2)—1 octet
  - SPar(2)—1 octet
  - NPar(3)—10 octets
- G.992.1 Annex B
  - NPar(2)—1 octet
  - SPar(2)—1 octet
  - NPar(3)—10 octets
- G.992.1 Annex C
  - NPar(2)—1 octet
  - SPar(2)—1 octet
  - NPar(3)—10 octets
- G.992.2 Annexes A and B
  - NPar(2)—1 octet
  - SPar(2)—1 octet
  - NPar(3)—8 octets
- G.992.2 Annex C
  - NPar(2)—1 octet
  - SPar(2)—1 octet
  - NPar(3)—8 octets
- G.992.1 Annex H
  - NPar(2)—1 octet
  - SPar(2)—1 octet
  - NPar(3)—8 octets
- G.991.2 Annex A
  - NPar(2)—1 octet
  - SPar(2)—2 octet
  - NPar(3)—63 octets
- G.991.2 Annex C
  - NPar(2)—1 octet
  - SPar(2)—2 octet
  - NPar(3)—64 octets
- ETSI MCM VDSL
  - NPar(2)—1 octet
  - SPar(2)—1 octet
  - NPar(3)—3 octets

5.9 NONSTANDARD INFORMATION FIELD

Messages that transport parameters, namely, CL, CLR, MP, and MS, may optionally contain nonstandard (NS) information. When the nonstandard information field is included, it may contain one or more blocks of data. The first octet in the NS field identifies the number of blocks that are contained in the field. Each nonstandard information block contains the following:

- A length indicator (one octet) of the block
- Country code per Recommendation T.35 (two octets)
- Provider code per Recommendation T.35 (four octets)
- Nonstandard information (M octets)

In summary, each block contains $M_i + 7$ octets; note that the value of M may be different for each block. Because there are a total of N blocks, the total number of octets in the nonstandard information field may be expressed as

$$\text{Total Number of NS Octets} = 1 + 7N + \sum_{i=1}^{N} M_i$$

where N is the number of blocks in the NS field, M_i is the number of vendor-specific information octets for information block i.

5.10 MESSAGE COMPOSITION

In the list of octets (p. 143), only the group of parameters that are enabled or identified in SPar(1) field are sent, depending on the command or message type being sent. Table 5.4 below shows required fields for inclusion in each of the defined message types. Note the messages requiring transmission of the parameter octets are CL, CLR, MS, and MP. The remaining messages do not pass any information parameter octets.

5.11.1 Transaction Types

In the preceding discussion, we looked at the data structure of the transmitted message types. For proper use of the above message types and structures, G.994.1 defines permitted transaction types for use in the preactivation (handshake) session. In this section, we provide a general description of the permitted transaction types defined in G.994.1; complete details are provided in Recommendation G.994.1 [1].

There are two types of transactions defined in G.994.1, namely, *basic* and *extended* transactions.

Table 5. 4 Required Field per Each Message Type

Message	Identification: Message Type and Revision ID	Identification: Vendor ID	Identification: ID Field Parameters	Standard Information Parameters	Nonstandard Information
MR	X	-	-	-	-
CLR	X	X	X	X	As Necessary
CL	X	X	X	X	As Necessary
MS	X	-	X	X	As Necessary
MP	X	-	X	X	As Necessary
ACK	X	-	-	-	-
NAK	X	-	-	-	-
REQ	X	-	-	-	-

5.11.1 Basic Transactions

Basic transactions are used to:

- Exchange and negotiate capabilities between the central office (CO) and customer premises (CP) units.
- Select a mode of operation.

The CP unit always initiates a basic transaction. Table 5.5 summarizes the permitted *basic transactions* as defined in G.994.1, and each transaction ends with an ACK(1). Transactions A, B, and D are all stand-alone transactions; at the successful end of any of these transaction types, the modems generally exit the handshake session. Otherwise the modems reenter the initial transaction state. Successful execution of Transaction C requires follow-up with either transaction type A, B, or D.

In transaction A, the CP unit selects a mode of operation and requests that the CO unit transition to this mode of operation. If the CO unit accepts this request, it responds with an ACK(1) message, and then the modems exit handshake and transition to the selected mode of operation.

In transaction B, the CP unit requests that the CO unit select the mode of operation. The CO unit then responds with the desired mode of operation; if the CP unit agrees, then it responds with an ACK(1). The modems then exit handshake and transition to the selected mode of operation.

Table 5.5 G.994.1 Basic Transaction Types

Transaction Identifier	CP Unit	CO Unit	CP Unit
A	MS _	ACK(1)	
B	MR _	MS _	ACK(1)
C	CLR _	CL _	ACK(1)
D	MP _	MS _	ACK(1)

In transaction C, the CP unit sends a list of capabilities to the CO unit who then responds with a capabilities list of the CO unit, which also serves as an acknowledgment. The CP unit responds with an ACK(1) upon successful reception of the CL message. In this transaction, both stations have exchanged and negotiated capabilities. For selection to be complete, follow-up with a transaction of type A, B, or D is needed.

In transaction D, the CP unit proposes a mode of operation to the CO unit and requests that the CO unit select the mode of operation. The CO unit responds with a MS message to select the mode of operation. The CP unit then responds with an ACK(1) message, and then the modems exit handshake and transition to the selected mode of operation.

In any of the above transaction types, if either the CP or CO units cannot determine an agreed common mode of operation (standard or nonstandard), then the corresponding unit sends a MS message to the other end with the nonstandard field bit and all information parameter and standard information parameter bits set to zero. When the other end receives the MS message, it responds with an ACK(1), and then the modems execute the clear-down procedure.

5.11.2 Extended Transactions

Extended transactions are essentially a concatenation of two basic transactions arranged in such a way that the CO unit is positioned to control the negotiation procedures. Recall that the basic transactions are initiated by the CP unit and so are the extended transactions. However, after the initial message of MS, MR, or MP is sent by the CP unit, the CO unit responds with a REQ-MR, REQ-MS, or REQ-CLR message, which positions the CO unit to control negotiation of operating mode. At the end of an extended transaction, the modems either exit handshake or go to the initial transaction state.

There are five extended transaction types. Table 5.6 below provides a list of the extended transactions, and a brief description of each is provided.

In transaction A:B, the CP unit selects a mode of operation, but instead of the CO unit responding with an ACK(1) as in basic transaction A, it responds with a request for mode request (REQ-MR), which requests that the CP unit directly transition to transaction B where the CO unit initiates a mode select request.

In transaction B:A, the CP unit requests that the CO unit select the mode of operation by issuing a mode request (MR) message, but instead of responding with a mode select (MS) as in basic transaction B, the CO unit sends a REQ-MS mes-

Table 5.6 G.994.1 Extended Transaction Types

Transaction Identifier	CP Unit	CO Unit	CP Unit	CO Unit	CP Unit
A:B	MS _	REQ-MR _	MR _	MS _	ACK(1)
B:A	MR _	REQ-MS _	MS _	ACK(1)	
A:C	MS _	REQ-CLR _	CLR _	CL _	ACK(1)
B:C	MR _	REQ-CLR _	CLR _	CL _	ACK(1)
D:C	MP _	REQ-CLR _	CLR _	CL _	ACK(1)

sage requesting the CP unit to make the mode selection. The CP unit then responds with the MS message as in basic transaction B.

In transaction A:C, the CP unit requests a specific mode of operation by issuing a MS message, but instead of responding with ACK(1) as in basic transaction A, the CO unit requests that the capabilities list be exchanged. Once the two units exchange capabilities and negotiate mode of operation, a follow-up transaction is needed for mode selection.

In transaction B:C, the CP unit requests that the CO unit select the mode of operation via a MR message, but instead of the expect MS response of basic transaction B, the CO unit requests a capabilities list request, where both units exchange their capabilities list and negotiate the mode selection. A follow-up transaction is needed for the mode selection.

In transaction D:C, the CP unit proposes a mode of operation via the MP message and requests that the CO unit select the operating mode. Instead of the expected MS message from the CO unit of basic transaction D, the CO unit requests a capabilities list request, where both units exchange their capabilities list and negotiate the mode selection. A follow-up transaction is needed for the mode selection.

5.11.3 Message Segmentation

Note that in the above transactions, some messages can become rather large when passing messages that contain the identification and standard information parameter fields. Excluding the two FCS octets and any octets inserted for transparency, the maximum message length in a frame is 64 octets. If a message is longer than 64 octets, then it must be segmented into two or more messages. The message types that can be segmented are those that contain the parameter octets, namely, CL, CLR, MP, and MS.

When a receiving station is parsing a segmented message, the receiving station sends an ACK(2), which indicates to the sending station that it is ready to receive the remainder of the message. Once the complete message is received, the receiving station responds with an ACK(1) or other appropriate response.

5.11.4 Example Transactions

In this section, we provide some example G.994.1 sessions to demonstrate the handshake process.

Example 1: Table 5.7 shows the use of a transaction sequence that combines basic transaction C with A. The CP unit first initiates a capabilities list request

Table 5.7 Example 1—Basic Transaction C Followed by Basic Transaction A

CP Unit	CO Unit	CP Unit	CP Unit	CO Unit
CLR _	CL _	ACK(1)	MS _	ACK(1)

Table 5.8 Example 2 – Extended Transaction A:C Followed by Basic Transaction A

CP Unit	CO Unit	CP Unit	CO Unit	CP Unit	CP Unit	CO Unit
MS _	REQ-CLR _	CLR _	CL _	ACK(1)	MS _	ACK(1)

where the two units exchange and negotiate capabilities via the CLR and CL messages. The CP unit then selects the mode of operation via the MS command (basic transaction A).

Example 2: Table 5.8 shows an example transaction that combines extended transaction A:C with basic transaction A. First the CP unit selects a mode of operation and requests that the CO select this mode. Instead the CO unit requests the CP unit for a capabilities list request. The two modems then exchange capabilities lists and negotiate operating modes. Once the exchange and negotiation are complete, the CP unit selects the mode of operation via the MS message.

5.12 G.hs START-UP/CLEAR-DOWN PROCEDURES

Finally, this section describes the start-up and clear-down procedures of a handshake session. In G.994.1, there are two types of start-up procedures: *duplex* and *half-duplex* start-up modes. Typically, the duplex start-up mode is used by ADSL transceivers, and the half-duplex start-up is used by SHDSL transceivers. Either the CO unit or the CP unit may initiate either type of start-up. Manufacturers of DSL modems are generally encouraged to support both modes of operation to provide interoperable handshake communication with other types of DSL equipment. The following subsections describe the duplex and half-duplex start-up procedures and the clear-down procedure as defined in G.994.1.

5.12.1 Duplex Start-up Procedures

As mentioned, the customer premises unit or central office may initiate duplex start-up. The sequences of each initiation are described below.

Customer Premises Initiated Duplex Start-up

Figure 5.8 shows the CP initiated duplex start-up sequence. The sequence begins with the CP unit (labeled HSTU-R) transmitting the R-TONES-REQ signal, which is the transmission of phase reversals every 16 ms in one or both of its signaling families. When the central office unit detects R-TONES-REQ, it responds with signal transmission (C-TONES) on one or both of its signaling families. When the CP unit detects C-TONES for a minimum of 50 ms, it transmits silence (R-SILENT) on the line for a minimum of 50 ms and a maximum of 500 ms, and then follows with a signal transmission (R-TONE1) on one or both of its signaling families. When the CO unit detects R-TONE1, it responds to the transmission of octet value 81_{16}, which is the one's complement of the HDLC Flag octet, on the modulated carriers (C-GALF1). The CP unit then responds with the transmission of HDLC flags on modulated carriers (R-FLAG1). Once the CO unit detects

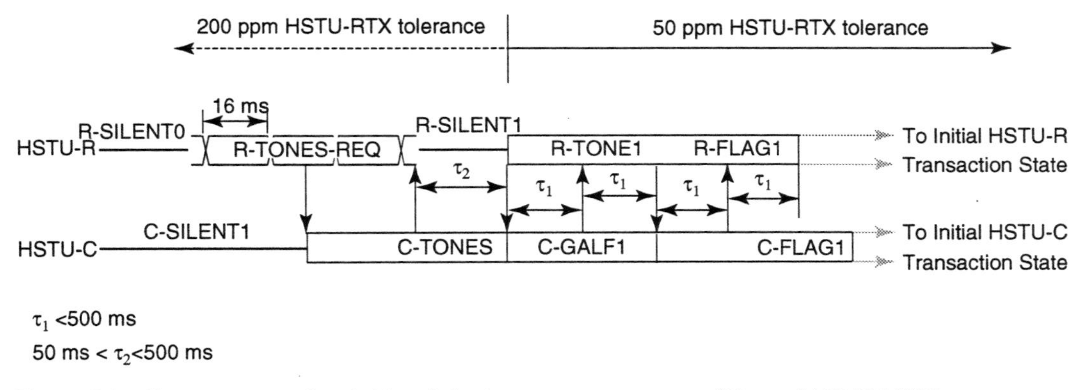

Figure 5.8 Customer premises initiated duplex start-up sequence (Figure 14/G.994.1[1]).

R-FLAG1, it responds by sending C-FLAG1 on its carrier(s) and then the units proceed to the initial transaction state of the handshake session. The timer values are shown in the figure.

Central Office Initiated Duplex Start-up

Figure 5.9 shows the CO initiated duplex start-up sequence. Initially both units are in start-up. In the CO initiated mode, the CO unit starts by sending C-TONES on its carriers. When the CP unit detects C-TONES, it transmits R-TONE1 on its carriers. The remaining sequence is the same as those CP initiated duplex start-up in Figure 5.8 from the R-TONE1 phase onward. The timer values are shown in the figure.

5.12.2 Half-Duplex Start-up

As with the duplex mode, the customer premises unit or central office may initiate half-duplex start-up. The sequences of each initiation are described below.

Customer Premises Initiated

Figure 5.10 shows the CP initiated half-duplex start-up sequence. This sequence is the same as the CP-initiated duplex sequence up to the transmission of C-TONES by the CO unit and the transmission of R-FLAG1 by the CP unit. Transition to the initial transaction state begins after the completion of C-TONES in the CO unit and R-FLAG1 in the CP unit. The timer values are shown in the figure.

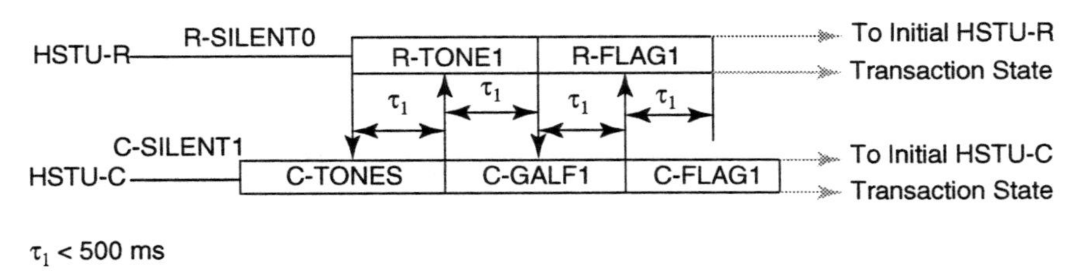

Figure 5.9 CO unit initiated duplex start-up (Figure 15/G.994.1[1]).

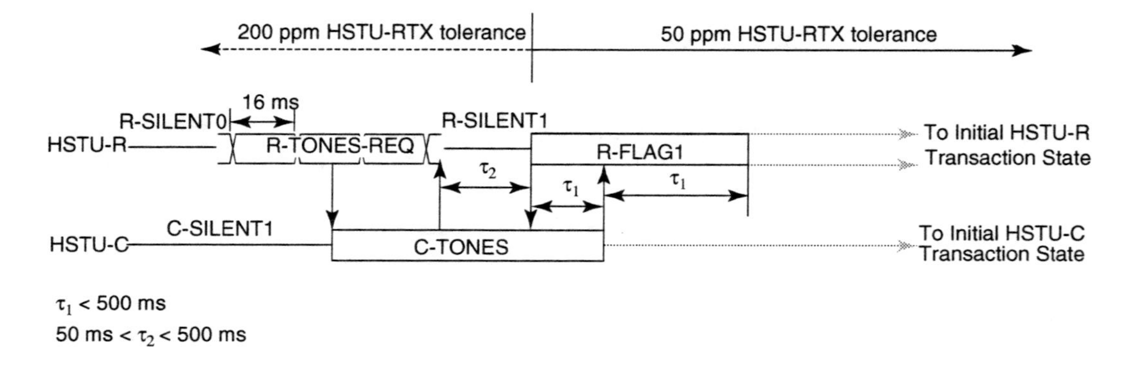

Figure 5.10 CP initiated half-duplex start-up sequence (Figure 16/G.994.1[1]).

Central Office Initiated

CO-initiated half-duplex start-up is derived from the CP initiated half-duplex start-up by eliminating the R-TONES-REQ and R-SILENT1 phases. The CP unit enters the initial transaction state upon completion of the R-FLAG1 transmission; the CO unit enters the initial transaction state upon completion of C-TONES. The timer values are provided in Figure 5.11.

5.12.3 Clear-down Procedure

Figure 5.12 shows the clear-down procedure for duplex mode. Either the CO or CP unit may initiate the clear-down procedure. The clear-down procedure is initiated by the unit that receives an ACK(1) after an MS message or a NAK-CD message: it transmits FLAGS for a period of less than 500 ms and then transmits four octets of GALF (referred to as R- or C-GALF2 in Figure 5.12) followed by silence. The unit that initiated the ACK(1) or NAK-CD message transmits C/R-FLAG2; the same unit continues to transmit the C/R-FLAG2 after it detects R/C-FLAG2 from the other unit for a period not to exceed 500 ms.

Figure 5.13 shows the half-duplex clear-down sequence, which may be initiated by either the CO or CO unit. Half-duplex clear-down begins with the unit that transmits ACK(1) from an MS message or a NAK-CD entering the silent state.

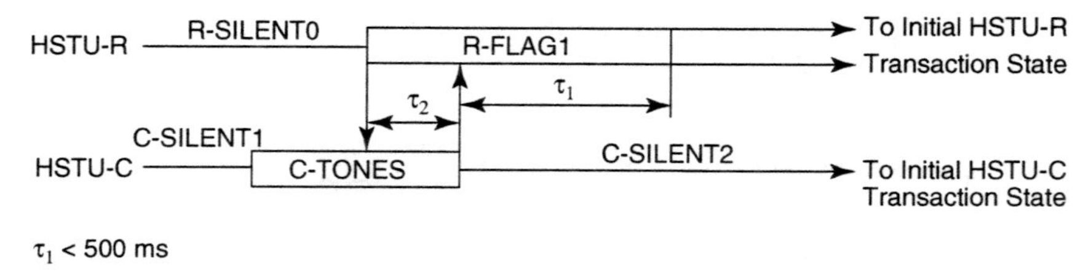

Figure 5.11 CO initiated half-duplex start-up sequence (Figure 17/G.994.1[1]).

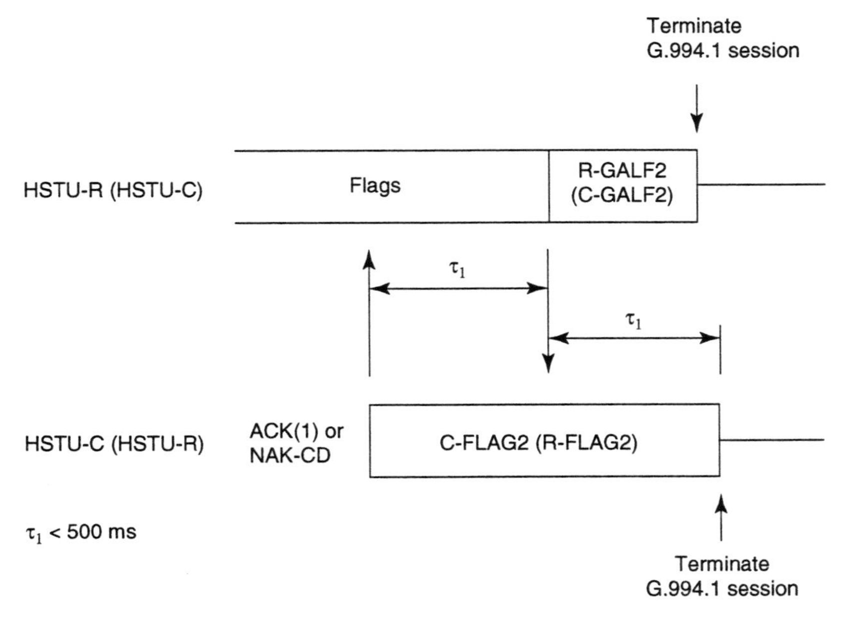

Figure 5.12 Duplex clear-down sequence (Figure 18/G.994.1[1]).

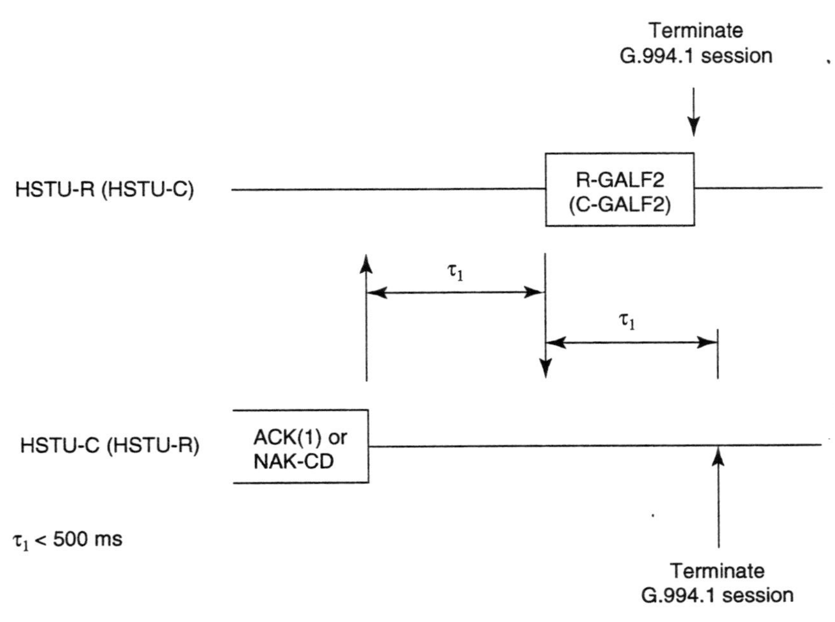

Figure 5.13 Half-duplex clear-down sequence (Figure 19/G.994.1[1]).

The other end detects the silent state for a maximum of 500 ms and then transmits four octets of R/C-GALF2, followed by silence.

REFERENCES

[1] ITU-T Recommendation G.994.1, "Handshake Procedures for Digital Subscriber Line (DSL) Transceivers," February 2001.

[2] ITU-T Recommendation G.992.1, "Asymmetrical Digital Subscriber Line (ADSL) Transceivers," July 1999.

[3] G.992.2, "ITU-T Recommendation G.992.2—Splitterless Asymmetric Digital Subscriber Line (ADSL) Transceivers," March 1999.

[4] G.991.2, "ITU-T Recommendation G.991.2: Single-Pair High-Speed Digital Subscriber Line (SHDSL) Transceivers," February 2001.

[5] ETSI TS 101 270-1 V1.1.2 (1998-06), "Transmission and Multiplexing (TM); Access Transmission Systems on Metallic Access Cables; Very High Speed Digital Subscriber Line (VDSL); Part 1: Functional Requirements," June 1998.

[6] ETSI TS 101 270-2 V1.1.1 (2001-02), "Transmission and Multiplexing (TM); Access Transmission Systems on Metallic Access Cables; Very High Speed Digital Subscriber Line (VDSL); Part 2: Transceiver Specification," February 2001.

[7] T1.424/Trial-Use, "Interface between Networks and Customer Installations—Very-High-Rate Digital Subscriber Line (VDSL) Metallic Interface," Committee T1 Standard, March 12, 2002.

CHAPTER 6

SINGLE-PAIR HIGH-SPEED DIGITAL SUBSCRIBER LINE (SHDSL)

HDSL applications have traditionally targeted provisioning of T1 (1.544 Mb/s) service in North America and E1 (2.048 Mb/s) service in Europe. As we have seen in Chapter 4, HDSL in North America was provisioned on two pairs with first generation HDSL that used 2B1Q technology. In Europe, first-generation HDSL used 2B1Q technology for provisioning 2.048 Mb/s service on 1, 2, and 3 wire pairs and CAP technology for provisioning 2.048 Mb/s service on 1 or 2 pairs. The higher number of pairs allows deployment to a greater distance because the bit rate on each line is lower, whereas deployment on a single pair allows a shorter maximum distance because the bit rate on the wire pair is larger. HDSL service was typically deployed to business customers addressing high-capacity access applications. The advantage to the service provider in deploying HDSL was lower provisioning costs and faster provisioning times compared with the original repeatered AMI or HDB3 access technologies.

In 1996, a proposal was made in ETSI TM6 to begin work on a multirate symmetric DSL transmission system to address business access applications delivering fractional E1 rate services, that is, $n \times 64$ kb/s service up to the maximum of 2.048 Mb/s. The deployable distance would be a function of the bit rate: the lower the bit rate, the greater the distance; the higher the bit rate, the shorter the distance. Although there was general interest for this new work, the project did not begin if full force until mid-1998. The project was entitled SDSL for symmetric digital subscriber line in ETSI TM6.

During this same time, T1E1 was developing second-generation HDSL also known as HDSL2 (see Chapter 4). HDSL2 targeted the transmission of T1 service on a single wire pair. Although there was not an equivalent HDSL2 project in ETSI TM6, the ETSI SDSL project covered the transmission of the E1 (2.048 Mb/s) payload on a single wire pair in its definition. During the discussion of the scope of ETSI SDSL, there was debate on the possible operation over baseband POTS. The result of the debates was that ADSL would target residential access applications with operation over baseband POTS and that SDSL would target business access applications without the use of baseband POTS.

Although never standardized, there were significant deployments of symmetric DSLs using 2B1Q and carrierless AM/PM (CAP) technologies. Primary deployments were by competitive access providers on local loops that were leased from the incumbent operators. As described in Chapter 4, the architecture of the CAP-based solution is closest to that of SHDSL. The 2B1Q SDSL solutions were extensions of the core first-generation HDSL technology, where the transceivers were configured to operate at select bit rates that were integer multiples of 8 kb/s. These solutions targeted business customers with high-capacity access applications.

In January 1999, a companion project to ETSI'S SDSL and to Committee T1's HDSL2 was begun in the ITU-T Study Group 15 Question 4 (i.e., ITU-T Q4/15) called single-pair high-speed digital subscriber line (SHDSL) and the corresponding Recommendation was referred to as G.shdsl. As with ETSI SDSL, SHDSL defines operation at bit rates ranging from 192 kb/s to 2312 kb/s in increments of 8 kb/s. The system defines transmission of only symmetric payloads, that is, the upstream and downstream bit rates are the same. Business applications are the primary target for SHDSL.

The core technology for SHDSL and ETSI SDSL is that of HDSL2, namely, 16-level trellis coded pulse amplitude modulation (16 TC-PAM). This chapter provides a description of SHDSL transceivers, some of their driving applications, and the manner in which SHDSL addresses these applications.

6.1 APPLICATIONS OF SHDSL

The definition of SHDSL was driven to support high capacity network access applications to business and small-medium enterprise customers. Such applications include T1 or E1 extension, fractional T1 or E1 access, wireless base station access to the central office, provisioning of multiple voice channels via a high-access channel (also referred to as a "pair-gain" access), work at home applications supporting data and digitized voice, campus applications, and others. The support of SHDSL for each of these applications is described below.

6.1.1 T1 or E1 Extension Application

A fundamental application of SHDSL is the equivalent to that of HDSL2. In North America, HDSL2 transports a DS1 (1.544 Mb/s) payload on a single twisted wire pair on loops that fall within the carrier serving area (CSA) limits of 9 kft of 26-gauge wire or 12 kft of 24-gauge wire. The block diagram in Figure 6.1 shows the use of SHDSL in the provisioning of a T1 access circuit, as is done using HDSL2. In the central office (CO), the T1 circuit is connected to the SHDSL unit via a DSX-1 interface. The T1 AMI interface circuit within the SHDSL central office unit converts the AMI signal into a 1.544 Mb/s bit-stream. At the customer premises, the

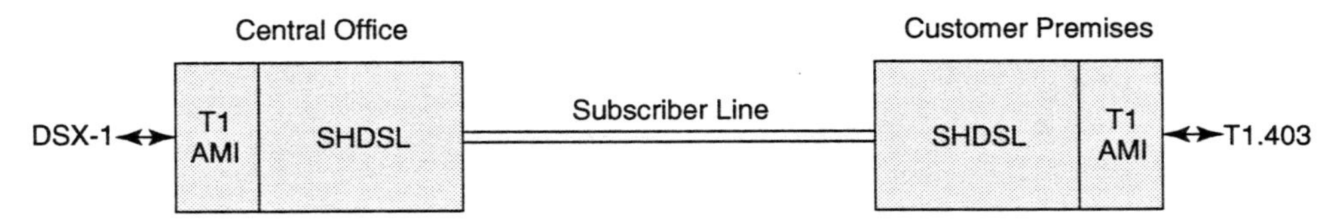

Figure 6.1 T1 extension application.

SHDSL unit provides a T1 AMI interface to the end customer per the T1.403 DS1 standard. For this application, the SHDSL transceiver is configured to the parameters defined by the HDSL2 standard, which is included in Annex A (North American annex) of the SHDSL Recommendation G.991.2.

SHDSL may also be used to provision an E1 (2.048 Mb/s) access circuit. In this case, the T1 AMI interface circuits shown in Figure 6.1 would be replaced by a G.703/G.704 interface circuit, and the SHDSL transceiver would need to be configured to support the payload bit rate of 2.048 kb/s (or 2.304 kb/s if the extra four 64 kb/s time slots are to be supported) [1, 2, 7] and the corresponding PSD in Annex B of G.991.2.

Fractional T1 access may be supported in the following manner. The DS1 circuit would contain predetermined time slots with active data, which the service operator provisions prior to service activation. The SHDSL transceiver takes only those active time slots from the DS1 circuit and transports them on the subscriber line at a bit rate that is less than the T1 rate of 1.544 Mb/s. The lower line bit rate allows for operation on longer loops than would be feasible with SHDSL transporting the full 1.544 Mb/s rate.

6.1.2 Connection of Wireless Base Station to Central Office

The application in Figure 6.2 shows SHDSL providing a transport of the voice channels in a wireless base station and feeding them to the nearby central office for switching or routing. Proper interface circuits would need to be defined for connecting the multiple voice channels into an SHDSL frame.

6.1.3 Campus Applications

SHDSL transceivers may also be applied to limited distance modem applications in a campus environment. In this case, the public network is not utilized. An example of campus application is shown in Figure 6.3. SHDSL modems may be configured with proper interfaces to interconnect local area networks and PBXs located within different buildings on a campus. The distance between buildings would determine the maximum bit rate configurable for the SHDSL transceivers.

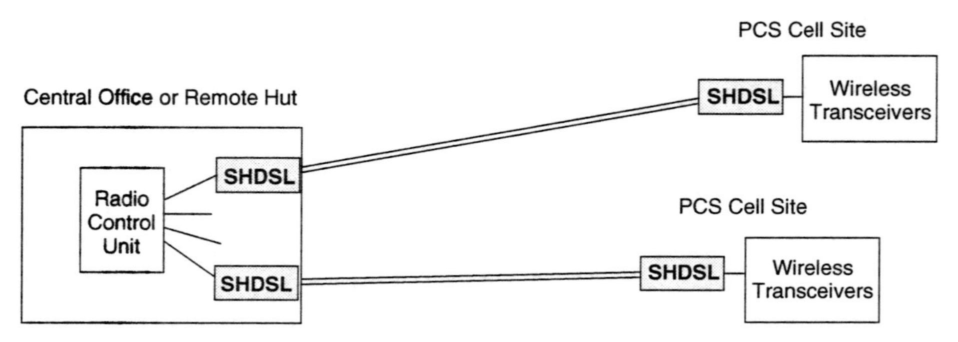

Figure 6.2 Transport application from wireless base station to central office.

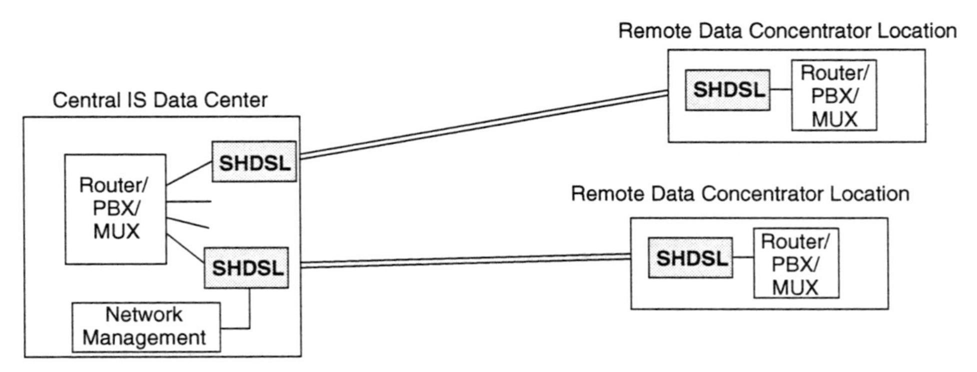

Figure 6.3 Campus private network application.

6.1.4 Provisioning of Multiple Voice Circuits

In situations where there may be a shortage of wire pairs, SHDSL transceivers may be used to transport multiple digitized voice channels on a single wire pair, as shown in Figure 6.4. At the remote location, the individual digitized voice channels would be converted back to analog voice signals and provisioned on individual wire pairs to the end subscriber. An example would be that one wire pair feeds a building that is a two-family residence. An SHDSL access line could be provisioned to transport four digitized voice channels where two voice channels could be made available to each unit in the two-family residence. Note that in such an application, to provide lifeline service the remote unit would need to be powered from the central office or have a battery backup capability at the remote location. Lifeline service provides operation of the analog voice service in the event of a local power failure.

6.1.5 Work-at-Home Application

Figure 6.5 shows a possible work-at-home scenario where an access circuit is used to deliver one or more pulse code modulation (PCM) voice circuits together with a

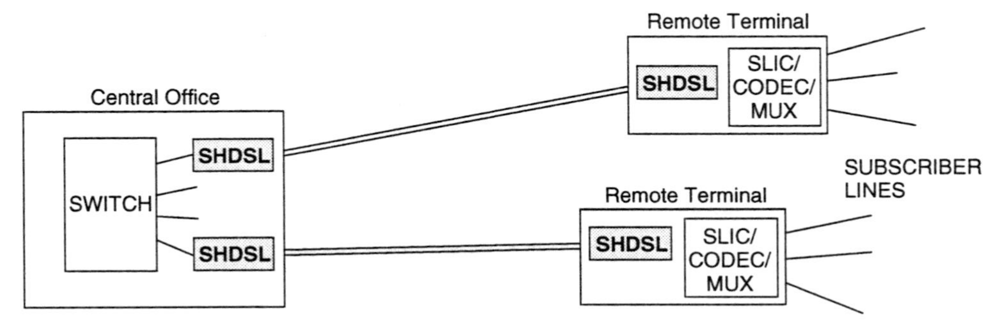

Figure 6.4 Provisioning of voice service using pair gain configuration.

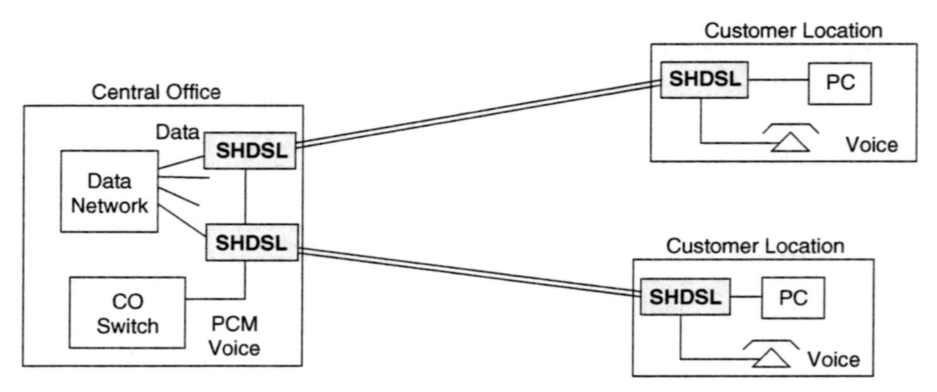

Figure 6.5 Work-at-home application—data plus PCM voice.

high-speed data channel. In this configuration, the SHDSL transceiver would be used to transport the multiplexed data and voice channels on a single high-speed access between the residence location and the CO. The distance between CO and customer location would determine the maximum bit rate that could be supported in the access line.

6.2 STANDARDS FOR MULTIRATE SHDSL

ITU-T Recommendation G.991.2 (also referred to as G.shdsl) defines the interface specification for SHDSL transceivers. G.shdsl defines a multirate single-pair symmetric DSL transport using 16-level trellis coded PAM. The main body of Recommendation G.991.2 defines the interworking specification that is independent of any regional requirements. Annex A of G.991.2 defines requirements and objectives that are specific to North America; similarly, Annex B of G.991.2 defines requirements and objectives that are specific to the European community.

Japan does not use TC-PAM for SHDSL. Because the Japan network has an abundance of time compression multiplexed ISDN (TCM-ISDN), the line code chosen for SHDSL applications in the Japan access network is DMT as defined in Annex H of G.992.1.

G.shdsl defines symmetric data rates from 192 kb/s up to 2,312 kb/s in increments of 8 kb/s for transmission on a single wire pair. The same Recommendation also defines an option for operation on two wire pairs. This option may be used to either increase the reach by operating on two wire pairs at half the payload rate on each pair, or to simply double the capacity on a given link.

ETSI TM6 has developed ETSI TS 101 524 [2] that defines SDSL. This regional standard is the primary source of content to Annex B of G.991.2.

Committee T1 has defined T1.422 [4], which is a regional standard that points to the ITU-T Recommendation G.991.2, with specific reference to Annex A, which contains regional specific requirements for North America.

6.3 SYSTEM FUNCTIONAL REFERENCE MODEL

Figure 6.6 shows the system reference model for SHDSL as defined in G.991.2. The reference model shows one unit located in the CO and the other unit located at the customer premises (CP) location; each unit terminates one end of the subscriber line. Each end unit is typically referred to as an SHDSL transceiver unit (STU). The unit at the central office is labeled STU-C and that at the customer premises is labeled STU-R, where R stands for "remote" location.

The reference model in Figure 6.6 shows the different layers of processing required in each STU. At the CO end, the STU provides connection between the subscriber line and the network interface(s); at the CP side, the STU provides connection between the subscriber line and the customer interface(s). The different layers of processing convert the signals from the network or customer interfaces into a form suitable for transmission on the digital subscriber line. The different layers of processing are identified as follows from highest to lowest layer:

- Network or customer interfaces
- TPS-TC: transmission protocol specific–transmission convergence layer
- PMS-TC: physical medium specific–transmission convergence layer
- PMD: physical medium dependent layer

The PMD layer is the lowest layer-processing block in the SHDSL transceiver unit; hence, it is the block that is least dependent on the supporting application. The PMD is the core modem of the STU in that it provides the modulation and demodulation operations at the bit level. The functions of the PMD layer include the following:

- Modulation and demodulation
- Bit clock and symbol clock generation and recovery
- Trellis coding and decoding
- Echo-cancellation
- Channel equalization
- Initialization and training

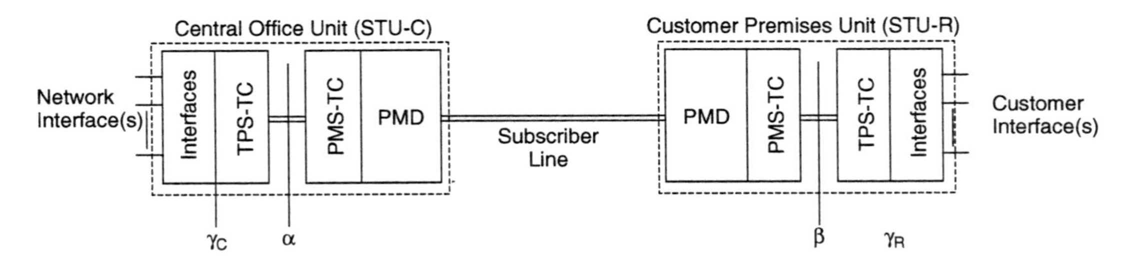

Figure 6.6 System reference model.

The PMD block processes the bit stream as though it is a random bit sequence in that it does not require knowledge of the meaning or relation of any of the bits that are transmitted. Any bit errors in the PMD are passed on to the higher layers.

One layer above the PMD is the PMS-TC layer, which contains the framing and frame synchronization functions. This layer needs to know the relation of the payload bits to each other for proper identification in the frame. The framing at this level primarily separates the payload bits from the overhead bits. The overhead channel in the PMS-TC provides the following functions: frame boundary identification, performance monitoring using a 6 bit cyclic redundancy check (CRC-6), function indicator bits, an embedded operations channel, and optional bit stuffing for support of synchronous timing or rate adaptation functions. So the PMD and PMS-TC layers together provide the capability of transmitting and recovering a payload bit stream and supporting the required operations and maintenance functions via processing of the overhead channel. These blocks together can support the widest range of applications and hence they are seen to be *application invariant* in the system reference model.

The TPS-TC is more application specific in that it provides any subchannel separation and identification needed in support of the SHDSL based service offering. To do so, the TPS-TC works together with interfaces block for transport of the different payload channels. For example, the SHDSL service may be configured to support one high-speed data channel and two digitized voice channels. The TPS-TC provides the framing of the three subchannels in the payload bit sequence that connects to the PMS-TC block. Correspondingly, the interfaces block provides the physical interfaces required in support of the data and digitized voice subchannels.

The connection between the TPS-TC and PMS-TC blocks in the STU-C is termed the α-interface. This is a logical interface that defines the subchannel frame structure of the payload bits to be transmitted in the PMS-TC. In the STU-R, the corresponding interface is termed the β-interface.

The connection between the interfaces and TPS-TC blocks is called the γ-interface. The γ-interface in the central office unit is referred to as the γ_C interface, whereas that in the customer premises unit is referred to as the γ_R interface. In general, the γ-interface is a logical interface, and its definition is totally dependent on the application being supported.

As mentioned earlier, the maximum distance that an SHDSL access circuit can be deployed depends on the bit rate of the channel. In some cases the desired range of deployment for the given line bit rate is greater than the specified range for that bit rate. To address these extended reach applications, the SHDSL recommendation defines two options: (1) the use of repeaters and (2) an alternative two-pair operation.

Figure 6.7 shows the reference model for deployment of SHDSL link with repeaters to extend the reach of operations. The repeaters are identified by the term

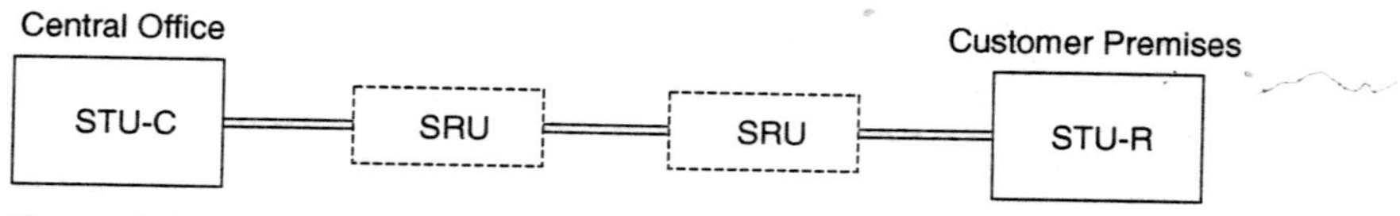

Figure 6.7 Reference model for deployment of SDHSL with repeaters.

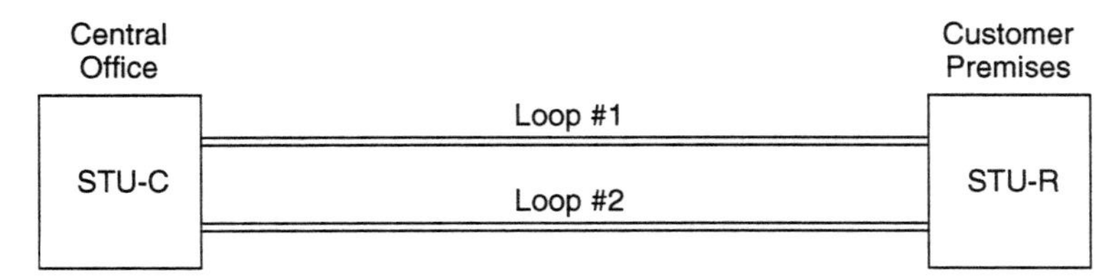

Figure 6.8 Reference model for alternative two-pair deployment.

SHDSL repeater unit (SRU). The issues generally associated with repeaters are the powering of the repeater units and the spectral compatibility with other service deployed in the same cable that are not served with repeaters but served directly from the CO.

Figure 6.8 shows the system reference model for deployment of SHDSL using two wire pairs. Because the reach of SHDSL is longer for smaller bit rates, this option may be used to increase the reach by provisioning the service on two wire pairs, where each wire pair transports half the payload rate. Alternatively, for a given reach, two-pair operation may be used to simply double the capacity on the given link.

6.4 HDSL4

An example of a system that uses the two wire pair option of SHDSL technology is HDSL4.

HDSL4 may seem to be an odd technology to develop following the success of HDSL2 because HDSL4 uses four wires for DS1 transmission while the primary goal of HDSL2 was to reduce cost by reducing the number of wires from four to two. Each technology has its place. HDSL2 is the technology of choice for DS1 (1.544 Mb/s) transmission for lines within the carrier serving area design rules (CSA, up to 9 kft of 26 AWG wire). HDSL2's use of only two wires reduces copper-line costs and also reduces the cost of the transceiver because a two-wire system requires one-half as many line interfaces. However, extending HDSL2 beyond CSA length lines is not attractive due to the high cost of midspan repeaters and spectral incompatibility with other systems.

HDSL4's role is to serve lines beyond HDSL2's reach. The use of four wires permits the transmitted signal PSD to avoid the higher frequencies used by HDSL2. This enables nonrepeatered HDSL4 to reach 11 kft of 26 AWG, whereas HDSL2 can reach only 9 kft. Also, the lower frequency characteristic of HDSL4 enables HDSL4 to remain spectrally compatible with other systems in the same cable even when HDSL4 is repeatered to virtually any distance. HDSL4 makes the trade-off of using an additional pair of wires to reduce the number of required repeaters, and provided that spare pairs are available, this is generally a very cost-effective trade-off. The combination of HDSL2 for shorter lines and HDSL4 for longer lines makes T1 and HDSL technology obsolete. For example, a 20 kft all 26

AWG loop would require two repeaters using first-generation HDSL technology, but HDSL4 would require only one repeater and the HDSL4 would have much better spectral compatibility.

6.5 SHDSL TRANSCEIVER OPERATIONS

6.5.1 SHDSL PMD Layer

The core technology for SHDSL is based on that of HDSL2, namely, 16-level trellis coded PAM (16TC-PAM). The fundamental difference is that HDSL2 was designed to optimize performance for the transport of 1.544 Mb/s in support of provisioning T1 service on a single wire pair. SHDSL uses 16TC-PAM technology for the provisioning of multirate symmetric services, where the loop distance and noise environment determines the maximum bit rate to be supported on a single wire pair. The payload bit rates defined by SHDSL range from 192 kb/s to 2.312 kb/s in increments of 8 kb/s. The overhead may be either 8 kb/s or 16 kb/s. Generally, the use of 16 kb/s as overhead is a carryover from HDSL. Because SHDSL accommodates all HDSL bit rates in addition to the other bit rates, the same 16 kb/s overhead option was preserved.

Figure 6.9 shows the general transceiver block diagram for SHDSL supporting 16-level trellis coded PAM. The upstream and downstream channels share same frequency components, so echo cancellation is the method used to separate the direction upstream and downstream channel transmission.

The scrambler randomizes the transmit data that is provided by the framer circuit. Randomized data are required for optimal performance of key transceiver algorithms such as the echo canceler and the equalizer. On the receive side, the descrambler provides the reverse operation of the scrambler and feeds the recovered data to the framer.

To provide greater immunity to crosstalk and other random background noise, the SHDSL transceiver incorporates trellis coding, which is the same trellis code as used in HDSL2. In the receiver, the trellis decoder is typically implemented, using the Viterbi algorithm to decode the received data, which provides

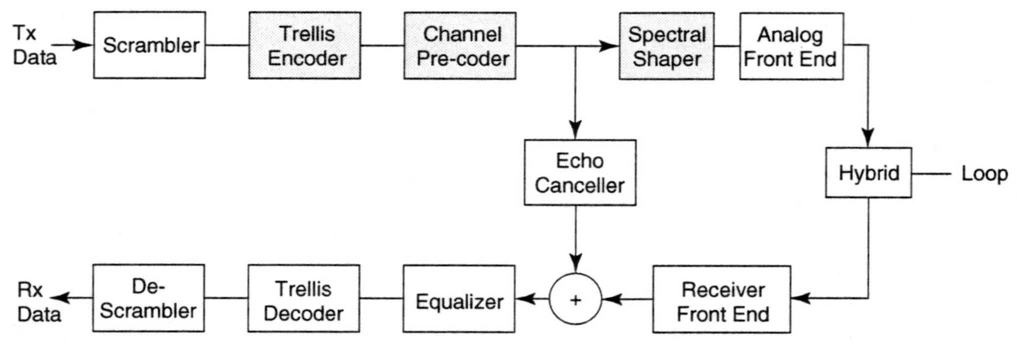

Figure 6.9 General transceiver functional block diagram for HDSL2.

near optimum maximum likelihood estimation of the transmitted symbols. To achieve the benefits of both the trellis coding and decision feedback equalization, channel precoding is applied. The channel precoder basically implements the equivalent of a decision feedback equalizer at the transmitter in which no incorrect symbols are ever transmitted in the feedback filter. This eliminates any error propagation from occurring, which would otherwise degrade performance, and eliminate the effects of the trellis coding gain. The details of the channel precoder are described in detail later in this chapter.

The spectral shaper is a high-order low-pass filter that shapes the signal spectrum in a way to provide good performance and assure spectral compatibility with other services in the cable. In SHDSL, there is a family of spectral shapes defined, one for each bit rate. Most of the spectral shapes are symmetric, in that both the upstream and downstream channels share the same spectrum and spectral shape. There are some selected bit rates that have asymmetric shapes defined. These rates are typically the line bit rates defined for classic HDSL, namely,784 kb/s, 1168 kb/s, 1544 kb/s, and 2048 kb/s. Asymmetric power spectral densities (PSDs) have different spectral shapes defined for the upstream and downstream channels. The asymmetric PSDs are specially engineered PSDs to meet a higher performance criteria in support of the service application while maintaining spectral compatibility with other signals in the cable.

The analog front-end block contains the digital-to-analog (D/A) conversion, analog filters, and line driver circuits. The hybrid circuit provides the two-to-four wire conversion for connecting the subscriber line to the transmitter output and the receiver input.

The receiver front-end block contains analog filters, an automatic gain control circuit that adjusts the total received signal to an optimum level within the circuit's dynamic range, and an analog-to-digital (A/D) converter for converting analog samples into digital samples for processing in the receiver.

Because upstream and downstream channels share the same bandwidth, the echo canceler is the device that removes the local echo from the total received signal and passes the desired far-end signal to the receiver for additional processing.

The equalizer compensates for any impairment introduced by the channel, such as amplitude and delay distortion. The equalizer is generally an adaptive equalizer, which learns and adapts to the characteristics of the channel during initialization and continuously adapts to any slowing changing line characteristics during data mode. Together with the channel precoder in the transmitter, the two collective algorithms produce the same performance improvements of a decision feedback equalizer (DFE). Because there is no error propagation, the transceiver achieves the performance improvements of both the trellis code and the DFE.

Prior to SHDSL, similar services were being provisioned using 2B1Q technology. The functional block diagram of 2B1Q is similar to that of SHDSL in Figure 6.9 except that 2B1Q does not use trellis coding and channel precoding. A bit-to-symbol map is used in place of these two blocks. The performance difference between the two systems is due primarily to the inclusion of a trellis code in SHDSL. The coding gain of the reference trellis code is 5.1 dB.

All of the blocks in Figure 6.9, relative to SHDSL, are described in more detail in the following sections.

6.5.2 Scrambler

The SHDSL Recommendation [1] allows for the use of different scrambler polynomials during the different phases of activation, including data mode. Table 6.1 shows a list of the scrambler polynomials of use in G.shdsl. The polynomials used during preactivation are selected during the handshake phase of initialization (see the discussion on initialization later in the chapter). The polynomials of index 0 are generally used in data mode and preactivation mode. The other polynomials are selected for use during the line probe phase of initialization, where the transceivers seek to learn the transmission capability limits of the given line.

Figure 6.10 shows a block diagram of the common scrambler structures used in data mode for the downstream and upstream transmitters (index 0), which is the same as those used in HDSL, HDSL2, and ISDN.

6.5.3 Trellis Encoder

In order to optimize performance for the different modes, trellis-coded modulation is used to maximize the immunity to crosstalk encountered in the loop plant. This is the same basic trellis code structure used in HDSL2 described in Chapter 4. To minimize the end-to-end latency, any types of block code or concatenated codes that require interleaving were ruled out because the latency introduced was too large. The use of interleaved block codes (such as Reed-Solomon codes), concatenated codes with interleaving, and turbo coding techniques proved to be difficult to use because they require latencies significantly greater than 500 μs to meet the performance objective. What remained was conventional trellis coding with channel precoding (such as the Tomlinson-Harashima precoding). Although multidimensional trellis codes were examined, it was determined that the simple one-dimensional Ungerboeck codes [10] were suitable, meeting up to 5 dB of coding gain (ideal asymptotic coding gain) within a latency of 500 μs.

For the one-dimensional Ungerboeck codes, thirty-two states were sufficient to achieve a code gain of 4.0 dB. To achieve 5 dB of coding gain would require implementation of a 512 state code; the challenge here is the design of a decoder such that the implementation loss is minimized and the latency requirement is still met. Two proposals were provided for 512-state trellis codes: one from Pairgain in [11] and the other from Adtran in [12]. Both codes were linear codes claiming coding

Table 6.1 Scrambler Polynomials

Polynomial Index	Downstream Polynomial $p(x)$	Upstream Polynomial $p(x)$
0	$1 + x^{-5} + x^{-23}$	$1 + x^{-18} + x^{-23}$
1	$1 + x^{-1}$	$1 + x^{-1}$
2	$1 + x^{-2} + x^{-5}$	$1 + x^{-3} + x^{-5}$
3	$1 + x^{-1} + x^{-6}$	$1 + x^{-5} + x^{-6}$
4	$1 + x^{-3} + x^{-7}$	$1 + x^{-4} + x^{-7}$
5	$1 + x^{-2} + x^{-3} + X^{-4} + x^{-8}$	$1 + x^{-4} + x^{-5} + X^{-6} + x^{-8}$
6	Reserved	Reserved
7	Not Allowed	Not Allowed

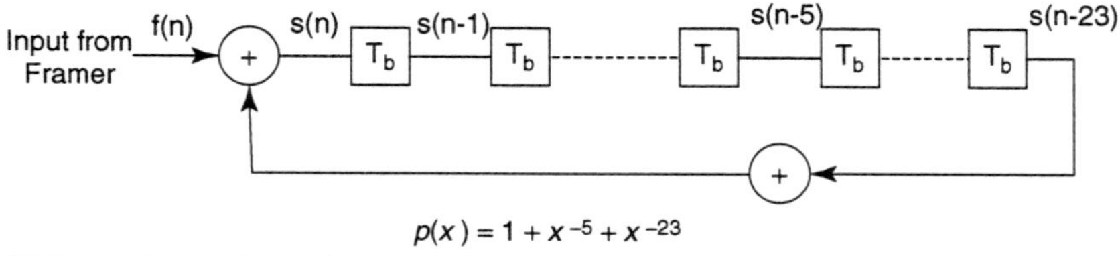

(a) Downstream channel scrambler

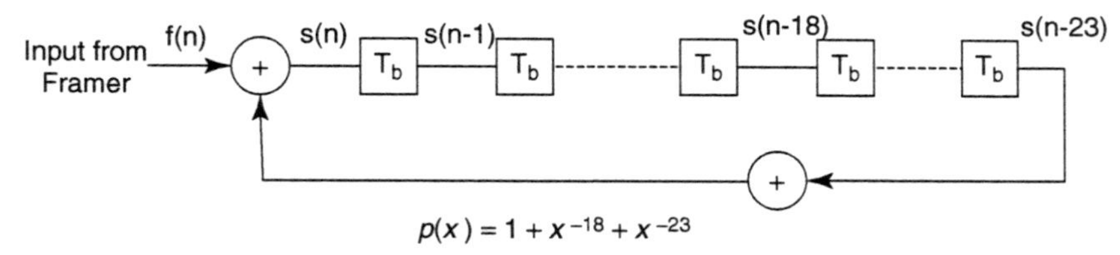

(b) Upstream channel scrambler

Figure 6.10 Example SHDSL upstream and downstream channel scrambler.

gains about 5 dB. The commonality is that the two proposed codes used rate one-half convolutional codes where the convolutional coding was performed on one information bit while the other information bits were passed uncoded. The general structure of the trellis is shown in Figure 6.11, where the value of k is 3 (i.e., for every 3 information bits, there are 4 coded output bits.

To address the numerous codes possible, the agreed trellis code structure includes a programmable non-systematic feed-forward convolutional encoder that codes the least significant bit of the 3-bit information symbol [13]. The structure of the programmable convolutional encoder is shown in Figure 6.12. The convolutional code is a nonsystematic rate one-half code, where for each input data bit there are two output bits. The generator polynomials of the two output paths of the convolutional encoder are

$$P_a(x) = a_0 + a_1x^{-1} + a_2x^{-2} + ... + a_{20}x^{-20}$$

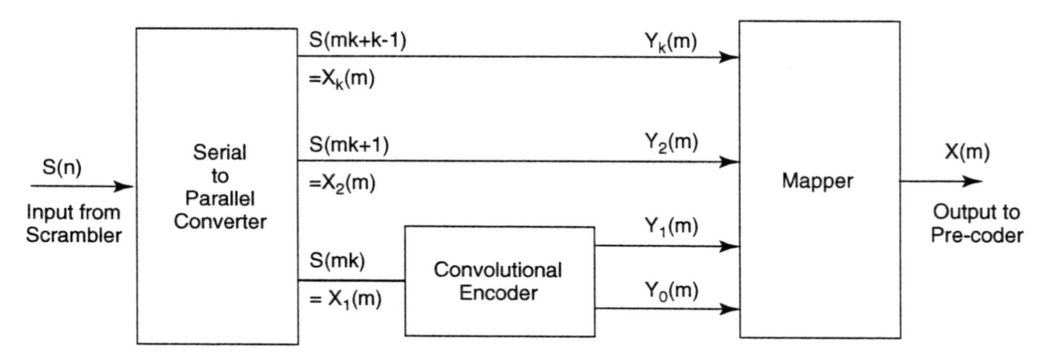

Figure 6.11 HDSL2 trellis encoder structure.

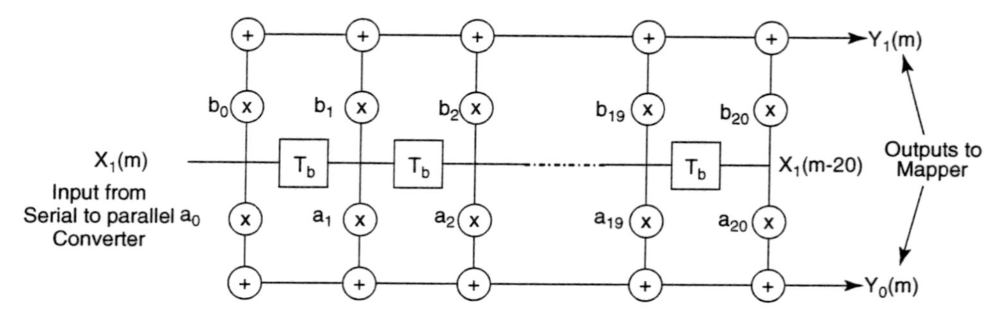

Figure 6.12 Convolutional encoder structure.

and

$$P_b(x) = b_0 + b_1x^{-1} + b_2x^{-2} + ... + b_{20}x^{-20}$$

for outputs Y_0 and Y_1, respectively. The coefficients $a_0, a_1, ..., a_{20}$ and $b_0, b_1, ..., b_{20}$ are binary coefficients and the operator '+' represents modulo-2 addition. A reference code having a coding gain of approximately 5 dB has coefficients $\{a_9, a_8, ..., a_0\} = \{0101101110\}$ and $\{b_9, b_8, ..., b_0\} = \{1100110001\}$; the same coefficients are represented in octal as A = 556 and B = 1461, where all of the digits are octal digits. Another code with approximately the same coding gain has coefficients A = 732 and B = 1063.

The entire trellis code in Figure 6.11 is a rate three-fourths code, where the encoder accepts 3 information bits and outputs 4 coded bits. Because the convolutional code that operates on the least significant bit of the input code word is nonsystematic, the trellis code is also nonsystematic. The remaining 2 information bits are passed to the symbol mapper uncoded. The coefficients of the convolutional encoder, defined by two 20th order polynomials, are provided by the manufacturer's equipment that contains the receiver. During initialization, the coefficients are passed from receiver to encoder. An advantage of this programmable approach is that manufacturers could provide codes that are suitable to the type of decoder that they have implemented (e.g., a Viterbi decoder or sequential decoder). Also note that with this configuration, it is possible that the upstream and downstream channel could have different trellis codes.

6.5.4 Bit-to-Symbol Mapping

The parallel bits at the output of the trellis encode must be mapped in symbols suitable for transmission on the line. The four bits at the trellis-encoder output are mapped into sixteen possible levels. The bit mapping of each of the levels are shown in Figure 6.13.

The input bits to the trellis encoder are $X_1(m)$, $X_2(m)$, and $X_3(m)$; the output bits are $Y_1(m)$, $Y_2(m)$, $Y_3(m)$, and $Y_4(m)$. $X_1(m)$ is the least significant bit of the trellis coder input bits, and $Y_1(m)$ is the least significant bits of the output symbol. The table in Figure 6.13 shows the mapping of the trellis encoder output bits to the output symbol. The mapping is also shown pictorially in the constellation diagram in the bottom of the figure.

Mapping Table

Trellis Encoder Output, $Y_3(m)\ Y_2(m)\ Y_1(m)\ Y_0(m)$	Level, $x(m)$
0000	−15/16
0000	−13/16
0010	−11/16
0011	−9/16
0100	−7/16
0101	−5/16
0111	−1/16
1100	1/16
1101	3/16
1110	7/16
1111	9/16
1000	11/16
1010	13/16
1011	15/16

Mapper Output Level

-15	-13	-11	-9	-7	-5	-3	-1	+1	+3	+5	+7	+9	+11	+13	+15
0000	0001	0010	0011	0100	0101	0110	0111	1100	1101	1110	1111	1000	1001	1010	1011

Trellis Encoder Output

Figure 6.13 SHDSL bit-to-symbol mapping.

6.5.5 Channel Precoder

The functional block diagram of the channel pre-coder is shown in Figure 6.14. The receiver computes the feedback filter coefficients (C_1, C_2, ..., C_N) during the training phase in the initialization process and then transfers the coefficients to transmitter during the parameter exchange phase. The input sequence $x(mT_s)$ is the output of the bit-to-symbol mapping in the trellis encoder. The output of the feedback filter, $v(mT_s)$, is computed by

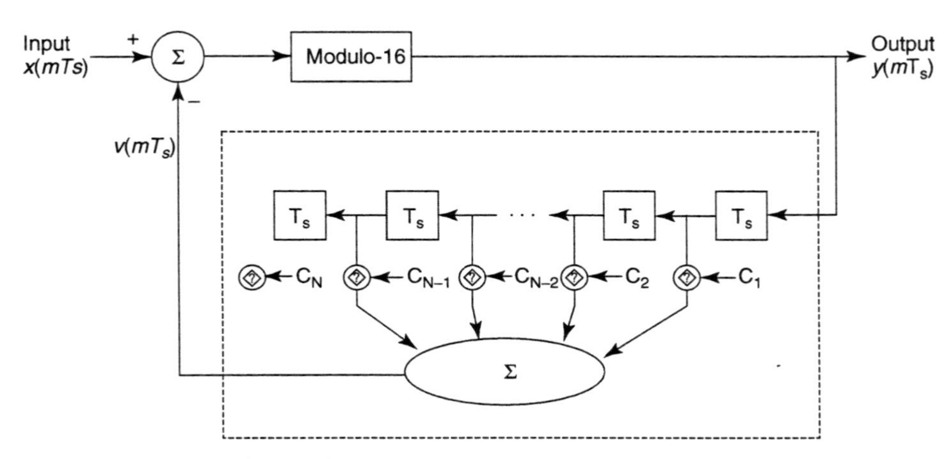

Figure 6.14 Channel precoder.

$$v(mT_s) = \sum_{k=1}^{N} C_k \cdot y(mT_s - kT_s)$$

where T_s is the symbol interval, $y(mT_s)$ is the precoded output sample, m is the sample time index, and N is number of coefficients in the feedback filter. A modulo-16 operator then operates on the difference between the input sample and feedback filter output sample $u(mT_s)$. Operation of the modulo-16 block is find an integer $d(mT_s)$ such that the sum $u(mT_s) + 2\, d(mT_s)$ falls between the values of -1 and $+1$. The resulting value of $y(mT_s)$ is then $u(mT_s) + 2\, d(mT_s)$.

In the HDSL2, the receiver determines the value of N and the value is passed on to the transmitter at initialization during the parameter exchange phase. The value of N has minimum value of 128 and a maximum of 180 samples; any value in between is valid.

6.5.6 Spectral Shaper

SHDSL uses a nominal sixth order roll-off filtering of NRZ (nonreturn to zero) pulses. The bandwidth of the main lobe is determined by the symbol rate (*fsym*) of the pulse sequence. The symbol rate is the ratio of the bit rate to the number of information bits per symbol interval, namely,

$$f_{sym} = \frac{R_b}{N_b}$$

where R_b is the bit rate of the line signal and N_b is the number of information bits per PAM symbol. For trellis coded 16-level pulse amplitude modulation (PAM), there are three information bits per symbol.

The following equation defines the nominal PSDs of SHDSL, which are common for both North America and Europe:

$$PSD_{SHDSL}(f) = \begin{cases} 10^{-\frac{PBO}{10}} \cdot \dfrac{K_{SHDSL}}{135} \cdot \dfrac{1}{f_{sym}} \cdot \dfrac{\left[\sin\left(\dfrac{\pi f}{N_{sym}}\right)\right]^2}{\left(\dfrac{\pi f}{N f_{sym}}\right)^2} \cdot \dfrac{1}{1 + \left(\dfrac{f}{f_{3dB}}\right)^{2-Order}} \\ \quad \cdot \dfrac{f^2}{f^2 + f_c^2}, \quad f < f_{intercept} \\ 0.5683 \cdot 10^{-4} \cdot f^{-3/2}, \quad f_{intercept} \leq 1.1\text{MHz} \end{cases}$$

where

- *PBO* defines the amount of power backoff in dB
- K_{SHDSL} is a PSD scaling coefficient

- f_{sym} is the symbol rate, which is one third the line bit rate
- N is a PSD shaping factor, set equal to 1 for all bit rates
- f_{3dB} is the shaping filter 3 dB cutoff frequency
- *Order* is the order of the low pass shaping filter, which is 6 for 16-level TC-PAM
- f_c is the cutoff frequency of the high coupling filter
- $f_{intercept}$ is the frequency where the two functions in the PSD equation intercept in the frequency range of f_{sym}

Table 6.2 summarizes the parameters for the TC-PAM spectral shaper. The parameters in this table are applicable to both North America and Europe.

Figure 6.15 shows plots of various SHDSL PSDs for payload bit rates of 256, 384, 512, 768, and 1152 kb/s. These plots do not include the noise floor portion defined in the nominal PSD equation above. The high pass coupling filter is set to 5 kHz, the order of the filter roll-off is 6, and the shaping filter 3 dB cutoff frequency is set to one-half the symbol rate in all cases.

For comparison purposes, Figure 6.16 shows a plot of the 2B1Q SDSL PSDs supporting for the same bit rates as those of the TC-PAM PSDs in Figure 6.15. The 2B1Q signals have 2 information bits per symbol as opposed to 3 information bits for the TC-PAM case; hence, the bandwidth of the TC-PAM spectrum will be narrower (better spectral efficiency) than the equivalent case for 2B1Q configuration.

2B1Q SDSL uses fourth order low-pass roll-off as opposed to sixth order roll-off in TC-PAM. As seen in the spectral plots of 2B1Q SDSL in Figure 6.16, the out of band energy is significantly greater for 2B1Q than for TC-PAM. The larger out of band energy is undesirable because of the added crosstalk that would be introduced to other systems in the cable, particularly that into ADSL.

In summary, the higher spectral efficiency and higher order filtering of 16-level TC-PAM makes G.shdsl a less disturbing signal to ADSL than 2B1Q SDSL when deployed in the same cable. Upon the adoption of the G.shdsl standard by the ITU-T, 2B1Q SDSL was removed from the basis systems list in the second issue of the spectrum management standard (T1.417) and replaced by G.shdsl (see Chapter 10 for more details on spectrum management).

Table 6.2 Parameters for TC-PAM Spectral Shaper

Payload Data Rate	K_{SHDSL}	f_{3db}	Transmit Power
$R_b < 1{,}536$ kb/s	7.86	$1.0\, f_{sym}/2$	13.5 dBm or less as determined by PBO
$R_b = 1{,}536$ or 1,544	8.32	$0.9\, f_{sym}/2$	13.5 dBm
$1{,}544 \le R_b \le 2{,}048$ kb/s	7.86	$1.0\, f_{sym}/2$	13.5 dBm or less as determined by PBO
$R_b \ge 2{,}048$ kb/s	9.90	$1.0\, f_{sym}/2$	14.5 dBm or less as determined by PBO

Notes:

- $N = 1$ for all cases.
- The amount of overhead may be 8 or 16 kb/s for any payload rate (depending on application).
- Filter order is 6 in all cases.
- The symbol rate is one third the line bit rate (payload + overhead).

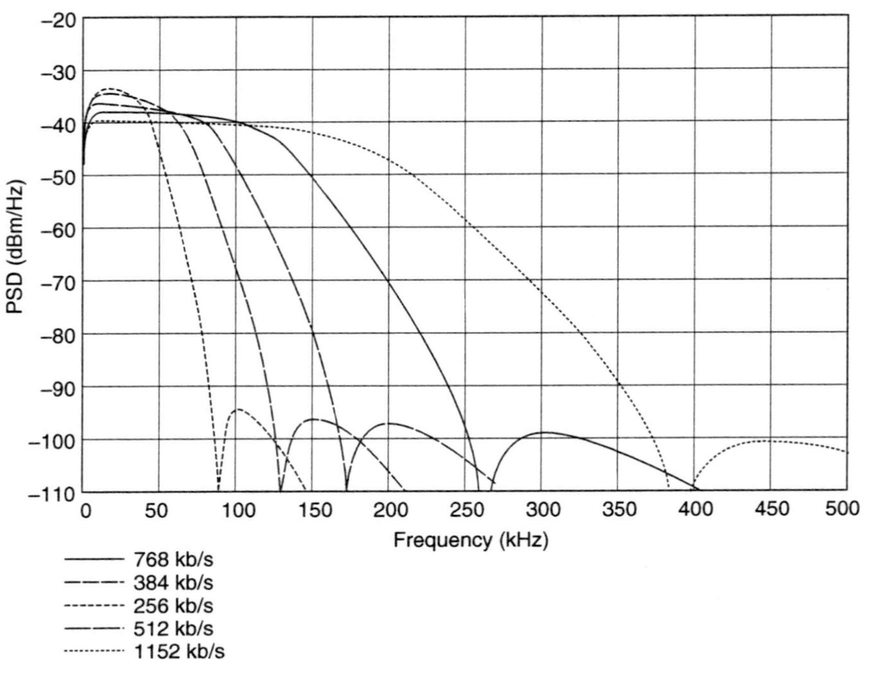

Figure 6.15 Nominal PAM PSDs without noise floor.

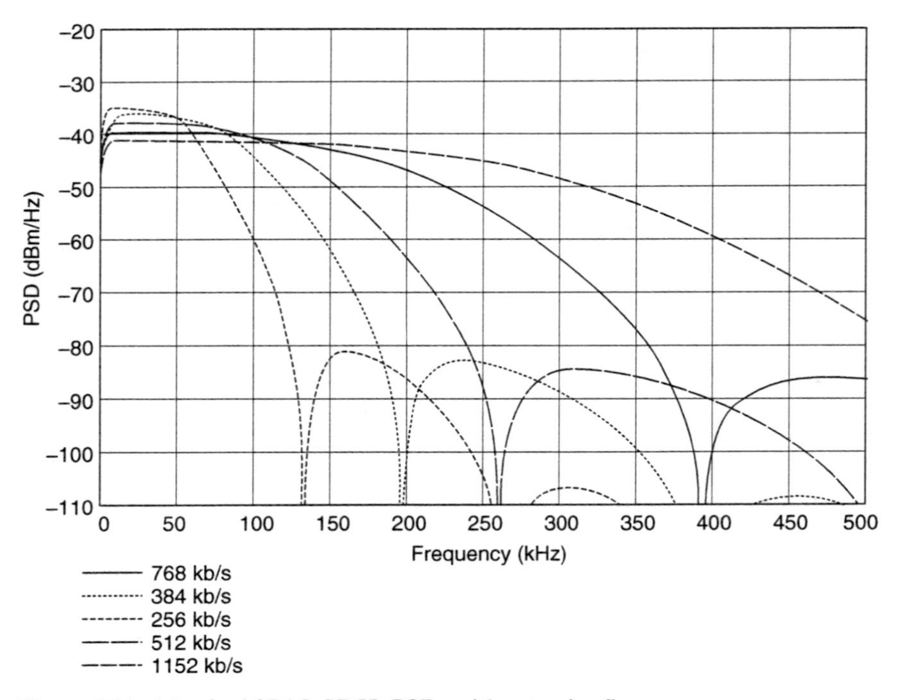

Figure 6.16 Nominal 2B1Q SDSL PSDs without noise floor.

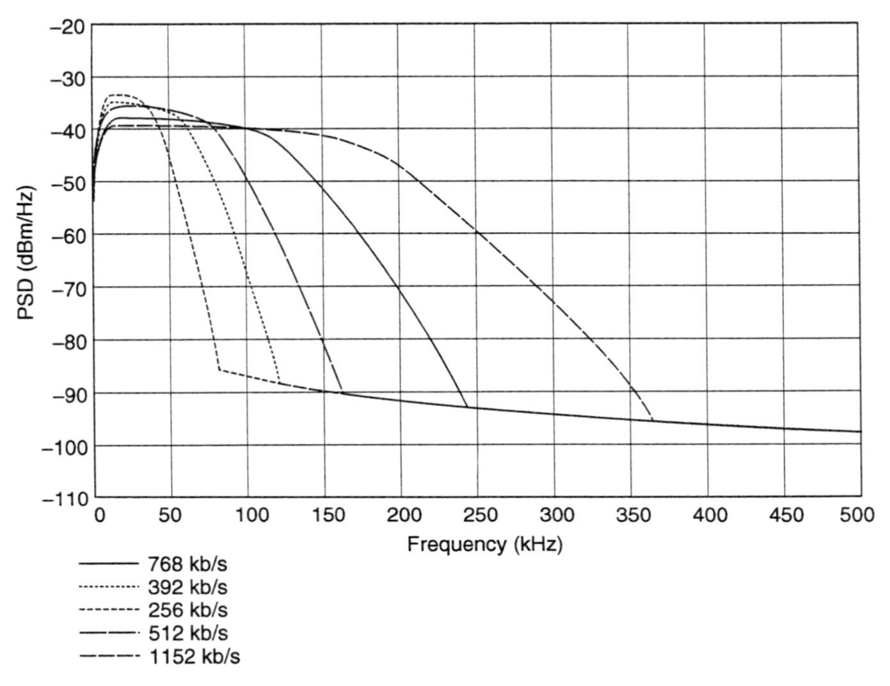

Figure 6.17 Nominal PAM PSDs with inclusion of out of band noise floor.

Implementations of PSDs may vary from different manufacturers. The out of band energy may vary depending on the specifics of implementations. To accommodate this variability, the G.shdsl standard [13] defines a noise floor function, which is shown in the nominal PSD equation above. The noise floor is applied at the frequencies above the intercept frequency, which is frequency where the main PSD and noise floor equations intercept in the vicinity of the symbol rate frequency f_{sym}. Figure 6.17 shows plots of the nominal TC-PAM PSDs, which include the out of band noise floor. Actual implementations would have the majority of the out of band PSD below the specified level.

In addition to the symmetric PSDs, G.shdsl defines numerous asymmetric PSDs. For North America, the asymmetric PSDs are taken from the HDSL2 standard, and they have been designed for optimal performance and minimal impact on spectral compatibility with other services deployed in the cable. Asymmetric PSDs are defined for 768 or 776 kb/s, and 1536 or 1544 kb/s configurations.

6.6 SHDSL PERFORMANCE

For SHDSL the PSD in the downstream direction is the same as that in the upstream direction. This is also true for 2B1Q SDSL. When the upstream and downstream channels use the same PSD shape, the PSDs are termed *symmetric* PSDs.

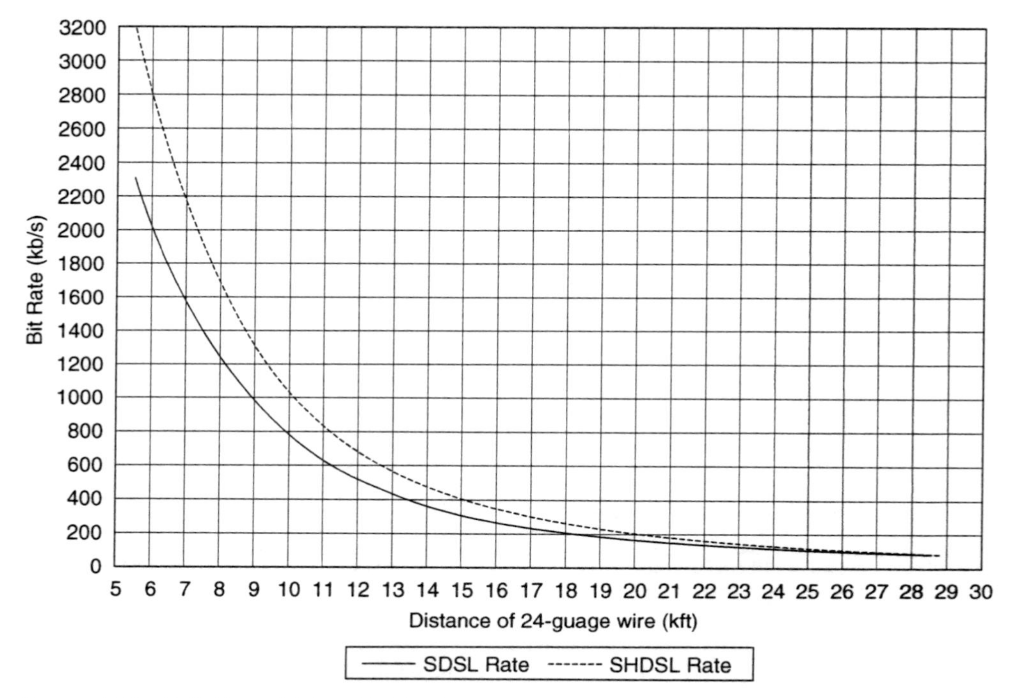

Figure 6.18 SHDSL and 2B1Q SDSL bit rates versus 24-gauge distance in presence of 49 SNEXT.

For the case of symmetric PSDs, the total transmit power is fixed to maximum value of 13.5 dBm. When these systems are deployed in the cable, the worst case disturbance is near-end crosstalk from the same system. Specifically, for an echo canceled system with symmetric PSD, assuming that all of the systems in the cable have the same transmit power, the worst-case disturbance is near-end crosstalk from other systems with exactly the same spectrum. Figure 6.18 shows the bit-rate versus distance of 26-gauge wire for both SHDSL using 16-level TC-PAM and 2B1Q SDSL in the presence of 49 self-near-end crosstalk (SNEXT) disturbers. These simulation results assume 6 dB of margin in all cases. The better performance of TC-PAM is due to inclusion of a trellis code with 5.1 dB of coding gain.

6.7 CORE SHDSL FRAMER (PMS-TC)

The structure of the SHDSL core frame is based on the superframe structure of HDSL2 as described in Chapter 4. This section briefly describes the SHDSL core frame per Recommendation G.991.2.

Figure 6.19 shows the core SHDSL physical layer superframe. The nominal frame period (T_{frame}) is 6 msec. There are four blocks for payload data bits identi-

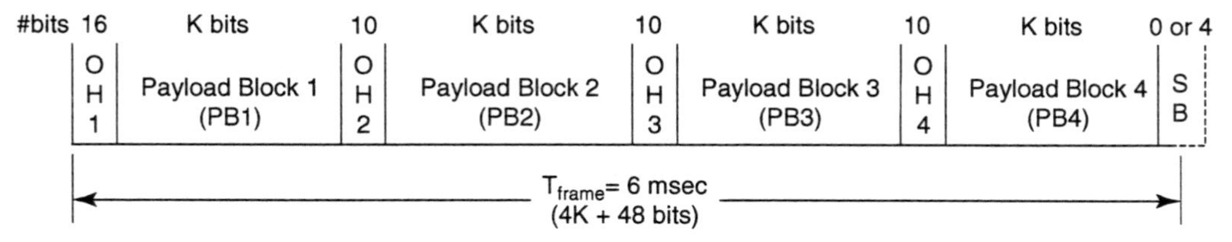

Figure 6.19 SHDSL core frame structure.

fied as PB1–PB4. Because the line bit rate is variable, the number of bits in an SHDSL superframe will depend on the payload bit rate.

There are also five subblocks allocated for overhead bits labeled as OH1–OH4 and block SB (stuff bits). OH1 has 16 bits, OH2–OH4 each has 10 bits, and the SB block has nominally 2 bits, but the size may vary between 0 and 4 bits. Note, however, that for the nominal superframe there are nominally 48 bits in the overhead channel.

The following is a summary of the overhead bit definitions.

- OH1 (16 bits): This block contains a 14-bit frame synchronization word (FSW). The definition of the FSW is vendor specific. During preactivation, the transmitter forwards its preferred FSW to the far-end receiver via the specified exchange protocol. The remaining two bits are fixed indicator bits. The first indicator bit identifies a loss of signal from an application interface. The second indicator identifies segment anomaly, which is generally a cyclic redundancy check (CRC) error determined on an incoming SHDSL frame.
- OH2 (10 bits):
  - Bits 1–4: EOC bits #1–4
  - Bits 5 and 6: crc-1 & crc-2
  - Bit 7: Fixed indicator bit #3, which defines the power status of the local power supply in the customer premises SHDSL unit (STU-R)
  - Bit 8: Stuff bit ID #1 (first copy). This bit is used when the pass-thru timing mode is enabled. Otherwise, this is a spare bit.
  - Bits 9 and 10: EOC bits 5 and 6.
- OH2 (12 bits):
  - Bits 1–4: EOC bits #7–10
  - Bits 5 and 6: bits crc-3 and crc-4
  - Bit 7: Fixed indicator bit #4 that defines segment defect. A segment defect indicates a loss of synchronization on the incoming SHDSL frame. Perhaps a regenerator has lost synchronization, and therefore the regenerated data is unavailable.
  - Bits 8–9: EOC bits #11 and 12
  - Bit 10: Stuff bit ID #2 (second copy). This bit is used when the pass-thru timing mode is enabled. Otherwise, this is a spare bit.

- OH3 (12 bits):
  - Bits 1–4: EOC bits #13–17
  - Bits 5 and 6: bits crc-5 and crc-6
  - Bits 7–10: EOC bits #18–20
- SB (0 or 4 bits):
  - Bits 1–2: Vendor depend definition.
  - Either 0 or 4: These stuff bits are not present during synchronous timing mode where the line clock is frequency locked to the payload clock.

The six CRC bits are used for performance monitoring of the subscriber line. The six CRC check bits, crc-1–crc-6, are those associated with the contents of the previous SHDSL superframe. The CRC is computed over all of the bits in the superframe except for the 14 sync word bits, the 6 CRC bits, and the nominally 2 SBs; hence, the data message contains 4K + 26 bits. The message polynomial is constructed such that first CRC computable bit is the coefficient of the term x^{4K+25}, and the last bit is the coefficient of the term x^0. The polynomial is then multiplied by a factor of x^6, and the result is divided (modulo-2) by the generator polynomial $x^6 + x + 1$. The coefficients of the remainder polynomial are used in the order of occurrence as the ordered set of check bits crc-1–crc-6. In the remainder polynomial, the coefficient of the term x^5 is crc-1 and that for the term x^0 is crc-6.

The embedded operations channel has 24 bits allocated in a 6 msec frame. This corresponds to a 4 kb/s clear data communications channel. This channel is used to pass operations and maintenance information between the CO and CP HDSL2 transceiver units.

6.8 TIMING CONFIGURATIONS

There are three mode timing operation defined in G.991.2, namely

- Plesiochronous mode
- Plesiochronous with network timing reference
- Bit synchronous mode

Each of these timing modes is described below.

Figure 6.20 shows the transceiver-timing configuration for the *plesiochronous* timing mode. This is the timing configuration used for classic HDSL operation. The transceiver line clock, that is, the transmit symbol clock, operates independent (free running) from the payload data clock. For this application, the payload timing reference needs to be transferred end to end. In this timing mode, the pulse stuffing operations are enabled, and the payload data clocks (i.e., Tx_Clock signals) are passed end to end. Note that due to the pulse stuffing and deleting operations, re-

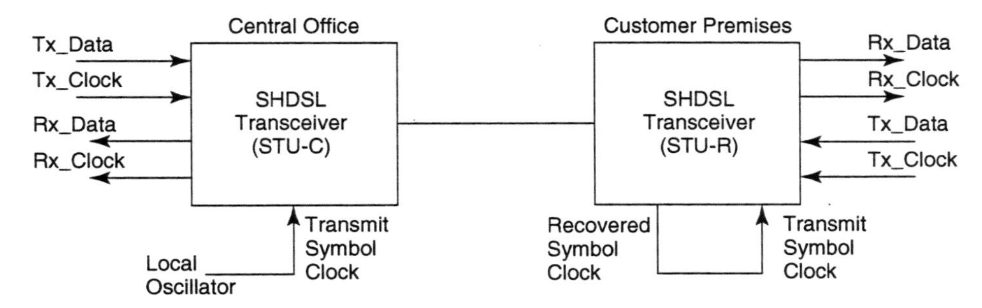

Figure 6.20 Plesiochronous timing mode.

covered payload data clock (Rx_Clock) will have some residual jitter. The amount of residual jitter depends on the circuit implementation.

In this timing mode, the central office unit (STU-C) serves as the master timing source for the line (symbol) clock, being that it is derived from a local oscillator. The customer premises unit (STU-R) derives the line (symbol) clock from the received line signal and uses this recovered clock at the transmit line (symbol) clock for the STU-R. This timing configuration at the customer premises unit is referred to as loop timing.

The *network timing* reference mode is similar to that of the plesiochronous timing mode. The exception is that the transmit symbol clock at the central office unit (STU-C) is derived from a local network timing reference as opposed to a local oscillator. In this mode of operation, the pulse stuffing operations are enabled as well. This mode of operation should allow for better residual output jitter on the recovered payload data clock because the timing sources are more closely matched.

The third timing mode operation is the *synchronous* timing mode, which is shown in Figure 6.21. In this mode of operation, the transmit symbol clock of the central office unit is frequency locked to timing source of the transmit data signal. Because the payload clock and the line clock are frequency locked, the stuff/delete operations in the core frame are disabled. Note that as in the other timing modes, the STU-R operates in the loop timing mode.

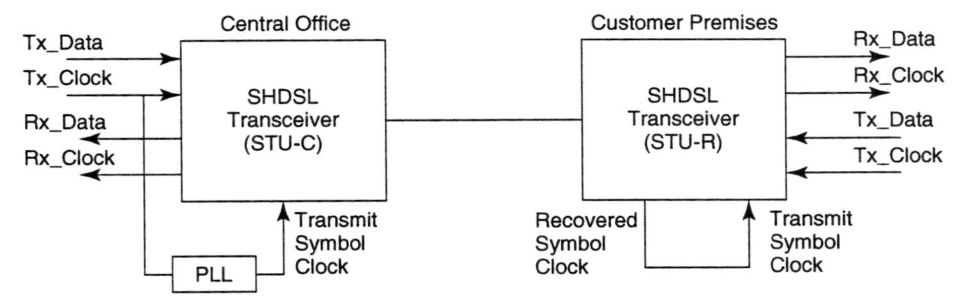

Figure 6.21 Synchronous timing mode.

6.9 APPLICATION SPECIFIC FRAMING (TPS-TC)

Earlier in the chapter, we described some possible applications for SHDSL transceivers. In order to support we need a TPS-TC and corresponding interface(s) with definitions that are in line with the application. Annex E of ITU-T Recommendation G.991.2 provides definitions of various types of TPS-TC that may be supported by SHDSL transceivers. Depending on the application being supported, other types of TPS-TCs may be defined as needed. Following are some example TPS-TC definitions.

6.9.1 Clear Channel Data

In transporting clear channel data, there is no special relationship between the structure of the user data and that of the core SHDSL frame. For this application, use of the core SHDSL frame as defined in Figure 6.19 is sufficient. The user payload data is inserted directly into the payload data blocks.

This same core frame structure may be used to transport T1 (1.544 Mb/s) or E1 (2.048 Mb/s) payloads. Although the T1 and E1 signals have a frame structure defined, the data may be placed directly into payload data blocks without regard for the boundaries of the T1 or E1 frames. External circuits may be used to determine the T1 or E1 frame boundaries carried in the payload bit sequences.

6.9.2 Fractional T1/E1 Transport

Figure 6.22 shows the subchannel frame structure for supporting fractional T1 or E1 services. Each active time slot from the T1 or E1 payload is assigned appropriate locations in the subblock data fields of the core framer's data block.

The transport of fractional T1 or E1 signals means that the SHDSL transceiver transports only the active time slots in the T1 or E1 payload. In a fractional T1 configuration, for example, both the customer and network interfaces in the system reference model of Figure 6.6 are T1 AMI signals. This interface operates at the full rate of 1.544 Mb/s; however, not all of the time slots are active. The TPS-TC block of the fractional T1 SHDSL transceiver takes only the active time slots from the T1 payload and inserts them into the core SHDSL payload data blocks as shown in Figure 6.22. At the receive end of the line, the TPS-TC block inserts the active time slots from the core frame into their corresponding time slot locations in the T1 frame. The operations are similar for the fractional E1 example.

6.9.3 Dual Bearer Transport

An example of a dual bearer application is in the simultaneous transport of a high-speed data channel together with one or more digitized voice channels. This application example is shown in the work-at-home application diagram of Figure 6.5. The customer interface for the high-speed data channel may be an Ethernet 10/100 Base-T interface, and those for the analog voice channels may consist of a codec and a subscriber line interface circuit (SLIC) for provisioning of analog/digital con-

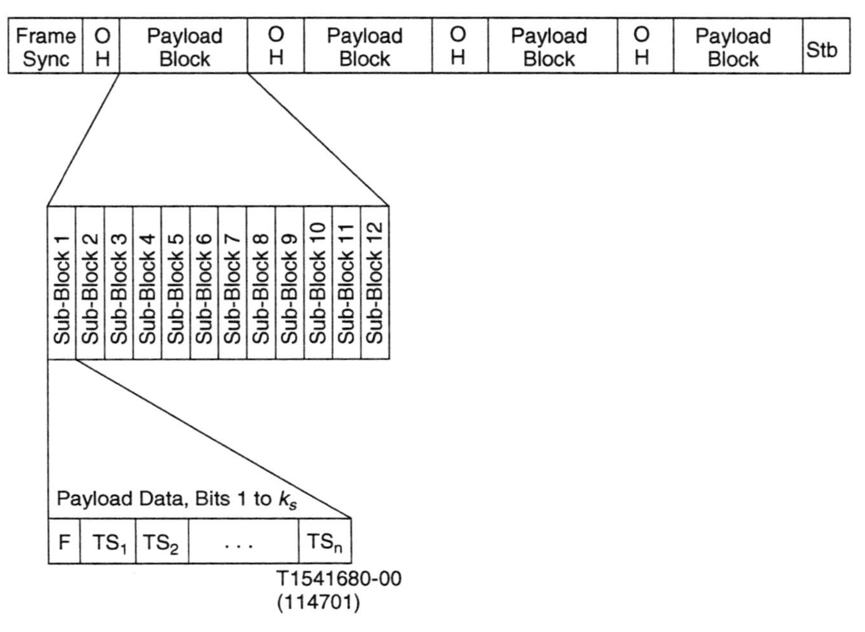

Figure 6.22 Fractional T1/E1 frame structure.

version and phone channel signaling. Details of the interface circuits are outside the scope of this chapter and are not described here.

The digitized voice channels may be transported through the network hierarchy in a bit synchronous T1 signal. The data signals would be transported through data network that is separate from the voice traffic.

Figure 6.23 shows the framing structure for dual bearer framing taken from Annex E.10 of G.991.2. Inside this frame, the digitized voice samples would be transported with proper time slot assignments in one of the bearers designated as the synchronous bearer. The data traffic would be transported without any special frame boundary alignment in the second bearer designated as the asynchronous data bearer channel.

Note that in the dual bearer mode, the transceiver would most likely need to operate in the plesiochronous timing mode with bit stuffing enabled. This may be required in support of the bit synchronous traffic of the voice channels. Alternatively, the transceiver line clock may be frequency lock to payload data clock of the digitized voice traffic.

6.10 INITIALIZATION

Initialization in SHDSL consists of two phases: the preactivation phase and the core activation phase. Figure 6.24 shows the total activation sequence for SHDSL link initialization. The preactivation phase consists of two handshake sequences

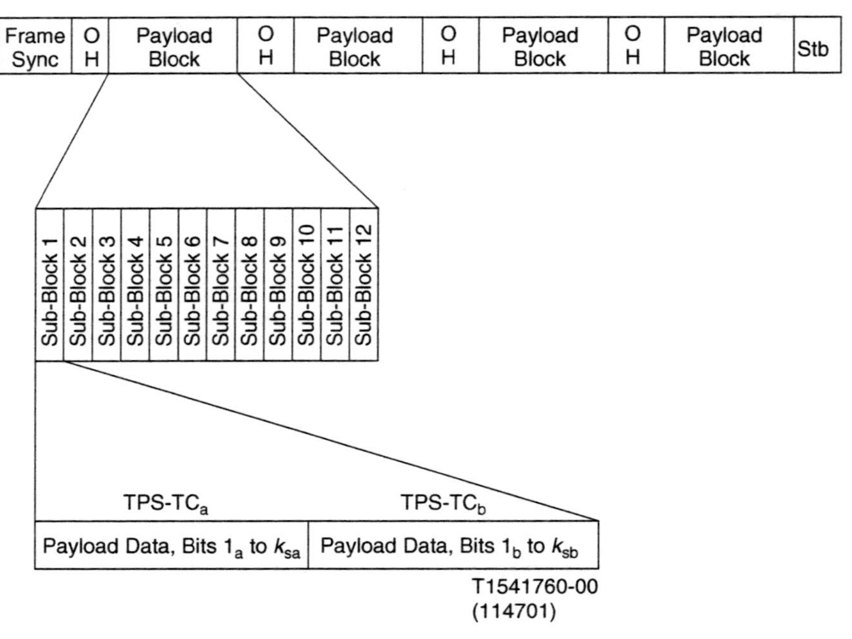

Figure 6.23 Dual bearer subchannel framing.

and a line probe session, whose total goal is to negotiate system parameters and learn the best SHDSL configuration that can be supported on the given channel. Once the system configuration parameters are determined from the preactivation phase, the core activation phase trains the modem for optimum performance when transitioning to data mode.

As seen in Figure 6.24, there are numerous timer values associated with the total activation time. The total activation time (t_{Act_Global}) is the sum of the preacti-

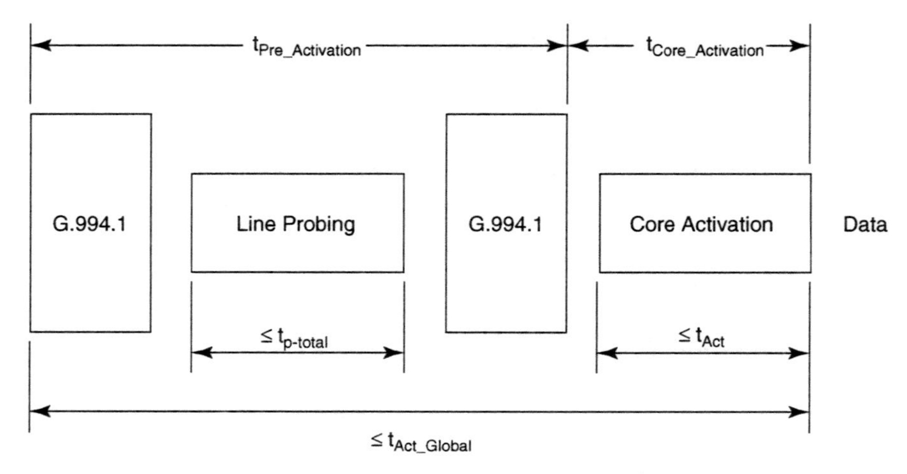

Figure 6.24 SHDSL total activation sequence.

vation and core activation times. The line probing session has a maximum duration time ($t_{p\text{-}total}$) of 10 seconds. The core activation time (t_{Act}) has a maximum duration of 30 seconds.

Note that in the preactivation phase, the handshake mechanism does not have a maximum time-out associated with it, in that the handshake mechanism is outside the scope of the SHDSL specification. However, reasonable designs of the handshake can limit the total time to less than 5 seconds for typical SHDSL configurations. In most cases, we can expect the total activation time to be less than 30 seconds total.

6.10.1 Preactivation Phase

As we mentioned earlier, the goal of the preactivation phase is to negotiate the selection of the transceiver parameter and determine (if necessary) the optimum transmission configuration using the optional line probe capability. Figure 6.25 shows the preactivation sequence of SHDSL.

In the first handshake (G.994.1) session, the transceivers negotiate the configuration of line probe sequences. The sequences P_{r1}–P_{rN} represent the line probe symbol sequence generated by the customer premises unit (STU-R). Similarly, the line probe sequences generated from the CO (STU-C) unit are labeled P_{c1}–P_{cN}. During this preactivation phase, the transceiver operates using 2-PAM. The transmitter components active during line probe are the scrambler, bit-to-symbol mapper, and the spectral shaper. The symbol rate, spectral shape, duration, and power backoff level are selected during the first handshake session.

6.10.2 Core Activation

Core activation (Figure 6.26) begins with the customer premises (CP) unit transmitting sequence C_r for a period of t_{cr} seconds. The value of t_{cr} depends on the bit rate: If the payload bit rate is 768 kb/s or less, then the nominal value of t_{cr} is 2 seconds; for the higher bit rates, the nominal value is 1 second. The CO unit may use the received sequence to perform timing recovery and train its equalizer. The CP unit may use this sequence to train it echo canceller.

Upon the completion of sequence C_r, the CO unit begins sending sequence S_c one-half second later. Sequence S_c is the 2-level PAM signal formed by inputting ones into the scrambler of the reference startup transmitter. Recall that the transmit power is set to the value specified in the preactivation frame sent dur-

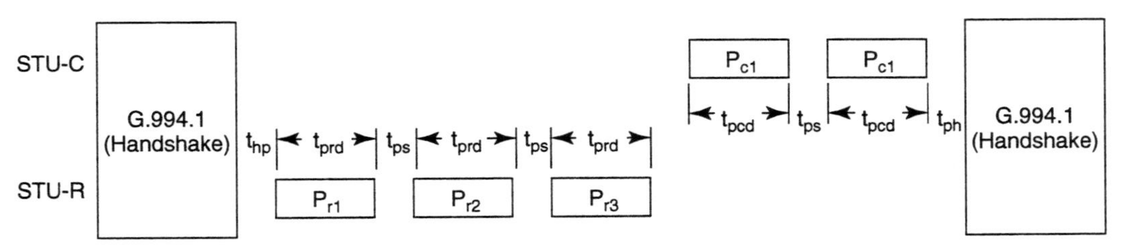

Figure 6.25 SHDSL preactivation sequence.

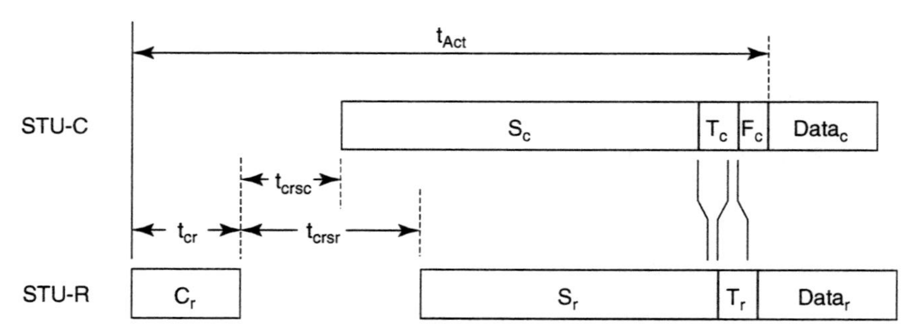

Figure 6.26 SHDSL core activation sequence.

ing the handshake sequence exchanges. For a duration t_{crsr}–t_{crsc} seconds, S_c is the only signal on the line. The duration of t_{crsr} depends on the bit rate: for payload rates of 768 kb/s and below, the nominal value is 3 seconds; otherwise the value of t_{crsr} is 1.5 seconds. During this time the CP unit may train its equalizer and timing recovery circuit from the received signal; during this same time, the CO unit may train its echo canceler. The CP unit begins transmitting Sequence S_r at t_{crsr} seconds after completion of C_r, and now there is simultaneous transmission of both upstream and downstream data. The sequence S_c is transmitted for a minimum of 5 seconds, which is the time required (timer T_{PLL}) for the CP unit to synchronize its phase locked loop.

Sequence S_r is also a two-level PAM signal resulting from an input of ones into the scrambler of the reference startup transmitter, and the transmit power is that specified in the handshake sequence exchange. The CP unit begins transmitting at t_{crsr} seconds after it concludes transmitting C_r. During the period of simultaneous S_c and S_r transmission, the transceivers continue training their equalizers, echo canceler and other necessary functions. If the transceiver functions have not converged by conclusion of S_c and S_r, then the transceiver enters an exception state. At that point the initialization process would need to be restarted.

After the CO unit transceiver has converged and it has been sending the S_c signal for at least 5 seconds (i.e., the value of the T_{PLL} timer), it transitions to sending signal T_c. During the transmission of T_c, the channel precoder coder coefficients and other system information is sent to the CP unit from the CO unit. Once the CP unit has converged and has begun detecting the T_c signal, it begins transmitting the T_r signal to the CO unit. As with T_c, the T_r signal passes the channel precoder coefficients and other signal parameters to the CO transceiver. The information transferred in the T_c and T_r signals are contained in a core activation frame.

Once the CO unit has detected the T_r signal and has completed transmission of the core activation frame, it begins sending signal F_c. Signal F_c sends the core-activation frame of T_c except that the frame sync word is reversed, and all of the remaining information bits are set to arbitrary values. Two of these frames are sent during F_c, and this can serve as an acknowledgment that the CO unit received T_r.

Upon conclusion of S_c transmission, the CO unit begins sending data and the CP unit begins sending data upon completion of T_r.

REFERENCES

[1] ITU-T Recommendation G.991.2, "Single-Pair High-Speed Digital Subscriber Line (SHDSL) Transceivers," February 2001.

[2] ETSI TS 101 524 V1.1.3 (2001-11), "Transmission and Multiplexing (TM); Access Transmission System On Metallic Access Cables; Symmetrical Single Pair High Bitrate Digital Subscriber Line (SDSL)," November 2001.

[3] Committee T1–T1.418-2000, "High Bit Rate Digital Subscriber Line—2nd Generation (HDSL2)."

[4] T1.422, "Single-Pair High-Speed Digital Subscriber Line (SHDSL) Transceivers," October 2001.

[5] Committee T1–T1.403-1999, "Network and Customer Installation Interfaces–DS1–Electrical Interface."

[6] ITU-T Recommendation G.703, "Physical/Electrical Characteristics of Hierarchical Digital Interfaces," November 2001.

[7] ITU-T Recommendation G.704, "Synchronous frame structures used at 1544, 6312, 2048, 8448 and 44,736 kbit/s Hierarchical Levels," October 1998.

[8] ETSI TS 101 135 V1.5.3 (2000–09), "Transmission and Multiplexing (TM); High Bit-Rate Digital Subscriber Line (HDSL) Transmission Systems on Metallic Local Lines; HDSL Core Specification and Applications for Combined ISDN-BA and 2,048 kbit/s Transmission," September 2000.

[9] T1E1.4/2001-006R2, "Draft Standard: High Bit Rate Digital Subscriber Line—2nd Generation (HDSL2/HDSL4) Issue 2," November 2001.

[10] G. Ungerboeck, "Channel Coding with Multilevel/Phase Signals," *IEEE Transactions on Information Theory* IT-28, no. 1, January 1982.

[11] Pairgain Technologies, "A 512-State PAM TCM Code for HDSL2," T1E1. 4/97-300, September 22, 1997.

[12] Adtran, "Performance and Characteristics of One-Dimensional Codes for HDSL2," T1E1.4/97-337, September 25, 1997.

[13] Adtran, Cicada, Siemens, Tellabs, and Westell, "Proposal to Break the FEC Logjam for HDSL2," T1E1.4/97-443, December 8, 1997.

CHAPTER 7

VDSL

Very-high bit-rate digital subscriber lines (VDSL) is currently the highest-speed DSL, with standardized symmetrical data rates as high as 26 Mbps and asymmetrical rates up to 52 Mbps downstream with 6.4 Mbps upstream. VDSL is an evolutionary successor to earlier DSLs in that VDSL increases

DSL data rates to support an ever-increasing customer demand for faster access to all types of information. Before exploring the technology, a brief history of VDSL is helpful:

The VDSL concept was first published in 1991 [1] and was a result of a joint Bellcore-Stanford research study into the feasibility of 10+ Mbps symmetric and asymmetric data rates on short phone lines. The study specifically searched for the potential successors to the then more prevalent 1.5 Mbps HDSL and the then relatively new (then also only 1.5 Mbps) ADSL.[1] The first serious suggestions that VDSL be standardized came almost simultaneously in the American T1E1.4 group from Amati Communications Corp. in [2] and in the ETSI group from British Telecom [3] as a function of the first ADSL trials in Britain at 2 Mbps and 6 Mbps (supplied by Amati to British Telecom) where discussions on the potential of higher speeds on shorter fiber-fed optical network unit (ONU)-based copper loops were active between the two companies. VDSL history also has connection to early high-speed ATM network studies in the ATM Forum [4] and DAVIC [5], which attempted to transmit 26 and 52 Mbps symmetrically on one or more twisted pairs over very short distances (less than 100 meters) for local area networks. Although the latter ATM and DAVIC efforts are somewhat forgotten, in that 100BASE-T and now Gigabit Ethernet became the methods of ubiquitous use for internal computer networks on twisted pair, these early ATM and DAVIC efforts did also provide useful information to the development of present-day VDSL standards. The DAVIC group went so far as to consider modification of the early ATM systems for very short-phone-line subscriber access.

Today, an architecture and transmission technology has emerged from years of VDSL study, and draft documents are just completing formal standardization. A special issue of the *IEEE Communications Magazine* has several articles on VDSL [35],[36]. Section 7.1 describes the basic VDSL concept and architecture as well as some important spectral issues. Section 7.2 then explores some applications for VDSL and the consequent implications on interfaces and basic architecture, of VDSL. Sections 7.3 and 7.4 then provide greater detail on the transmission details of VDSL implementation. The reader is also referred to Chapter 11, which looks at advances projected in DSL. Section 7.5 provides some early information on the area of VDSL known as Ethernet in the first mile (EFM). Recent interest in 100 Mbps symmetric transmission is discussed as VDSL enhancement in Chapter 11.

7.1 BASIC VDSL CONCEPTS

Basic VDSL architecture presumes a shorter twisted-pair transmission line than earlier DSLs. VDSL line lengths are typically between 150 meters (500 feet) and 2000 meters (6600 feet). On such short phone lines, the very high-speed digital transmission of tens of megabits/second is possible. The existence of shorter phone lines implies a telecommunications network that increasingly uses fiber, or perhaps

---

[1]The author (Cioffi) would like to acknowledge gratefully encouragement and support from then-retiring Dr. Joseph Lechleider of Bellcore, who encouraged and financially supported this early VDSL study.

also wireless transmission, in that the nearest central-office portion of the access network previously carried all data via twisted-pair. All DSL technologies prior to VDSL had served the large majority of customers via copper wire pairs directly from the CO to the customer location. VDSL's short line reach, in contrast, requires the large majority of customers to be served via copper wire pairs running from the customer site to a nearby network node that is then linked to the central office via fiber or radio. Thus, this architecture could be described as *hybrid fiber-copper*.

Section 7.1.1 describes the basic VDSL combination of twisted pair with fiber and relates speed/length trade-offs for VDSL, whereas Section 7.1.2 then further details consequent architecture for VDSL. An important area of VDSL performance and installation is spectrum compatibility with an array of existing DSL and home phone–line signals, on both twisted pair and in the wireless ether surrounding telephone lines. Section 7.1.3 then illustrates some of the complexities of spectral design and describes and evaluates current VDSL spectrum recommendations. A brief synopsis of what we here call "the grand debate" is provided in Section 7.1.4 and then is validated by the two transmission methods specified in the draft standards to be described later in Sections 7.3 (DMT) and 7.4 (QAM).

7.1.1 ADSL Extension

ADSL is now acknowledged as a successful telecommunications service with tens of millions of lines in deployment, and hundreds of millions hoped to be deployed in the next decade or two. However, in its earliest days of standardization, ADSL faced the severe criticism that even its greatest standardized speed of 8 Mbps was too slow to match the data rates possible on what are called hybrid fiber-coax (HFC) networks. HFC networks upgrade existing unidirectional cable TV networks in two ways:

1. The cable TV networks are made two-way in HFC by replacing upstream-blocking unidirectional amplifiers in the cable plant by more sophisticated two-way non-upstream-blocking bidirectional amplifiers.
2. The cable TV networks are increased in bandwidth in HFC by replacing coax near the TV head-end by fiber, multiplexing multiple separate coax signals on the same fiber in different bands, and thus rendering fewer subscribers per coaxial branch.

Phone companies believed in 1994 and 1995 that they must replace their existing phone-line networks with HFC, and several attempted to do so, only to find later that the costs were prohibitive.

ADSL was already bi-directional,[2] but with limited speeds downstream and even lower speeds upstream. In 1994 and 1995, ADSL was perceived as unable to support the full set of video, voice, and data services necessary to compete with HFC. VDSL was proposed by ADSL proponents as a next higher-data-rate step

---

[2]The asymmetry in ADSL allowed a longer line length for reliable transmission of a given data rate [6].

for ADSL: If fiber can be installed in HFC, then why not install it in existing networks when there are customers ready to pay for higher speeds than ADSL, and instead use fiber-based-loop-shortening to increase the speed and symmetry of ADSL? The initial VDSL architecture of Figure 7.1 ensued as the future of DSL deployment when (and if) customers were willing to pay for more and more fiber. VDSL is an incremental alternative that leverages existing phone lines in contrast to network replacement mandated by HFC. In 1995 and beyond, VDSL's incremental deployment won increasing favor with telephone service providers and is the actual mode of choice today. Cable suppliers continue to upgrade their TV networks to HFC at significant cost, but it becomes increasingly clear that the merits of DSL will prevail for nearly all services other than (unidirectional) analog and newer digital television delivery, for which the plant architecture of cable seems still to be well suited.[3]

The optional splitter of ADSL is preserved in VDSL so that network-powered analog voice service can be delivered normally on the same line as VDSL. The cost of the fiber section is high, but can be divided by the number of customers served. As fiber penetrates closer and closer to the customer, that cost is shared by a smaller number of customers. Thus an important trade-off in VDSL is the length of the fiber versus the length of the remaining copper. There is no single good answer to this trade-off as it depends on applications, customer willingness to pay, transmission method, and of course cost of the fiber—however, VDSL allows a wide range of trade-offs, as this chapter illustrates.

Figures 7.2(a) and 7.2(b) illustrate data rates for both upstream and downstream VDSL transmission on 24-gauge twisted pair (.5 mm European) versus loop length for the United States, Japan, the United Kingdom, and some other countries (998 curve) and some European countries (997 curve) DMT VDSL draft standards.[4] Clearly the data rates are quite high on short loops, ensuring a large individual bandwidth per user (often higher than cable networks, which customers must share due to the cable plant architecture). Although still evident, the premium paid in range loss for symmetry of data rate is less as loop lengths get shorter, and then VDSL also offers a way to offer increasingly symmetric individual service to customers. As the number of small businesses worldwide explodes, most often in urban areas where line lengths are short, the potential for symmetric support of the voice, conferencing, peer-to-peer gaming or working, "home" Web server upstream bandwidths is then evident with VDSL. Today, an increasing number of service providers consider exploring early VDSL deployment for business services, particularly, EFM support (see Section 7.5). Korea Telecom has committed to deploy DMT VDSL service starting in 2003 and is the first telco to offer commercial VDSL service that is not a trial. Five percent of Korean DSL is mandated to be DMT VDSL by 2005 (and Korea has more DSL than any country in 2002–2003). In 2002, only a tiny fraction of the nearly 10 million businesses in the United States were connected by fiber (and a yet smaller fraction in other countries) [7]. Tele-

---

[3]It is conceivable that Internet-based approaches to TV may allow an opportunity for DSL.

[4]Achievable data rates for the other "single-carrier" standard will be less, with these curves of Figures 7.2(a) and (b) as an upper bound. (See [6].)

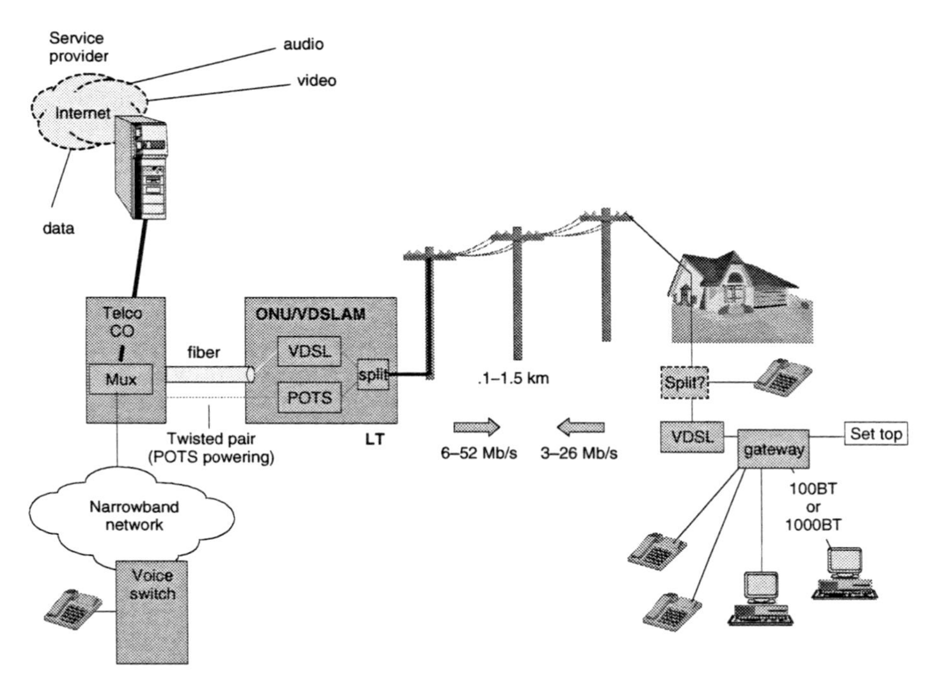

Figure 7.1 VDSL system architecture with ONU/fiber-loop-carrier system.

phone companies are installing more fiber to businesses every year. With the advent of ATM passive optical network (APON) access, the deployment of fiber is expected to accelerate. One large fiber installer and service provider estimates they will increase this number by 2,000 business in the next two years at a cost of $1 billion [7]. In 2003, construction of residential fiber-to-the-home (FTTH) began on a small scale. Thus, VDSL will play a major role in the future service offerings to small and many large businesses before fiber connection is financially viable or completed. Other service providers still believe that support of video and television may also be viable in the future, although the economics of this application may be harder to justify versus cable.

The wealth of ADSL installations also mandates another practical requirement that VDSL service must be compatible in many respects with existing ADSL. An existing customer with an ADSL modem on his premises (perhaps in his/her portable computer) may move or travel into another area, or may live in an area where VDSL arrives, and will still want his/her ADSL modem to function as it always has. Thus, the ONU-side modem in Figure 7.1, often called the LT (line termination) in VDSL would need to support ADSL service, but would of course also allow higher speed service if/when that customer decides to purchase a higher-speed VDSL modem and the higher-speed VDSL service. Also, a customer who buys a VDSL modem will certainly want that modem to work with an existing ADSL connection at lower speed if that is all that is available. In addition to interoperation with ADSL modems, VDSL modems must also be compatible spectrally with ADSL modems that may share the same binder and with existing home-

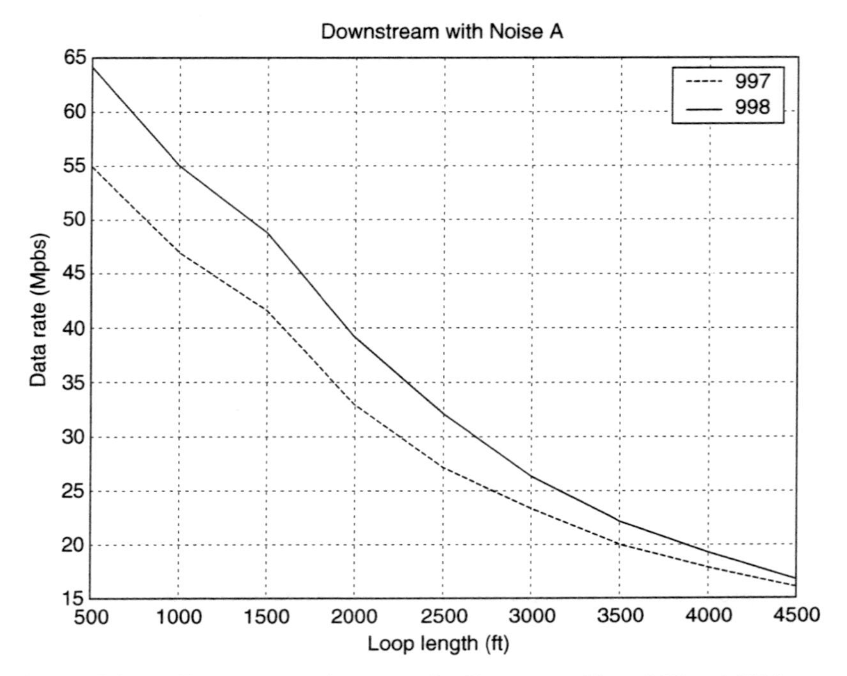

Figure 7.2(a) Downstream data rates for Frequency Plans 997 and 998 in draft standard DMT VDSL [17].

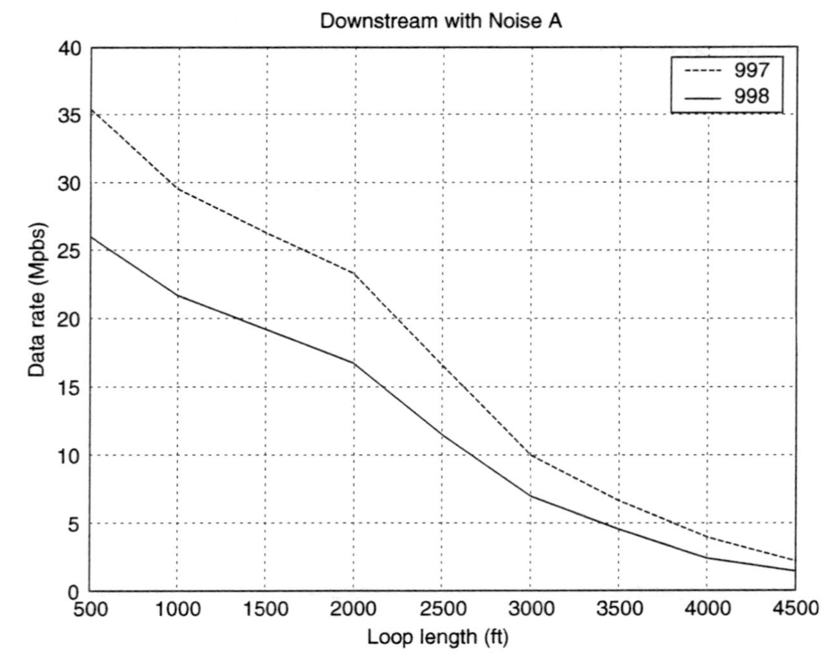

Figure 7.2(b) Upstream data rates for draft standard DMT VDSL Frequency Plans 997 and 998.

premises networks, as well as with perhaps a plethora of other standard or nonstandard systems that exist in/near the cable (for instance, HDSL, SHDSL, or nearby HAM radio). Section 7.1.3 deals more directly with this spectrum issue, whereas Chapter 8 explores the area of regulation and various newer forms of spectrum management that VDSL may exploit appear in Chapters 11 and 12.

Overall, VDSL offers a mechanism for service providers to upgrade their networks incrementally and with continued profitability to include increasing amounts of fiber, approaching an ultimate goal of a network based entirely of fiber. Telephone-line service providers have a very powerful story and future with VDSL, now following ADSL.

7.1.2 VDSL Architecture

With the VDSL incremental growth opportunity for DSL for the next century, the questions are "where, when, and how" to insert the fiber into the network. Clearly the installation of fiber should usually start closer to the service provider where the cost of a fiber can be shared over the many customers served by that fiber. Further from the central office, the fiber would service fewer customers and thus appear more costly to install per customer. Indeed today, 15 percent of the American network has what is called fiber feeder as shown in Figure 7.1. Initially, with the fiber loop carrier system being deployed today, usually only the POTS/voice connections in Figure 7.1 exist. However, VDSL or any DSL is added by placing the DSLAM at the end of the fiber, sometimes known as an optical network unit (ONU). Many phone companies have massive expenditures and fiber deployments underway to augment their ADSL service deployments so that line lengths are shortened and ensure higher ADSL speeds more reliably. It is also possible to see an ADSL DSLAM also at the ONU, which is sometimes also called a remote terminal (RT) or LT, depending on the country and the exact type of DSL deployed (but unfortunately the use of these terms is inconsistent). We use the term LT in the ensuing part of this chapter. The trend toward more fiber and shorter twisted pairs will continue with time and VDSL. Splitter circuits can be used at both ends to protect and preserve the existing analog POTS service. Additionally, the high speeds of VDSL allow multiple digital voice signals to also be carried to the customer. Within the customer premises (home or business, home is shown), a gateway is used to demultiplex the various VDSL signals and route them to the appropriate applications device, which could be a phone, computer, or television/entertainment system.

Within the central office, another demultiplexer/multiplexer can be used to extract application signals and route them appropriately. Figure 7.1 presumes a heavy use of Internet delivery, as well as Ethernet distribution within the home, but other mechanisms for such multiplexing are possible and discussed. In particular, wireless local area networks (LANs) [8] and/or home-phone distribution systems [9] have been considered.

Telephone lines often are designed to have a few intermediate points to which fiber or connection of any device is simpler than any other position along the line. Since the early 1970s, phone companies have averted deploying phone lines of length greater than 2.4 miles, a length often called a "carrier service area" or CSA. Early versions of such deployment often used what is called digital loop carrier (DLC) or subscriber loop carrier (SLC). SLCs were originally served with T1 links, and later HDSL, where fiber is today.[5] DLCs or SLCs multiplex several voice signals onto a single twisted pair to the RT before digital signals reconvert to analog and traverse up to the last 2.4 miles as analog POTS. Such RT points are strained in terms of the physical space available for new equipment, thermal temperatures allowed (often having no fans or air conditioning, unlike a telephone company central office[CO]), and available power to energize DSL modems. Thus, as ADSL migrates into VDSL, DSL modems need to be smaller and consume less power per line, constraints that have been facilitated by continued advances in VLSI technology since VDSL's inception. Wider bandwidths also increase radiation levels from twisted pair, mandating lower power spectral densities at higher frequencies than used in earlier DSLs. The CSA interface point typically allows speeds of up to 6 Mbps/640 kbps in either ADSL or VDSL to be deployed to all customers, a factor of four (or more) increase with respect to most ADSL rollouts of 2001, which target phone line lengths of up to four miles (the latter thus covering theoretically about 90 percent of the existing network).

Another point of service potentially is the so-called "distribution point" (or sometimes carrier-service interface [CSI]), which typically is within 3000 feet of the customer. This point is where larger cables are terminated and smaller cables servicing up to a few hundred customers begin. Usually the box at the CSI basically serves as a cross-connection point for twisted pairs. However, the entire distribution-point box can be replaced if fiber feeds this CSI point. VDSL modems placed in such an enclosure then energize the subscriber-side twisted pairs that emanate. Power and size constraints are at least as difficult at this point as at the CSA remote terminal, usually with only a small area (few square inches) and about 1 watt of available power per DSL customer.

Another intermediate point yet closer to individual customers is often called the "cabinet" or "pedestal." Usually only four to sixteen customers are served from the pedestal with individual twisted pairs emanating to these customers. The pedestal again is normally a cross-connection point for telephone lines, but fiber can be deployed to this physically accessible point, and a VDSL modem deployed there. Very high speeds are possible on the resulting phone lines of 100 meters or less, potentially 100s of Mbps or more, higher yet than current VDSL (see Section 7.5).

Placing fiber to each successive point is increasingly costly because the cost of the fiber per subscriber necessarily increases as the number of customers decreases. Considerable cable-trenching, physical labor, and placement of many ONUs may be necessary as the fiber deployment extends closer to the customer. However, in the future if the customer demands higher bandwidth, then potentially higher revenue is possible also to pay for the fiber-deployment costs. Ultimately, fiber might be run to the home or even into the home to the desk/TV-top. The key to VDSL is the incremental deployment if and where customers are willing to pay for more fiber. The cost of deploying fiber can be from $100,000 to $1 million per half mile in

---

[5]One might argue "where fiber is supposed to be today," as many DLCs still use T1s or HDSLs in the digital segment where fiber may eventually be placed. This process of fiber installation to the DLC is actually occurring slower than phone companies projected, and thus copper reuse, perhaps in some coordinated overall fashion to the DLC from the central office, is more prevalent than most telcos care to admit.

areas of reasonable customer density. The placement of an ONU can cost an additional $250,000. Fiber to the premises ultimately eliminates active electronics intermediate to customer and telco, a potential maintenance advantage that should be figured also.

Realistically, though some service providers hesitate in admission, fiber-fed terminals will see mixture of lines from 12,000 feet down to 100 feet, and VDSL and ADSL modems will be operating from the same remote terminal (in practice, even if not in theory).

HFT Concept

Alternatively, a concept that is analogous to HFC networks has often been promoted for VDSL. In hybrid-fiber twisted pair or hybrid fiber-coax, the fiber carries fully modulated analog DSL signals from the CO on different wavelengths, which are then demodulated and optical-to-electrical converted for final transmission on the phone line at the fiber/twisted-pair interface. This particular configuration could have considerably less power consumption and size required at the fiber/tp interface than the usual VDSL configuration in Figure 7.3. However, it may be wasteful of optical bandwidth—with the state-of-the-art wavelength-division-multiplexing technology today perhaps limiting the number of wavelengths on the fiber (with sufficient linearity for DSL) to less than 100. However, most of the digital complexity (signal processing and multiplexing at various levels) is then remoted to the CO where it can be more easily and cost-effectively implemented. This technique remains a research subject presently.

Unbundling Issue

Colocation of VDSL modems is yet more difficult when the VDSL modems are not in a CO. This is because sharing of space by different service providers at the cabinet, CSA, or distribution point is physically difficult (there is not enough

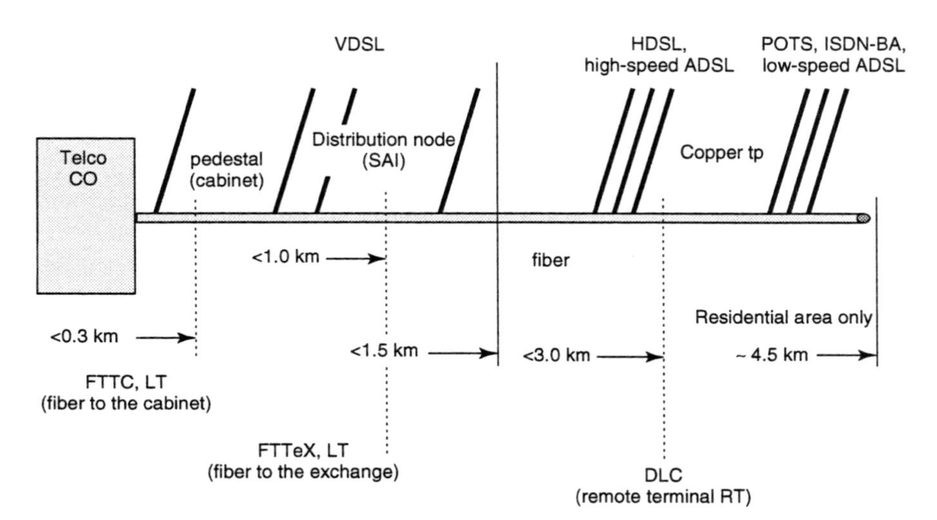

Figure 7.3 Illustration of intermediate points to which fiber (shaded line) may traverse.

space). Today this is a hotly debated issue in DSL deployment, and a single solution has not yet emanated. Some service providers accuse incumbent local exchange carriers of installing more fiber just to complicate collocation. Potential solutions for VDSL collocation are to:

1. Standardize the backplane interface and card size(s) of VDSL so that many service providers may plug into an ONU. (This is still problematic in that someone must go to the ONU and do it, which is costly and difficult in buried ONUs)
2. Provide separate fibers to the ONU for each service provider and divide the existing small space according to the fibers that enter.
3. Use the HFT concept and collocate at the central office where more space is available.
4. Use higher-level unbundling at layer two or three in the protocol stack (see Chapters 9 and 11).
5. A universal line card that can be sold/leased by any service provider.

Other solutions may evolve. VDSL standards to date have only encompassed collocation by mandating that a single-spectrum type shall be used in all VDSL transmission types to minimize crosstalk between VDSLs. Largely, current VDSL standards are just beginning to address the intricacies of the VDSL collocation issue.

POTS Splitters in VDSL

Splitter circuits for ADSL and VDSL are described in basic detail and design in [6], Chapter 3. For VDSL, the necessity of a splitter continues to receive attention. The rising use of splitterless ADSL suggests that perhaps splitterless VDSL is also advisable for compatibility and volume deployment reasons. The first splitterless VDSL proposals appeared in [11], [12]. Although these proposals in standards met with minority opposition (which is sufficient to block standardization), most advocates of the design in Section 7.3 are pursuing various forms of splitterless operation as an additional feature and option, albeit proprietary. The VDSL technology in Section 7.4 will not operate without splitters because of the consequent home-wiring effects.

In addition to simple separation of POTS and DSL signals, splitter circuits separate internal customer wiring from DSL, typically running new CAT five twisted-pair wiring on the DSL splitter port to the DSL customer modem (without bridged taps or other internal wiring issues), while the POTS signals remain on the existing wiring. This splitter can considerably simplify transmission problems because many customers have flat (untwisted) wiring, multiple taps, and other internal wiring deficiencies that degrade VDSL transmission without splitters. At the higher transmission rates of VDSL, all these internal effects become increasingly important. A splitterless DMT VDSL modem operates in the presence of these internal-wiring VDSL effects, albeit with degradation. Nonetheless, many internal networks (particularly those in service for small businesses, but also many residences) have good quality internal wiring. Furthermore, POTS signals may be carried far more economically via digital encapsulation in the VDSL service itself.

Literally, tens to hundreds of voice signals may augment high-speed data service for a customer with several telecommunications users. With such POTS delivered digitally by DSL, there is little need or use for an analog POTS line. In some high-end residences, if there is a power-failure with the internal wiring, analog POTS may still (for a single user) be returned to the phone lines automatically as long as the ONU remains capable of supporting (as always) a single line–powered analog POTS service on that same line. The potential for improved data rates, as well as easier installation, without splitters is the attraction of such operation. Such voice-over DSL has become increasingly popular, and VDSL is a logical extension of this popular nonanalog POTS DSL service. In this splitterless configuration, the VDSL customer modem becomes a gateway to both data and voice services within the premises, which may then have additional wiring for distribution.

VDSL transmission designs on a splitterless channel will need to be robust to bridged taps, increasing amounts of radio interference, potential crosstalk (on same or other lines) from home services already present on the phone lines, and further signal attenuation. Such modems may also have need for control of power-spectral density masks also to avoid excessive emission on the customer's premises that might interfere with local HAM operators, emergency radio, or other appliances.

This area is likely to be one of controversy in the future (see Section 7.1.4) because it does distinguish technologies in Sections 7.3 and 7.4 substantially. The home architectures that emerge from ADSL (see Chapter 3) will likely be those of choice for VDSL.

7.1.3 VDSL Spectrum Issues

As the highest speed DSL yet, VDSL uses the greatest amount of spectrum. Thus, it has the greatest concerns for spectrum compatibility. The issues of crosstalk and emissions from VDSL into surrounding telephone lines and radio receivers are more important and complicated in particular. Also, the crosstalk from existing services affects VDSL spectrum design and performance. Furthermore, increasingly popular home LANs (e.g., HomePNA) on twisted pair within customer premises will also complicate issues and trade-offs, as there is both spectrum overlap on the same line with VDSL as well as crosstalk issues into other VDSL lines from the home LANs. Considerable debate occurred for the design of VDSL spectrum, and there are correspondingly three internationally approved spectrum plans (presumably one selected for any specific geographic region). Unfortunately, the competitive interests between different service providers and the competitive interests between different transmission techniques have not worked to VDSL's best spectrum advantage so far, as significant compromise has occurred, sometimes for nontechnical reasons or rationale. However, this area may be reopened as VDSL spectrum-management standardization continues and as advanced transmission enhancements alter basic parameters and issues. Fortunately the three options do provide considerable flexibility for the future, as issues are revisited by technologists, marketing persons, and national regulators. This uncertainty makes this section at this time a bit difficult to write, but this book focuses on technical issues and describes the three plans, illustrating the various trade-offs. In the long-term massive deployment of DSL, the service providers and equipment/chip vendors who best comprehend all the aspects of this area will be able to garner the best business advantages in massive DSL deployment. These spectral options appear in Figures 7.4(a), (b), and (c), and are discussed in the next subsection.

Spectral Plans

The need for a fixed spectrum plan is only necessary for compatibility of the SCM plans of Section 7.4, whereas the digital duplexing of the DMT spectrum allows arbitrary placement of band edges without spectral roll-off penalty (although a 7.8 percent cyclic prefix penalty is necessary, see Section 7.3). It is possible that the plans in Figures 7.4(a) and 7.4(b) will be replaced by those that fully consider all aspects of applications and deregulation in the future. See Chapter 11 for more on spectrum management and the future of DSL. The third international spectral plan in Figure 7.4(c) encompasses the possibility of spectrum flexibility. This option is accommodated only in the DMT VDSL standard of [17]. (It originally appeared of interest only in Sweden, but at the time of this writing appears it will be selected in several Asian VDSL deployments.)

Robustness

VDSL must be able to accommodate frequency-selective disturbances, the most well known of which are bridged taps of different lengths. In this chapter, VDSL performance in the presence of bridged taps and other frequency-selective disturbances is evaluated in terms of robustness. A robust system on a line with a bridged tap shares the unavoidable degradation equitably between the downstream

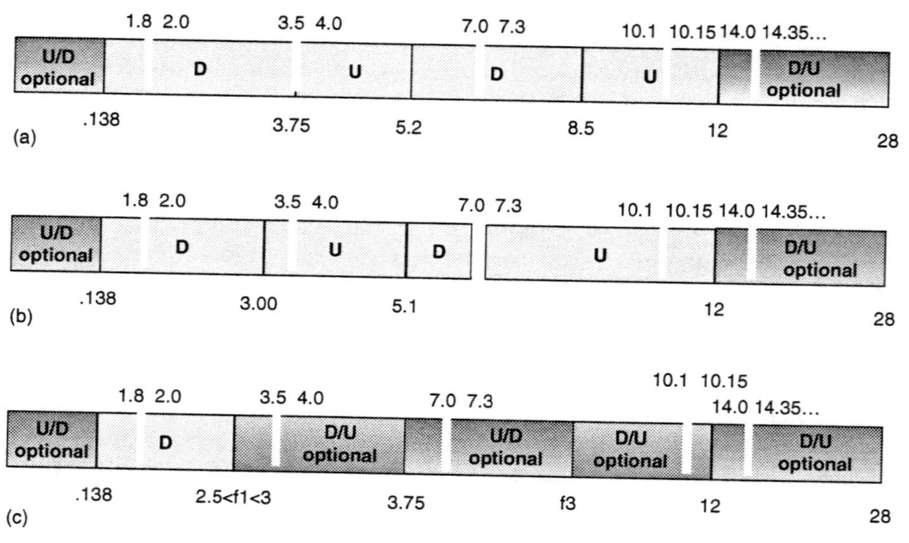

Figure 7.4 **(a)** Plan 998—North American, United Kingdom, and Japan VDSL spectrum. (Additional radio bands notched when used at 18.068–18.168, 21–21.45, 24.89–24.99 MHz). **(b)** Plan 997—VDSL Spectrum for some European countries. (Same unshown notched bands as Figure 7.4(a). **(c)** International flexible VDSL Spectrum Plan (f1, f2, f3, f4 determined programmably).

and upstream directions. A robust system on a line subject to mobile radio interference adapts to cope with changes in the noise profile.

Bridged Tap Robustness. Bridged taps occur in the loop plants of all operators (including countries where the operator claims to have no bridged taps) and in particular are extremely pronounced in occurrence when splitterless designs are used. Although it is impossible to avoid the effects of bridged taps completely, it is highly desirable for VDSL performance to degrade somewhat gracefully on lines with bridged taps. For a symmetric service, graceful degradation occurs if the ratio of upstream rate to downstream rate remains close to one, even if total sum data rate up and down decreases slightly. In this symmetric case, huge rate loss in one of the directions because of bridged taps is highly undesirable. In asymmetric transmission, it is desirable to maintain the ratio of asymmetry under different bridged-tap configurations.

Figure 7.5 illustrates the adverse effects of bridged taps on transmission performance. The figure shows the transfer function (in dB) of a 4050-foot loop, with bridged taps (66, 56, 46, and 36 feet long) and without bridged taps. The bridged taps cause the transfer function to exhibit notches periodically in frequency. As the bridged taps get shorter, the notches become deeper, and they move to higher frequencies. The existence of such notches (10–20 dB deep) can seriously harm transmission.

Below the graph, two different four-band frequency plans are drawn. Note that this specific loop has very large attenuation at frequencies above 7 MHz, so the spectrum above 7 MHz is unsuitable for data transmission. Therefore, only the lower two bands would actually be used. Each frequency plan copes differently with this kind of disturbance. If plan A were used in the presence of a bridged tap 36 to 66 feet long, then upstream transmission performance would be degraded significantly, although the downstream direction would be minimally affected. If plan B were used, then the downstream transmission performance would instead be degraded. Both plans fail to be robust.

One might argue that some other fixed four-band plan would actually show more immunity to such situations. However, the bridged-tap length is unknown; it may vary from 10 to more than 100 feet. Thus, the notches may appear in almost any frequency of the VDSL spectrum. For any four-band plan spanning the VDSL frequency range, there will always be a bridged tap with such a length that performance will be degraded in only one direction. The greatest loss for such unidirectional loss with asymmetric transmission often occurs in the upstream direction because the lower upstream bandwidth loses a greater fraction of its data-carrying capability to a notch when it occurs solely in an upstream band. However, for any desired ratio of downstream and upstream data rates, there is one optimal frequency-division duplexing scheme, which one can prove attains the maximum possible robustness. The solution is to partition the spectrum into infinitesimally small bands and alternatively assign them to upstream and downstream transmission. Then any frequency-selective disturbance (such as a bridged tap) will have an equal impact on both directions of transmission. Figure 7.6 shows such a frequency plan and illustrates why symmetric service is maintained.

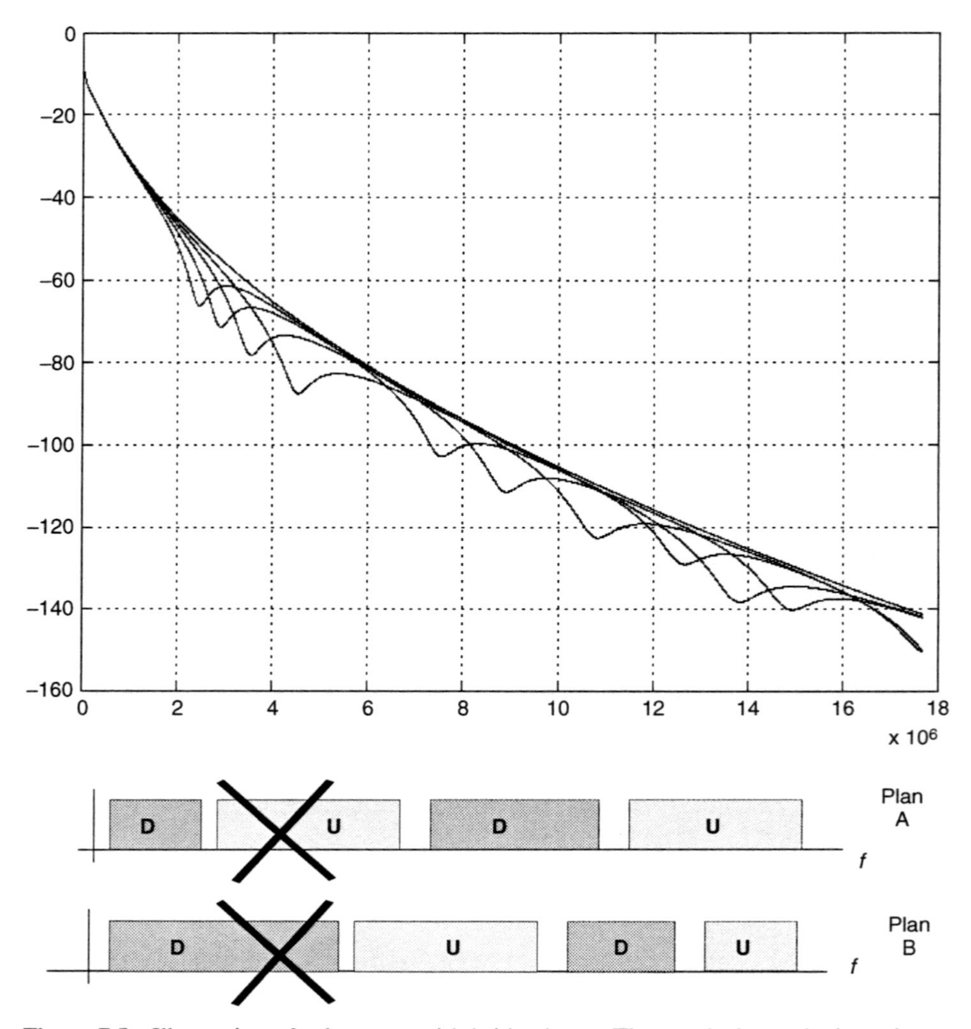

Figure 7.5 Illustration of robustness with bridged taps. The graph shows the insertion loss (in dB) of a 4050′ loop with bridged-taps of length 66′, 56′, 46′, and 36′ (20 m, 17 m, 14 m, and 11 m respectively). Below the graph, two different frequency plans are shown. When plan A is used, only upstream transmission is affected. When plan B is used, only downstream transmission is affected. In both cases, symmetric service is disabled.

The implementation of this optimal scheme may prove too complex,[6] so suboptimal schemes with somewhat lower but still adequate robustness may have to be used instead. By interpolating between the four-band plan and the optimal plan, we deduce that a number of bands as large as possible is highly desirable. As the number of bands increases, the data rate loss caused by a bridged tap will be dis-

[6]Recent demonstrations of full "zippering" have been able to suggest that, at least in some situations, large numbers of alternating up/down bands are indeed feasible with acceptable implementation.

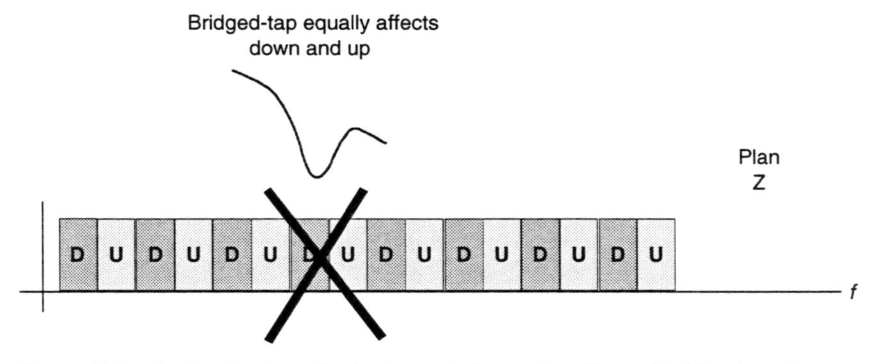

Figure 7.6 Optimal robust duplexing solution—the effect of bridged tap is shared between upstream and downstream.

tributed more evenly between the two directions of transmission. The simulations to follow demonstrate this fact.

Simulations

The simulation results that are shown below were obtained using a popular telco simulation tool. The four different frequency plans that were evaluated are shown below (numbers refer to MHz):

998 band plan A in g.993.1

Up = (3.75–5.2, 8.5–12)

Down = (0.138–3.75, 5.2–8.5)

997 band plan B in g.993.1

Up = (3.25–5.1 , 7.1–12)

Down = (.138–3.25, 5.1–7.1)

Digital duplexing 5-band plan

Up = (0.03–0.138, 3.08–4.78, 10.242–17.66)

Down = complement of up

Digital duplexing 7-band plan

Up = (0.03–0.138, 2.5–3.5, 4.5–5.5, 11–17.66)

Down = complement of up

Digital duplexing 15-band plan

Up = (0.03–0.138, 2.1–2.5, 2.75–3, 3.25–3.5, 4–4.25, 4.5–4.75, 5–5.5, 10.5–17.66)

Down = complement of up

Bandplan C in g.993.1 is generally intended to be variable in frequency cut-offs and allows for the needed flexibility in future VDSL systems. See, for instance,

Chapter 10 or Section 7.5. An early option has up = (0.03–.138, 2.5–3.75, Fx-12) and down = (.138 to 2.5, 3.75-FX, 12 and up). This is effectively a six-band plan. Presently Bandplan C has an ambiguity as to the upstream/downstream cutoff at 2.5 as to whether it is 2.5 or 3. The author interprets this to mean that this frequency can only be chosen between 2.5 and 3 MHz.

The services evaluated were for both the noise A and noise D environment of [13]. For various band plans, the reach in meters was computed both with and without bridged taps. Table 7.1 shows the resulting data rates for a 4500-foot loop.

We immediately see that using a larger number of bands always improves upstream data rate. A four–band plan has upstream data rate annihilated with the 998 or 997 plans, a particularly concerning problem for those desiring symmetric service. It is worth noting that the reach of the extra long symmetric service is improved by more than 35 percent (300–500 meters) when more than seven bands are used. This service represents an important market segment, and might be the first VDSL service to be deployed. A more detailed set of results appears in [14]. In particular, four lines to one customer could offer 10 Mbps symmetric "10BT" Ethernet service if seven or more bands were used. Such a service, plus downstream video, might be desired by a small business (see Section 7.5).

Mobile Radio (Ingress) Noise Robustness

Mobile radio ingress noise robustness is important also in VDSL. The area is sensitive because ingress transmissions can include those used in defense of various countries. Roughly though, individuals without security clearances in various countries can learn of two types of disturbances:

1. Several narrowband analog voice signals in a band of a few hundred kilohertz that can be anywhere between 1 and 20 MHz and can move (or hop) in their general band location
2. Direct-sequence spread spectrum signals of 300–500 kHz bandwidth (spread voice or data) with center frequency that hops throughout the 1 to 20 MHz band

Even when frequency bands are known, non-linear/non-ideal channel/receiver elements can lead to RF in other bands through various sum and difference effects on single or multiple RF interferers.

One cannot specify in advance those bands to be annihilated by radio noise, and the joint operation of VDSL and of emergency and/or defense systems might become highly preferred, especially in emergency situations or disaster areas.

Again the solution is as in Figure 7.6 to be robust with VDSL duplexing. This section models a single data signal of width 500 kHz coupling into a phone line at a

Table 7.1 Bridged-tap Robustness Results for 4500 foot 26-Gauge Loop

ANSI Noise-A	998	997	5-band	7-band	15-band
Upstream	40 kbps	260 kbps	1.69 Mbps	2.68 Mbps	3.37 Mbps
Downstream	15.6 Mbps	15.4 Mbps	15.6 Mbps	14.7 Mbps	14.0 Mbps

level sufficient to cause loss of use of the same band on that line. This level can vary from –80 dBm/Hz to –110 dBm/Hz, depending on the length of line and other system parameters. In this case, the duplexing plans were again evaluated. The loop simulated is again a 1.1 km .4 mm loop, and Noises A and D [15] were used in addition to the radio noise, along with 20 VDSL FEXT. The worst-case position of the radio noise actually was the same for both plans, 4–4.5 MHz. The loss is again less for the plan with more sub-bands.

The seven-band plan again robustly achieves over 7 Mbps symmetrically in all cases for Noise A whereas the four-band plan is only 1.4 Mbps. The relative drop in performance for the four-band plan is also larger (even though absolute data rates are smaller because of analog duplexing. Although analog duplexing is a separate effect, it is fair to include in four-band plans because these plans were advocated by proponents of systems that must use analog duplexing, whereas the larger number of bands were advocated by proponents of systems that could implement the multiple bands without exess bandwidth. For Noise D, the data rates are 2.5 Mbps for the seven-band plan and only 575 kbps for the four-band. Again the relative as well as absolute loss is larger for four-bands because it is not robust to radio noise. A larger number of bands (more than eight) can actually increase the Noise A result to over 8 Mbps symmetric.

HomePNA

The Home Phoneline Networking Alliance (HomePNA) and in-progress standard draft G.989.1 (G.pnt) of the ITU [9] specify a use of the telephone line bandwidth between 5 and 10 MHz for internal home phoneline networking. HomePNA spectrum of course overlaps VDSL, leading to signal disruption of VDSL on the same line as well as generating large NEXT into neighboring VTU-Rs of customers other than the HomePNA user.

Despite the G.vdsl and G.pnt standards being developed within the same standards committee, the frequency bands for VDSL and G.pnt overlap. This course was justified by assuming the installation of unidirectional low-pass filters for every user of a HomePNA network (presuming that the customer takes the time to locate his network interface somewhere outside his home and then prop-

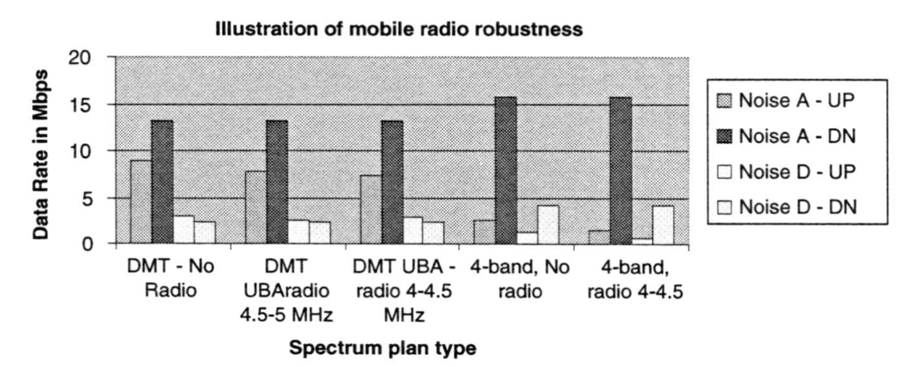

Figure 7.7 Illustration of mobile radio noise robustness for two plans with 1.1-km loop, 20 VDSL FEXT, noise A and D, .4 mm line.

erly installs the filter), even though that filter is not necessary for his internal computers to talk to one another. The authors of this book doubt the practicality of requiring all HomePNA users to install such filters before the very first VDSL could be installed in the same distribution area.

A second, more elegant solution, which does involve complexity increase for the VTU-R, appears in [23] where G.pnt signals could be canceled from a VDSL signal when on the same line or on neighboring lines as long as there were only one or two of significant amplitude. However, the method will not work for larger numbers.

Note this spectral incompatibility concern does NOT apply to Ethernet, which is typically installed on category 5 wiring which is separate and isolated by nature and design from the telco network. Thus, even though Ethernet uses the same band, there is no actual spectral overlap on physically collocated wires.

7.1.4 The Grand Debate

Dating to the days of ISDN and ADSL standardization, there has been a debate over the best transmission technology to use for DSL. Although ADSL standards have universally selected the specific multicarrier transmission method known as DMT after considerable deliberation and testing, a few CAP and QAM proponents nevertheless marketed nonstandard ADSL modems for a significant time period, before most switched to and supported standardized DMT. Subsequent debate was heated in the marketplace, and there were abortive attempts to reverse ADSL standards (from DMT to CAP/QAM). The supposed threat of fundamentally high complexity of DMT leading to high prices eventually was unequivocally refuted in the ADSL marketplace, where low-cost interoperable components abound today. Early proposals for VDSL recommended an extension to the ADSL DMT standard (see [2]). Later, other VDSL proposals argued for standardization of the QAM/CAP technologies. A protracted and complex debate emerged. Should the VDSL standard specify DMT or SCM? Should the standard specify both techniques or only one? Should a compromise such as filtered multitone (FMT) be adopted? The chairpersons of T1E1.4, ITU-Q4/15, IEEE 802.3, and ETSI TM6 coordinated their efforts with firm determination to converge on a standard specifying one technique. Despite the best efforts of the combined leadership of all three committees, a few key companies opposed the selection of any one technique. This denied the industry a timely standard that would have fostered interoperability and would have assured both vendors and carriers that the technology they pursued was not a dead-end technology. Because a definitive standard was not possible, T1E1.4 chose to the develop a trial-use VDSL "standard" that was essentially a temporary catalog of two leading VDSL proposals.

This newer debate has continued in VDSL for many years with two large industrial consortia emerging with two complete transmission specifications for VDSL:

1. VDSL Alliance—Discrete Multitone transmission (DMT) with digital or analog duplexing
2. VDSL Coalition— "Single" Carrier modulation (SCM) with analog duplexing

(The single is in quotes because the VDSL Coalition actually advocates a solution with two carriers in each direction, which might possibly be augmented an optional third carrier upstream below 138 kHz.) Both groups have contributed a draft trial-use standard to T1E1.4, which appear in three documents: A common reference document [15], an SCM document [16], and a DMT document [17]. The DMT specification supports up to 4096 4.3125 kHz ADSL-style tones, and any number of up/down stream transmission bands. The lower 256 tones are exactly the same as ADSL, facilitating backward compatibility. Both groups have about fifty companies in them, with about ten common members. The companies in both groups represent an enormous cross-section of the telecommunications world.

The VDSL Alliance solution is backward interoperable with ADSL and has taken greater time to develop into its present converged state [17], but offers some outstanding flexibility and performance features in a large number of possible configurations. The VDSL Coalition specification [16] is slightly simpler to understand (although more pages in reality) and to design to, but not interoperable with ADSL. The Coalition specification had the advantage of earlier availability of transmission components that were partially compliant with it. The T1E1.4 group will revisit the trial-use standard for permanency in 2003, when it is likely only one solution will survive; thus, the grand debate will likely continue for some time.

There are a number of differences in markets addressed, data rates and capabilities, and potential chip complexities (although experts on both sides are now agreeing even at VDSL speeds, transmission modems are becoming a negligible fraction of overall DSL implementation complexity in VLSI). This area is controversial, and so the reader is better left to evaluate the two approaches themselves rather than for these authors to here opine the eventual surviving method.

7.2 APPLICATIONS AND THEIR INTERFACES

Applications for VDSL are manifold, with the consequent relative impacts yet indeterminate. This section begins by some degree of analysis of different data-rate combinations and applications for VDSL in terms of the presently evolving views of broadband access. Generally, this section argues that speeds for DSL access will steadily increase from the current installed asymmetric 100s of kilobits to 1 Mbps-plus to 10s of Mbps of symmetric transmission speed as fiber reaches closer to the DSL customer.

The application types of audio, video, and data as usual will guide vision of the requisite speeds of future broadband DSL access here, followed by an example progression of DSL applications that could, for instance, motivate the speed combinations illustrated by the VDSL transmission methods discussed in Sections 7.3 and 7.4.

7.2.1 VDSL Applications and Speed Types

This section will individually consider the three application types of audio, video, and data/computing. Each area shows considerable need from a very realistic viewpoint for the increasing DSL speeds of VDSL. In combination, the different application types further suggest that within a reasonable time period, DSL will be choking for yet greater bandwidth, a need to which the DSL industry will need to respond.

Audio

"Audio" to telecommunications engineers often means a voice signal, although clearly today high-fidelity audio and music are of increasing interest. Good quality voice signals require 64 kbps (although some networks reduce that via compression to 32 kbps or 16 kbps). Small businesses and high-end residences are using increasing numbers of phone channels, multiplexed on a single line. The recent success of the ADSL area known as VoDSL (voice-over DSL) suggests that 16 voice channels on a single DSL link (1 Mbps) is highly desirable—indeed symmetric 1.5 Mbps service (24-channel T1) has long been a staple of new telco business. Clearly transmission of this or higher symmetric rates allows more economic service of a mutiline business. One could argue that symmetric data rates from 1 to 150 Mbps are clearly of immediate interest, but are limited in their use to places where those speeds can be implemented. Clearly, extension over phone lines to a greater number of termination points is of interest.

However, new audio applications also abound. The success of Napster and other similar applications admits the desirability of downloading CD-quality audio to a variety of home storage devices for immediate or delayed playback. The best high-quality compression methods today can achieve this quality with no less than 64 kbps. Thus, Table 7.2 suggests some transfer speeds for perhaps a teenager using DSL to download new music, both an audio single and an entire album. Clearly, 56 kbps is unacceptable, but even current ADSL speeds of today (1.5 Mbps) lead to annoying wait times. Speeds of 6 Mbps (asymmetrically) do lead to reasonable transfer times.

Data

Anyone familiar with e-mail attachments (often documents) or downloads of documents or software applications from the Web is aware that many files today can easily be 2 MB in size. That size has increased accordingly to Moore's law (even faster in the storage industry) by a factor of 2 every 2 years, meaning that in 2–5 years, file sizes of 10 MB might be common (and in fact can exist today, albeit rare in everyday use). Ideally, one would like to purchase a computer (perhaps a laptop) and connect it to a phone line and then have file transfer at the highest speeds that the connection of the computer DSL modem to the network DSL modem would allow. Some parts of the industry have already found that current

Table 7.2 Music Transfer Times with Various DSL Speeds

	56 kbps	1.5 Mbps	6 Mbps	26 Mbps	150 Mbps
Single	4.25 min	10 sec	2.5 sec	590 ms	100 ms
Album	1 hr, 11 min	2 min, 40 sec	40 sec	9.5 sec	1.6 sec

One song is 4 minutes of 64 kbps audio.

Table 7.3 10 MB Data-File Transfer Times with Various DSL Speeds

	1.5 Mbps	6 Mbps	26 Mbps	150 Mbps
10 MB	1 min	13.3 sec	3 sec	500 ms

DSL speeds are not yet sufficient for best e-mail/surfing file transfer. Such file transfers may be in either direction, especially for someone working at home who has extracted a file, altered it, and then wants to restore it to the corporate server with changes, suggesting symmetric transmission may also become desirable. Table 7.3 illustrates file transfer speeds for a 10 MB file.

Video (and Image)

A single photo image from a digital camera today can easily consist of 1 MB, whereas an entire 50-photo album would take then 50 MB. Table 7.4 lists DSL speeds and times for transferring such a photo or album. Again bidirectional transfer could be common, lending to increasing demand for symmetric DSL services.

For video, real-time video takes from 1.3 to 4 Mbps for a single good quality channel. HDTV (high-definition TV) requires 20 Mbps, but perhaps a more realistic DSL application would be the transfer of a DVD movie over the Internet to a residential display device, for subsequent viewing (perhaps with stops and starts while the viewers take leave of the viewing room for personal needs). A DVD two-hour movie requires about 2 GB to store, leading to the transfer times in Table 7.5.

Future HDTV movies could occupy as much as 130 GB. The tables clearly indicate a desirability for faster DSL speeds, even when real-time video (TV) is ignored.

Distributed Computing

As a final example, view the Internet as an example of distributed computing. The parallel bus transfer speed of a 1.5 GHz Pentium 4 processor is 48 Gbps, symmetric. Clearly in the limit, anything less slows the processor capability in a fully distributed environment. Thus, higher and higher broadband access speeds are motivated by the basic applications of which we know today. DSL broadband access speeds will become (if not already) the bandwidth limiter, and thus VDSL seeks to address this limitation.

For this reason, VDSL standards have outline a series of increasing speeds and symmetries as objectives of VDSL, as summarized in Table 7.6.

Table 7.4 Single Photo and Album Transfer Times with Various DSL Speeds

	1.5 Mbps	6 Mbps	26 Mbps	100 Mbps
Single Photo	5 sec	1.25 sec	300 ms	75 ms
Album	4 min	1 min	1.9 sec	450 ms

Table 7.5 DVD Movie Transfer Times with Various DSL Speeds

	1.5 Mbps	6 Mbps	26 Mbps	100 Mbps
DVD	4 hours	1 hour	14 min	3 min

Short-speeds have been tending toward 100 Mbps symmetric, whereas 10 Mbps symmetric on one or more coordinated lines is also of newer interest.

Clearly, not all of the speeds can be achieved at all the ranges in all environments, but Table 7.6 gives an idea of the types of applications, audio, data, and video above that might be enabled and over what lengths, basically consistent with Figures 7.2(a) and 7.2(b).

As an example, the most ambitious of ADSL plans today use fiber to bring all line lengths to less than 4 km, allowing a maximum ADSL speed of about 6 Mbps downstream. Thus, a possible progression of VDSL speeds as line lengths get shorter might follow Figure 7.8's progression of bandwidth use [18]:

The use of bandwidth from DC to 20 MHz would allow successively, 4 km ADSL service (up to 6 Mbps) at 4km, to be followed by an increase of the upstream data rate to an equivalent 6–8 Mbps by using spectrum from 1–5 MHz judiciously to create equal upstream and downstream data rates of 6–8 Mbps over 2 km or perhaps also to allow some increase in the downstream data rate. The band from 5.3 MHz to 10 MHz would be used to increase downstream data rates to 26 Mbps at a range of about 1 km, and the band above 10 MHz subsequently used to restore symmetry at 26 Mbps or higher on line lengths below 300 m where that upper bandwidth is useful. In Figure 7.8, the U/D areas mean that the band could be assigned to either direction. One can readily note that trade-offs in symmetry level, range, and data rate will easily produce other possibilities. The reality of VDSL application is that the exact spectrum use and data rates are still not carefully evaluated. In an attempt to define spectrum usage, a group of phone companies known as the Full Service Access Network (FSAN) tried to specify a reduced set of data rates and associated spectrum use, but the outcome was controversial with independent frequency plans in Europe and the United States as in Figures 7.4(a) and (b) depending on the belief in real-time TV as a DSL application (or lack thereof) and a third programmable international plan (Figure 7.4[c]) that may have been the only sanguine output from the study because it encompassed the reality that the application base was not yet well-enough understood to standardize. It appears at time of writing that efforts in EFM (see Section 7.5) and also new understandings of spectrum compatibility (see Chapter 11) may merit adjustment again to the bandplans. The U.S. group standardized on 22 Mbps down and 3 Mbps upstream to be supported over a target range of 1 km, and then specified an incompatible 6/6

Table 7.6 VDSL Speeds and Ranges

VDSL Class (Range)	Long (1.5 km)	Medium (1 km)	Short (< 300 m)
Downstream Speed	13 Mbps	26 Mbps	52 Mbps
Upstream Speed	1.5 Mbps	3-13 Mbps	6-52 Mbps

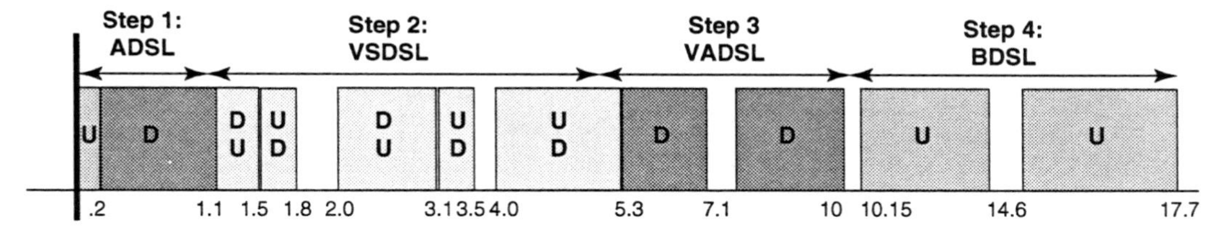

Figure 7.8 Possible VDSL rate progression and bandwidth use.

and 13/13 Mbps target objective at the same and shorter ranges—some Europeans preferred a more symmetrical set of data rates. The basic problem is that the exact driving economics and applications of VDSL were not yet known at the time of specification, thus the third programmable alternative that allows some degree of ability to change as necessary (see Section 7.3 on digital duplexing).

Regulators in all countries of the world have only just begin to investigate the spectrum use in future VDSL. Indeed it may be that the packet-level unbundling in Chapter 11 becomes the regulator choice (as indeed has happened already for one major operator in the United States [19]), opening the issue of spectrum use in VDSL once again. The overall conclusion here is that the only sanguine solution may be one that offers considerable flexibility to adjust to market and regulatory requirements as broadband DSL access develops. Section 7.3 and Chapter 11 further investigate such flexibility.

7.2.2 Common VDSL Reference Configurations

In an attempt to keep the two standardized transmission methods discussed in Section 7.1.4 somewhat in common application compliance, a common reference model was adopted and illustrated in Figure 7.9. The interfaces and functionality associated with the two interfaces are common to both VDSL transmission meth-

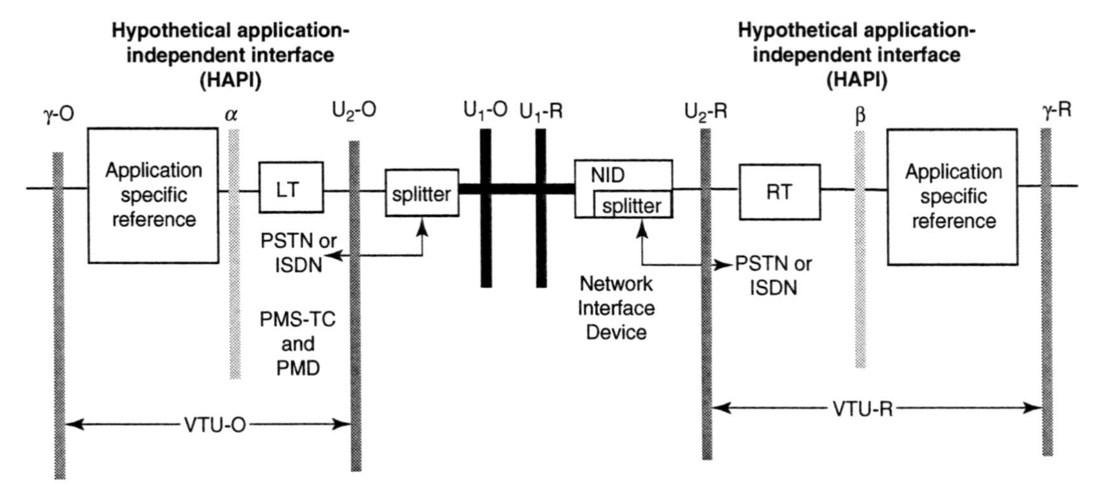

Figure 7.9 VDSL reference model from [15].

ods. The PMS-TC (physical medium-specific transmission convergence) and PMD (physical-medium dependent) interfaces are specified for each transmission method, while the spectra of the U interfaces and the splitter are again specified in common for the two transmission methods. Like ADSL, VDSL also specifies two paths, a slow or interleaved path and a fast or non-interleaved path as in Figures 7.10 and 7.11. The former undergoes interleaving as well as forward-error correction to allow for maximum impulse-noise protection while the latter allows for minimum delay (no more than 1 ms in VDSL). The application-specific reference is the DSLAM device that basically makes use of a subset of the functionality for a given PMS-TC interface, converting from the γ interface. The γ interface can be an ATM or STM interface, for instance at speeds well above those of an individual DSL modem and perhaps is the interface to the DSLAM itself. The application specific layer then extracts the pertinent bits (presumably set by the normal ATM method for setting permanent or virtual channels) for each of the fast and slow data paths through the VDSL modem, formats those bits into a known and reversible format within each stream, and then forwards fast and slow bits to the PMS-TC interface. Theoretically, these two bit streams could be visualized as the same for the two line codes; in reality, the DSLAM manufacturer and application-module manufacturer would know which of the two line codes it is using and then make the appropriate translations. Thus, even though both might interface to a common fiber interface (for instance OC-12 or Gig Ethernet), the ensuing DSLAMs will be quite different depending on the line code used.

Indeed the greatest challenge of VDSL, presuming a working modem, may be the high-speed extraction and identification of the individual application signals, likely sent through an ATM switching system. Later chapters deal with that process, but this section notes a few characteristics of interest for transmission.

One can envision the application devices as multiplexing and demultiplexing the applications signals discussed in Section 7.2.1, of which there may be many simultaneously present in both the fast and slow buffers, and formatting them for/from the modem itself. The high-bandwidth data channel created by VDSL may allow for numerous applications to simultaneously flow.

7.2.3 Operations

The VDSL standard Part I [15] has an elaborate list of operational and maintenance capabilities. As with previous DSLs, the main parameters of interest are the state of the modems, the likelihood or presence of errors on the link, and the synchronization of network functions. VDSL allows passage of the 8 kHz Network Timing Reference.

7.3 DMT PHYSICAL LAYER STANDARD

Discrete multitone transmission (DMT) is the only worldwide standard for ADSL, and VDSL transmission using DMT is the most efficient of the high-performance transmission methods that allow a transmission system to perform near the funda-

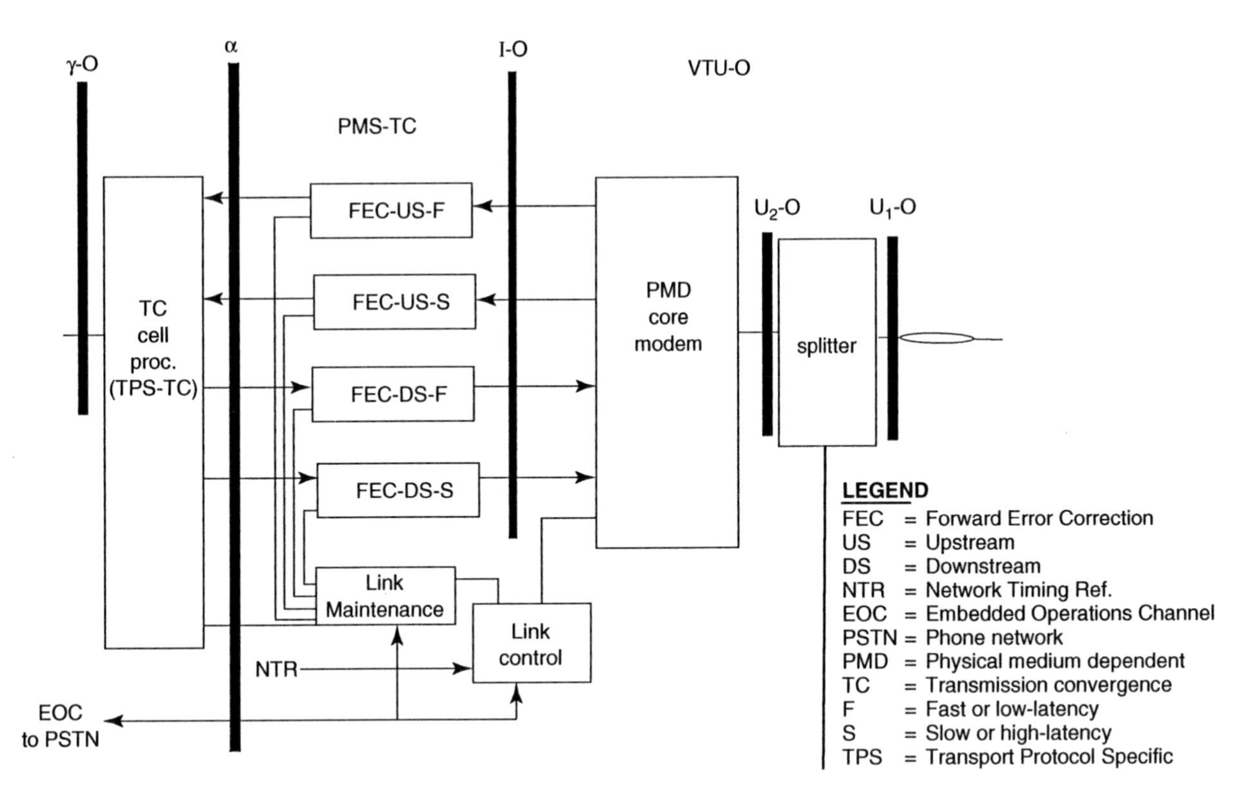

Figure 7.10 VTU-O reference model.

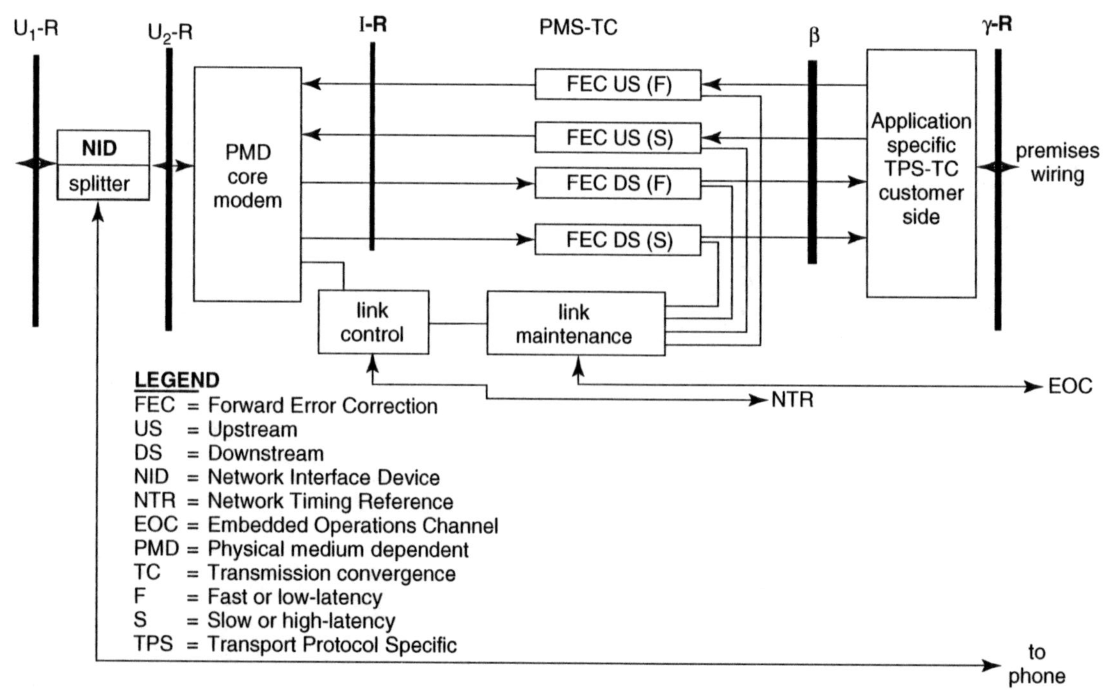

Figure 7.11 VTU-R reference model.

mental limit known as capacity [6]. For difficult transmission lines, there is no other cost-effective high-performance alternative presently, and VDSL has the most difficult transmission environment of all DSLs. Variants of DMT for wireless transmission (known as OFDM) have also come into strong use in the areas of wireless local area networks (IEEE 802.11(a), [8]), wireless broadband access (IEEE 802.16 [20]), as well as digital terrestrial television (HDTV and digital TV) and radio broadcast [21] in most of the world. All are known to also be particularly difficult transmission problems. The standardized VDSL DMT method is a natural extension of the method used for ADSL and backward compatible with it, as described in Section 7.3.1. Section 7.3.2 describes the "zipper" duplexing method that is also called "digital duplexing," which is an enhancement to the original DMT method that allows the upstream and downstream transmissions to be compactly placed in the limited transmission bandwidth of a telephone line. Section 7.3.4 investigates initialization.

7.3.1 Basic Multicarrier Concept

The basic multicarrier transmission concept of dividing a transmission band into a large number of subcarriers and adaptively allocating fractions of total energy and data rate to each to match an individual line characteristic is explained in [6]. It is this adaptive loading feature that sets multicarrier methods in a higher league of performance than other DSL transmission methods. VDSL presents a highly variable transmission environment with bandwidths that can vary from a few MHz to nearly 20 MHz, with intervening radio interference in several narrow bands, with huge spectrum notching effects from bridged taps, and with a variety of crosstalking situations. VDSL is undoubtedly the most difficult and highly variable transmission problem yet faced by DSL engineers. Multicarrier transmission is the most practical method to meet that challenge.

The first challenge for VDSL is interoperability with existing ADSL. One of the applications for VDSL is simply a speed extension of ADSL, meaning that it is possible for an existing ADSL customer to have that service provided by a new ONU in his neighborhood that is VDSL ready. A VDSL modem in the ONU that will interoperate with that existing ADSL modem is highly desirable so that no extra labor or purchases are necessary at the customer's premises should that customer elect to keep his current ADSL service for a period of time before electing to move to VDSL to increase the speeds of his ADSL service. Similarly, an existing ADSL customer may elect to purchase a VDSL modem (or may have one from a previous residence or business address) and then need to interoperate with an existing ADSL CO modem. Thus, a requirement for incremental DSL roll out to higher speeds and increasing use of fiber is that VDSL interoperate with existing ADSL, meaning the lower 256-down/32-up DMT tones of an ADSL modem, must also be implemented by an interoperating VDSL modem. For this reason, the DMT VDSL standard [17] uses the same tone spacing of 4.3125kHz that was used in ADSL. The VDSL standard allows for the DMT VDSL modem to use numbers of tones of 256, 512, 1024, 2048, and 4096 or, 2^{n+8}, $n = 0,1,2,3,4$ in either an asymmetrical or symmetrical allocation. Support of 2048 tones is considered a default for full compliance with other VDSL modems, but interoperation with smaller numbers of tones is illustrated in Figure 7.12.

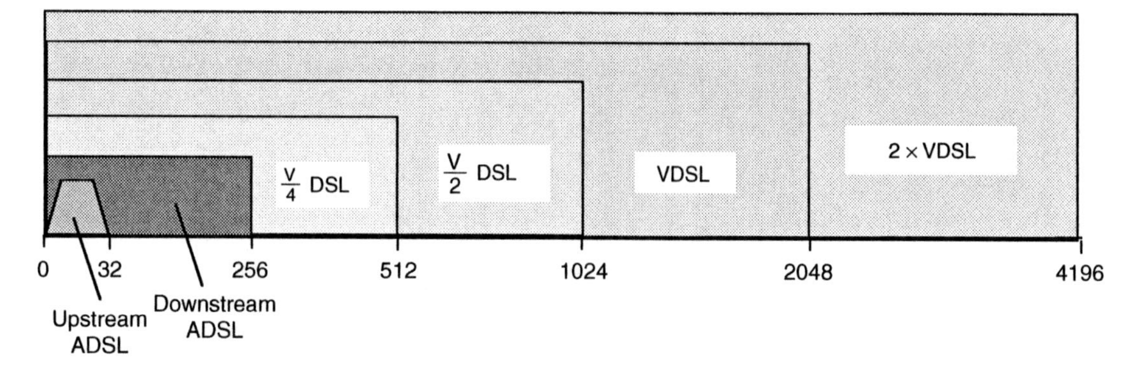

Figure 7.12 Interoperability diagram for standardized DMT modems of increasing speeds.

From Figure 7.12, one can determine upon close inspection that the modem (ONU or customer premises) with the smallest number of DMT tones then dictates the maximum that can be used (in either an upstream or downstream direction). Such elimination of superfluous tones can occur naturally during training, or may be selectively programmed during special initialization exchanges, with the latter usually the preferred implementation.

We now proceed to characterize some details of the DMT VDSL standard.

Cyclic Extension: The cyclic extension [6] size for DMT VDSL is optionally programmable to sizes $m \times 2^{n+1}$, where m is an integer. The modem must be able to implement at least the default of $20 \times 2^{n+1} = 40 \times 2^n$ for the case of $n = 0$, which corresponds to interoperability with ADSL.[7] Longer cyclic prefixes on shorter channels might allow some additional performance-enhancing features to be added for DMT VDSL that are not implemented in ADSL, which are described more in Section 7.3.2.

Encoder: The constellation encoder for DMT VDSL and tone-ordering procedures are identical to those of ADSL standards [6], although implemented perhaps over a larger set of tones for VDSL. There are two such ordering procedures inherited from ADSL.

Pilot: The pilot of ADSL has been made optional and generalized in VDSL. In early ADSL, the pilot was always sent downstream on tone number 64 (276 kHz). In VDSL, the VTU-R can decide to use a pilot on any (or no) tone in initialization. If a pilot is used, the 00 point in the standardized 4-point QAM constellation on that selected tone is sent in all symbols. The synchronization symbol of ADSL has been eliminated in VDSL, except when interoperating with an ADSL modem.

Timing Advance: The VTU-R is capable of changing the symbol boundary of the downstream DMT symbol it receives by a programmable amount, which is communicated during initialization to the VTU-O modem. This is also a new feature for VDSL that is used for implementation of digital duplexing as described in Section 7.3.2.

Power Back Off (PBO): PBO has been studied by standards groups as a way to prevent upstream FEXT from a customer closer to an ONU from acting as a large noise for a VDSL customer further away. The basic problem is that the closer user could operate at a higher data rate or with better performance than is necessary or fair to other customers. Various methods for reducing the disparity among lines vary from introducing a flat PBO at all frequencies as a function of measured received signal (see [16]) to spectrally shaped methods. Two of the shaping methods are:

a) Reference noise method—force all upstream transmissions to have an FEXT of common spectral shape (and thus harm)

[7]Actually, in this case, the cyclic prefix is reduced by 8 samples if the synchronization symbol of ADSL is to be inserted every 69 symbols or 17 ms.

b) Reference length method—force all upstream transmissions to have an FEXT the same as that of a nominal "reference" length VDSL line.

These two methods, particularly (b), seem to have won favor with standards groups, but the area is still debated at the time of writing. A method making use of actual line measurements in [24] was introduced for future spectrum management and essentially eliminates the PBO issue, but it came after VDSL standards had entered final voting and could not then be standardized. Chapter 11 shows several far superior methods in dynamic spectrum management (DSM).

Express Swapping: The standardized bit swapping of ADSL is mandatory also in VDSL [6]. However, VDSL also offers a highly robust and high-speed optional capability of altering the bit distribution all at once (instead of one tone at a time as in the older bit swapping). This is known as *express swapping*. The additional commands are described in the VDSL standards documents [17] and in Chapter 3. Express swapping allows the new bit table to be sent in one command and protected by a CRC check. If correctly received, all tones are replaced with the new bit distribution at an immediately succeeding point specified in the commands and protocol. This allows a system to react very quickly to abrupt transients caused by excitation of crosstalkers or RF interferers, or perhaps an off-hook line change in splitterless operation. It also can enable advanced spectrum management features that may occur in the future (see Chapter 11).

7.3.2 Digital Duplexing

This subsection describes *digital duplexing,* which is a method for minimizing bandwidth loss in separating downstream and upstream DMT VDSL transmissions. Originally, this method was introduced by Isaksson, Sjoberg, Nilsson, Mesdagh, and others in a series of papers that refer to the method as "Zipper" in [22] and [29–32]. General principles, as well as some simple examples, illustrate how digital duplexing works and why it saves precious bandwidth in VDSL. This description is intended for readers familiar with basic DMT, although a brief DMT review is included with emphasis on DMT features that are enhanced with digital duplexing. Thus, readers can use their DMT knowledge and the examples and explanations of this section to understand the relationship of the cyclic suffix to the cyclic prefix, and thus consequently to comprehend the benefits of *symbol-rate loop timing* and to appreciate the use of windows without intermodulation loss.

Excess bandwidth is a term used to quantify the additional dimensionality necessary to implement a practical transmission system. The excess-bandwidth concept is well understood in the theory of intersymbol interference where various transmit pulse shapes are indexed by their percentage excess bandwidth. In standardized and implemented DMT designs for ADSL, for instance, the symbol rate is 4000 Hz while the tone width is 4312.5 Hz, rendering the excess bandwidth

(.3125/4) = 7.8%. In ADSL, additional bandwidth is lost in the transition band between upstream and downstream signals when these signals are frequency-division duplexed. In VDSL, this additional bandwidth loss is zeroed through an innovation [22] known here as digital duplexing, which particularly involves the use of a "cyclic suffix" in addition to the well-known "cyclic prefix" of standardized DMT ADSL. This subsection begins with a review of basic DMT and of its enhancement through the use of the cyclic suffix, including a numerical example that illustrates symbol-rate loop timing. This discussion illustrates why the time-domain overhead is all that is necessary to allow full use of the entire bandwidth without frequency guard bands in a very attractive and practical implementation.

This section proceeds to investigate crosstalk issues both when other VDSL lines are synchronized and not synchronized. Windowing and its use to mitigate crosstalk into other DSLs or G.pnt are also discussed, as are conversion-device requirements.

This section then continues specifically to use a second example that compares a proposed analog duplexing plan for VDSL with the use of digital duplexing in a second proposal. In particular, 4.5 MHz of excess bandwidth is necessary in the analog duplexing whereas only the equivalent of 1.3 MHz is necessary with the more advanced digital duplexing. The difference in bandwidth loss of 3 MHz accounts for at least a 6–12 Mbps total data rate advantage for digital duplexing in the example, which provides a realistic illustration of the merit of digital duplexing.

Basic DMT

Figure 7.13 illustrates a basic DMT system for the case of baseband transmission in DSL [2], showing a transmitter, a receiver, and a channel with impulse response characterized by a phase delay Δ and a response length υ in sample periods. *N/2* tones are modulated by QAM-like two-dimensional input symbols (with appropriate N-tone conjugate symmetry in frequency) so that an N-point Inverse Fast Fourier Transform (IFFT) produces a corresponding real baseband time-domain output signal of N real samples.

For basic DMT, the last $L = \upsilon$ of these samples are repeated at the beginning of the symbol (packet of samples) of transmitted samples so that $N + L$ samples are transmitted, leading to a time-domain loss of transmission time that is ${}^{L}/_{N}$. This ratio is the excess bandwidth. The minimum size for $L = v$ is the channel impulse response duration (in sampling periods) for basic DMT (later for digitally duplexed DMT, L will be the total length for all prefix/suffix extensions and thus greater than v). Sometimes DMT systems use receiver equalizers [6] to reduce the channel impulse response length and thus decrease v. Such equalizers are common in ADSL. The objective is to have small excess bandwidth by decreasing the ratio ${}^{v}/_{N}$. In both DMT ADSL and VDSL, the excess bandwidth is 7.8 percent.

The IFFT of the DMT transmitter implements the equation

$$x_k = \frac{1}{N}\sum_{n=0}^{N-1} X_n \cdot e^{j\frac{2\pi}{N}nk}$$

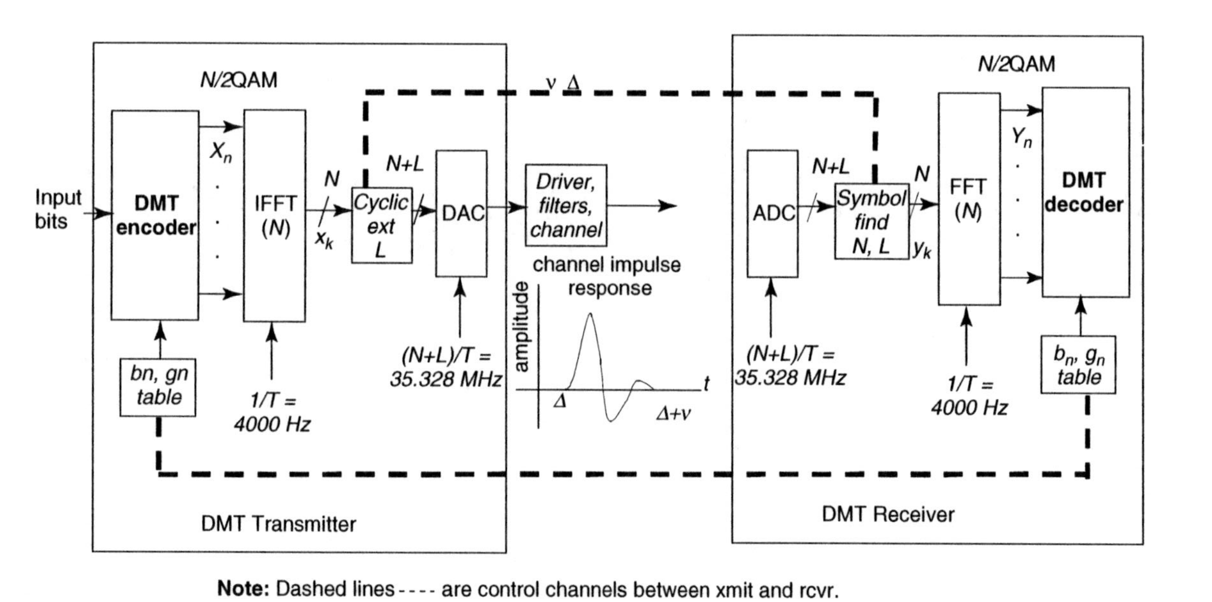

Note: Dashed lines - - - - are control channels between xmit and rcvr.

Figure 7.13 Basic DMT transmission system.

where x_k, $k = 0,\ldots,N-1$ are the N successive time-domain transmitter outputs (the prefix is a trivial repeat of last v). The values X_n are the two-dimensional modulated inputs that are derived from standard QAM constellations (with the number of bits carried on each "tone" adaptively determined by loading as in [6] and stored in the b_n, g_n tables at both ends). To produce a real time-domain output, $X_n = X^*_{N-n}$ in the ubiquitous case that N is an even number. In ADSL, $N = 512$, while in VDSL, $N = 512 \cdot 2^{n+1}$, $n = 0,1,2,3,4$. One symbol of $N + v$ samples is transmitted every T seconds for a sampling rate of $N + {}^{L}/_{T}$. With ${}^{1}/_{T} =$ 4000 Hz in ADSL and VDSL, the sampling rates are 2.208 MHz and up to 35.328 MHz, respectively, leading to a cyclic prefix length of $v = L = 40$ samples in ADSL and a cyclic-extension length of up to $L = 640$ samples in VDSL. The extension length in VDSL also includes the cyclic suffix to be later addressed.

When the cyclic extension length is equal to or greater than the impulse-response length of the channel ($L \geq v$), DMT decomposes the transmission path into a maximum of $N/2$ independent simple QAM-like transmission channels that are free of intersymbol interference and can be easily decoded. The receiver in Figure 7.13 extracts the last N of the $N + L$ samples in each symbol at the receiver (when no cyclic suffix is used) and forms the FFT according to the formula

$$Y_n = \sum_{k=0}^{N-1} y_k \cdot e^{-j\frac{2\pi}{N}nk}$$

where the reindexing of time is tacit and really means samples corresponding to times $k = L + 1 ,\ldots, L + N$ in the receiver. *The receiver must know the symbol alignment* and cannot execute the FFT at any arbitrary phase of the symbol clock. If the receiver were to be somehow offset in timing phase, then time-domain samples from another adjacent symbol would enter the FFT input, displacing some corresponding time-domain samples from the current symbol. For instance, suppose the FFT executed m sample times too late, then the output would be

$$\widetilde{Y}_n = \sum_{k=0}^{N-m-1} y_{k+m} \cdot e^{-j\frac{2\pi}{N}nk} + \sum_{k=N-m}^{N-1} u_{n-m+k} \cdot e^{-j\frac{2\pi}{N}nk}$$

where u_k are samples from the next symbol that are unwanted and act as a disturbance to this symbol. Furthermore, m samples from the current symbol were lost (and the rest offset in phase). Thus,

$$\widetilde{Y} = Y_n + E_n$$

where E_n is a distortion term that includes the combined effects of u_k, the missing terms y_k, and the timing offset in the symbol boundary. $E_n = 0$ only (in general) when the correct symbol alignment is used by the receiver FFT. DMT systems easily ensure proper phase alignment through the insertion of various training and synchronization patterns that allow extraction of correct symbol boundary.

It is important to note that the FFT of any other signal with the same N might also have such distortion unless the symbol boundaries of that signal and the DMT signal were coincident. In the later case of time coincidence, the FFT output is simply the sum of the two signals' independent FFTs. Indeed the receiver would have no way of distinguishing the two signals and would simply see them as the sum in the time-coincident case.

The second signal could be the oppositely directed signal leaking through the imperfect hybrid in VDSL. If the transmitted and received symbols are aligned in time and frequency-division duplexing is used, tones are zeroed in one direction if used in the other. Then, the sum at the FFT output is simply either the upstream or the downstream signal, depending on the duplexing choice for the set of indices n. No zeroed tones are necessary between upstream and downstream frequency bands as that is simply a waste of good undistorted DMT bandwidth. No analog filtering is necessary—the IFFT, cyclic-prefix, and the FFT do all the work *if* the system is fortunate enough to have the time coincidence of the two DMT signals traveling in opposite directions. The establishment of time coincidence of the symbols at both ends of the loop is the job of the cyclic suffix, which the next subsection addresses. In other words, the FFT works on any DMT signal of symbol size N samples, regardless of source or direction as long as the symbol is correctly positioned in time with respect to the FFT. This separation is not easily possible unless the DMT signals are aligned—thus VDSL DMT systems ensure this alignment through a cyclic suffix to be subsequently described.

Table 7.7 provides a comparison and summary of DMT use in ADSL and in VDSL. Note that DMT VDSL uses digital frequency-division multiplexing (FDM) and spans at most sixteen times the bandwidth of ADSL at its highest bandwidth use. This full bandwidth form is actually optional and the default values are shown in parenthesis on the right, with the default actually being exactly one-half the full.

Cyclic Suffix

The cyclic suffix is appended to the end of a DMT symbol (the opposite side of the cyclic prefix) and could repeat for instance the first 2Δ samples of the DMT symbol (not counting the prefix samples) at the end as in Figure 7.14, where Δ is

Table 7.7 Comparison of DMT for ADSL and VDSL.

	ADSL	VDSL (default)
FFT Size	512	8192 (4096)
# of tones	256	4096 (2048)
Cyclic ext length	$L = \nu = 40$	$L = 640 \geq \nu + \Delta$
Sampling rate	2.208 MHz	35.328 (17.664) MHz
Bandwidth	1.104 MHz	17.664 (8.832) MHz
Duplexing	Analog FDM	Digital FDM
Tone width	4.3125 kHz	4.3125 kHz
Excess bandwidth	86 kHz (prefix) +40 kHz (filters)	1.3 MHz (650 kHz) (no filters)

The 40kHz EB in ADSL is only if it's a true FDD system—echo cancellation doesn't have this penalty, even if the spectra are overlapped.

the phase delay in the channel (phase delay or absolute delay, not group delay, which is related to ν). A symbol is then of length $N + L$. The value for L must be sufficiently large that it is possible to align the DMT symbols, transmit and receive, at *both* ends of the transmission line. Clearly, alignment at one end of a loop-timed line[8] is relatively easy in that the line terminal (LT) need only wait until an upstream DMT symbol has been received before transmitting its downstream DMT symbols in alignment on subsequent boundaries. Such single-end alignment is often used by some designers to simplify various portions of an ADSL implementation. The alignment is not necessary unless digital duplexing is used. However, alignment at one end almost surely forces misalignment of the symbol boundary at the VTU-R as in Figure 7.14.

In Figure 7.14, let us suppose that $N = 10$, $\nu = 2$, and that the channel phase delay or overall delay is $\Delta = 3$. The reader can imagine a master time clock and that the LT begins transmitting a prefixed DMT symbol at time sample 0 of that master clock. The first two samples at times 0 and 1 are the cyclic prefix samples and are followed by 10 samples of the DMT symbol at times 2 through 11. See the vertical line at the left in Figure 7.14 that denotes the time axis (increasing time going down). In the second line from the left in Figure 7.14, the downstream (ds) symbol and prefix are shown on the scale of the timeline. At the receiver, all samples undergo an absolute delay of 3 samples (in addition to the dispersion of $\nu = 2$ samples about that average delay of 3). Thus, the cyclic prefix's first sample appears in the VTU-R at time 3 of the master clock, and the DMT symbol then exists from time 3 to time 14 as in the third vertical line from the left in Figure 7.14. The samples used by the receiver for the FFT are samples 5 through 14, whereas samples 3 and 4 are discarded because they also contain remnants from a previous DMT symbol. The LT has DMT symbol alignment in Figure 7.14, so that it also received the upstream prefix first sample at time 0 and continued to receive the corresponding samples of the upstream DMT symbol until time 11. Thus, the DMT symbols are aligned at the LT. However, in order for the upstream DMT symbol to arrive at this time, it had to begin at time –3 in the VTU-R (fourth vertical line from the left in Figure 7.14). Thus, the upstream symbol transmitted by the VTU-R occurs in the VTU-R at times -3 through 8 of the reader's master clock. Clearly the DMT symbols are not aligned at the VTU-R even though they are aligned at the VTU-O.

In Figure 7.15, a cyclic suffix of 6 samples is now appended to DMT symbols in both directions, making the total symbol length now 18 samples in duration (thus slowing the symbol rate or using excess bandwidth indirectly). The LT remains aligned in both directions and transmits the downstream DMT symbol from master-clock times 0–17. These samples arrive at the NT at times 3–20 of the master clock, with valid times for the receiver FFT now being 5–14, 6–15, ..., 11–20. Each of these windows of 10 successive receiver points carry the same information from the transmitter and differ at the FFT output by a trivial phase rotation on each tone that can easily be removed. The upstream symbol now transmits corre-

---

[8]Loop-timing of the sample clock means that the NT (VTU-R) uses the derived sample clock from the downstream signal as a source for the upstream sample clock in DMT.

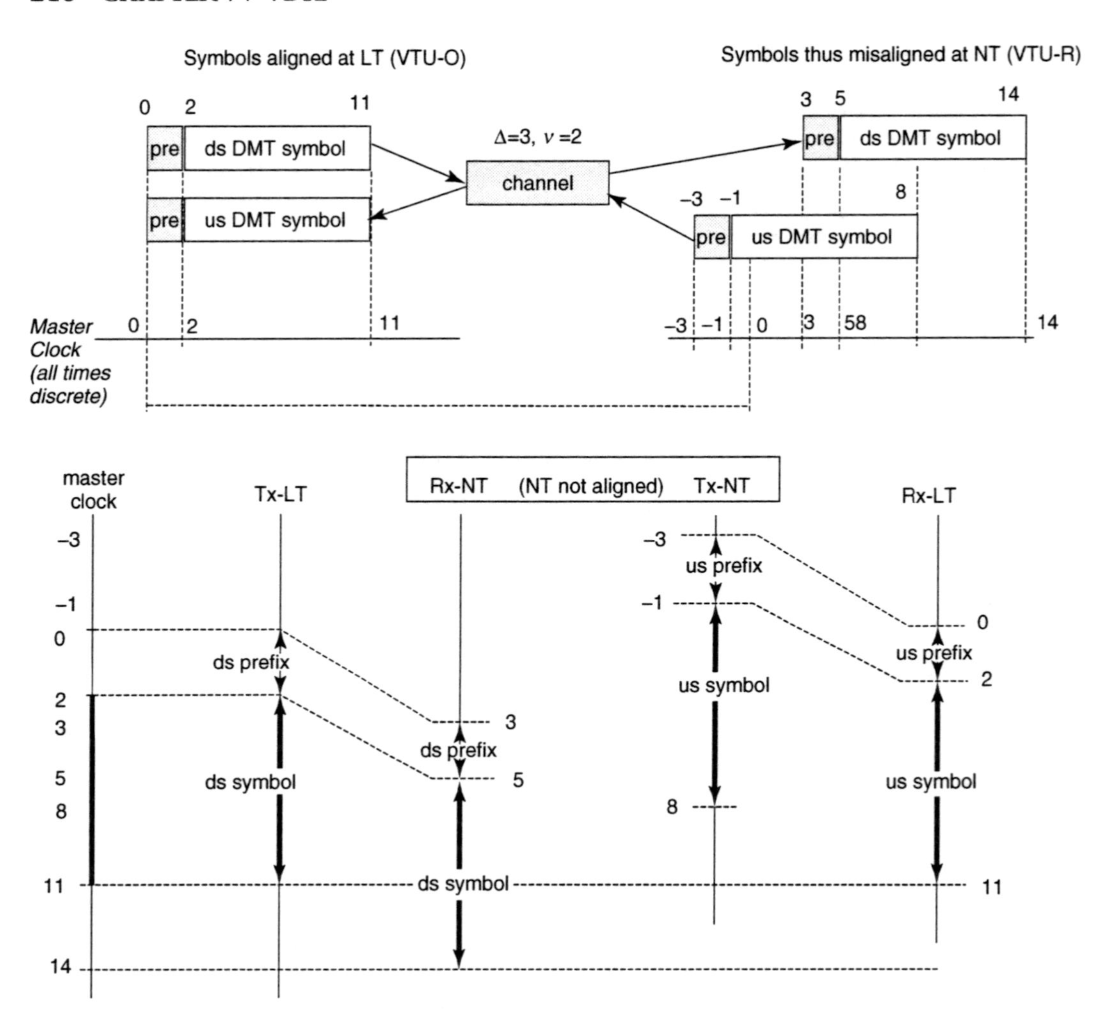

Figure 7.14 DMT symbol alignment without cyclic suffix.

sponding valid DMT symbols from time −1 to time 8, also 0–9, 1–10, and the last valid upstream symbol is time 5–14. At the VTU-R, the first downstream valid symbol boundary from samples 5–14 and the last upstream valid symbol boundary, also at samples 5–14, are coincident in time. Thus, the receiver's FFT can correctly find both LT and VTU-R transmit signals by executing at sample times 5–14 without distortion of or interference between tones. At this time only (for cyclic suffix length 6), the receiver FFT and IFFT can be executed in perfect alignment, and thus, downstream and upstream signals are perfectly separated. This DMT system in Figure 7.15 has *symbol-rate loop timing,* and, thus, a single FFT can be used at each end to extract both upstream and downstream signals without distortion. (There is still an IFFT also present for the opposite-direction transmitter.) This loop is now digitally duplexed.

Note the cyclic suffix has length double the channel delay (or equivalently was equal to the round-trip delay) in the example. More generally, one can see that from Figure 7.15 that VTU-R symbol alignment will occur when the equivalent of time 5, which is generally $\upsilon + \Delta$, is equal to the time $-3 + 2 + L_{suf}' = -\Delta + \upsilon + L_{suf}$. Equivalently, $L_{suf} = 2\Delta$. In fact, any

$$L_{suf} \geq 2\Delta$$

is sufficient with those cyclic suffix lengths that exceed 2Δ just allowing more valid choices for the FFT boundary in the VTU-R. For instance, if the designer had chosen a cyclic suffix length of 7 in our example, then valid receiver FFT intervals would have been both 5–14 and 6–15. This condition can be halved using the timing advance method in the next subsection.

Timing Advance at LT

Figure 7.16 shows a method to reduce the required cyclic suffix length by one-half to $L_{suf} \geq \Delta$. This method uses a *timing advance* in the LT modem where

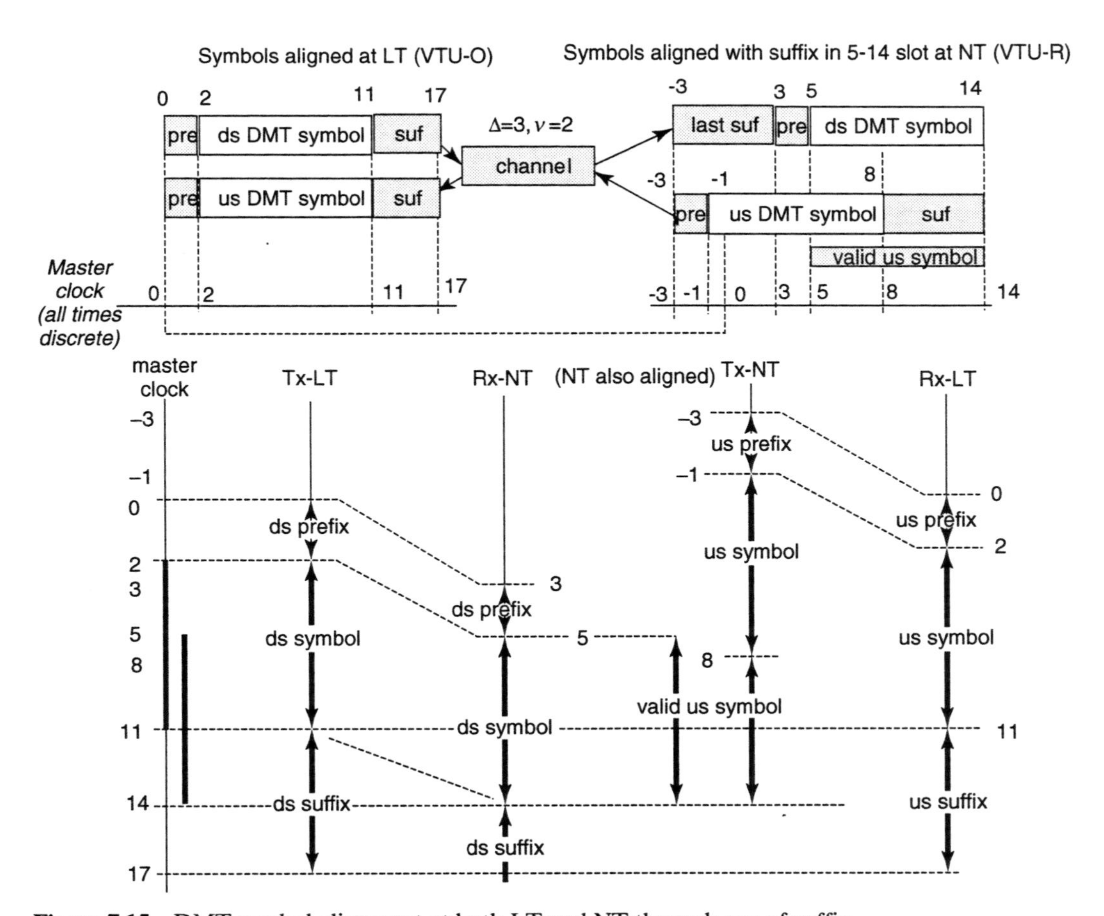

Figure 7.15 DMT symbol alignment at both LT and NT through use of suffix.

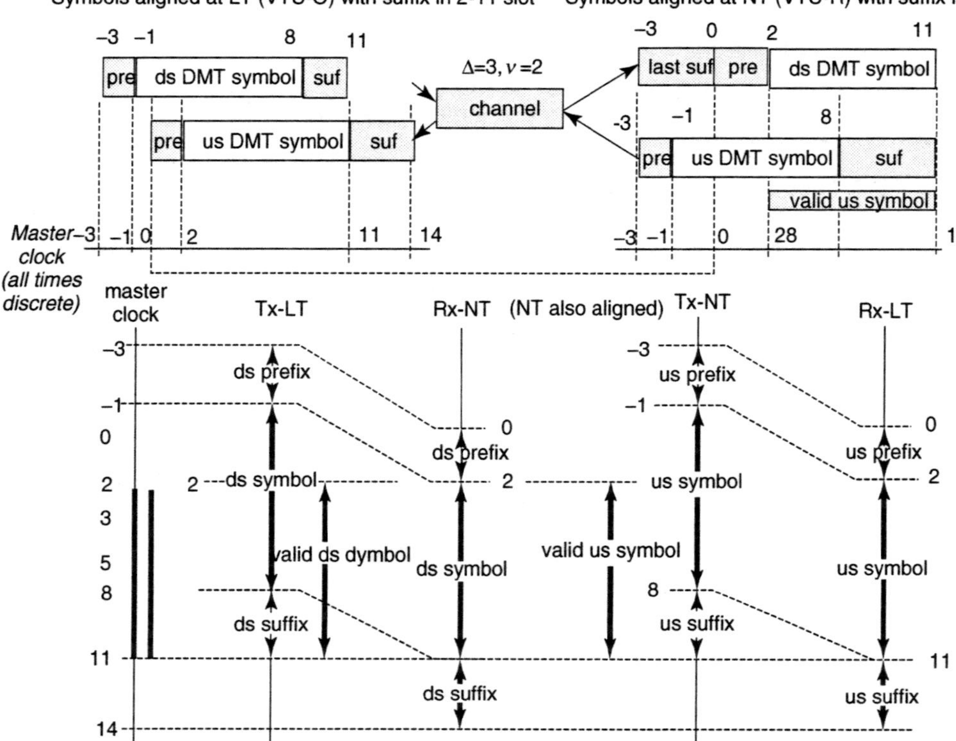

Figure 7.16 Illustration of cyclic suffix and LT timing advance.

downstream DMT symbols are advanced by Δ samples. The two symbols now align at both ends at time samples 2–11, and the cyclic suffix length is reduced to 3 in the specific example of Figure 7.16. Thus, for VDSL with timing advance, the total length of channel impulse-response length must be less than 18 μs (the same requirement as if there were no digital duplexing), which easily is achieved with significant extra samples for the suffix in practice. In VDSL, the proposal for L is 640 when $N = 8192$. The delays of even severe VDSL channels are almost always such that the phase delay Δ plus the channel impulse-response length ν are much less than the 18 μs cyclic extension length (if not, a time equalizer can be used, see [6]). One notes that the use of the timing advance causes the transmitters and receivers at both ends to all be operational at the same phase in the absolute time measured by the master clock.

Digital duplexing thus achieves complete isolation of downstream and upstream transmission with no frequency guard band—there is, however, the 7.8 percent cyclic extension penalty (which is equivalent to 1.3 MHz loss in bandwidth in full VDSL and 650 kHz loss in the default or "lite" VDSL). Thus, it is not correct, nor appropriate, to place frequency guard bands in studies of DMT performance in VDSL. Some papers on this subject have reached misleading conclusions due to the unnecessary inclusion of guard bands for DMT digital duplexing.

Digital duplexing in concept allows arbitrary assignment of upstream and downstream DMT tones, which with FDM VDSL means these two sets of upstream and downstream tones are mutually exclusive. On the same line, there is no interference or analog filtering necessary to separate the signals, again because the cyclic suffix (for which the equivalent of 1.3 MHz of bandwidth has been paid or 650 kHz in default VDSL) allows full separation even between adjacent tones in opposite directions via the receiver's FFT. However, while theoretically optimum, one need not "zipper" the spectrum in extremely narrow bands, and instead upstream and downstream bands consisting of many tones may be assigned as described earlier. Some alternation between up and down frequencies is universally agreed as necessary for reasons of spectrum management and robustness, although groups differ on the number of such alternations, which will depend on assumed loop topologies.

Crosstalk

Analysis of NEXT between neighboring VDSL circuits needs to consider two possibilities: synchronization of VDSL lines and asynchronous VDSL lines. The first case of synchronous crosstalk is trivial to analyze and implement with FDM. Adjacent lines have exactly the same sampling clock frequency (but not necessarily the same symbol boundaries). There is no NEXT from other synchronized VDSLs with FDM as will be clear shortly. The second case of asynchronous self-VDSL NEXT is more interesting. In this case, the sidelobes of the modulation pulse shapes for each tone are of interest.

A single DMT tone consists of the sinusoidal component

$$x_n(t) = \left[X_n \cdot e^{j\frac{2\pi}{N}\cdot\frac{N+v}{T}\cdot nt} + X_{-n} \cdot e^{-j\frac{2\pi}{N}\cdot\frac{N+v}{T}\cdot nt} \right] \cdot w_T(t)$$

$$= 2|X_n| \cdot cos(2\pi f_0 nt + \angle X_n) \cdot w_T(t),$$

where $f_0 = \frac{2\pi}{N} \cdot \frac{N + v}{T}$ or 4.3125 kHz in ADSL and VDSL, and $w_T(t)$ is a windowing function that is a rectangular window in ADSL but is more sophisticated and exploits digital-duplexing's extra cyclic suffix and cyclic prefix in VDSL. Figure 7.17 shows the relative spectrum of a single tone with respect to an adjacent tone for the rectangular window.

Note the notches in the crosstalker's spectrum at the DMT frequencies: all are integer multiples of f_0. Thus, the contribution of other VDSL NEXT will clearly be zero if all systems use the same clock for sampling, regardless of DMT symbol phase with respect to that clock. This is an inherent advantage of DMT systems with respect to themselves because it is entirely feasible that VDSL modems in the same ONU binder group could share the same clock and thus have no NEXT into one another at all. Indeed, this is a recommended option in [17]. When this occurs, performance enhancement occurs automatically.

When the sampling clocks are different, however, the more that sampling clocks of VDSL systems differ, the greater the deviation in frequency in Figure 7.17 from the nulls, allowing for a possibility of some NEXT. Studies of such NEXT for DMT digital duplexing are highly subjective and depend on assumptions of clock accuracy, number of crosstalkers with worst-case clock deviation, and the individual contribution to the NEXT transfer function of each of these corresponding worst-case crosstalkers. Nonetheless, reasonable implementation renders NEXT of little consequence between DMT systems.

If the VDSL PSD transmission level is S = -60 dBm/Hz, and the crosstalk coupling is approximated by $(m/49)^{.6} \cdot 10^{-13} \cdot f^{1.5}$ for m crosstalkers the crosstalk PSD level is

$$S_{xtalk}(f) = S \cdot (1/49)^{.6} \cdot f^{1.5} \cdot \left| \sin c\left(f \Big/ f_0 \right) \right|^2 .$$

The peak or sidelobes can be only 12 dB down with such rectangular windowing of the DMT signal, as in Figure 7.17. Asynchronous crosstalk may be such that especially with misaligned symbol boundaries, a really worst-case crosstalker could have its peak sidelobe aligned with the null of another tone (this is actually rare, but clearly represents a worst case). To confine this worst case, the following methods are used.

Windowing of Extra Suffix and Prefix

This section explains how windowing can be implemented without the consequence of intermodulation distortion when digital duplexing is used. Windowing in digitally duplexed DMT exploits the extra samples in the cyclic suffix and cyclic

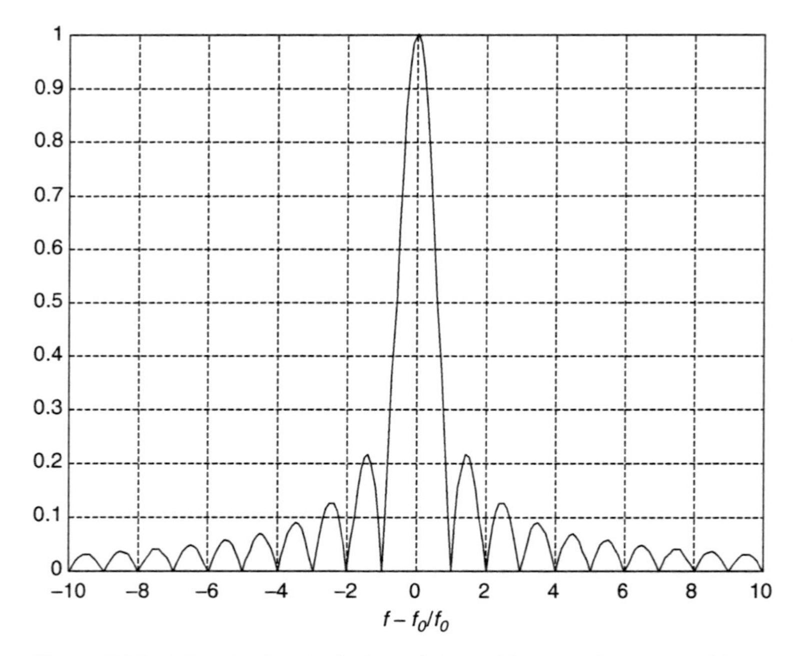

Figure 7.17 Magnitude of windowed sinusoid versus frequency. Note notches at DMT frequencies.

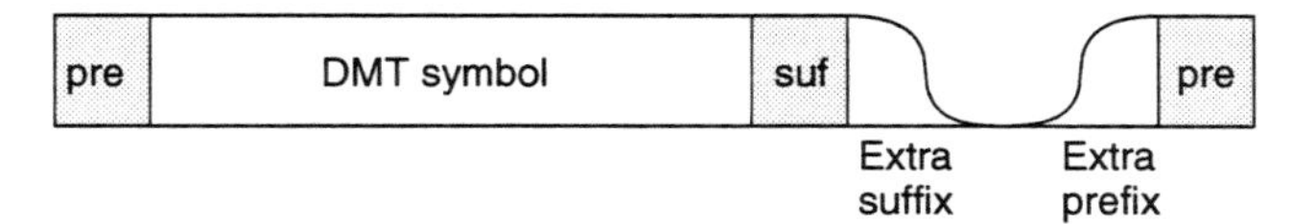

Figure 7.18 Illustration of windowing in extra suffix/prefix samples—smooth connection of blocks without affecting necessary properties for digital duplexing.

prefix beyond the minimum necessary. Because the cyclic extension is always fixed at $L = 640$ samples in VDSL, there are always many extra samples. Figure 7.18 shows the basic idea: the extra suffix samples are windowed as shown with the extra extension samples now being split between a suffix for the current block and a prefix for the next block. The two are smoothly connected by windowing, a simple operation of time-domain multiplication of each real sample by a real amplitude that is the window height. The smooth interconnection of the blocks allows for more rapid decay in the frequency domain of the PSD, which is good for crosstalk and other emissions purposes. A rectangular window will have the per tone (baseband) roll off function given by

$$W_T(f) = \frac{sin\left(\pi f / f_0\right)}{\left(\pi f / f_0\right)},$$

or the so-called "sinc" function in frequency. Clearly, a smoother window could produce a more rapid decay with frequency. A logical and good choice is the so-called raised-cosine window. Let us suppose that the extra cyclic suffix contains $2L' + 1$ samples in duration, an odd number.[9] Then the raised cosine function has the following time-domain window (letting the sampling period be T') with time zero being the first point in the extra cyclic suffix and the last sample being time $2L'T'$ and the center point thus $L'T'$:

$$w_{T,rcr}(t) = \frac{1}{2}\left\{1 + cos\left(\pi \cdot \left[\frac{1}{L'T'}\right]\right)\right\} \quad t = 0,\ldots,L'T',\ldots,2L'T'$$

One notes the window achieves values 1 at the boundaries and is zero on the middle sample and follows a sampled sinusoidal curve in between. The points before time $L'T'$ are part of the prefix of the current symbol, whereas the points after $L'T'$ are part of the suffix of the last symbol.

[9]If even, just pretend it is one less and allow for two samples to be valid duplexing end points.

The overall window (which is fixed at 1 in between) has Fourier transform (let $\alpha = \frac{L'}{L - L'}$, ignoring an insignificant phase term)

$$W_T(f) = \frac{L'}{\alpha} \cdot \frac{\sin\left(\pi f / f_0 \right)}{\left(\pi f / f_0 \right)} \cdot \frac{\cos\left(\alpha \pi f / f_0 \right)}{1 - \left(2\alpha \pi f / f_0 \right)^2}.$$

Larger α means faster roll-off with frequency. This function is improved with respect to the sinc function, especially a few tones away from an up/down boundary. Reasonable values of α corresponding to 100–200 samples will lead to even the peaks of the NEXT sidelobes below -140 dBm/Hz at about 200 kHz spacing below 5 MHz. The reduction becomes particularly pronounced just a few tones away, and so at maximum, a very small loss may occur with asynchronous crosstalk. Thus, signals other than 4.3125kHz DMT see more crosstalk, but within 200 kHz of a frequency edge, such NEXT is negligible. This observation is most important for studies of interference into home LAN signals like G.pnt, which at present almost certainly will not use 4.3125 kHz spaced DMT.

Overlapped Transmitter Windows. Figure 7.19 shows *overlapped windows* in the suffix region. The smoothing function is still evident and some symmetrical windows (i.e., square-root cosine) have constant average power over the window and the effective length of the window above L' can be doubled, leading to better sidelobe reduction. This overlapping requires an additional $2L'+1$ additions per symbol, a negligible increase in complexity.

Receiver Windowing. Receiver windowing can also be used to again filter the extra suffix and extra prefix region in the receiver, resulting in further reduction in sidelobes. Figure 7.20 shows the effect on VDSL NEXT for both the cases of a transmitter window and both a transmitter and receiver window. Note the combined windows has very low transmit spectrum (well below FEXT in a few tones) and below -140 dBm/Hz AWGN floor by forty tones. If it is desirable to further reduce VDSL NEXT to zero, an additional small complexity can be introduced as in the next subsection with the adaptive NEXT canceler.

Adaptive NEXT Canceler for Digital Duplexing. Figure 7.21 illustrates an adaptive NEXT canceler and its operation near the boundary of up and down frequencies in a digitally duplexed system. Figure 7.21 is the downstream re-

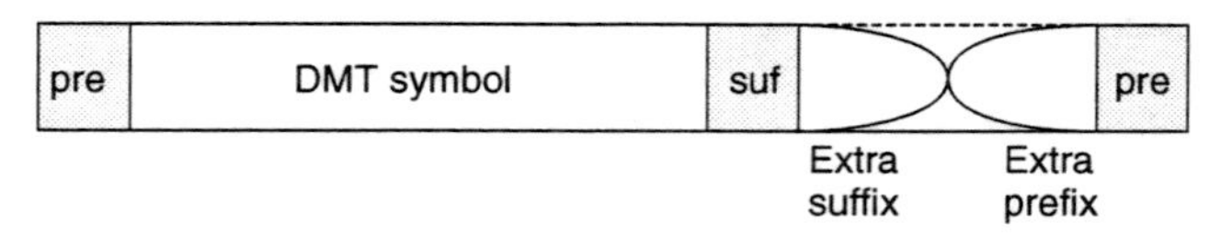

Figure 7.19 Illustration of overlapped windowing.

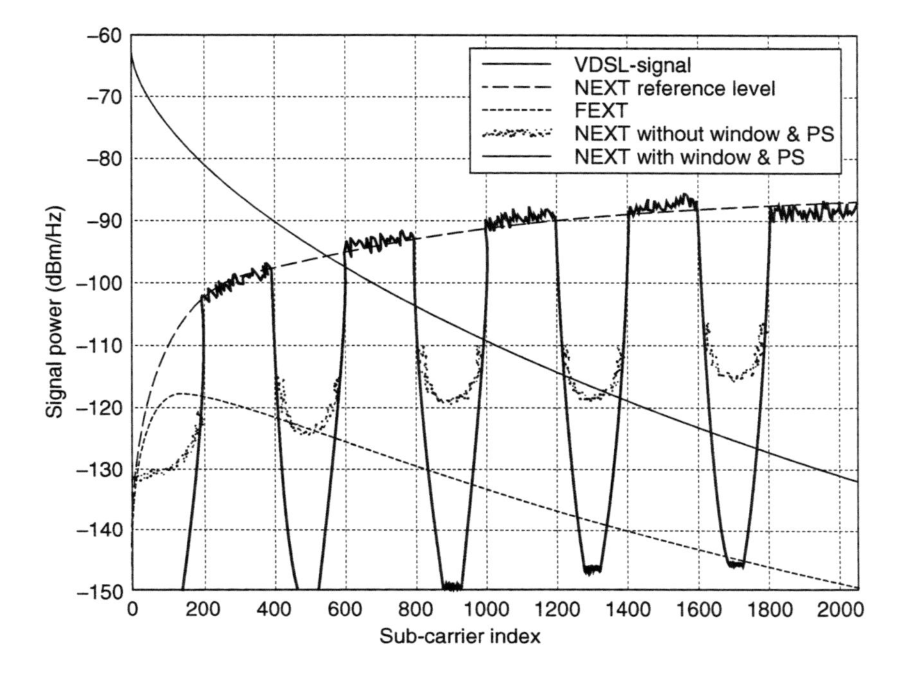

Figure 7.20 PS = transmitter windowing (pulse shaping), and window here means receiver window. This simulation is for a 1000 m loop of .5 mm transmission line (24 gauge).

ceiver, but a dual configuration exists for the upstream receiver. Note any small residual upstream VDSL NEXT left after windowing in the downstream tone n (or in tones less than n in frequency index) must be a function of the upstream signal extracted at frequencies $n + 1, n + 2, \ldots$ at the FFT output. This function is a function of the frequency offsets between all the NEXTs and the VDSL signal. This timing-clock offset is usually fixed but can drift with time slowly. An adaptive filter can eliminate the NEXT as per standard noise cancellation methods [6]. A very small number of tones are required for the canceler per up/down edge if transmit windowing and receiver windowing are used. Adaptive noise cancellation can be used to make VDSL self-NEXT negligible with respect to the -140 dBm/Hz noise level. This allows full benefit of any FEXT reduction methods that may also be also in effect (note the NEXT is already below the FEXT even without the NEXT canceler, but reducing it below the noise floor anticipates a VDSL system's potential ability to eliminate or dramatically reduce FEXT). See, for instance, Chapter 11. Note this noise canceller is an early version of the per-tone equalizer concept in [37].

7.3.3 DMT VDSL Framing

The DMT transmission format supports use of Reed-Solomon forward error correction (see [6]) and convolutional/triangular interleaving. The RS code is the same as that used for ADSL with up to 16 bytes of overhead allowed per code word.

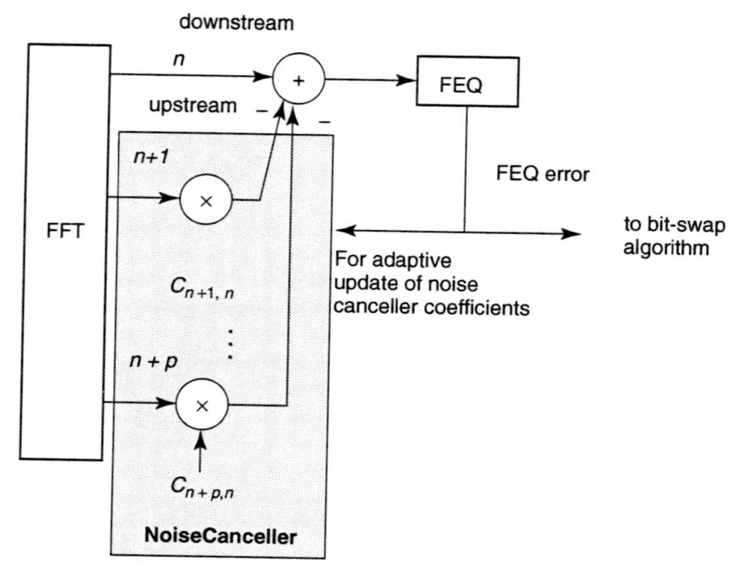

Figure 7.21 Adaptive noise canceler for elimination of VDSL self-NEXT in asynchronous VDSL operation. Shown for one upstream/downstream boundary tone (can be replicated for each up/down transition tone that has NEXT distortion with asynchronous VDSL NEXT). No canceler necessary if NEXT is synchronous.

There is no fixed relationship between symbol boundaries and code word boundaries, unlike ADSL.

Instead, any payload data rate that is an integer multiple of 64 kbps (implying an even integer multiple of payload bytes on average per symbol) with dummy byte insertion where necessary and as described in [17] is allowed. Triangular interleaving that allows interleaving at a block length that is any integer sub-multiple of a code word length (in bytes) is allowed (ADSL forced the block size of the interleaver to be equal to the code word length). Given the high speeds of VDSL, the loss caused by dummy insertion is small, compared with the implementation advantage of decoupling symbol length from code word length. Triangular interleaving was described in [6] and again in [17], so it is not described here.

Latency can take on any value between 1 ms (fast buffer requirement) and 10 ms (slow buffer default) or more. The latency is determined according to code word, data rate, and interleave-depth parameter choices as in [17]. Fast and slow data are combined according to a frame format that no longer includes the synchronization symbol of ADSL and has updates of the fast and slow control bytes with respect to ADSL. Superframes are no longer restricted to just sixty-nine symbols as in ADSL.

Framing can be implemented as in Chapter 9 of [17] and need not be repeated here.

7.3.4 Initialization

An earlier text describes in detail the various aspects of training of a DMT modem [6]. The VDSL training procedure is described in [17] and compatible with the popular g.hs (g. 994) of ITU. The fundamental steps of training are the same as in [6] with the LT now being expected to set a timing advance and measure round-trip delay of signals so that the digital-duplexing becomes automatic. The length of cyclic prefix versus suffix and other detailed framing parameters are set through various initialization exchanges.

One feature of digital duplexing is that it does allow very simple echo cancellation if there is band overlap. With synchronized symbols, there is only one tap per tone to do full echo cancellation where that may be appropriate. However, the NEXT generated by overlapping bands at high frequencies might discourage one from trying unless NEXT cancellation (coordinated transmitters and receivers) can also be used, which would only be one tap per tone per significant crosstalker.

The reader is otherwise referred to [17] for more details.

7.4 MULTIPLE-QAM APPROACHES AND STANDARDS

The VDSL system specified in [16] uses either CAP or QAM as a modulation scheme [6], and analog frequency division duplexing (FDD) to separate the upstream and downstream channels. There are two carriers or equivalently center frequencies, both with 20 percent excess bandwidth raised cosine transmission in each direction, following frequency plans 997 (Europe) or 998 (North America). The symbol rate of each of the signals is any integer multiple of 67.5 kHz, and the carrier/center frequencies can also be programmed as any integer multiple of 33.75 kHz. This allows for the receiver for each signal to estimate signal quality and request an appropriate center frequency and symbol rate, as well as corresponding signal constellation, which can be any integer QAM constellation from four points to 256 points as described in detail in [16]. Radio frequency emission control occurs through programmable notch filters in the transmitter, for which a decision feedback equalizer in the receiver can partially compensate.

With overhead included, data rates are certain integer multiples (not all) of 64 kbps up to 51.84 Mbps downstream and 25.92 Mbps upstream. SCM advocates sometimes cite "blind operation"—that is, no training, as an advantage, but of course one data rate must be fixed for all systems for true blind operation.

7.4.1 Profiling in SCM VDSL

To accommodate short (< 1 kft), medium (1000–3000 feet) and long (> 3000 feet) transmission at both asymmetric and symmetric rates, SCM VDSL can transmit up to four QAM signals, two in each direction. For long loops, only one carrier downstream and one carrier upstream (just above the downstream band) are permitted. For medium range, a second downstream carrier is permitted in addition to the two carriers of long loops, while the short loops can use all four carriers. It was basically this four-max carrier restriction of SCM that forced the number of bands in the 997 and 998 frequency plans. Using four bands is otherwise very suboptimal and unnecessary as mentioned earlier.

Figure 7.22 depicts the concepts of the profiles that are created by altering the center frequencies and symbol rates chosen. Notches at radio bands must be inserted to ensure emissions meet radio requirements; however, there are not a sufficient number of carriers to simply achieve this notching by reversing direction, although the 998 spectrum plan does leave one radio band as a reversal point near 4 MHz between upstream and downstream transmission.

As Figure 7.22 illustrates, larger symbol rates will likely be accompanied by larger center/carrier frequencies.

Decision feedback equalization (DFE) is presumed in the receiver and no Tomlinson precoding is used, even though FEC is used. The FEC in [16] and interleaving are basically identical to those used in the DMT standard. The DFE is described in the next subsection.

7.4.2 Operation of the DFE

The way a DFE handles RF interference (RFI) and notching is briefly discussed with reference to Figure 7.23.

The analog front end (AFE) of any VDSL system needs to do some analog processing to reduce very strong RFI to an acceptable level and avoid overloading of the receiver's A/D and other input circuitry. This issue is not discussed any further here. The feed-forward filter of the DFE creates a notch around the frequency at which the RF interferer is located, so that very little RFI is present at the output of this adaptive filter. The energy that is removed from the received signal by this notch is then restored by the feedback filter in such a way that the folded spectrum

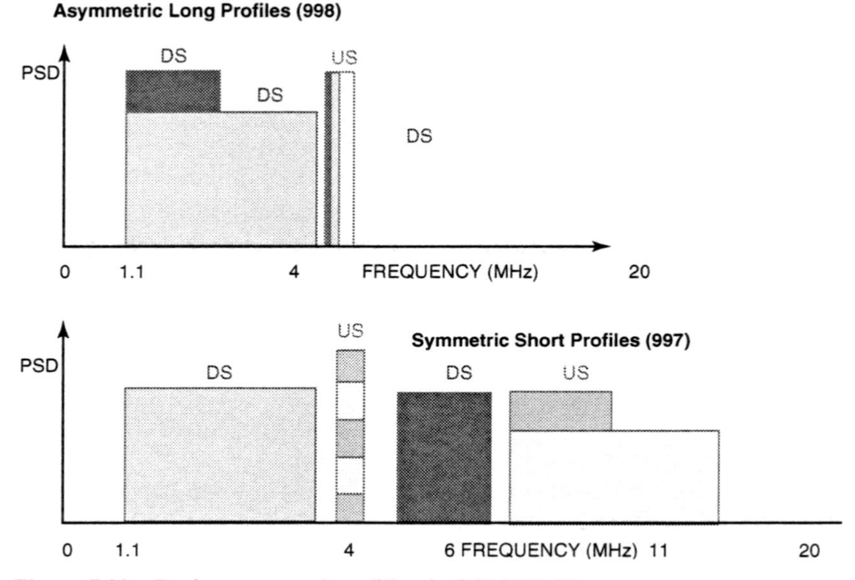

Figure 7.22 Basic concept of profiling in SCM VDSL.

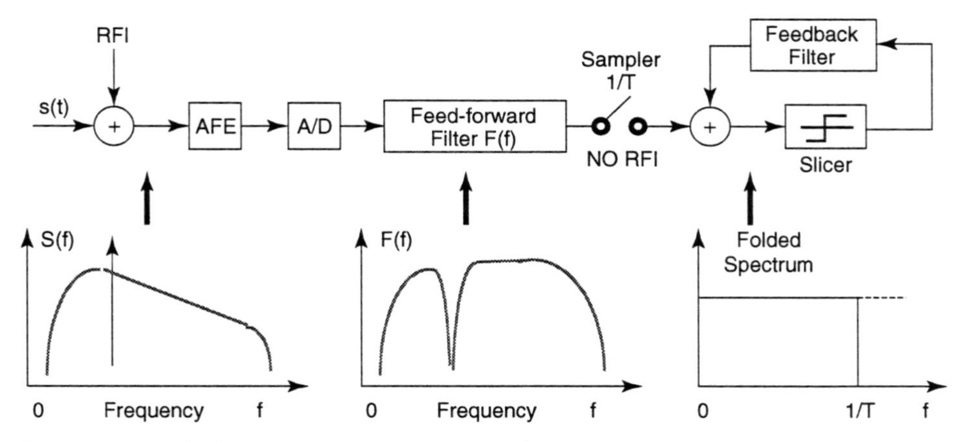

Figure 7.23 Principle of operation of the DFE in the presence of RFI.

at the input of the slicer is flat. Actually, this is often claimed to be optimum performance by SCM advocates, but it is not. Optimum performance can only occur when the transmit band is silenced, and more carriers are used to have a QAM signal on each side of the notch [25]. DFEs with multiple or single deep notches can require high-precision implementation and must execute at the symbol rate, leading to billions of operations per second being required at VDSL speeds. Most QAM designers reduce the number of taps from the levels needed for excellent performance because of this complexity problem. Instead, designers hope that the difficult notching is not required often and that the small number of taps in the equalizer is sufficient.

In a way, SCM VDSL designers reduced system complexity in early chips by sacrificing performance on difficult channels and hoping that the low-complexity chips would be consequently attractive. QAM is well suited with the DFE on channels that have continuous transmission bands and mediocre distortion, where it can eliminate generation of multiple carriers. However, as the channel distortion grows, the complexity of the DFE quickly overtakes the complexity of the multiple carrier generation, and QAM cannot handle channels with the severest distortion of VDSL well. SCM designers hope that an early low-cost solution can be replaced by increasing complex QAM components of the future that gradually address an increasing number of severe distortion situations in VDSL.

7.4.3 FMT

An informative annex of [17] contains a filtered multitone (FMT) system [26], which is actually based on the principle of profiling and QAM transmission in [16]. This approach of specifying 8 to 256 carriers according to profiling that is enhanced with respect to [16] had been acceptable to some DMT and SCM advocates as a compromise to the grand debate. However, elements of both sides found the proposal either too complex (DMT advocates) or resembling DMT too much (SCM advocates), and so the proposal was provided for information only in the DMT draft standard [17], although it more appropriately fits the SCM model and so is listed in this section here. For more details, the reader can refer to the annex of [17]. Rather here we pursue a general theory of equivalence in transmission in the hope to add to what is in [17] in terms of understanding.

Reference [26] describes a transmission method that is intermediate to multitone and QAM/CAP transmission methods in that both types of signals can be constructed using elements in that contribution. This section contains an example that illustrates the construction of either type of signal.

The following section on equivalence simplifies the basics of so-called "equivalence theory" [27], which outlines the conditions for equivalent performance of QAM/CAP methods and multichannel (or DMT) transmission methods. The areas of bandwidth optimization and interpolation/decimation are also addressed in the section below on equivalence. The section on paradox provides a second example called the "Bandwidth Optimization Paradox" in equivalence theory that illustrates the effect of incorrectly satisfying the equivalence conditions. This situation happens often in the VDSL transmission area, almost always when RF ingress/egress constraints are present and also commonly when bridged-taps or frequency-selective large-amplitude crosstalk is present.

Equivalence

References [27] and [28] are sometimes cited as noting performance equivalence of various types of modulation methods, including, but not limited to, DMT and CAP/QAM/SCM. The methods in [26] are generally within the main consideration in [27] with the finite-delay packet realizations considered in [28]. The conditions for equivalence are sometimes also not well understood nor well translated. To simplify understanding of the conditions, this section illustrates equivalent transmitters and receivers for infinite-length filters. After viewing these examples, this section progresses to stating the general equivalence conditions, but in simple non-mathematical terms, along with a "CDEF" (authors of reference [27] in alphabetical order) formula that allows computation of system performance.

Transmit Equivalence. Figure 7.24 illustrates equivalent transmitters based on a bank [26] of 8 equal bandwidth filters for modulation. Channel characteristics are such in this system that the optimum bandwidth use (which can be found through a procedure known as "water-filling" [27]) uses only 6 of the filter bands. Four of the used bands (set A) are adjacent, and the other 2 (set B) are also adjacent. Sets A and B, however, are separated by an unused band.[10]

The bands in Set A can carry 2, 3, 5, and 6 bits, respectively, corresponding to increasing signal-to-noise ratios in the corresponding channel, while the two bands in Set B can carry 4 and 6 bits, respectively. The unused bands carry 0 bits. The average number of bits per band in Set A is 4, and the average number in Set B is 5. If the symbol rate for the system is 1 MHz, then the data rate is 26 = (2 + 3 + 5 + 6 + 4 + 6) bits transmitted 1 million times per second for a total data rate of 26 Mbps.

[10]An unused band corresponds to high noise, perhaps RF or crosstalk, or can correspond to excessive channel attenuation, for instance, called by bridged taps or just high-frequency attenuation on a twisted pair.

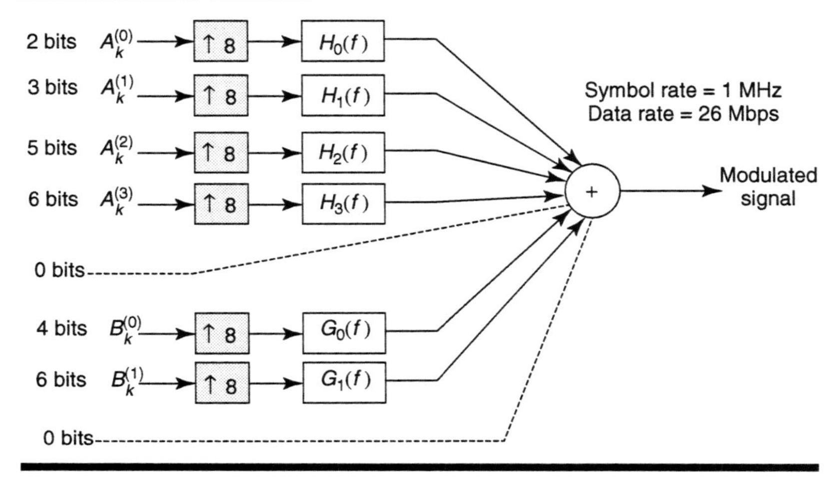

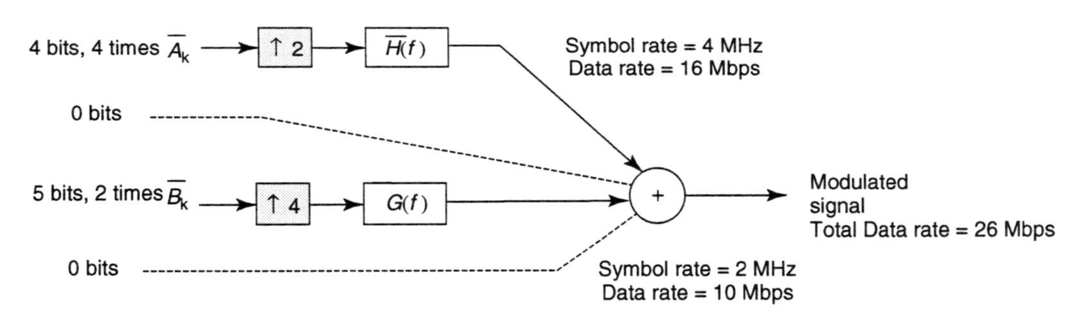

Figure 7.24 Transmitter equivalence example.

For equivalent performance, a QAM system with two bands of different width may be used instead as also shown in Figure 7.24. The first A band has a symbol rate of 4 MHz and carries 4 bits on all symbols for a data rate of 16 Mbps. The B band carries 10 Mbps at a symbol rate of 2 MHz with 5 bits on all symbols. The total data rate is also 26 Mbps. With the correct choice of receiver for each system (see Figure 7.25), the performance of the two systems in Figures 7.24 and 7.25 is identical.

Note that the carrier frequencies and symbol rates for the CAP/QAM system have been chosen so that the used bands exactly correspond to those in the multitone transmission system. Further, the average number of bits in any set of multitone bands must equal the fixed number of bits used in the corresponding QAM/CAP band. The example could be extended to any number of sets of bands in a straightforward manner following the same rules of using the same bandwidth and average bits/band for all separate bands. In VDSL, seven or more disjoint bands are common because there are many gaps in used spectrum caused by RF ingress/egress constraints (there are five amateur radio bands alone that overlap VDSL), bridged-tap notches, and frequency selective large crosstalk (e.g., a home LAN signal).

Receiver Equivalence. Figure 7.25 illustrates the corresponding receivers for the transmitters of Figure 7.24. The multitone system chooses an appropriate set of matched filters and directly decodes the outputs of each band. No equalizer is necessary within any band because each band is free of intersymbol and intra-band interference. The complexity of modulation is contained within the transmit filters and corresponding receiver filters entirely: DMT systems use a cyclic prefix and FFT to implement the filters, but there are other approaches to filter implementation as in [26]. The QAM/CAP system uses a decision feedback equalizer (DFE) for each band. The DMT and QAM systems will be equivalent if the transmit spectra are the same as in Figure 7.26. Reference [27] shows that the MMSE-DFE converts each band into the equivalent of an intersymbol-interference-free additive white Gaussian noise channel and the signal-to-noise ratio of this channel is computed according to

$$SNR_{dfe} = 2^C - 1$$

where C, the capacity in bits per Hz for the particular band and DFE under study, is a function of the channel. This result presumes an infinite number of DFE taps

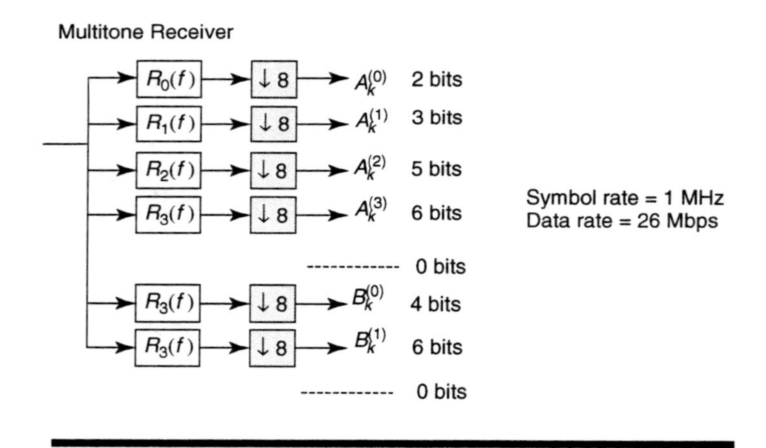

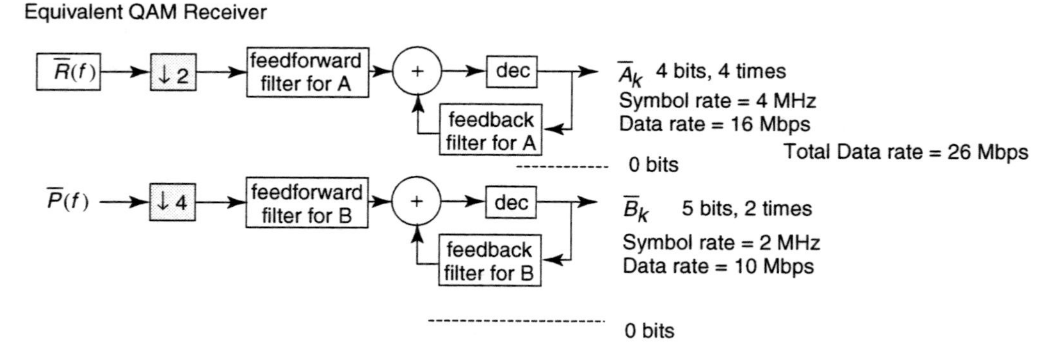

Figure 7.25 Receiver equivalence example.

and may be difficult to approach in practical implementation. An equivalent SNR for any correctly designed multitone system is well known to follow the same expression. The formula does not mean that either system operates at capacity unless extremely powerful codes with no extra margin are used. However, as long as the same code of the same power is used in both systems, then the suboptimum performance is the same in each band for both the multitone and the infinite-DFE-equalized QAM system.

For the QAM system in Figures 7.24 and 7.25, let us assume that uncoded QAM is used on both the 4-bit and the 5-bit subchannels and the performance is measured to be at probability of symbol error approximately 10^{-6}. The capacities are approximately 3 bits per Hz higher than what uncoded QAM transmission can achieve and so are equal to 4 + 3 = 7 and 5 + 3 = 8 bits/Hz, respectively. Thus the DFE and MT SNRs are

$$SNR_{dfe,A} = 2^7 - 1 = 127 \quad (21\text{ dB})$$

for band A, and

$$SNR_{dfe,B} = 2^8 - 1 = 255 \quad (24\text{ dB})$$

for band B. An overall SNR can be computed as in [28], but is not necessary for the present discussion.

Conditions for Equivalence. Simply stated and paraphrased from [27] and [28], the QAM/CAP systems and multitone system have the same performance when:

1. Each continuous set of adjacent used bands in the multitone system can be replaced by a single band in the QAM system whose symbol rate is equal to the width of the set of used adjacent multitone bands and whose carrier/center frequency is exactly in the middle.
2. The average number of bits/Hz must be the same in the two systems in each used band.

When these conditions are met, the CDEF result can be used to compute the SNR level achievable by either multitone or infinite-DFE-equalized QAM(S) in each band as long as the capacity in bits/Hz for each band is known (the capacity is fundamental to each channel and can be computed through well-known water filling methods that are also summarized in [27] and the references therein).

It should be noted that a single used band replacing several separated bands does not perform as well regardless what receiver structure is used. There must be a separate modulated signal and separate DFE for each separate band. Intuitively, this means that when the best spectrum has gaps that a transmit band-reject, or equivalently a "notch" filter is not sufficient to reach best performance, multiple QAM signals with optimized center and carrier frequency are then necessary. The following gives an example of performance loss when this rule is not observed. When capacity-achieving codes are used (never occurs in practice because of 6 dB margin necessary in DSL), it is possible to mislead oneself using the CDEF result into thinking that for this one situation, notch filtering is sufficient. However, one will find the corresponding transmit filters are not realizable (i.e., do not satisfy the Paley-Wiener criterion) and thus only a mathematical aberration. Thus, separate QAM signals for each optimum band are necessary in practice, whatever the applied code.

A Misinterpretation of CDEF. Some authors assume the "SNR is high" and derive equivalence results under this condition. Unfortunately, a high-SNR assumption although certainly applicable to many voiceband modems does *not* apply to all channels, and in particular does not ever apply to DSL channels. The high-SNR assumption means that the so-called optimum water-filling band of Shannon and capacity is the entire transmission band—clearly this is not true in DSL. One can easily show all methods are equivalent with sufficiently high SNR, but not at the practical SNR levels of DSL. For such an assumption to be true in DSL, kilowatts of power would have to be delivered to the transmission line.

Thus, one needs to be careful with statements such as "It performs the same, but is less complex" because the "it" may not be a system that has equivalent performance in all situations, so the statement is highly misleading. A well-designed multiband system, whether using QAM or DMT, can perform the same, but then and only then should complexity be compared. There have been many studies of complexity with various assumptions and the consequent conclusions. In applications such as voiceband modems, the complexity comparison can be much closer, and on channels with relatively less linear distortion and nearly white noise, the QAM design can be simpler. An example in the following clearly shows a VDSL situation where the single-carrier QAM system actually *"performs worse, but costs more"* and corresponds to a well-known RF-noise and emissions constraint of VDSL.

Interpolation and Decimation. Interpolation (transmitter) and decimation (receiver) to the master clock speed, corresponding to the $\uparrow 8$, $\uparrow 4$, and $\uparrow 2$ and $\downarrow 8$, $\downarrow 4$, and $\downarrow 2$ in Figures 7.24 and 7.25, respectively, become more sophisticated as the number of filter elements increases. A simple solution is to use Fast Fourier Transform (FFT) methods to interpolate from any number of used carriers to the full-time domain sampling rate, which occurs with the use of one inverse transform in the transmitter and one forward transform in the receiver. Almost all multicarrier and filter-bank approaches use some kind of fast transform to implement the interpolation and corresponding decimation operations. However, other interpolation methods are possible, including multirate filters. Some single-carrier proponents go as far as to compute performance with continuously variable symbol rate, without consideration of the corresponding interpolation problem. Tacit in such theoretical calculation is the assumed solution of the interpolation problem, which is more than likely most efficiently implemented with two fast transforms in each of the transmitter and the receiver. These two transforms are double the complexity of the straightforward use of a direct modulation in the frequency domain. Straightforward filter implementation of the interpolation/decimation is sometimes not feasible unless a considerable amount of performance is sacrificed. Because the exact bands to use may vary from channel to channel, interpolation can become more difficult for single-carrier systems, leading to some design trade-offs that sacrifice performance in certain situations, but not in others in order to achieve a tolerable level of complexity.

Thus, complexity evaluations of different transmission systems should include the decimation/interpolation problem. For difficult DSL channels, the trade-off is often better for multicarrier than for single carrier. For easier single-optimum-band channels, a QAM signal with a few choices of a single symbol rate may be easier to implement. The latter is the case, for instance, in v.34 and v.34.bis voiceband modems where a symbol rate and carrier frequency are selected from a menu of possible choices during training. Thus, complexity of transmission method comparison depends on how difficult the channel is. When the channel is very difficult, then multicarrier modulation is often easier to implement. It is the difficulty of the DSL transmission channel in ADSL and VDSL that attracted multicarrier specialists to this problem over the past decade (these specialists do not advocate multicarrier use everywhere, just where appropriate and simpler).

Paradox

Let us return to the example of Figures 7.24 and 7.25 in Figure 7.26, but suppose that only the three subchannels are to be used because the channel SNR is too poor on the other five. Let us also suppose that these three subchannels each carry 3, 4, and 5 bits (or an average of 4 bits/Hz) for a total data rate of 12 Mbps. The

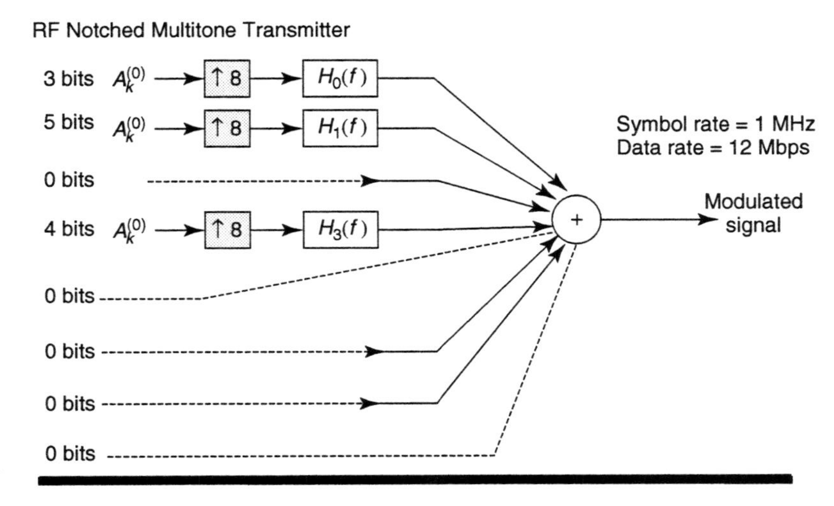

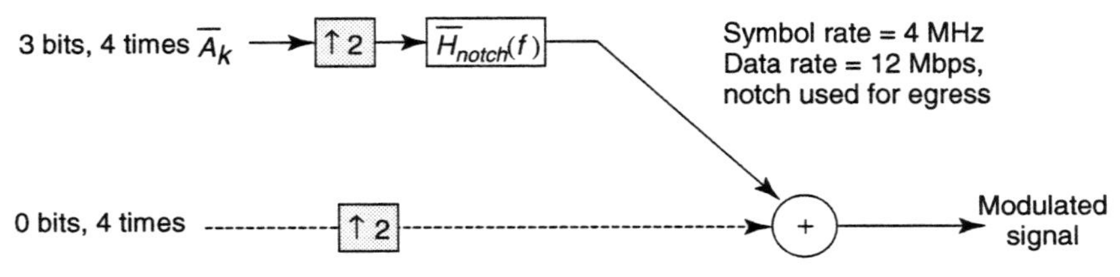

Figure 7.26 Example of the common paradox.

empty band in the middle of the upper 4 on tone 3 would likely correspond to a situation where energy needs to be silenced for RF egress/ingress reasons, or potentially for a bridged-tap or other effect that causes frequency selective band loss. An optimum QAM system with a DFE would actually be two QAM signals that each would in this example carry 4 bits at symbol rates of 2 MHz and 1 MHz, respectively, also for a total of 12 Mbps. This optimum QAM system is not shown in Figure 7.26, but should be obvious to construct if the reader understands the early examples in Figures 7.24 and 7.25.

QAM proponents often state that they can implement the situation as shown in the lower half of Figure 7.26, with a 4 MHz symbol rate, 3 bits/Hz, for again a total of 12 Mbps. An appropriate transmit notch filter is used to silence the bands silenced in waterfill and creates a power spectrum that is the same as the optimum spectrum for this channel. This section will investigate the performance of this alternative "single-carrier" approximation to the correct "two-carrier" solution for the case of an infinite-complexity MMSE-DFE receiver.

Let us suppose further that the transmission system uses a code (like Reed-Solomon forward error correction, which is for instance used in all proposals for VDSL) that improves performance by 3 dB, but that a 6 dB margin must also be ensured at the desired error probability of 10^{-6}. A simple method for quantifying bit-rate for a coded QAM system is to use the "gap" [27] approximation:

$$b = \log_2\left(1 + \frac{SNR_{dfe}}{\Gamma}\right)$$

where $\Gamma =$ 6 dB for QAM with FEC at symbol error probability of 10^{-6}. Thus, the number of bits transmitted at this error rate is computed as the capacity for a channel with 6 dB less SNR. When margin is included (as mandated in all DSL standards, including VDSL), the formula generalizes to

$$b = \log_2\left(1 + \frac{SNR_{dfe}}{\Gamma \cdot \gamma_m}\right)$$

where γ_m is the margin. This means that achievable data rate for 6 dB margin corresponds to a channel with 12 dB less SNR, or equivalently a 4-bit loss using the well-known 3 dB/bit rule in QAM transmission (4 bits × 3dB/bit = 12 dB).

For the case of the correctly optimized TWO-QAM-DFE system, the necessary SNR_{dfe} can be computed for $b = 4$ in the above formula as 24 dB. Thus, each of the MMSE-DFE receivers would attain probability of error 10^{-6} with 6 dB margin and 3 dB of coding gain at total data rate of 12 Mbps when the $SNR_{dfe} = 24$ dB. The capacity for such a system is thus $c = \log_2(1 + 10^{2.4}) = 8$ bits/Hz by the CDEF result, or the channel's capacity is then 8 bits/Hz (3 MHz) = 24 Mbps. A DMT or multicarrier system with the correct bands would also achieve this same performance level.

For the single-carrier QAM system with infinite DFE of Figure 7.26, the channel's capacity in total bits/second is invariant and remains equal to 24 Mbps. For the single QAM system, this means that this capacity in bits/Hz is 24/4 =

6 bits/Hz for the suboptimally selected symbol rate of 4 MHz. Thus, an infinite-complexity MMSE-DFE receiver will achieve performance of

$$SNR_{dfe,4MH:} = 2^6 - 1 = 63 \quad (18\,\text{dB})$$

The new margin of this system operating at the same probability of error and same data rate of 12 Mbps (again 4 Msymbols/sec × 3 bits/symbol = 12 Mbps) has a gap-plus-margin of

$$\text{gap} + \text{margin (dB)} = 10 \cdot \log\left(\frac{\text{SNR}_{\text{dfe, 4MHz}}}{2^3 - 1}\right) = 9\ \text{dB}.$$

Because the gap for the QAM+FEC is fixed at 6 dB, the margin has reduced to 3 dB. Thus, incorrect selection of the symbol rate and use of a single carrier, even with infinite-complexity notch filters in the transmitter and infinite-complexity DFE in the receiver, is 3 dB worse in performance than the use of the correct symbol rate. The greater the mismatch of DFE symbol rate to correct bandwidth, the greater the loss in margin. Thus, it is essential to use the correct symbol rate, or truly set of symbol rates, to get equivalence of the two methods. For VDSL, this often means several DFEs are necessary because waterfilling optimized spectra often consist of several disjoint frequency bands, especially when RF emissions or ingress are of concern. Although the DFE feedforward section may notch the RF (or match to an egress filters notch), it corresponds to using 4 MHz instead of the correct 3 MHz in the simple example here. Even the best DFE with infinite length filters will perform worse than a set of DFEs corresponding to the used bands (which are probably simpler to implement anyway).

Indeed, the complexity of the notch filters and DFE is much higher when only a single QAM signal is used because the notches need to be synthesized, and the performance is worse. This is clearly a poor design choice, and in fact "costs more and performs worse." Unfortunately, this result has been overlooked by QAM VDSL proponents. Simply put, if one has a hard design problem with multiple bands because of RF notches, bridged taps, crosstalk noise, and line distortion—a single carrier approach does not perform well and may be more complex than a well-designed transceiver with multiple carriers. This is an axiom of transmission and cannot be circumvented. However, on a simple transmission channel where only one band need be used (for instance in voiceband modems or HDSL), multiple carriers are almost never necessary, and so a single carrier approach may then offer some complexity reduction. VDSL and ADSL abound with practical examples where multiple bands must be used.

7.5 ETHERNET IN THE FIRST MILE (EFM)

Figure 7.27 shows the basic concept of EFM [34]. One, two or four phone lines may be coordinated to deliver 10 or 100 Mbps symmetric VDSL service in EFM. Even though "Ethernet" physical-layer copper twisted-pair standards have long been es-

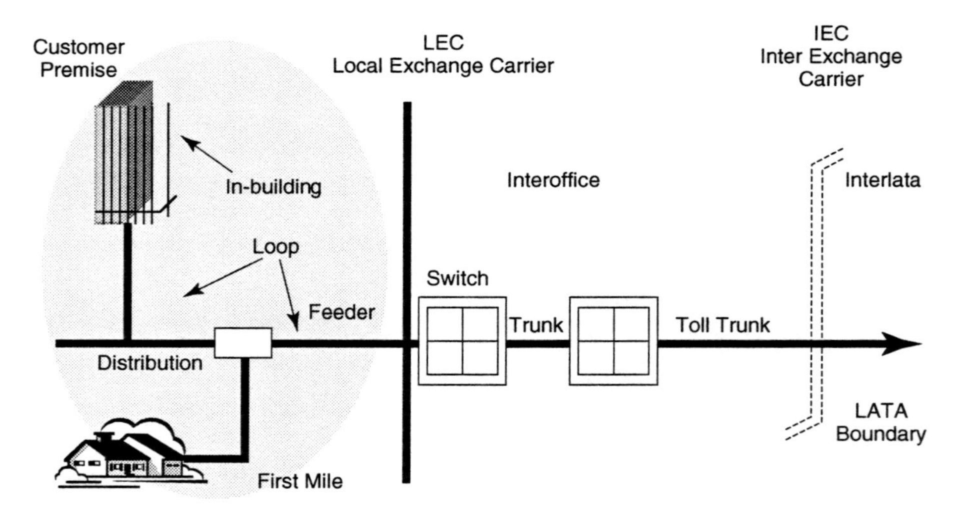

Figure 7.27 Ethernet in First Mile Illustration (shaded area is EFM interest). (Courtesy of H. Barass).

tablished, they are restricted to a length of 100 meters for 10 Mbps ("10BT"), 100 Mbps ("100BT") and 1000 Mbps ("1000BT"). Each uses a category-5, 5e, or 6 set of four twisted pairs in synchronous point-to-point transmission. Longer-length transmission fundamentally requires a different physical-layer modulation, while the upper layer "Ethernet/TCP/IP" functionality can be maintained so that the DSL line essentially looks like a "long-range Ethernet." The current physical-layer Ethernet use of three or four twisted-pairs in coordination admits that same possibility for EFM (clearly longer length for any given speed can be attained by sharing the transmission bandwidth of several lines to achieve the desired rate of 10 or 100 Mbps), as well as the possibility of carrying the entire data rate on just one line also (over a shorter distance than two or four coordinated lines).

Presently, the IEEE 802.3 standards committee is studying EFM possibilities, and VDSL proponents have suggested this is an excellent match to VDSL capabilities. However, EFM appears interested in only symmetric transmission where VDSL under Plans 997 and 998 are clearly designed for asymmetric transmission. The flexible VDSL spectra under the DMT standard in [17] clearly does allow different band use for EFM where appropriate.

Some documents on EFM line modeling occur in [33],[34], but models are not fully accepted nor standardized presently for long-length lines. In particular, FEXT modeling at high frequencies becomes very important, especially with the use of multiple lines at wide coordinated bandwidths. Chapter 11 enumerates several EFM range results using dynamic spectrum management concepts.

7.5.1 Multiline FEXT Modeling

The multiple-input multiple-output (MIMO) characterization of a cable of twisted pairs merits attention and measurement for studies in Ethernet in first-mile (EFM) efforts. As noted, groups of twisted pair within a cable may be combined for better transmission/duplexing: The interaction between lines within a subgroup or the entire cable can be exploited to improve performance and reduce transceiver complexity, motivating a model. Reference [33] suggests a temporary model for MIMO FEXT that can be used to evaluate/test EFM.

MIMO FEXT Channel

Figure 7.28 illustrates the matrix or MIMO FEXT twisted-pair channel. Each of the M inputs to this matrix channel may produce a component of the signal at each of the K outputs. Usually, $M = K$. For instance, a quad (four twisted pairs tightly packed together) has $M = 4$ inputs, $K = 4$ outputs, and a total of $16 = M \cdot K$ transfer functions of interest. These $M \cdot K$ transfer functions can be summarized in a $K \times M$ matrix $\mathbf{H}$. The $K \times 1$ vector of channel outputs $\mathbf{Y}$ is then related to the $M \times 1$ vector of channel inputs $\mathbf{X}$ by

$$\mathbf{Y} = \mathbf{HX}.$$

In some cases, four pair is really eight separate wires, allowing a 7×7 matrix channel.

Ultimately, the designer would desire the exact $\mathbf{H}$ for each binder of wires: Any FEXT information is contained within this matrix. Approximate models are of interest in evaluating the various EFM opportunities in terms of range, rates, and service applications/market. In recognition that such transfer matrices are not well known, Reference [33] suggests an $\mathbf{H}$ model for temporary use in EFM studies. NEXT matrix models are of less MIMO interest because NEXT is either avoided by duplexing choice or by echo/NEXT cancellation between lines.

The kmth element of $\mathbf{H} = [H_{km}(f)]_{\substack{k=1,\ldots,K \\ m=1,\ldots,M}}$ is the transfer function from input m to output k. When $m = k$, $H(f)$ is simply the transfer function of the kth line, $H_{kk}(f)$, and can be determined from basic transmission line theory, given the length and RLCG parameters of the line [10]. Reference [10] also models FEXT power transfer of the off-diagonal terms as proportional to the line transfer function $|H_{kk}(f)|^2$, the square of frequency f^2, and the length of the line (in meters), d, which is explained on page 90 of [6]. This corresponds to a crosstalk-insertion loss transfer path of

$$H_{km}(f) = h_{fext} \cdot H_{kk}(f) \cdot (if) \cdot \sqrt{d} \qquad (*)$$

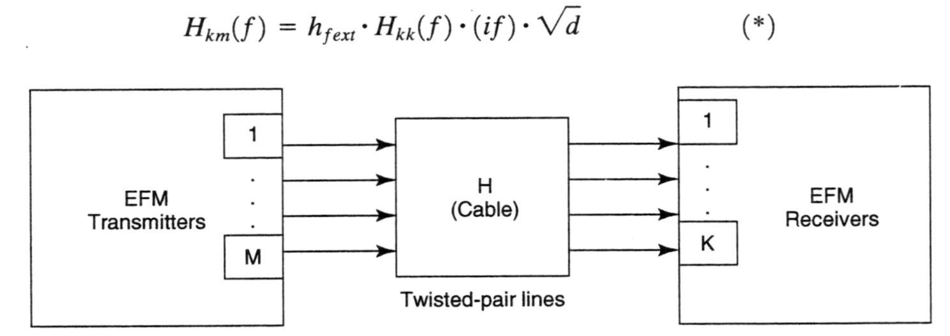

Figure 7.28 Matrix channel.

with a worst-case value of $h_{fext} = \sqrt{7.74 \times 10^{-21} \cdot (.3048m/ft)} = 4.8 \times 10^{-11}$ for two adjacent category 3 telco-plant crosstalking lines. Equation (*) is for one crosstalking line—thus an extra multiplicative factor of $(K - 1)^{0.6}$ used in models that average several lines is not used in (2) because that factor is for more than just two adjacent crosstalking lines. The factor h_{fext} is reduced nominally by a factor of 10 (or 20 dB) for category-5 wiring with tighter twisting. However, quads in the telephone plant that instead sometimes twist all four lines in an ensemble, may have a higher value than the value above, as much as an increase by a factor of 20 dB. Thus, a range of h_{fext} may be given by (category 5 independent twists) $4.8 \times 10^{-12} \le h_{fext} \le 4.8 \times 10^{-10}$ (category 3 ensemble-twisted quads).

This same FEXT model is often seen in a form that describes only energy transfer, in other words the squared magnitude of Equation (*). Here, the model is converted to voltage transfer because EFM studies may desire phase information also. The equation above may sometimes be augmented by a linear phase term $e^{j2\pi f t}$ where τ is chosen to make the corresponding crosstalk impulse response causal. The matrix **H** can then be formed by finding the insertion-loss function for any twisted pair in the bundle and inserting this insertion loss along the diagonal terms of the matrix **H**. The off diagonal terms are equal to Equation (*) with possibly randomly chosen phase offsets and/or linear phase. Some schemes that coordinate the lines may find the details of the off-diagonal terms increasing important.[11] In all cases, simple squaring of Equation (*) and treating the FEXT like Gaussian noise (which is what current transceiver designs and implementations do) leads to the lowest possible (i.e., uncoordinated) performance.

However, actual individual FEXT insertion losses do not follow such a smooth characteristic with frequency and indeed vary up and down with respect to the model in Equation (*), see Figure 7.29. The variations can vary with the individual pairs as in Figure 7.29.

The inaccuracy of the Equation (*) model is increasingly evident at higher frequencies. There is a raised sinusoidal appearance to the magnitude transfer. This plot shows coupling functions between a pair on a 500 meter .5 mm cable with 50 pairs. Cosine terms can be added to the model to closely approximate the location of the dips and peaks in frequency seen in the measured FEXT insertion loss functions. Thus, the proposed model is

$$H_{km}(f) = \begin{cases} n_{lines} \cdot \sqrt{h_{fext} d} \cdot (j2\pi f) \cdot \\ H(f) & m = k \\ \left[1 + 0.3 \cdot \cos\left(\frac{2\pi f d}{c_{line}}\right) - 0.3 \cdot \cos\left(\frac{4\pi f d}{c_{line}}\right)\right] \cdot H(f) \cdot e^{j\varphi} & m \neq k \end{cases}$$

[11]When the off-diagonal terms are significantly smaller than the diagonal, the best coordinated schemes all converge to making the line appear as if there were no FEXT. Although the details of the transfer function are then not of consequence to performance, the implementation still though depends on knowing the phase.

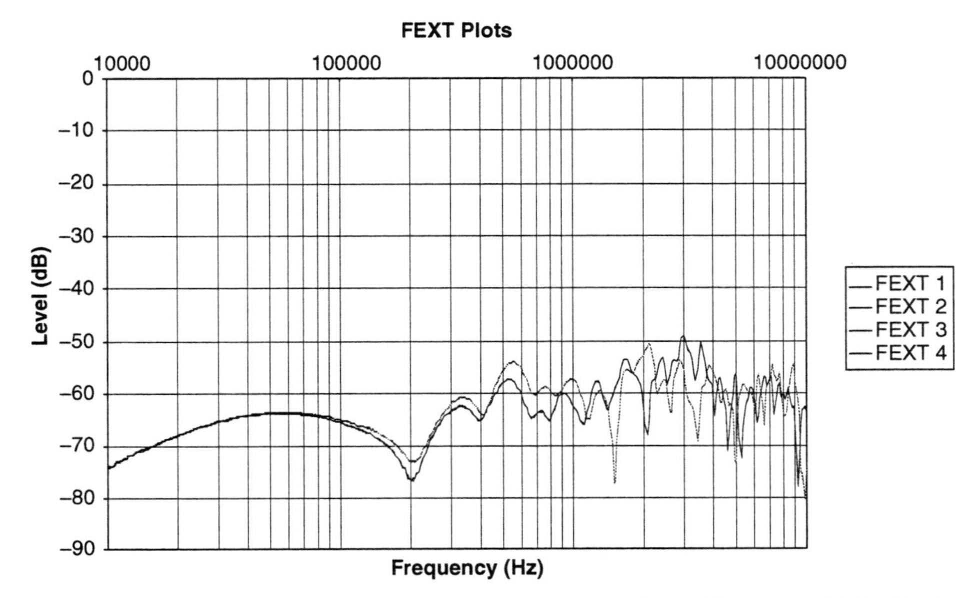

Figure 7.29 500-meter example FEXT insertion loss functions. (Courtesy of John Cook of BTexact).

where $H(f)$ is as above derived from standard transmission theory. c_{line} is the speed of light on the media (often just use 300 Mmeters/sec), and φ is a phase term (i.e., $\varphi = 2\pi f\tau + \phi_{km}$ where τ makes the response causal and ϕ_{km} is chosen from a uniform distribution over $(0,2\pi)$ independently for each pair of indices m and k). Normalizing each off-diagonal entry by the factor $n_{lines} = (K-1)^{0.6/2}/\sqrt{(K-1)} = (K-1)^{-0.2}$ on average can account for the fact that distant lines have less crosstalk than close lines within the bundle and produces a slightly more accurate model when $K = 25$ or 50.

The following table suggests values for h_{fext} and n_{lines}:

	Category 5 Quad	Category 3	Telco Quads
h_{fext}	4.8×10^{-12}	4.8×10^{-11}	4.8×10^{-10}
n_{lines}	$3^{-0.2} = .803$	$49^{-0.2} = .459$	$3^{-0.2} = .803$

This proposed MIMO FEXT model attempts to augment well-known existing models to include:

Quads: which may have better or worse coupling depending on the type of quad and associated independent (cat 5) or ensemble (cat3 telco quad) twisting

Phase: phase coupling between lines that can be important for implementation of coordinated transmission schemes

Notches: the length-dependent frequency variation not included in earlier models, but which may become important in EFM studies

This model is in somewhat of a state of improvement at time of writing and likely readers would want to pursue dynamic spectrum management and EFM standards of future to get any updates occurring after time of writing.

REFERENCES

[1] P. S. Chow, J .S. Tu, and J. M. Cioffi, "Performance Evaluation of a Multichannel Transceiver for ADSL and VADSL Services," *IEEE Journal on Selected Areas in Communications* 9, no. 6 (August 1991): 909–919.

[2] J. Cioffi and J. A. C. Bingham, "A Proposal for Consideration of a VADSL Standards Project," ANSI Contribution T1E1.4/94-183, December 5, 1994; see also J. Cioffi and K. Jacobsen, "15 Mbps on the Mid-CSA," ANSI Contribution T1E1.4/94-088, April 18, 1994, Palo Alto, CA, and "Range/Rate Projections for NxDMT" ANSI Contribution T1E1.4/94-125, June 6, 1994.

[3] K. Foster, "A Proposal to Study the Feasibility of High-Speed Copper Drop Standardization," *British Telecom contributions TD 56 and WD 62 to ETSI RG12 subgroup of TM3,* October 10, 1994, Rome, Italy.

[4] ATM Forum, ATM User-Network Interface Specification, V.3.0, Physical Layer UNI Interfaces STS-1 for 51.84 Mbps 100 meter transmission on Unshielded twisted-pair, see *http://www.atmforum.com/atmforum/library/53bytes/backissues/others/53bytes-0795-5.html,* May 1995.

[5] DAVIC (Digital Audio Visual Council) "Specification 1.0—Part 8 Lower-Layer Protocols and Physical Interfaces," *http://www.ccett.fr/dam/dvc_spec.htm,* Short-range baseband asymmetrical PHY on copper and coax, circa 1995.

[6] T. Starr, J. M. Cioffi, and P. Silverman, *Understanding Digital Subscriber Lines,* Upper Saddle River, NJ: Prentice Hall, 1999.

[7] Singh, Jagdeep (CEO, OnFiber), "Presentation on future of Fiber Transmission and Broadband Access," to Computer Science and Telecommunications Board of United States National Research Council, January 26, 2001, Palo Alto, CA.

[8] *IEEE Standard 802.11a-1999,* "Wireless LAN Medium Access Control (MAC) and Physical Layer (PHY) Specification: High-Speed Physical Layer in the 5 GHz Band," issued by IEEE Computer Society, September 16, 1999.

[9] International Telecommunications Draft Standard G.pnt, Study Group SG15/Q4, 2001.

[10] Spectrum Management for Loop Transmission Systems, ANSI Standard T1.417-2001, New York, NY.

[11] T. Riley, " Proposal to Add a New Work Program to the VDSL Project," *T1E1.4 Contribution 99-,* Clearwater, FL, December 5, 1999.

[12] J. Cioffi, G. Ginis, and W. Yu, "Performance of splitterless DSL at higher speeds," *ANSI Contribution T1E1.4/99-558,* Clearwater, FL, December 5, 1999.

[13] ETSI TS 101 270-2 v1.1.5, (2000-12) TM "VDSL Part 2 Transceiver Specification," European Telecommunications Standards Institute Report on VDSL, TM (transmission and multiplexing), Access transmission systems on metallic access cables, ETSI: Sophia Antipolis, 2000.

[14] J. Cioffi and W. Yu, "G.vdsl: Robust VDSL in the Presence of Bridged Taps," ITU Contribution NG-039, Nashville, TN, November 2, 1999.

[15] Q. Wang, ed., "Very-High-Speed Digital Subscriber Line (VDSL) Metallic Interface, Part 1: Functional Requirements and Common Specification," *ANSI T1.424–2002 Standard.*

[16] V. Oksman, ed., "Very-High-Speed Digital Subscriber Line (VDSL) Metallic Interface, Part 2: Single-Carrier Modulation Specification," *ANSI T1.424–2002 Standard.*

[17] S. Schelestrate, ed., "Very-High-Speed Digital Subscriber Line (VDSL) Metallic Interface, Part 3: Multicarrier Modulation (MCM) Specification," *ANSI T1.424–2002 Standard.*

[18] J. Cioffi et al., "North American Prioritization of High-Speed Asymmetric VDSL Service with Respect to Other VDSL Services in Response to G.vdsl Issue Item 2.7," *ANSI Contribution T1E1.4/99-393,* Baltimore, MD, August 23, 1999.

[19] FCC Order 00-336, Second Memorandum Opinion and Order. Adopted: September 7, 2000; released: September 8, 2000, Ameritech and SBC Corporations.

[20] V. Erceg, et al., "Channels or Fixed Wireless Applications," *IEEE 802.16 Standards Contribution,* Xc-00/NNr0, February 23, 2001, Tampa, FL. See also The IEEE 802.16 Working Group on Broadband Wireless Access Standards, fixed-wireless access standard and activities; see Web page *http://wirelessman.org/,* January 16, 2001.

[21] ETSI EN 300 429 V1.1.2 (1997-08): Digital Video Broadcasting (DVB); Framing Structure, Channel Coding and Modulation for Cable Systems.

[22] M. Isaksson, P. Deutgen, F. Sjöberg, P. Ödling, S. K. Wilson, and P. O. Börjesson, "Zipper—A Flexible Duplex Method for VDSL," Proceedings of International Conference on Copper Wire Access Systems (CWAS 1997), pp. 95–99, Budapest, Hungary, October 1997. See also M. Isaksson, et al., "Zipper—A Duplex Scheme for VDSL Based on DMT," ANSI T1E1.4/97-016, 3–7, 1997.

[23] K. Cheong, J. Choi, J. Fan, R. Negi, N. Wu, and J. Cioffi, "Soft Cancellation via Iterative Decoding to Mitigate the Effect of Home-LANs on VDSL," *ANSI Contribution T1E1.4/99-333R2,* Baltimore, MD, August 23, 1999.

[24] J. Cioffi, G. Ginis, W. Yu, and C. Zeng, "Example Improvements of Dynamic Spectrum Management," *ANSI Contribution T1E1.4/2001-089R1,* Costa Mesa, CA, February 23, 2001.

[25] J. Cioffi, G. Ginis, W. Yu, G. Cherubini, E. Eleftheriou, and S. Olcer, "Construction of Modulated Signals from Filter Bank Elements and Equivalence of Line Codes," *ANSI Contribution T1E1.4/99-395,* Baltimore, MD, August 23, 1999.

[26] G. Cherubini, E. Eleftheriou, S. Oelcer, and J. Cioffi, "Filtering Elements to Meet Requirements on Power Spectral Density," *ANSI Contribution T1E1.4/99-429,* Baltimore, MD, August 23, 1999.

[27] J. M. Cioffi, G. P. Dudevoir, M. V. Eyuboglu, and G. D. Forney, Jr., "MMSE Decision Feedback Equalizers and Coding: Parts I and II," *IEEE Transactions on Communications* 43, no. 10 (October 1995): 2582–2604.

[28] J. M. Cioffi and G. D. Forney, Jr., "Generalized Decision-Feedback Equalization for Packet Transmission with ISI and Gaussian Noise," *Communication, Computation, Control, and Signal Processing,* eds. A. Paulray, V. Roychowdhury, and C. Schaper (a tribute to Thomas Kailath). Boston: Kluwer, 1997.

[29] F. Sjöberg, M. Isaksson, R. Nilsson, P. Ödling, S. K. Wilson, and P. O. Börjesson, "Zipper—A Duplex Method for VDSL Based on DMT," *IEEE Trans on Communications* 47, no. 8, (August 1999): 1245–1252.

[30] J. G. Denis, M. I. Mestdagh, and P. Ödling, "Zipper VDSL: A Solution for Robust Duplex Communication over Telephone Lines," *IEEE Communications Magazine* 38, no 5 (May 2000): 90–96.

[31] F. Sjöberg, R. Nilsson, M. Isaksson, P. Ödling, and P. O. Börjesson, "Asynchronous Zipper," Proceedings of IEEE International Conference on Communications (ICC'99), vol. 1 (Vancouver, Canada), June 1999, pp. 231–235.

[32] F. Sjöberg, M. Isaksson, P. Deutgen, R. Nilsson, P. Ödling, and P. O. Börjesson, "Performance Evaluation of the Zipper Duplex Method," Proceedings of IEEE International Conference on Communications (ICC'98), Atlanta, GA, June 1998, pp. 1035–1039.

[33] J. Cioffi and J. Fang, "A Temporary Model for EFM/MIMO Cable Characterization," *IEEE 802.3 Standards Contribution,* Los Angeles, CA, October 18, 2001.

[34] IEEE 802.3 Draft "Ethernet in the First Mile Copper Standard Test and Data Rate Project," *IEEE 802.3 Standards Contribution,* Los Angeles, CA, October 18, 2001.

[35] J. Cioffi, P. Spruyt, T. Pollet, V. Oksman, J. J. Werner, K. Jacobsen, and J. Chow, "VDSL," *IEEE Communications Magazine* (May 2000).

[36] S. Olcer and J. Cioffi, guest eds., "VDSL Special Issue," *IEEE Communications Magazine* (May 2000): entire issue.

[37] K. Van Acker, G. Leus, M. Moonen, O. vad de Weil, and T. Pollet, "Per-tone Equalization for DMT-based Systems," *IEEE Transactions on Communications,* vol. 49, n. 1: 109–119. January 2001.

CHAPTER 8

UNBUNDLING AND LINE SHARING

8.1 OVERVIEW

DSLs operate by transmitting data over copper telephone lines.[1] Each line traverses between a customer premises and a telephone company central office (CO) or a digital loop carrier remote terminal (DLC-RT). These facilities are owned by the incumbent local exchange carrier (ILEC, the traditional phone company) and prior to deregulation, the ILEC operated as a regulated monopoly and was not required to make its facilities available to competing service providers such as competitive local exchange carriers (CLECs). Without access to these facilities, competitive local services were not possible. With the

[1]Industry and regulatory documents use the terms "line" and "loop" to refer to the twisted pair of wires connecting the customer to the network. In this book, we primarily use the term "line."

1984 Modified Final Judgement divestiture of AT&T, the telecommunications regulatory climate began its gradual change from a regulated monopoly to a mostly deregulated competitive market. The competitive market was seen as an environment that would accelerate innovation and reduce the price of services. To avoid abuse, the activities of a monopoly are tightly restricted by government regulations that require lengthy public review and approval of every new aspect of service. The regulators operated in the mode of telling the monopoly what it could not do, and there was little incentive for the service provider to reduce the price of services. For the competitive market model to succeed, the CLECs must have access to necessary facilities owned by the ILEC. This access must be provided at a reasonable cost, availability, and reliability. CLECs offer their services via unbundled lines leased from the ILEC. Alternatively, a CLEC may offer ADSL service via line sharing where the ILEC uses the voice frequency band for its regulated telephone service, and the CLEC uses the higher frequencies for its services. During the late 1990's it became apparent that local service competition could be further facilitated by the use of alternative local access media in addition to unbundled telephone lines. The alternative media includes coaxial cable, terrestrial radio, direct satellite links, and fiber owned by the CLEC.

Despite the business failure of some CLECs, the creation of a competitive telecommunications market has seen some success. There are more than 300 CLECs in the United States providing service to millions of customers. It appears that the ILECs, CLECs, and ISPs (internet service providers) are learning that cooperation with each other is necessary for their mutual benefit. Figure 8.1 shows the types of companies that provide portions of the total service to the end customer. Sometimes more than one of these functions will be provided by one company or by company affiliates.

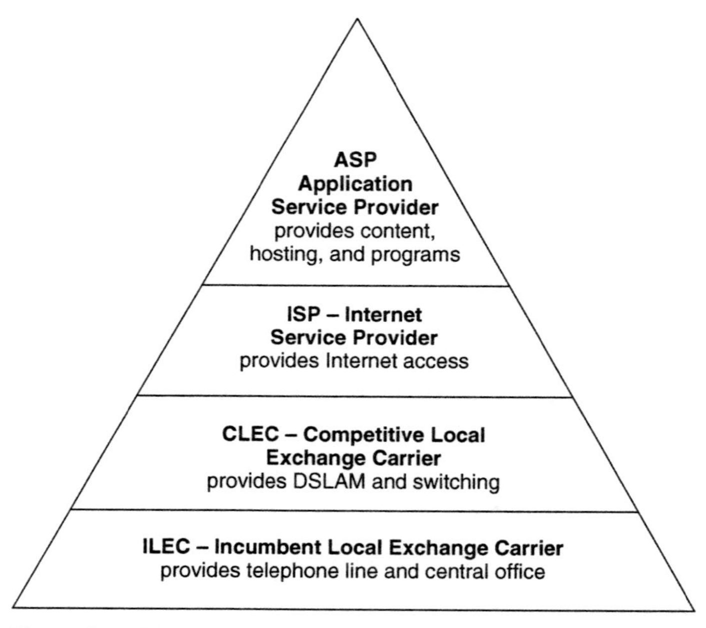

Figure 8.1 Hierarchy of service providers.

8.2 U.S. REGULATORY PROCESS

For an engineer, it is quite astonishing to realize that the chief architects of our networks are regulatory lawyers who have little understanding of layered protocols, statistical processes, or Maxwell's equations. Any company in the telecommunications industry should maintain a current understanding of the regulatory environment. Some major companies have large staffs devoted to regulatory activities, and at times may contract for more regulatory resources. It seems that every few years new regulatory policy causes fundamental changes to occur in the industry that can make or break the fortunes of certain companies. Equipment that once belonged to the service provider suddenly is classified as customer equipment, causing equipment vendors to develop different distribution channels for their equipment. Portions of the network infrastructure that were controlled solely by one company suddenly must be manageable by a multitude of service providers, causing sweeping changes in operational support systems. The opening of network interfaces have threatened equipment interoperability and reliability, requiring the industry standards bodies and the National Reliability and Interoperability Council (NRIC) to scramble to find ways to minimize service problems. Even for devices as simple as analog telephones, the quality of service has been compromised by some types of customer equipment in an unregulated environment. The regulator who does not understand the engineering implications of his or her policies is as foolhardy as the engineer who does not understand the implications of regulatory policy on his or her designs.

This book focuses on the regulatory environment in the United States. The regulatory situation for each nation is unique. The legacy telecommunications infrastructure, societal norms, government, and legal systems differ. The telecommunications deregulation trend began in the United States, but many countries are following this trend with a delay ranging from one to ten years.

The foundation of U.S. regulatory policy is the laws passed by Congress (see Figure 8.2). The Telecommunications Act of 1934 established a telephone service as a regulated monopoly. Later, the Telecommunications Act of 1996 established the basis deregulation with local service competition, including collocation in COs and line unbundling. The Federal Communications Commission (FCC) implements the laws by developing national rules and enforcing these rules. The FCC rule making involves a public review process that typically takes two to four years from the first notice of inquiry to an issued Report and Order. Following the Report and Order, parties can request reconsideration, and if the FCC deems there is sufficient merit, the FCC may issue an order on Reconsideration or a Memorandum Modifying an Order. Ultimately, FCC and state orders may be challenged in judicial courts. In some cases, FCC orders have been invalidated by court order.

Each state has a Public Utility Commission (PUC) that regulates services within the state, whereas the FCC regulates interstate services. State PUCs usually follow the policies developed by the FCC. Voice service providers must be certified in each state where they operate for the type of service offered. Regulated services must be tariffed in each state that the service is offered. The tariff process involves advance disclosure of the intended service, public comment, and review by the state PUC.

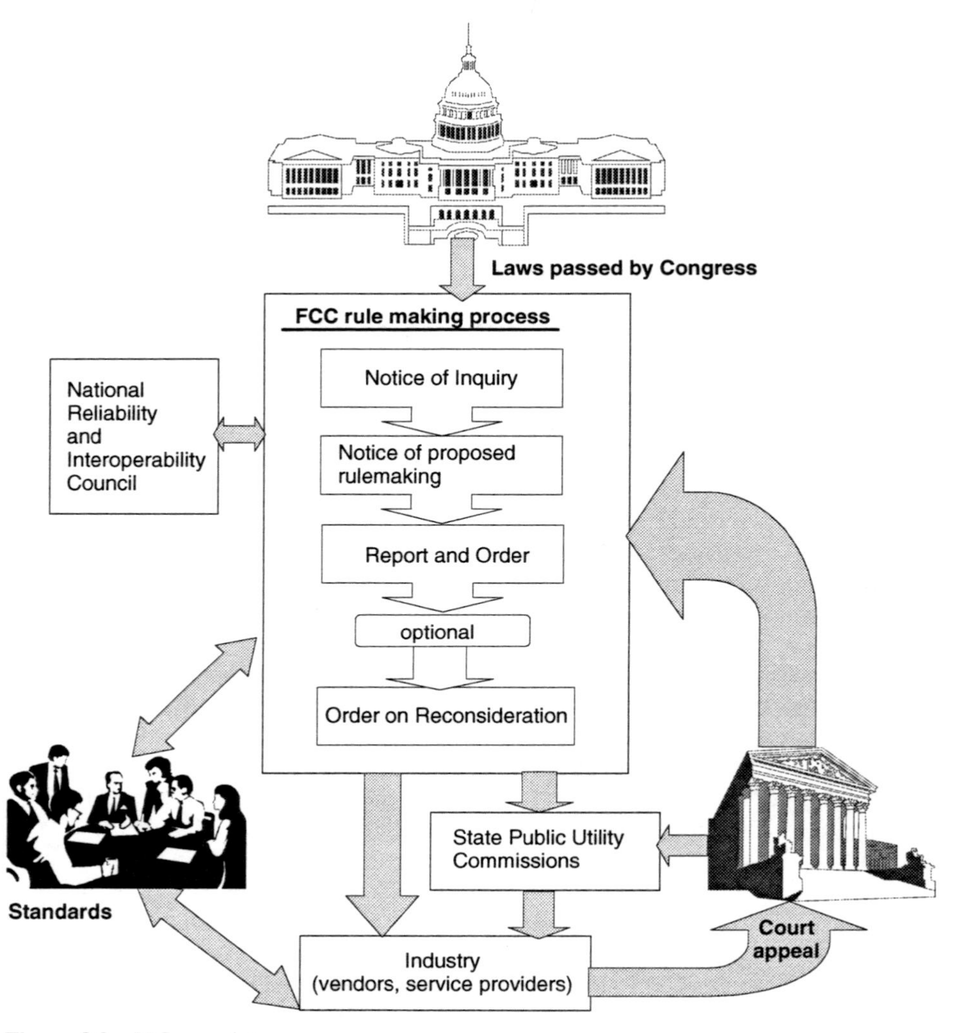

Figure 8.2 U.S. regulatory process.

The telecommunications deregulation and unbundling process has generally been led by the United States, with many other countries following a similar course a few years later. The European Union required all its member countries to implement local access unbundling during 2001.

8.3 UNBUNDLING

In the United States, the Telecommunications Act of 1996 required ILECs to lease unbundled network elements (UNEs) to certified CLECs. The UNEs include:

- Lines from the CO to the customer
- Lines from accessible remote network sites to the customer (subloops)
- Spare fiber facilities (sometimes called "dark fiber")
- Inside wire owed by ILEC
- Interoffice transmission facilities (trunks), including shared transport
- Floor space within the CO where a CLEC may place transmission, switching, and other types of equipment provided it meets the safety, fire, and heat dissipation requirements known as NEBS (Network Equipment Building Standard)
- Telephone numbers, including number portability database
- Network interface device (NID) demarcation point at customer site
- Access to operational support systems (OSSs) and network data bases, including subscriber line qualification data from LFACS (loop facilities assignment system) (see Section 8.6)
- SS7 signaling facilities and advanced intelligent network databases
- Local circuit switch line ports

On March 31, 1999, the FCC released the First Report and Order CC Docket 98-147 with detailed rules for collocation, and gave notice of future proposed rule making relating to spectrum compatibility and line sharing. Each of the UNEs must be available separately and at a price that is approximately equal to the cost to the ILEC (know as TELRIC pricing—total element long range incremental cost). In the event that the requested facilities are not available, the ILEC is not required to build additional facilities unless the CLEC pays for the additional cost. Safeguards exist to allow verification of ILEC claims that spare facilities do not exist, in particular CO collocation space. At the CLEC's preference, the ILEC must provide collocation space with or without locked equipment cages that would restrict access to the CLEC equipment. Collocation may be physical (CLEC maintains the equipment) or virtual (ILEC is paid to maintain the equipment on behalf of the CLEC). In both cases, the CLEC must pay to lease the space. The ILEC may apply environmental and safety standards (generally known as NEBS) to the CLEC equipment that are no stricter than the ILEC applies to its own equipment. CLECs may also locate equipment in a site outside the CO. For DSL systems, it is desirable for the DSLAM to be as close as possible to the MDF (main distributing frame) in the CO to minimize additional line length.

The unbundled subscriber lines are provided for the exclusive use of the CLEC; in the next section, line sharing is discussed. Unbundled lines typically are categorized by transmission technology, for example:

- Telephone service capable lines
- ISDN capable lines
- ADSL capable lines
- HDSL capable lines
- Lines capable of SDSL or SHDSL at certain bit rates

Carriers who connect equipment to unbundled lines are required by the ILEC to meet spectrum management specifications that limit the effects of crossstalk caused by the transmission system. These requirements apply to equipment at both ends of the line, and include total transmitted power, power spectral density, transverse balance, and longitudinal output voltage. Many types of transmission systems are limited to a maximum line length because the effects of crosstalk are more critical for systems that must also deal with a large amount of attenuation. Chapters 10 and 11 discuss spectrum management in depth.

Unbundled DSL capable lines consist of a direct copper path from the central office main distributing frame (MDF) to the point of network demarcation at the customer premises (generally the NID), and do not include electronics such as line-terminating equipment, repeaters, or digital loop carrier. Telephone-grade unbundled lines may include loading coils, and may be either a direct copper line or a DLC channel connected to a copper sub-loop. Unbundled access via digital loop carrier is discussed further in Section 8.5.

The Telecom Act of 1996 and subsequent FCC Orders prohibit ILECs from directly providing high-speed data services (other than ISDN) to customers. However, the ILEC may create separate subsidiary companies that provide high-speed data service via DSLAMs and data switches owned by the subsidiary. These subsidiaries are treated much like a CLEC; they are not required to provide unbundled DSLAM access or other elements to CLECs.

8.4 LINE SHARING

Line sharing unbundles a portion of the frequencies conveyed by a telephone line. The frequency band above approximately 4 kHz is used by the data service provider while the frequency band below 4 kHz is used by the voice service provider. Thus, two different service providers simultaneously share the same line. A splitter device at the network end of the line combines the data and voice-band signals while preventing interference between the two transmission bands. The splitter consists of a high-pass filter leading to the DSLAM and a low-pass filter leading to the voice switch. The high-pass filter is most often implemented as a pair of dc (direct current) blocking capacitors whose principal function is to avoid disruption of the voice service if the wire pair was accidentally shorted at the DSLAM.

As shown in Figure 8.3, line sharing implies the sharing of a telephone line by an ILEC and a data service provider, and not the sharing of a phone line by multiple end-user customers such as an old-time party line. There is a configuration where multiple DSL customers do share one or more common telephone lines: a remote DSLAM located in a multiple dwelling unit (MDU). This scenario is not called line sharing, even though the multiple customers served by the remote DSLAM do share the same line(s) that connect the remote DSLAM to the network.

In theory, there are many ways that multiple service providers could share a telephone line:

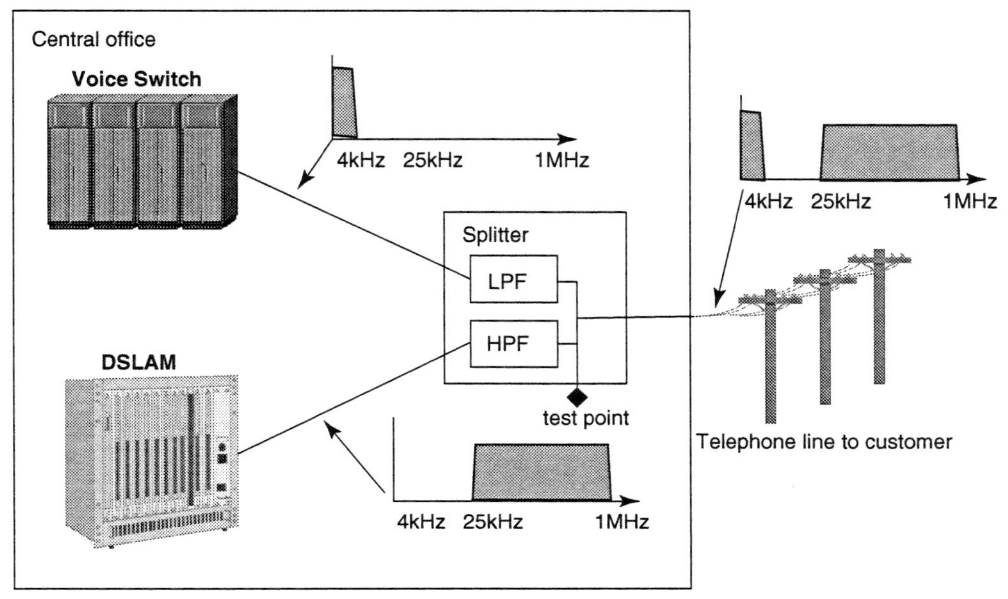

Figure 8.3 Line sharing splitter at network end of line.

- *Two fixed frequency bands:* One band for traditional voice service, and one band for DSL data transmission. This is the method defined by the FCC for line sharing. This method works well for traditional voice services and for standard ADSL-based service. However its inflexible structure may be poorly suited to future technologies and services (e.g., voice-over-DSL).
- *More than two fixed frequency bands:* More than two service providers could share a line by dividing the line into multiple frequency bands. One example that could use three bands would be a customer subscribing to three service providers for voice, data, and alarm service. This method would suffer the inefficiencies of multiple frequency-guard-bands. Because some security/fire alarm services use signaling below the voice-band (below 200 Hz), the current practice of line sharing for voice and data can be considered a three-band method of line sharing.
- *Flexible frequency bands:* The frequency dividing the two bands could be variable depending on the needs of the two service providers. For example, one provider could use a frequency band based upon the number of derived voice channels provided, and the data service provider could use the remaining frequency band. In the extreme, the frequency band separation could change dynamically. This method would be complex to administer, could require a complex splitter, and would also complicate spectrum management.
- *Time division sharing:* Burst-mode transmission systems from multiple providers could be synchronized to allow more than one DSL modem on the

line. The time division could be fixed or variable. This method trades the capacity loss of guard-time periods, and the complexity of modem synchronization for the capacity loss due to frequency guard-bands and splitting filters. This method could complicate spectrum management.

- *Code division multiple access:* The cost and technical feasibility of CDMA-type line sharing is uncertain.
- *Higher-layer bit stream sharing:* An access provider with a DSL modem at each end of the line would provide higher-layer multiplexing of packet/cell streams from multiple service providers. The multiplexing could be performed at layer two via ATM virtual circuits (PVCs or SVCs), or at layer three via IP paths. This method has the following benefits:
  - Any number of service providers may have simultaneous access to the customer.
  - Service providers may be changed without wiring changes in the central office.
  - Full flexibility of the division of bandwidth between service providers.
  - Full flexibility for types of services (voice, data, alarm, and video).
  - Statistical multiplexing permits a bandwidth overbooking advantage.
  - No capacity loss due to frequency guard-bands.
  - Greater aggregate capacity for all DSL lines may be achieved by the mutual coordination of all transmitted signals; this is most feasible if all lines are controlled by the same entity.
  - DSL modem cost shared by multiplex providers.
  - Advances in DSL technology could be applied without changes to architecture or regulatory structures.
  - High reliability because a single entity has full administrative and diagnostic control of the line. There would be no need for multi-entity coordination of diagnostic testing.
- *Hybrid of higher-layer bit stream sharing and fixed frequency division:* One frequency band for traditional voice service, and DSL operation in the upper frequency band with multiplexed bit-streams from multiple service providers.

Line sharing is a boon to service competition. CLECs typically pay $5 to $7 per month to lease a line to carry data with line sharing, whereas non-shared, unbundled lines often cost at least twice as much. More important, line sharing permits the data service to customers where no spare wire pairs are available. With line sharing, the data service provider will almost never be unable to serve a customer due to a lack of spare pairs in a cable. Carrying two services on a line improves cable utilization and reduces cable plant exhaustion.

The FCC released the line sharing order December 9, 1999, as the combined Third Report and Order in CC Docket 98-147 and Fourth Report and Order in CC Docket 96-98. The line sharing order required ILECs to provide a new UNE to certified CLECs. The new UNE was termed the high-frequency portion loop (HFPL), and the order describes it as residing above the traditional voice band. To avoid excessive constraints on technology, the FCC purposely did not define the dividing line between the high- and low-frequency portions of the line, though the order does mention that the voice-band is commonly considered to reside from 300 Hz to 3.4 kHz and that the HFPL should not interfere with voice band services. This would suggest that ILECs might not be required to provide line sharing on lines with P-phone type service that incorporates signaling at 8 kHz.[2] The FCC order does not require that any specific type of DSL technology be used on the high frequency portion of the line, but mention is made of ADSL and MVL[3] type technologies. ILECs are required to provide line sharing only on lines that are already used for traditional voice service. Thus, the ILEC is not required to provide line sharing on currently unused lines or nonvoice lines. However, the CLEC still can lease dedicated lines for nonshared use. The price for the line sharing UNE is set by the states using a process called TELRIC; this price is approximately equal to the telephone company's cost. The line sharing UNE consists of the high-frequency portion of a line from the MDF in a CO to the point of network demarcation at the customer's premises, or from a DLC-RT to the network demarcation at the customer's premises. The copper line from the DLC-RT to the customer is sometimes called a subloop.

As with dedicated unbundled lines, line sharing uses a telephone line that is owned by the ILEC and leased to the CLEC. The ILEC retains maintenance responsibility for the line. If requested by the CLEC, the ILEC must perform line conditioning provided that the line conditioning would not degrade the voice-band service. Line conditioning generally consists of the removal of loading coils or removal of bridged taps. Loading coils are necessary for good voice-band performance on lines longer than 18 kft. Loading coils must be removed for DSL operation. Bridge tap remove is necessary for less than 5 percent of lines. The ILEC may charge a reasonable fee for line conditioning.

The line sharing order requires that the DSL line sharing transmission not interfere with voice service. The two services are combined by use of a splitting filter (see description earlier in this section) that may be owned by either the ILEC or the CLEC. The splitter ownership, physical location, technical characteristics, installation process, and equipment cabling arrangements are negotiated jointly by the ILEC and CLEC. Both ILEC- and CLEC-owned splitters are in practice (Figures 8.4 and 8.5). Factors in the positioning of the splitter are the amount of central office cabling required, the number of main distributing frame (MDF) wiring jumpers, and the provisions for direct access to the line for testing. As shown below, the CLEC owned splitters could eliminate one MDF jumper and shorten some CO cabling. Whereas, the ILEC-owned splitters could permit a pool of splitters to be shared by many CLECs, the CLEC-owned splitter would also allow the splitter to be integrated with the DSLAM.

The line sharing splitter at the central office restricts the bandwidth available to the DSLAM (low frequencies eliminated) and to the voice switch (high frequen-

---

[2]Germany, and possibly some other countries in Europe, may implement line sharing of ADSL over basic rate ISDN using 4B3T line coding with approximately 0 to 100 kHz for ISDN and ADSL occupying the band above this.

[3]MVL is multiple virtual line, a proprietary DSL technology.

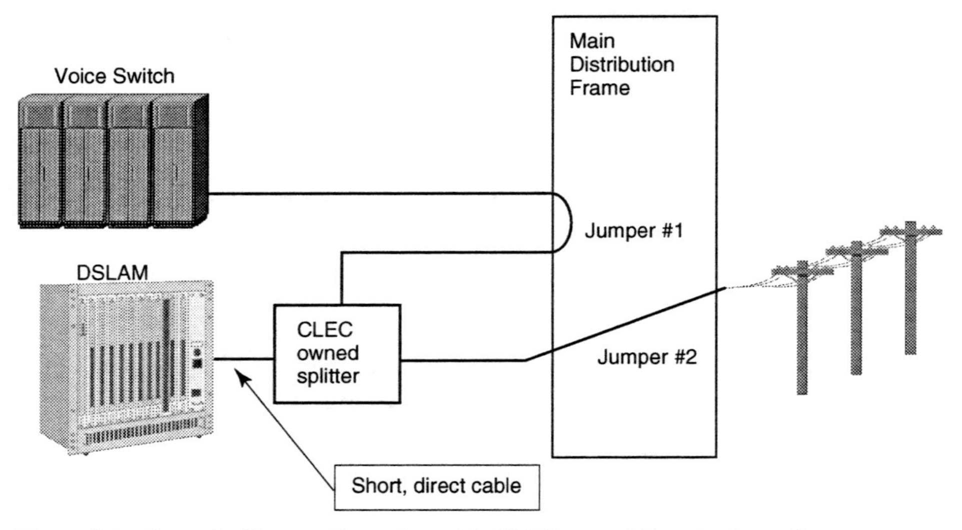

Figure 8.4 Central office configuration with CLEC-owned line sharing splitter.

cies eliminated). Neither the DSLAM nor the voice switch could test the full bandwidth of the shared line. This restricts the ability to fully diagnose the line characteristics. Figure 8.6 shows four alternative locations for line test systems:

1. *Inside or in front of the DSLAM.* This position could be used for a line testing system owned by the data carrier. Testing would be limited to frequencies above 4 kHz unless a bypass function was included with the splitter. Only lines that had already been connected to the data carrier's equipment could be tested; thus this configuration is of little value for preservice line qualification.
2. *Before the voice switch.* This position could be used for a line testing system owned by the voice carrier. Testing would be limited to frequencies below

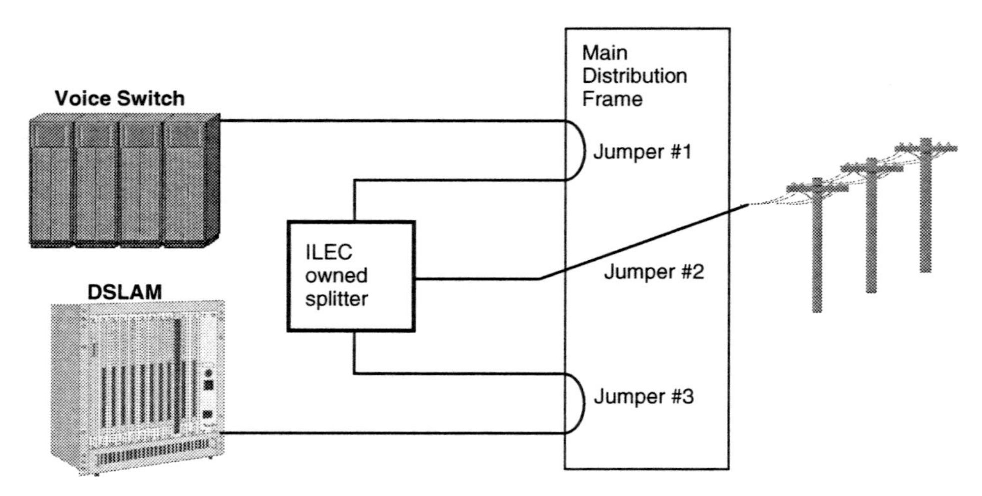

Figure 8.5 Central office configuration with ILEC-owned line sharing splitter.

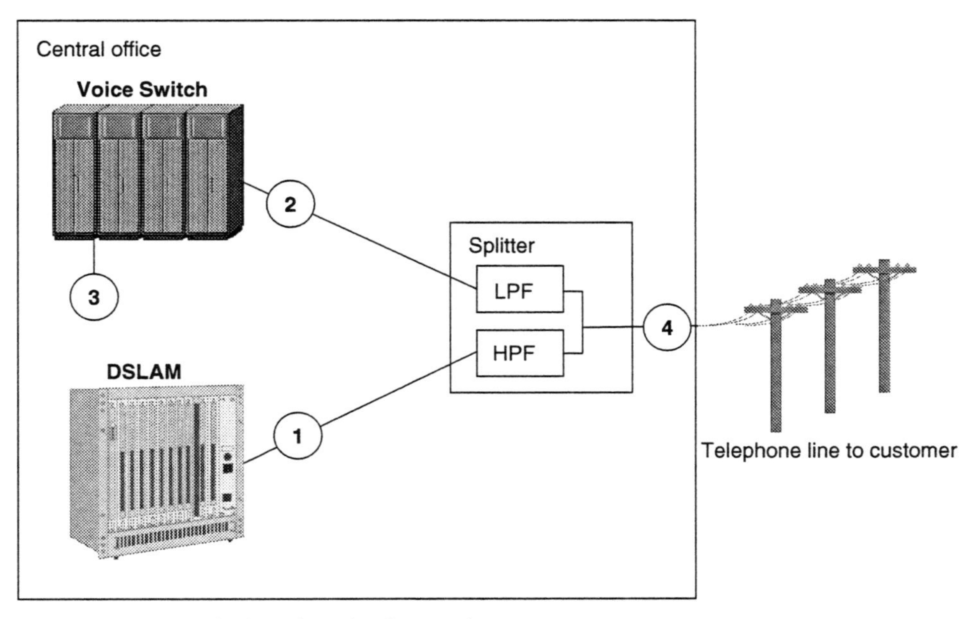

Figure 8.6 Potential locations for line testing systems.

4 kHz unless a bypass function was included with the splitter. Pre-service line qualification could be performed on any line with voice service.

3. *Behind the voice switch.* This position could be used for a line testing system owned by the voice carrier. This takes advantage of the existing metallic test access network within the voice switch. However, the bandwidth of this test bus is less than 100 kHz on some voice switches, and a splitter bypass function would be needed for testing beyond about 4 kHz. Testing within the 4 kHz voice band, such as performed by MLT (mechanized loop test system), is sufficient to detect gross line faults (e.g. line short, open, foreign voltage) and to make a gross estimate of line length. A much greater bandwidth measurement is necessary to generate a moderately accurate estimate of DSL bit-rate capacity. ILECs are providing CLEC access to MLT to perform tests on lines used for line sharing.
4. *Line side of the splitter.* This position could be used by the ILEC or the CLEC. This test access location avoids the need to bypass the splitter for full-band line testing. It is unlikely that this could be used for pre-service line qualification since it would be too expensive to connect all voice lines to the test access system. Integration of the splitter and the test access would help the economics of this configuration.
5. *Manual test access (not shown in the figure).* A technician could connect a portable test set at the MDF or possibly at the splitter.

A combination of methods 1 and 3 above, could be used to test the high band via functions built into the DSLAM, and the low band via automated access to the ILEC's voice-band line test system.

The FCC line sharing order also expanded the charter of the NRIC to advise the FCC on policy for spectrum management of local lines based on technical standards developed by the T1E1.4 national standards working group.

8.5 UNBUNDLING AND DLC-FED LINES

Nearly one-third of subscriber lines in the United States were served from digital loop carrier remote terminals (DLC-RTs) or fiber-fed next generation DLC (NGDLC) in the year 2000. DLC and NGDLC shorten the copper telephone line by using fiber, T1, or HDSL to link a remote line multiplexer (the DLC-RT) to the CO, and then using individual copper telephone lines to connect each customer to the DLC-RT located within their neighborhood. The Telecommunications Act of 1996 required the unbundling of lines served from DLC-RTs, and it further required the unbundling of spare fibers connecting the CO to the remote sites. Using these facilities, a CLEC could place their own NGDLC-RT or remote DSLAM at the remote site once they had obtained the rights to locate equipment on the property. The high cost of outdoor cabinets, construction, power access, and the difficulty of obtaining public easement rights make CLEC overbuilds generally unattractive for neighborhood remote serving sites. A more attractive opportunity that is widely practiced is the placement of a remote DSLAM within an office building or multiple dwelling unit (MDU, e.g., an apartment building) to serve many customers within the building.

Being less than 12,000 feet in length and lacking loading coils, the lines fed from DLC-RT sites are ideal for DSL transmission. However, each DLC-RT must first be upgraded with the addition of DSL line cards, new common equipment plug-ins to perform statistical multiplexing, and, in some cases, an additional fiber-optic link to the central office.

CLECs can lease unbundled DLC transport for POTS and ISDN that provides a path from a DLC-COT (central office terminal) port to an RT-fed telephone line connecting to the customer's premises. The FCC's Second Memorandum Opinion and Order released September 8, 2000, permitted SBC Corporation to provide DSL access via ILEC owned NGDLC equipment (including the DSL line cards) and the associated optical concentration devices (OCDs) in the central office that aggregate the traffic from all the NGDLC systems. The FCC's Order required SBC to lease unbundled DSL access to CLECs at a price approximately equal to its cost. With this, the CLEC can lease access at the OCD to the DSL data bit-stream to and from a customer. This can be provided in a line sharing arrangement where the ILEC provides POTS on the same line. Though this FCC Order was specific to SBC, it clearly establishes a precedent for similar arrangements by other ILECs.

With line sharing and unbundled access to NGDLC-fed DSL lines, data carriers can provide service to most customer lines. The principal exceptions are lines exceeding 18,000 feet and lines using digital added main line (DAML, also know as universal digital carrier) technology. The attenuation of the lines beyond about 18,000 feet prohibit DSL operation at data rates needed for many applications.

High bit-rate operation on these long lines could be enabled by a midspan repeater, but ILECs have not made a practice of providing repeaters for the use of CLECs. Midspan repeaters are expensive, primarily due to the labor cost of splicing the repeater into the cable and the equipment cost for the apparatus case to protect the repeater from the harsh outside environment. The cost of the repeater can be reduced by encapsulating the repeater in epoxy with a wire stub enclosed; this configuration is known as a repeater brick. Mounting repeater bricks can be a problem if there are more than a few of them.

DAML systems convey two or more digitally derived POTS channels via one copper telephone line. DSL-like transmission is used on the line, with the traditional analog voice interfaces re-created at both the customer and the central office ends of the line. Unlike voice-over DSL systems, traditional DAML systems convey only derived voice channels and do not also convey data (except for data modulated within the voice band using V.34 or V.90 modems). Most DAML systems use 2B1Q modulation in a band from zero to a few hundred kHz . Because most DSL transmission systems also operate within this frequency band, DSL may not be used on lines with traditional DAML systems. Some new DSL products combine a DAML-like multiline voice function and a high speed data channel in an integrated system.

8.6 UNBUNDLING AND NETWORK OPERATIONS

Line sharing posed a huge challenge to telephone company methods and operations systems. It had never been imagined that two service providers might share a line. Revisions were necessary to keep track of services from different providers on the same line, and new methods were needed so that the entire line was not disconnected when a customer asked for one service to be disconnected. Also, new provisions were needed to deal with the line sharing splitter in the central office. Nevertheless, the changes proved to be feasible.

CLECs are provided direct access to the telephone company operations systems that keep track of telephone lines (such as LFACS, loop facilities assignment system) and test the lines (such as LMOS-MLT, loop maintenance operations system—mechanized loop test). LFACS contains information about what services are assigned to each line, and length of the line.

When a CLEC orders an unbundled line, the ILEC assigns a spare wire pair, and places the necessary jumper wires at the MDF and cross-box at the FDI (feeder distribution interface). The ILEC will remove loading coils and bridged taps if the CLEC requests and pays for the special conditioning. Usually, the ILEC does not perform preservice tests of the line unless requested by the CLEC. One benefit of line sharing is that more than 95 percent of lines work the first time because the line was already working for voice service, whereas about one-third of dedicated unbundled lines (not line shared) do not work the first time due to bad wire-pairs, and various line provisioning errors.

The early stipulations from the FCC indicated that a DSL transmission system was acceptable for general use if the system had been successfully used for service without reported harm to other services. This policy was motivated by the desire to spur competition and innovation. However, it highlighted the urgent need for a technically sound basis for determining spectral compatibility. "Harm" in the original policy was not well defined, and a system that caused no harm on short lines, or in small numbers, could easily cause a total loss-of-service for other systems within the same cable where the conditions were different. By the time such problems were widely reported and the fault proven, corrective measures could be difficult and expensive. The situation would be like the Food and Drug Administration approving a new drug for general use after reviewing only one clinical trial involving two persons. Realizing this in 1998, the FCC held an industry forum on spectrum management of DSL systems, and then requested T1E1.4 to develop a standard to define the technical specifications for DSL spectrum management.

The T1.417 spectrum management standard was completed by T1E1.4 at the end of the year 2000. It provides a set of predefined spectrum management classes that are suitable for DSL systems operating over smaller and larger bandwidths. Each spectrum management class has specified limits on total transmitted power, power spectral density (PSD), output voltage, transverse balance, and a deployment guideline. The deployment guideline specifies the maximum length of line the system may be used on. The spectrum management standard also stated technology-specific criteria for SDSL and SHDSL type transmissions systems, as well as providing an open-ended Method B that permits the demonstration of spectral compatibility for future technologies. Chapters 10 and 11 provide a more detailed discussion of spectrum management.

A carrier (ILEC or CLEC) using a line should identify the class of signal being transmitted on the line. This may be the T1.417 spectrum management class, or some other equivalent indication of the type of signal bandwidth, power, and deployment restriction. This information is needed for the administration of the deployment restrictions, and to assist trouble resolution.

DSL equipment at the customer's premises must be certified to not harm the network by the FCC Part 68 process or a new DSL specific process that may be developed in the future. Similarly, equipment at the network-end should also be addressed by DSL certification in the future. It is expected that this certification will require that signals transmitted by sample equipment be measured to verify that the signals are within the limits specified in the T1.417 spectrum management standard. These measurements and the corresponding interpretation of the measurement results should be verified by an entity with the necessary test systems, expertise, and impartiality.

It is unlikely that spectrum management will be proactively enforced. Carriers will be contractually bound to adhere to the requirements for the type of signal being sent, and to use certified equipment. Corrective action will be taken only if service trouble is detected. The FCC rules do not specify the process for resolving interference troubles. When excessive crosstalk is suspected to be the cause of service trouble, the carrier of the victim system may request the ILEC to investigate the trouble. If the ILEC finds that a transmission system is transmitting a signal in excess of the contractual terms for the use of the line, the ILEC has the right to disconnect that line. If all carriers adhere to the T1.417 spectrum management standard, this should not occur. The diagnosis of trouble resulting from excessive crosstalk will likely require lengthy efforts by highly qualified technicians. The degree of investigation will depend on the contractual agreements between the ILEC and CLEC. If the source of the trouble is not agreed to, then the dispute between carriers may need to be resolved by the state PUC.

Because there is 10 dB less crosstalk coupling between pairs in different binder groups, better DSL performance can be achieved when certain types of transmission systems are segregated into dedicated binder groups. This is true for systems whose performance is not limited by self near-end crosstalk, such as ADSL. Regardless of the number of ADSLs within a binder group, ADSL will perform better if there are no HDSL, SDSL, SHDSL, or T1 carries in the same binder group. Proposals to segregate ADSL from other services were rejected by the FCC because the spare pairs in these binder groups would be unavailable for other types of DSL service. The FCC line sharing order does permit the segregation of T1 carrier into dedicated binder groups because the FCC identified T1 carrier as a *known disturber*. T1 carrier is the only *known disturber* at this time, and indeed, T1 carrier is one of the worst crosstalkers found in telephone cables. The T1 carrier can even cause substantial crosstalk into the lines in adjacent binder groups. However, the T1 carrier does not always cause trouble for DSLs in the same binder group. ADSL, for example can operate reliably at 1.5 Mb/s on lines less than 9 kft long with a T1 carrier disturber in the same binder group. Nevertheless, ILECs have realized that removing T1 carrier will greatly expand the capacity for other services. As a result, new T1 carrier lines were rarely installed after the year 2000, and some ILECs are proactively replacing a large portion of their existing T1 carrier lines by moving the DS1 service to fiber-to-the-business, or HDSL2 lines. A 4-wire 1.5 Mb/s transmission system using TC-PAM modulation is expected to assure spectral compatibility for all lines, including long lines with repeaters.

It is tempting to try to expand the bit rate or line length permitted for a certain DSL technology by placing a restriction on the number such systems within a binder group, for example, extending the spectrum management deployment guidelines for SDSL technology by an extra 1000 feet, if only five SDSLs were permitted within a binder group. Such methods for binder group management are impractical for several reasons:

- It is difficult to administer limitations on the number of certain types of DSLs in a binder group.
- Such population restrictions would reduce the number of wire pairs available for DSL service.
- Since the pair-to-pair crosstalk coupling with a binder group varies by about 12 dB, very little would be gained by assuming the worst few pairs versus all pairs in the binder group.
- The effects of mixed types of disturbers on other pairs would negate any gain.
- Binder group integrity (the grouping of the set of pairs) is often not maintained throughout the length of the cable.

CHAPTER 9

SPECTRAL COMPATIBILITY OF DSL SYSTEMS

One of the driving elements of DSL technologies is to use the same cables that provide plain old telephone service (POTS) to customers for providing high-bit rate digital services to the end users. However, the key difference from analog voices services is that the bandwidths used in DSLs are significantly greater than that of the analog voice channel. In the cable, these wide-band signals crosstalk into other wire pairs in the same cable and may cause interference into their corresponding signals. Hence, crosstalk is one of the key impairments that limit the performance of digital subscriber line systems.

Spectral compatibility is the quantification of the level of disturbance that one DSL system has on other systems deployed in the same cable. A clear understanding of the interactions that the different types of systems have on each other in the cable is the foundation for managing the deployment of DSL systems for maintaining a specified quality of service.

In the mass deployment of digital subscriber line services, we can expect the cable to be filled with many different types of signal spectra. Given the wide variety of signals in the cable, there can be a wide range of interference between the different systems and a set of rules or guidelines are needed to manage the deployment of these systems to assure that a minimum quality of service. These rules or deployment guidelines is called *spectrum management.* The rules of spectrum management are based on a working knowledge of the spectral compatibility of the various DSL systems and an agreement on a desired quality of service.

This chapter presents an overview of the spectral compatibility fundamentals and demonstrates the effects that different DSL services deployed in the loop plant have upon each other. Included are discussions on the loop plant environment, cable characteristics, near-end crosstalk and far-end crosstalk, a comparison of key selected DSL spectra, and the compatibility of these spectra with others in the cable. The next chapter describes spectrum management, which is based on the principles described here.

9.1 THE LOOP PLANT ENVIRONMENT

POTS is provisioned to customers by routing twisted wire-pairs between the telephone network equipment at a central office (CO) or remote terminal (RT) and the customer premises (CP) location. The twisted wire pair that connects the CO to the CP is the subscriber loop, which may consist of sections of copper twisted wire pairs of one or more different gauges.

Twisted wire pairs are contained in cables that have large cross sections near the central office. Within each cable, the twisted wire pairs are grouped into binders of 10, 25, or up to 50 wire pairs, and there could be up to 50 binder groups per cable.

Figure 9.1 shows the basic architecture of the loop plant, in which subscriber loops are provisioned. Each subscriber loop can be divided into three portions of cable: feeder cable, distribution cable, and drop wire. Feeder cables provide links from the CO to a connection point or feeder distribution interface (FDI) in a concentrated customer area. Distribution cables provide links from the FDI to customer locations. The FDI provides connections of the wire pairs in the feeder cable

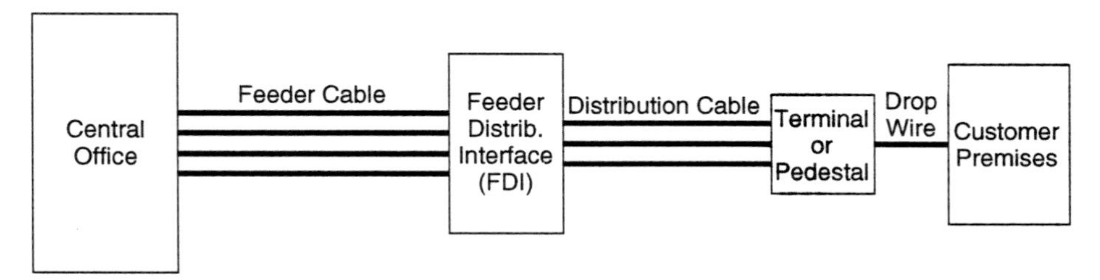

Figure 9.1 Architecture of the loop plant.

with those in the distribution cable. The drop wires represent the portion of wire that extends from the terminal on a telephone pole to the customer premises or underground from the pedestal to the customer premises.

Because loop plant construction is completed prior to service request, distribution cables are made available to all existing and potential customer sites. It is common practice to connect a twisted pair from a feeder cable to more that one wire pair in the distribution cable. Multiple connections from a feeder or distribution cable to more than one-customer premises form bridged taps. Typically, only one customer is serviced at one time while the other bridged taps are unterminated. Bridged taps provide the ability for cable pairs to be connected to any customer site passed by the cable, and they exist on about 80 percent of loops in the United States and Canada. Bridged taps also exist in the loop plant of many countries, though generally at a much lower proportion than found in North America.

The loop plant was originally designed to provision customers with POTS. To ensure proper quality of service, design rules were defined for subscriber loop provisioning. One set of rules that governs distribution of twisted wire pairs for voice service from the CO to the CP is the resistance design rules. Implemented in the 1980–1981 time frame (a very small portion of the current loop plant), the design rules are summarized as follows:

1. Loop resistance is not to exceed 1500 Ω.
2. Inductive loading is to be used whenever the sum of all cable lengths, including bridged taps, exceeds 15 kft.
3. For loaded cables, 88 mH loading coils are placed at 3 kft from the CO and thereafter at intervals of 6 kft.
4. For loaded cables, the total amount of cable, including bridged taps, in the section beyond the loading coil furthest from the CO should be between 3 kft and 12 kft.
5. There are to be no bridged taps between loading coils and no loaded bridged taps.

Revised resistance design (RRD) rules are defined in [11]. The majority of CO-fed loops in the United States follow the RRD rules. These rules require that the maximum loop resistance on an 18 kft twisted wire pair between the central office to the customer premises must be less than 1300 Ohms and on loops between

18 kft and 24 kft, the maximum resistance is 1500 Ohms. Loading coils are applied to all loops greater than 18 kft or have loop resistance greater than 1300 Ohms.

Table 9.1 provides an example of the typical RRD design loops for various distances; loops shorter than 15.3 kft are entirely 26 AWG wire. The numbers in the wire-gauge columns indicate the length of that type of wire in kilofeet.

Telephone cables are designed with different wire gauges ranging from 26 AWG (thin diameter resulting in higher resistance per unit length) to 19 AWG (thick diameter resulting in lower resistance per unit length). Because distances from the CO to each customer are different, distribution cables are equipped with different wire gauges to meet the resistance design guidelines and provide service to the maximum number of customers. On long loops, the distribution cables tend to use thicker gauge wire in the regions closer to the subscriber location in order to minimize the total loop resistance. At the CO, the feeder cables tend to use fine diameter gauges (typically 26 AWG) to maximize the number of wire pairs being served by the central office.

Some customers may be located so far away from the CO that it may not be possible to meet the resistance design rules. If a subscriber loop is provisioned with a length greater than the maximum defined by the RRD rules, then loading coils must be inserted in the loop to achieve proper voice quality. Note, however, that subscriber loops provisioned with loading coils are not suitable for support of wide-band DSL services because loading coils introduce too much attenuation of the frequencies above the voice channel, which are required by the DSL system. In short, loading coils must be removed from any subscriber loop that is to support a DSL based service.

Another set of rules, called carrier serving area (CSA) rules, define the distribution of twisted wire pairs from digital loop carrier systems. The radius covered by CSA rules are up to 9000 ft of 26-gauge wire and up to 12,000 ft on 24-gauge wire. The concept of CSA rules were originally developed for provisioning loops from digital loop carrier (DLC) systems in support 56 kb/s digital data service (DDS) and later slightly modified for the support of POTS from a DLC system. The CSA rules are defined as follows:

1. Nonloaded cable only.
2. Multigauge cable is restricted to two gauges.
3. Total bridged tap length may not exceed 2.5 kft, and no single bridged tap may exceed 2.0 kft.

Table 9.1 Example RRD Loops

Loop length (kft)	26AWG (kft)	24 AWG (kft)	22 AWG (kft)
16	14.5	1.5	
17	13	4	
18	11.5	6.5	
19	9	10	
20	7.5	12.5	
24	1	23	
25	0	24	1
26	0	22	4
27	0	20	7

4. The amount of 26 AWG cable may not exceed a total length of 9 kft, including bridged taps.
5. For single gauge or multigauge cables containing only 19, 22, or 24 AWG cable, the total cable length may not exceed 12 kft including bridged taps.
6. The total cable length including bridged taps of a multi-gauge cable that contains 26 gauge wire may not exceed

$$12 - \frac{3(L_{26})}{9 - L_{Btap}} \text{kft}$$

where L_{26} is the total length of 26 gauge wire in the cable (excluding any 26 gauge bridged tap) and L_{Btap} is the total length of bridged tap in the cable. All lengths are in *kft*.

The above CSA guidelines do not include any wiring in the CO nor any drop wiring and any wiring in the customer premises.

As the transport medium for wide-band signals, the twisted wire pairs introduce linear impairments into the signal. Specifically, the linear impairments are propagation loss, amplitude distortion, and delay distortion. The propagation loss in dB is directly proportional to the distance of the loop. Amplitude distortion results from the fact that signal components at higher frequencies experience more amplitude loss than the components at low frequencies. To a first order, the amplitude response of the twisted wire pair channel rolls off at roughly the square root of frequency. Finally, the delay is such that at low frequencies (less than approximately 10 kHz) there is very sharp variation in the group delay and at higher frequencies the delay response is relatively constant [1].

Test loops have been developed for American Wire Gauge (AWG) in T1E1.4 and metric cables in ETSI for ISDN [3–4], HDSL [5–7, 14], and ADSL [2, 15, 16] systems. These test loops provide an industry-accepted basis for evaluating the performance of the various DSL systems in the presence of different crosstalk scenarios. Reference [1] provides a detailed description of cable modeling. References [2–7] provide a comprehensive list of the test loops used for the North American and European loop environments.

9.2 CROSSTALK IN THE LOOP PLANT

Crosstalk generally refers to the interference that enters a communication channel, such as twisted wire pairs, through some coupling path. The diagram in Figure 9.2 shows two examples of crosstalk generated in a multi-pair cable. On the left-hand side of the figure, signal source $V_j(t)$ transmits a signal at full power on twisted wire pair j. This signal, when propagating through the loop, generates two types of crosstalk into the other wire pairs in the cable. The crosstalk that appears on the left-hand side, $x_n(t)$ in wire pair i, is called near-end crosstalk

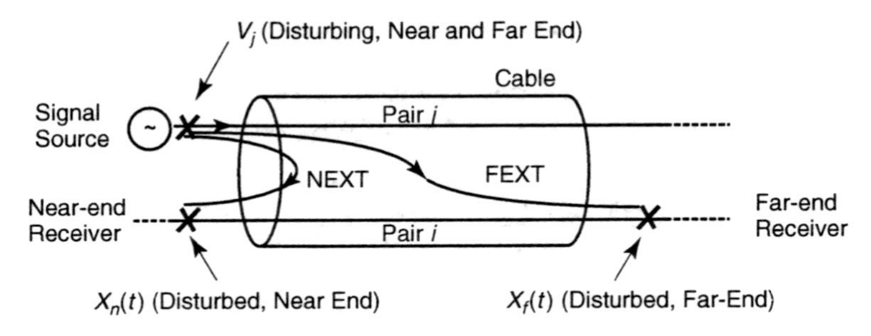

Figure 9.2 NEXT and FEXT in a multipair cable.

(NEXT) because it is at the same end of the cable as the crosstalking signal source. The crosstalk that appears on the right-hand side, $x_f(t)$ in wire pair i, is called far-end crosstalk (FEXT) because the crosstalk appears on the end of the loop opposite to the reference signal source. In the loop plant, NEXT is generally far more damaging than FEXT because NEXT has a higher coupling coefficient and, unlike far-end crosstalk, near-end crosstalk directly disturbs the received signal transmitted from the far end after it has experienced the propagation loss from traversing the distance from the far end down the disturbed wire pair.

In a multipair cable, relative to the desired receive signal on one twisted wire pair, all the other wire pairs are sources of crosstalk. For DSL systems, the reference cable size for evaluating performance in the presence of crosstalk is a 50-pair cable [2]. So by reviewing the example shown in Figure 9.1, we see that relative to the received signal on wire pair i, the other 49 wire pairs are sources of crosstalk (both near-end and far-end).

In the United States, 25-pair binder groups are most common; however the T1.417 spectrum management standard employs the 50-pair binder model for conservatism. The difference between a 50-pair and a 25 pair model is approximately 1.8 dB. This is shown in Figure 9.3.

9.2.1 Near-End Crosstalk Model

As described in references [2, 3, 5 and 6], for the reference 50-pair cable, the near-end crosstalk *coupling* of signals into other wire pairs within the cable is modeled as

$$|H_{NEXT}(f)|^2 = X_{49} \times \left(\frac{N}{49}\right)^{0.6} \times f^{\frac{3}{2}}$$

where $x_{49} = \dfrac{1}{1.13 \times 10^{13}}$ is the coupling coefficient for 49 NEXT disturbers, N is the number of disturbers in the cable, and f is the frequency in Hz. Note that the maximum number of disturbers in a 50-pair cable is 49. A signal source that out-

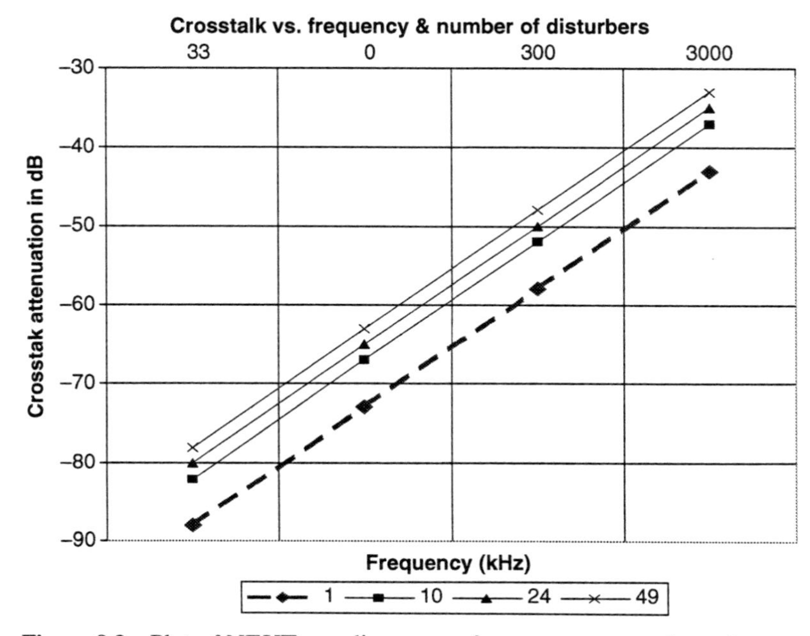

Figure 9.3 Plot of NEXT coupling versus frequency per number of disturbers.

puts a signal with power spectral density $PSD_{Signal}(f)$ will inject a level of NEXT into a near-end receiver that is

$$PSD_{NEXT}(f) = PSD_{Signal}(f) \times |H_{NEXT}(f)|^2.$$

So as illustrated in Figure 9.2, if there are N signals in the cable with the same power spectral density $PSD_{Signal}(f)$, the PSD of the NEXT at the input to the near-end receiver on wire pair i is $PSD_{NEXT}(f)$.

Note from the above expressions that the crosstalk coupling is very low at the lower frequencies and the coupling increases at 15 dB per decade with increasing frequency. For example, at 80 kHz, the coupling loss is 57 dB for 49 disturbers. The loss (in dB) for 49 disturbers at other frequencies may be computed using the following formula:

$$L_{49} = 57\text{ dB} - 15 \cdot \log\left(\frac{f}{80\text{ kHz}}\right),$$

where L_{49} is the near end crosstalk coupling loss in dB and f is the frequency in kHz.

9.2.2 Far-End Crosstalk Model

Correspondingly, in the same 50-pair cable, the far-end crosstalk coupling of signals into other wire pairs is modeled as

$$|H_{FEXT}(f)|^2 = |H_{Channel}(f)|^2 \times \left(\frac{N}{49}\right)^{0.6} \times k \times d \times f^2$$

where $H_{Channel}(f)$ is the channel transfer function, $k = 8 \times 10^{-20}$ is the coupling coefficient for 49 FEXT disturbers, N is the number of disturbers, d is the coupling path distance, and f is the frequency in Hz.

Note that the coupling is small at low frequencies and large at higher frequencies. The coupling slope increases at 20 dB/decade with increasing frequency.

9.3 NEXT VS. FEXT

If we compare the coupling coefficient for NEXT relative to that of FEXT, we see that the near-end crosstalk coupling is approximately six orders of magnitude greater than that for far-end crosstalk. On the other hand, notice that the coupling for NEXT increases at 15 dB per decade with increasing frequency, whereas the coupling for FEXT increases at 20 dB per decade with increasing frequency.

FEXT is an important consideration for ADSL crosstalk into other ADSLs. This is especially important for crosstalk from RT-fed ADSLs; this is a principal topic for work on the Issue 2 spectrum management standard.

A comprehensive study of loop plant cable characteristics, linear impairments, and crosstalk may be found in [1].

9.3.1 Channel Capacities in the Presence of NEXT and FEXT

Let us assume that a 50-pair cable is filled completely with signals that have the same power spectral density.

Figure 9.4 contains an example of NEXT and FEXT on a 9000 foot 26-gauge loop. The figure plots the power spectral density (in dBm/Hz) of the transmit signal, the insertion loss of the 9000 foot 26-gauge loop (in dB), and the PSDs of the received signal, the 49 near-end crosstalk disturbers, and the 49 far-end crosstalk disturbers. In this example, the transmit signal is an arbitrary pass-band signal whose bandwidth is approximately 700 kHz between its half-power points (−3 dB frequencies). Its power spectral density in the pass-band region has a level of –40 dBm/Hz. The power spectral density of the received signal is shaped by the insertion loss of the channel as shown in the figure. For example, at 500 kHz the insertion loss of the channel is 50 dB, so the resulting receive signal PSD is –90 dBm/Hz, which is 50 dB below its transmit level.

For the special case where the crosstalk comes from signals with the same PSD as that of the corresponding transmitter, near-end crosstalk is more specifi-

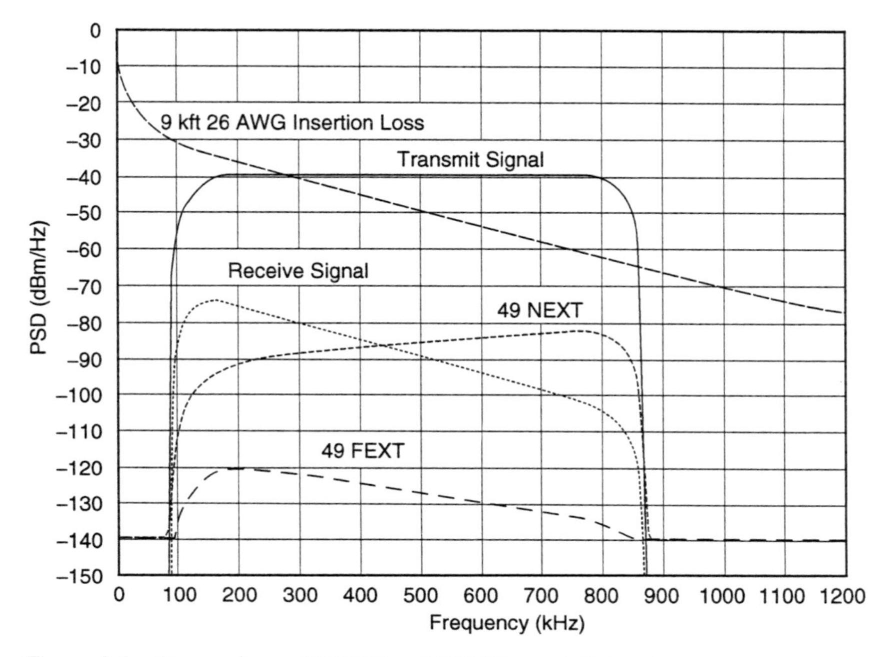

Figure 9.4 Comparison of NEXT and FEXT crosstalk levels.

cally referred to as self-NEXT (or simply SNEXT) and far-end crosstalk is referred to as self-FEXT (or simply SFEXT). Such is the case in the example of Figure 9.4.

In addition to the magnitude of the received signal, Figure 9.4 also shows the magnitudes of the NEXT and FEXT crosstalk power spectral densities. In each case, we assume 49 disturbers in the 50-pair cable. Also note that the noise floor for the signals shown in the figure is –140 dBm/Hz.

The area between the "Receive Signal" and "49 SNEXT" curves in Figure 9.4 represents the signal-to-noise ratio (SNR) at the receiver input relative to near-end crosstalk. For this near-end crosstalk case, the SNR is approximately 3.7 dB on the 9 kft 26 AWG loop. Correspondingly, the area between the "Receive Signal" and "49 SFEXT" curves in Figure 9.4 represents the SNR relative to far-end crosstalk. For this far-end crosstalk case, the SNR is approximately 40 dB on the 9 kft 26 AWG loop. Note that the SNR due to NEXT is significantly lower than that due to FEXT; hence, a FEXT limited environment is much more desirable than a NEXT limited environment.

To further quantify the effects of NEXT and FEXT, we compute the resulting Shannon (or channel) capacities, which defines the theoretical maximum bit rate that can be transmitted over a given channel with a small error rate. In general, the Shannon capacity for a given channel is

$$C = \int_0^\infty \log_2[1 + SNR(f)] \cdot df$$

where C is the channel capacity in bits/second, $SNR(f)$ is the signal-to-noise ratio density at the receiver input, and f is frequency in Hz. For the case of SNEXT, the channel capacity is 1.7 Mb/s and that for SFEXT is approximately 8.7 Mb/s; a difference of approximately 7 Mb/s.

Clearly, as shown in the above example, near-end crosstalk strongly dominates over the effects of far-end crosstalk in the digital subscriber loop environment.

9.4 THE "PRIMARY" DSL SIGNAL SPECTRA

This section describes power spectral densities of different DSL transmit signals that are deployed in the network. The label "primary" means that the DSL systems described here were the early generation to be deployed; hence they form the *basis* for definition of spectral compatibility. The standards based DSL include ISDN (integrated services digital network), HDSL (high-rate digital subscriber line), and DMT-based ADSL (asymmetric digital subscriber line). Other DSLs signals include 2B1Q based SDSL (symmetric digital subscriber line), CAP-based MSDSL (multirate symmetric digital subscriber line), and CAP-based RADSL (rate adaptive digital subscriber line) per Committee T1 TR-059. The signals targeting symmetric transmission applications include ISDN, HDSL, SDSL, and MSDSL, and those targeting data access applications are ADSL and RADSL. Following are descriptions of the signal power spectral densities for each of the above signals.

9.4.1 ISDN

ISDN basic rate provides symmetrical transport of 160 kb/s on the subscriber line. The line code is 2B1Q, and the corresponding transmit signal power spectral density is expressed as [3]

$$PSD_{ISDN}(f) = \frac{5}{9} \cdot \frac{V_p^2}{R} \cdot \frac{\left[\sin\left(\frac{\pi f}{f_0}\right)\right]^2}{\left(\frac{\pi f}{f_0}\right)^2} \cdot \frac{1}{1 + \left(\frac{1}{f_{3dB}}\right)^4}, \; f_{3dB} = 80 \text{ kHz}, 0 \leq f < \infty$$

where $f_o = 80$ kHz, $V_p = 2.5$ Volts and $R = 135$ Ohms. The transmit power of ISDN is 13.5 dBm. In Figure 9.5, the solid line plots the power spectral density of the ISDN transmit signal. Included in the plot is the effect of a high-pass filter with a cutoff frequency at 2 kHz, which models the transformer coupling of the signal onto the line.

Also shown in the figure (dotted plot) is the PSD of 49 near-end crosstalk disturbers from ISDN. As defined in references [2, 3, 5, 6, and 12], the noise floor of the system (modeled at the receiver input) is – 140 dBm/Hz.

Both the upstream and downstream signals occupy the same frequency band, so an echo canceler is used to separate the two directions of transmission on the

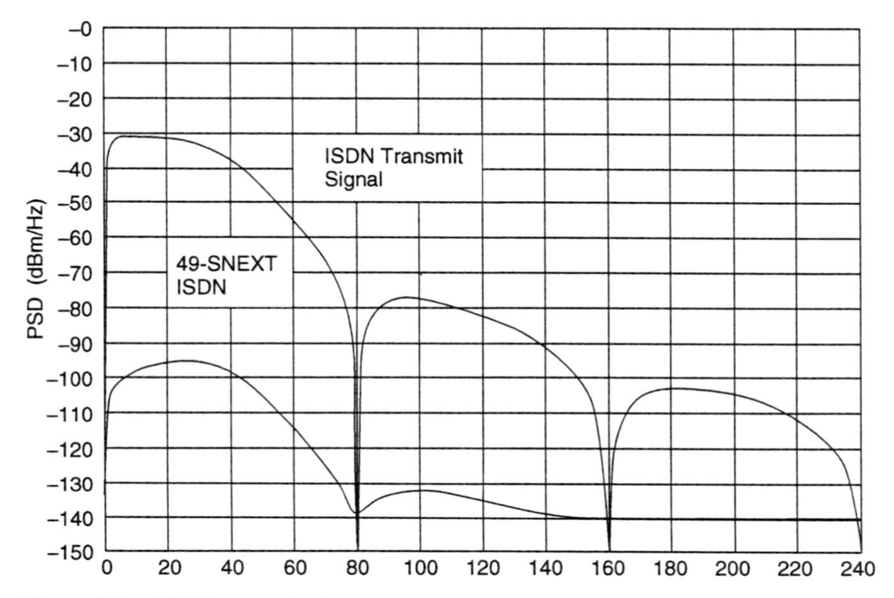

Figure 9.5 ISDN transmit signal and 49-NEXT spectra.

subscriber line. In such a system, ISDN is subject to self-near-end crosstalk (SNEXT).

The deployment objective for ISDN is to operate on nonloaded loops that range up to 18 kft in the presence of SNEXT. These include all loops that meet RRD rules.

9.4.2 HDSL

In North America, HDSL is a service that provides the transport of T1 (1.544 Mb/s) signals between the CO and the CP. This service is deployed using two subscriber lines (e.g., four wires), where the bit rate is 784 kb/s on each wire pair and half of the T1 payload is carried on each pair. The line code is 2B1Q, and the corresponding transmit signal power spectral density is expressed as

$$PSD_{HDSL}(f) = \frac{5}{9}\cdot\frac{V_p^2}{R}\cdot\frac{\left[\sin\left(\frac{\pi f}{f_0}\right)\right]^2}{\left(\frac{\pi f}{f_0}\right)^2}\cdot\frac{1}{1+\left(\frac{1}{f_{3dB}}\right)^8}, f_{3dB} = 196\text{ kHz}, 0 \le f < \infty$$

where $f_o = 392$ kHz, $V_p = 2.7$ Volts and $R = 135$ Ohms [5, 6, 13]. The transmit power of the HDSL signal is 13.5 dBm. In Figure 9.6, the solid line plots the power spectral density of the HDSL signal. Included in this plot (but not shown in the preceding equation) is a high-pass filter with a 2 kHz cutoff frequency to model the effects of transformer coupling. The dotted plot is the PSD of 49 near-end crosstalk disturbers from HDSL.

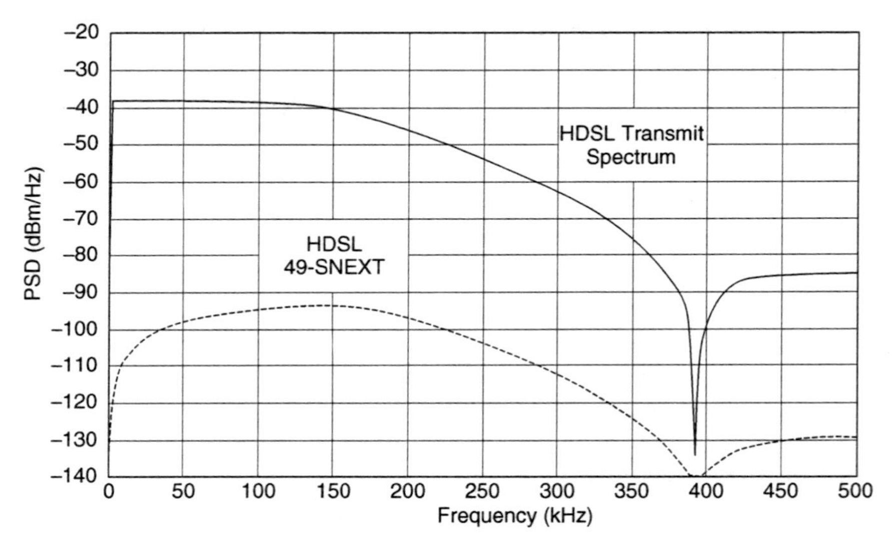

Figure 9.6 HDSL transmit signal and 49-NEXT spectra.

Both the upstream and downstream signals occupy the same frequency band, so an echo canceler is used to separate the two directions of transmission on the subscriber line. In such a system, HDSL is subject to SNEXT.

The deployment objectives for HDSL are operation on loops that meet CSA requirements while operating in the presence of SNEXT.

9.4.3 Symmetric DSL Technologies

SDSL technologies define the transport of symmetric DSL services on a single twisted wire pair. SDSL solutions deployed are echo cancellation based and are implemented using CAP and 2B1Q technologies. To distinguish between the two technologies, we refer to the 2B1Q solutions as SDSL and to the CAP-based solutions as MSDSL.

The MSDSL bit rates considered here are 160 kb/s, 272 kb/s, 400 kb/s, 784 kb/s, and 1560 kb/s. Each of the CAP-based systems assumes a two-dimensional eight-state trellis code to provide additional immunity to crosstalk. Figure 9.7 shows the spectral plots of the MSDSL spectra. The bandwidth of each spectrum is directly proportional to the bit rate, so the 160 kb/s signal has the narrowest bandwidth and the 1560 kb/s signal has the widest bandwidth. The line codes for each system are as follows: 160 kb/s uses coded 8-CAP (2 data bits per symbol), 272 kb/s and 400 kb/s each use coded 16-CAP (3 data bits per symbol), and 784 kb/s and 1560 kb/s each use coded 64-CAP (5 data bits per symbol). The spectral shaping of each transmit signal is square-root raised cosine with a roll-off factor of 15 percent. For each bit rate, the spectrum is scaled and shaped for a transmit power of 13.5 dBm.

Figure 9.8 shows the SDSL power spectral density plots for four data rates: 160 kb/s, 384 kb/s, 784 kb/s, and 1560 kb/s. Each system uses NRZ (nonreturn to zero) shaping followed by an N-th order Butterworth filter. The 160 kb/s spectrum

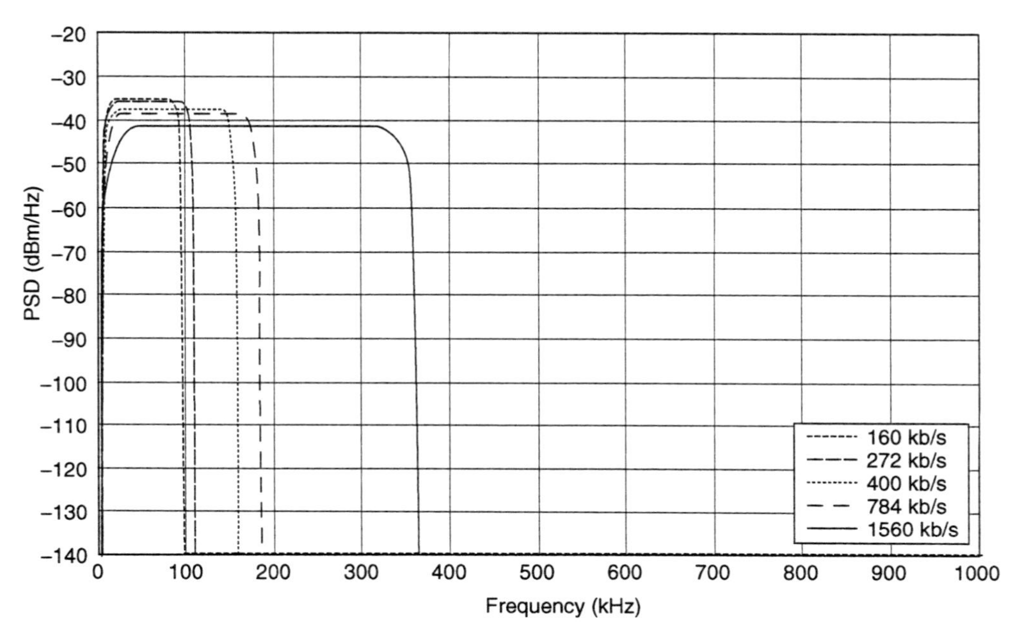

Figure 9.7 MSDSL transmit signal spectra.

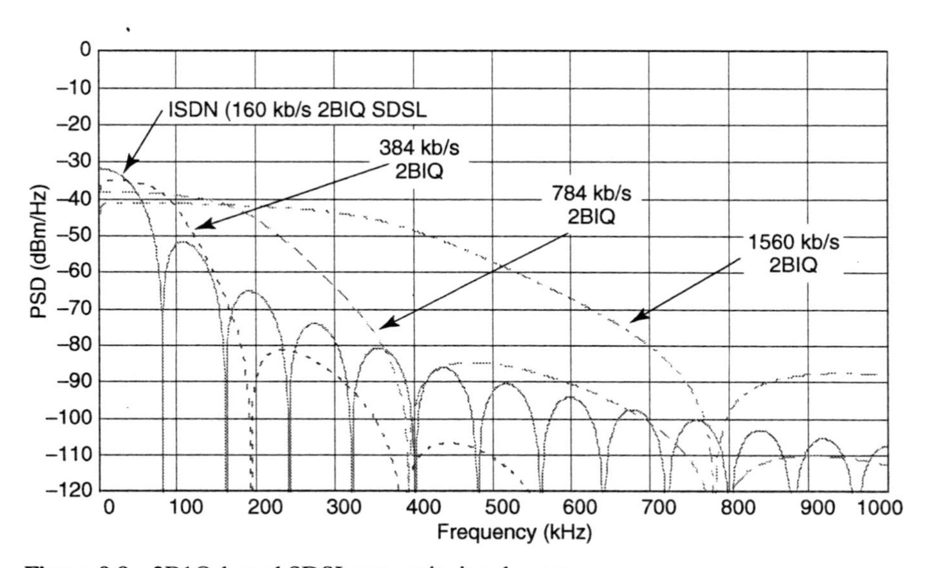

Figure 9.8 2B1Q-based SDSL transmit signal spectra.

is the same as that defined for ISDN in T1.601, which defines a second-order Butterworth filter for out of band energy attenuation. The remaining signals use a fourth-order Butterworth filter for out of band energy attenuation. The first null in each spectrum defines the signal bandwidth.

If we compare the PSDs of Figure 9.7 with those of Figure 9.8, the bandwidths of the multilevel CAP systems are narrower than those of the four level (2B1Q) systems for the same bit rate. Also, the spectral efficiency of the raised cosine shaping has significantly less energy in the high frequency out-of-band region that the NRZ shaping of the 2B1Q signals. From these figures, it is clear that the spectra with the more efficient spectral shaping will provide less interference into other systems at the higher frequencies.

For the same transmit power, the performance of both the CAP and 2B1Q systems are dominated by SNEXT. If the 50-pair cable is deployed with signals having different bandwidths, then consideration must be given to the effect that these different bandwidths have in producing NEXT onto other services. Recall that the NEXT coupling is greater at higher frequencies than at lower frequencies. As described earlier, the NEXT coupling increases at 15 dB/decade with increasing frequency.

9.4.4 ADSL and RADSL

Figure 9.9 shows transmit and near-end crosstalk spectral plots of the upstream and downstream ADSL channels. The same upstream and downstream bands are used for both ADSL [T1.413, G.dmt, and G.lite) and single-carrier RADSL [TR-059]. In general, ADSL and single-carrier RADSL are variable bit rate systems and the actual bandwidths of the upstream and downstream channels may vary depending on the bit rate and noise environment.

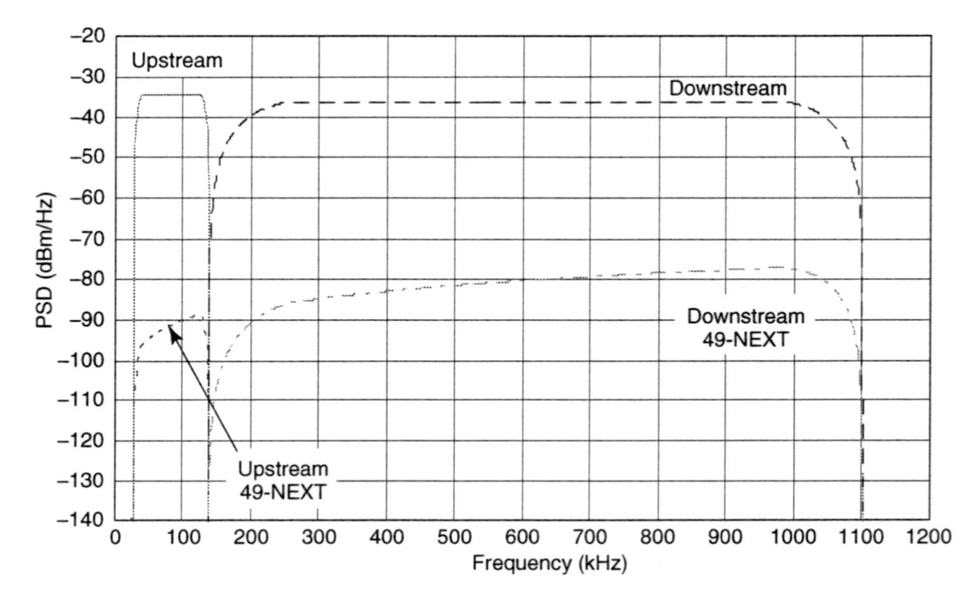

Figure 9.9 ADSL and RADSL FDM signal spectra.

The plots shown in Figure 9.9 are idealized PSDs displaying the maximum possible bandwidth of the upstream and downstream channels. Not shown in the figure are the out-of-band energy levels. Usually the PSD levels represent the RMS values. However, for ADSL the PSD levels shown in the figure are peak values as opposed to RMS values [12] to allow for varying gains at different carrier frequencies. No guard band is specified between the upstream and downstream channel; details of the DMT PSD masks are given in the ADSL standard [T1.413, G.dmt, and G.lite].

The spectra shown in Figure 9.9 use frequency division multiplexing (FDM) to separate the upstream and downstream channel. If the cable has only FDM-based ADSL systems deployed, there is no SNEXT; system performance would be limited by SFEXT. T1.413 and G.dmt also defines an echo-canceled version of ADSL, where the downstream channel completely overlaps the upstream channel. In this case, NEXT will be injected between the upstream and downstream channel. The channel most impacted is the upstream channel, where the NEXT from the downstream channel completely covers the upstream band.

RADSL systems per TR-059 have only been deployed using FDM for separation of the upstream and downstream channels; hence there is no NEXT generated between the upstream and downstream channels. The complete definition of the RADSL PSD masks is given in [9].

9.4.5 T1 AMI

The PSD of the T1 line signal is assumed to be a 50 percent duty-cycle random alternate mark inversion (AMI) code at 1.544 Mb/s. The single-sided PSD is represented by the following expression:

$$PSD_{T1} = \frac{V_p^2}{R} \cdot \frac{2}{f_o} \cdot \left[\frac{\sin\left(\frac{\pi f}{f_o}\right)}{\left(\frac{\pi f}{f_o}\right)} \right]^2 \cdot \sin^2\left(\frac{\pi f}{2f_o}\right) \cdot \frac{1}{1 + \left(\frac{f}{f_{3dB-Shaping}}\right)^6} \cdot \frac{f^2}{f^2 + f_{3dB-Xfmr}^2},$$

$0 \leq f < \infty$ where $V_P = 3.6$ V, $R_L = 100\ \Omega$, $f_o = 1.544$ MHz, $f_{3dB\text{-}Shaping} = 3.0$ MHz is the 3 dB frequency of a third-order Butterworth low-pass shaping filter, and $f_{3dB\text{-}Xfmr} = 40$ kHz is the high-pass transformer coupling cutoff frequency. Figure 9.10 shows a plot of the T1 AMI transmit signal PSD along with a 49-disturber near-end crosstalk PSD.

Unlike most DSLs, T1 carrier does not scramble the transmitted bits. As a result, the actual PSD at any time is highly dependent on the payload bit pattern.

9.5 COMPUTATION OF SPECTRAL COMPATIBILITY

To compute the resulting SNR margin, we first need to compute the resulting output signal-to-noise ratio (SNR_{out}) and then take the difference from the reference SNR value (SNR_{ref}) that corresponds to a bit error rate (BER) of 10^{-7}. Table 9.2

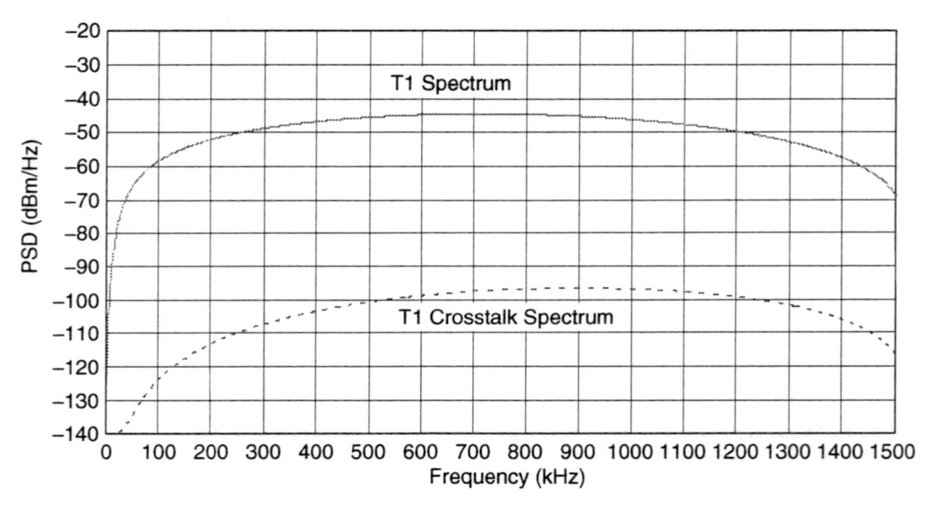

Figure 9.10 T1 AMI signal and crosstalk spectra.

lists the SNR_{ref} values [8] for the uncoded signal constellations considered in this study. To include the gain effects of the trellis code, the coding gain is subtracted from the uncoded signal's SNR_{ref}.

The output signal-to-noise ratio (SNR_{out}) of the decision feedback equalizer in the CAP/QAM receiver is computed by [8]

$$SNR_{out} = \exp\left[\frac{1}{f_B}\int_0^{f_B} \ln\left(1 + \sum_{n=0}^{\infty}\frac{s_R(f + nf_B)}{N_X(f + nf_B) + N_0}\right)\cdot df\right].$$

The output signal-to-noise ratio of the DFE in the PAM receiver is computed by [8]

$$SNR_{out} = \exp\left[\frac{1}{f_B}\int_{-\frac{f_B}{2}}^{\frac{f_B}{2}} \ln\left(1 + \sum_{n=\infty}^{\infty}\frac{s_R(f + nf_B)}{N_X(f + nf_B) + N_o}\right)\cdot df\right].$$

Table 9.2 List of Reference Signal-to-Noise Ratios of Various Line Codes

Line Code (without Trellis Coding)	$SNR_{ref\text{-}uncoded}$ (dB)
4-CAP/2-PAM	14.5
8-CAP	18.0
16-CAP/4-PAM	21.5
32-CAP	24.5
64-CAP/8-PAM	27.7
128-CAP	30.6
256-CAP/16-PAM	33.8

The SNR margin (M) is computed as follows:

$$M = SNR_{out} + G_{TC} - SNR_{ref\,,uncoded}$$

where G_{TC} is the coding gain of the trellis code and $SNR_{ref,uncoded}$ is the reference SNR of the uncoded line code (second column in Table 9.2). Alternatively, this equation may be expressed as

$$M = SNR_{out} - SNR_{ref\,,coded}$$

where $SNR_{ref,\,coded} = SNR_{ref,\,uncoded} - G_{TC}$.

9.6 SPECTRAL COMPATIBILITY OF THE DSL SYSTEMS

This section describes the spectral compatibility of the echo canceled (EC) transmission DSL systems relative to other DSLs deployed in the loop plant. The transmission DSLs included are ISDN, HDSL, and SDSL. For all spectral compatibility studies done in this chapter, only 50-pair cables with 26-gauge wire are assumed in each case.

9.6.1 ISDN

The transmit spectrum of ISDN is shown in Figure 9.5. Because ISDN is an echo-canceled symmetric transport system, we need to consider the effects of SNEXT. Because the SNEXT spectrum completely overlaps the ISDN transmit spectra, we expect this disturber to dominate over other disturbers whose spectra only partially overlap. Although the HDSL spectrum may fully overlap that of ISDN, the PSD level of HDSL will be lower than that of ISDN because the two systems have the same total transmit power.

In the evaluation of ISDN transceiver performance, echo canceler performance is considered for the ISDN transceiver. Seventy dB of echo cancellation have been achieved in practical ISDN transceivers. If there is no crosstalk in the cable, then the performance of the ISDN transceiver is limited by the performance of the echo canceler. Specifically, consider the scenario where we have a 50-pair cable of 26-gauge wire. If this cable has a single ISDN transmission system deployed and the remaining 49 wire pairs are not used, then the reach of an ISDN transceiver operating at a bit error rate (BER) of 10^{-7} with 6 dB of margin is 20.5 kft on 26-gauge wire. When we add one additional ISDN signal into the cable, the added single SNEXT disturber reduces the reach to 20 kft. With ten SNEXT disturbers, the reach is 19.1 kft and with twenty-five disturbers, the reach is 18.6 kft. Finally, if the whole 50-pair cable is filled only with ISDN systems, the maximum achievable reach of an ISDN system would be 18 kft, limited by SNEXT. Figure 9.11 shows a summary of the the ISDN reach as a function of SNEXT level.

Figure 9.12 shows the spectral plots of the transmit and receive signals of an ISDN system operating on an 18 kft 26 gauge loop. Also shown in the figure are

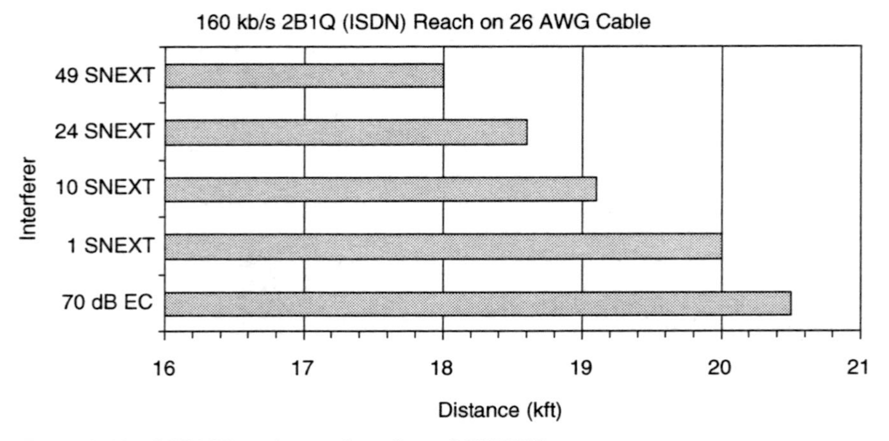

Figure 9.11 ISDN Reach as a function of SNEXT.

the insertion loss of the 18 kft loop and the 49-SNEXT plus the amount of echo that the echo canceler does not eliminate. The area between the received signal and crosstalk curves defines the received SNR.

We now consider the case when the cable includes a mixture of other DSL services. Other DSLs considered are HDSL, SDSL, and CAP RADSL upstream and downstream. For each case we consider the worst case scenario measuring the reach of ISDN in the presence of forty-nine disturbers from the "other" DSL in question. Figure 9.13 shows a comparison of the ISDN reach on 26-gauge wire in the presence of forty-nine disturbers from each of the "other" DSLs. Because of the total spectral overlap, SNEXT is a worse disturber to ISDN than any of the

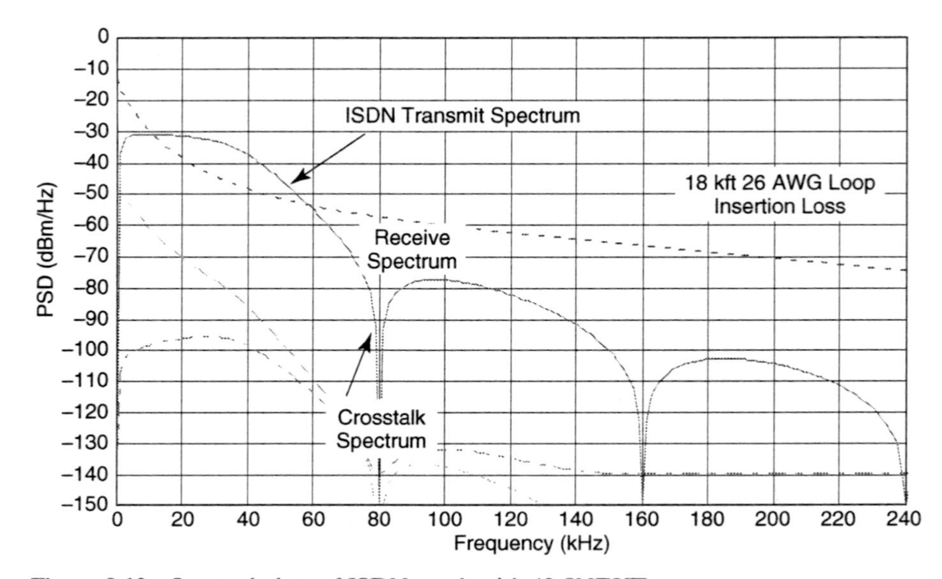

Figure 9.12 Spectral plots of ISDN reach with 49-SNEXT.

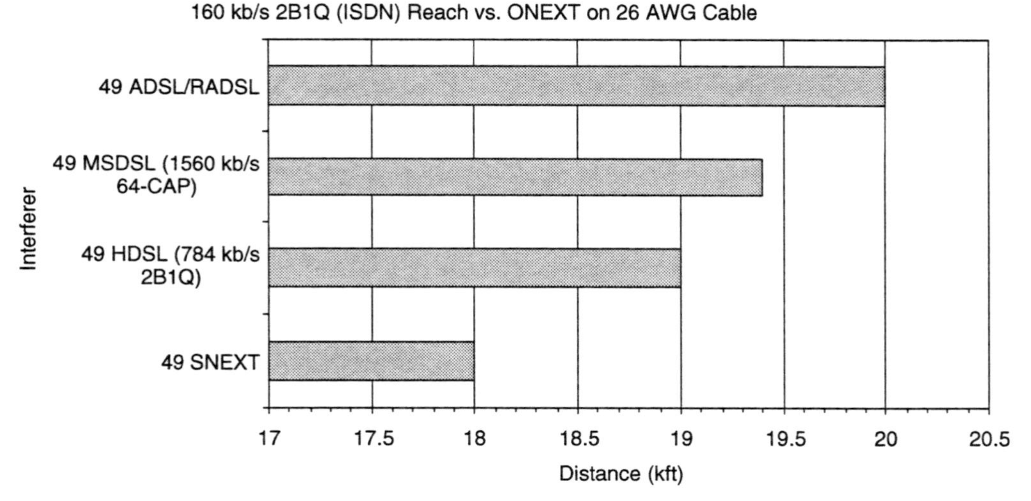

Figure 9.13 ISDN Reach as a function of other NEXT disturbers.

"other" DSLs, because SNEXT only has a partial overlap of its spectra with that of ISDN. Although the HDSL spectrum fully overlaps the ISDN spectrum, the PSD level of HDSL is lower than that of ISDN because the two systems have the same transmit power. The lower PSD level of HDSL therefore introduces less crosstalk into the ISDN band than does SNEXT.

In summary, SNEXT is the worst case disturber to ISDN basic rate. Deploying "other" services in the same cable with ISDN will have less impact on the performance of ISDN than if only ISDN was deployed in the cable.

9.6.2 HDSL

The transmit spectrum of HDSL is shown in Figure 9.6. As with ISDN, HDSL is also a symmetric echo canceled system. In the evaluation of HDSL transceiver performance, echo canceler performance is considered for the HDSL transceiver. Seventy dB of echo cancellation have been achieved in practical HDSL transceivers.

If there is no crosstalk in the cable, then the performance of the HDSL transceiver is limited by the performance of the echo canceler. Specifically, consider the scenario of a 50-pair cable containing 26-gauge wire. If there is only a single HDSL transmission system deployed and the remaining forty-nine wire pairs are not used, then the reach of an HDSL transceiver operating at a BER of 10^{-7} with 6 dB of margin is 13.7 kft on 26-gauge wire (Figure 9.14). When we add one additional HDSL system into the cable, the added single SNEXT disturber reduces the reach by 1.7 kft to 12 kft. With ten SNEXT disturbers, the reach is 10.6 kft, and with twenty-five disturbers, the reach is 10.1 kft. Finally, if the whole 50-pair cable is filled only with HDSL systems, the maximum achievable reach of an HDSL system would be 9.5 kft, limited by SNEXT.

Consider now the cases where HDSL is deployed in the cable with a mixture of other DSL services. Relative to HDSL, the "other" DSLs considered are ISDN,

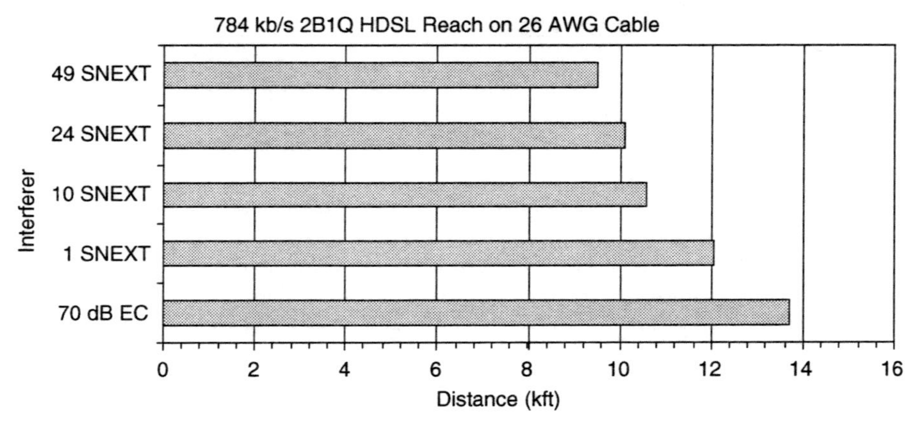

Figure 9.14 HDSL reach as a function of SNEXT.

SDSL, and CAP RADSL upstream and downstream. For each case we consider the worst case scenario measuring the reach of HDSL in the presence of forty-nine disturbers from the "other" DSL in question. Figure 9.15 shows a comparison of the HDSL reach on 26-gauge wire in the presence of forty-nine disturbers from each of the "other" DSLs. Because of the total spectral overlap, SNEXT is the worst disturber to HDSL than any of the "other" DSLs. The other DSL spectra only have a partial overlap with that of HDSL.

In summary, SNEXT is the worst case disturber into HDSL. Because the other spectra in the cable have near-end crosstalk spectra that do not fully overlap with the HDSL transmit spectrum, the overall interference will be less than the NEXT from other HDSL signals.

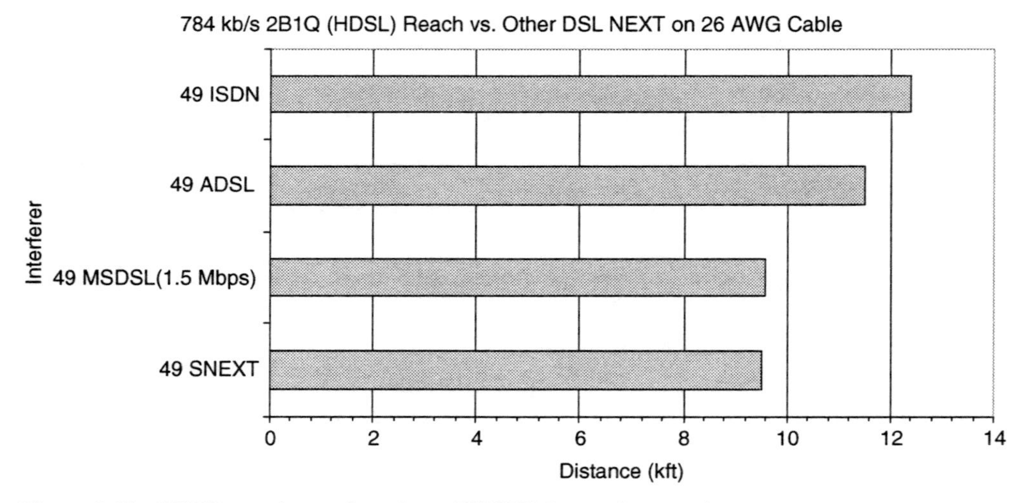

Figure 9.15 HDSL reach as a function of NEXT from other services.

9.6.3 SDSL

SDSL systems are echo-canceled systems and, therefore, are disturbed the most by SNEXT. In fact, SNEXT is the dominating disturber to SDSL. The wider the signal bandwidth, the greater the level of SNEXT. The bit rate of the signal is directly proportional to the signal bandwidth, so the reach of SDSL systems decreases with increasing bandwidth (hence, increasing bit rate). Figure 9.16 shows the reach of SDSL systems relative to SNEXT. Note the decrease in reach with increasing bit rate.

Because SDSL performance is limited by SNEXT, NEXTs from other DSLs do have as much impact on SDSL reach. However, depending on the signal bandwidth, SDSL systems may impact the performance of other DSL systems such as ADSL or CAP RADSL. The spectral compatibility of SDSL into other systems is discussed in the subsequent sections.

In summary, SNEXT is the dominant disturber into SDSL.

9.6.4 ADSL

In this study, we consider the spectral compatibility of FDM-based ADSL. Figure 9.9 shows the transmit and near-end crosstalk spectral plots of the upstream and downstream DMT ADSL channels. The spectral compatibility of DMT ADSL and CAP RADSL are similar in that neither system has SNEXT to deal with because the FDD transmission scheme places the transmitted energy in a frequency band separate from the receive band for the same end of the line. They both have SFEXT, and they must deal with NEXT from other DSL services in the same cable.

As with CAP RADSL, DMT ADSL is a variable bit rate system, and the actual bandwidths of the upstream and downstream channels may vary depending on the bit rate and crosstalk. Shown in Figure 9.9 is the maximum possible useful bandwidth for the upstream and downstream channels.

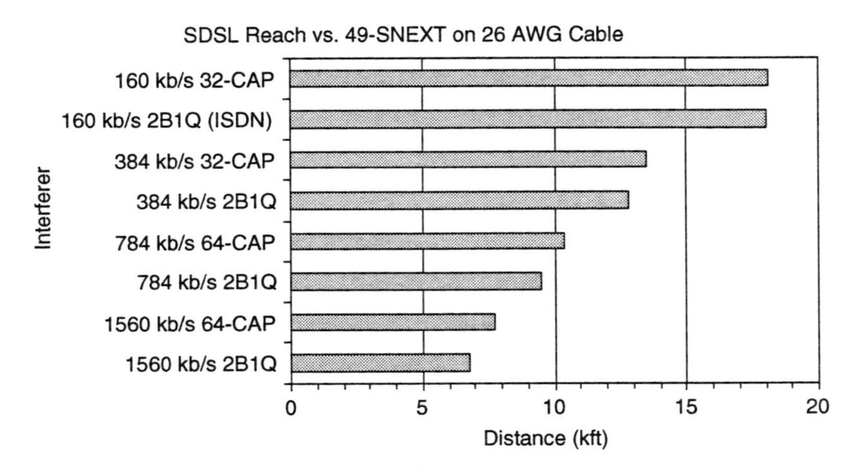

Figure 9.16 SDSL reach versus 49-SNEXT.

To evaluate the spectral compatibility of the DMT upstream channel with other services, we compute the reach of 272 kb/s DMT upstream channel in the presence of crosstalk from other DSL services. Figure 9.17 shows a comparison of the reach of a 272 kb/s upstream DMT system in the presence of NEXT from HDSL, T1 AMI, ISDN, 784 kb/s SDSL, and SFEXT. Clearly, SFEXT is the best noise environment providing the least amount of interference. T1 AMI also provides low interference into the upstream channel because the AMI signal energy is very low in the DMT upstream channel band. The dominant disturbers into the upstream channel are HDSL and SDSL because the NEXT from these services provides full bandwidth overlap with the DMT upstream channel. The ISDN spectrum has partial overlap with the DMT upstream channel, and therefore has less impact on upstream channel reach than does HDSL and SDSL.

To evaluate the spectral compatibility of the DMT downstream channel with other services, we compute the reach of a 680 kb/s DMT downstream channel in the presence of crosstalk from other DSL services. Figure 9.18 shows a comparison of the reach of a 680 kb/s downstream DMT system in the presence of NEXT from HDSL, T1 AMI, ISDN, 784 kb/s 64-CAP/QAM SDSL, and SFEXT.

As with the upstream channel, SFEXT is the best noise environment providing the least amount of interference; however, its reach is lower that the upstream channel because the loop has higher loss in the frequencies of the downstream channel. Contrary to the upstream, T1 AMI provides the dominant level of interference into the downstream channel because the AMI signal energy is highest in the DMT downstream channel band. Because of the significant bandwidth overlap with the downstream channel, HDSL is the next dominant disturber in line. ISDN and SDSL have the least impact of NEXT into the DMT downstream channel. The degradation in reach from T1 AMI versus the best case of SFEXT is approximately 6000 ft; the corresponding in reach from HDSL is approximately 5000 ft.

In summary, HDSL and SDSL are dominant disturbers into the upstream channel of ADSL. T1 AMI is the dominant disturber into the ADSL downstream channel. The best case for deployment of FDM ADSL services is to fill the cable

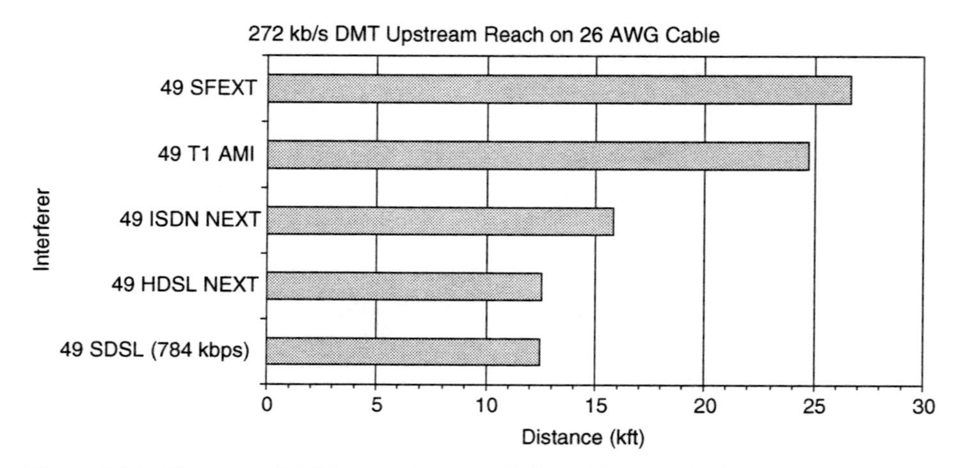

Figure 9.17 Upstream DMT spectral compatibility with other DSLs.

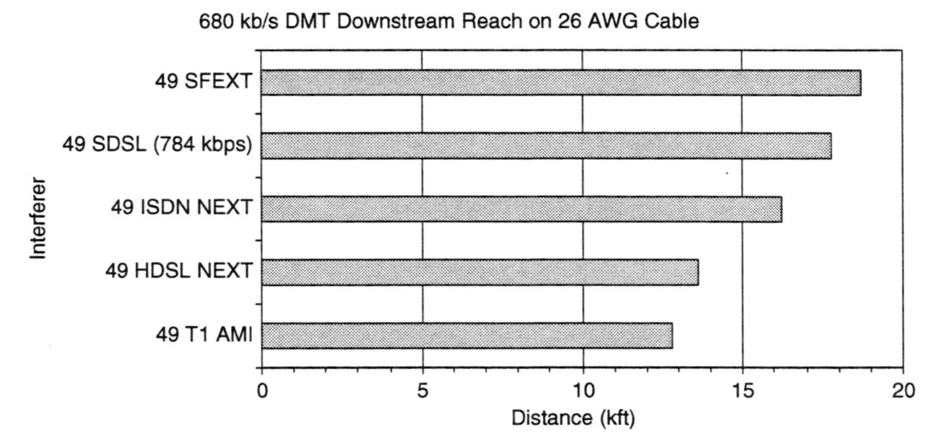

Figure 9.18 Downstream DMT spectral compatibility with other DSLs.

completely with ADSL and eliminate all NEXT. If the cable contains a mixture of DSLs, then NEXT from HDSL and SDSL are the dominant disturbers into the upstream channel, and T1 AMI and HDSL are the dominant disturbers into the downstream channel.

9.6.4 T1 AMI

Figure 9.19 shows the system model for determining the spectral compatibility of the downstream ADSL channel into T1 AMI. Because T1 AMI is a repeatered link, there are numerous points to consider observing the crosstalk, and they are labeled crosstalk points #1 and #2 in the figure.

In the conventional provisioning of T1 links, the first repeater is placed at a maximum of 3 kft from an end-point and a maximum of 6 kft between repeaters. Note, however that T1 lines were originally designed as trunk lines to interconnect COs and the wire gauges were 22 AWG or 19 AWG. Because the distribution plant usually uses 26 AWG wire (thinner than 22 or 19 AWG) directly out of the CO, the provisioning rules used in the distribution plant are not ubiquitously known. However, in this study, we will assume a worst case scenario of 26-gauge wire using the same repeater spacing rules for the trunk plant.

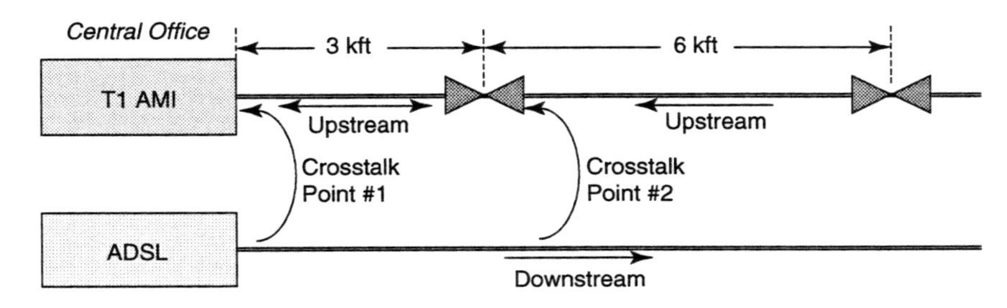

Figure 9.19 Crosstalk scenarios for DSLs into T1 AMI.

The ADSL downstream transmit signal is strongest at the CO transmitter output. So at the CO (crosstalk point #1 in Figure 9.19) the ADSL downstream signal introduces the strongest level of crosstalk into the T1 AMI receiver. The T1 AMI signals have maximum energy at the transmitter output of both the end units and repeaters. In the first repeater span, the loop segment length is 3 kft and so the received AMI signal would not be attenuated as much as it would be in a midspan repeater spacing of 6 kft. At crosstalk point #2, the downstream ADSL signal is attenuated by 3 kft of cable, and so the crosstalk level into the first repeater would be attenuated by that amount.

To estimate the performance of the AMI signal, we compute the signal-to-noise ratio (SNR) at the AMI receiver input and crosstalk points #1 and #2. In each case, the number of downstream ADSL disturbers assumed is twenty-four. The SNR is measured in two ways: (1) at the T1 AMI center frequency of 772 kHz, and (2) averaged through out the entire T1 AMI band to the first null, that is, 0 to 1.544 MHz. In the AMI receiver, we assume that the receiver provides automatic gain adjustment, no equalization, and ideal time sampling. To achieve a BER of 10^{-7}, we assume a 17.5 dB SNR is required for the three-level signal at the input to the AMI receiver. The margin achieved is the difference in SNR at the receiver input and the 17.5 dB reference value.

Table 9.3 below shows the spectral compatibility computation results with twenty-four ADSL downstream channel disturbers. The third column shows the SNR margin seen at the AMI receiver inputs measured at the AMI center frequency of 772 kHz. For both crosstalk points #1 and #2, the SNR margin is roughly 1 dB so the T1 AMI system should still provide service with better than 10^{-7} BER. The last column shows input SNR averaged over the entire T1 AMI band. At crosstalk point #1, the averaged SNR is roughly the same as the SNR at the center frequency. For crosstalk point #2, the averaged SNR is greater than center frequency SNR because of the greater attenuation suffered by the AMI signal on the 6 kft loop.

In summary, an estimation of spectral compatibility of ADSL with T1 AMI was provided using very pessimistic assumptions. Specifically, the same provisioning rules of repeater spacings on 22- and 19-gauge wire was applied to 26-gauge wire and the losses seen on 26-gauge wire were significantly greater. In all cases of twenty-four ADSL disturbers into a T1 AMI receiver, the input SNR is at least 1 dB greater than that required for achieving 10^{-7} BER performance.

If the repeater spacings are shorter than those assumed here, then the margins will improve. Based on this data, we expect ADSL to not degrade T1 AMI service in the distribution plant. As shown earlier, T1 AMI is the dominant disturber into the ADSL downstream signal.

Table 9.3 Spectral Compatibility Computation Results with 24 ADSL Disturbers

Crosstalk	**Center Frequency (772 kHz)**		**Averaged SNR (0 to 1544 kHz)**	
Point	**SNR (dB)**	***Margin (dB)***	**SNR (dB)**	***Margin (dB)***
#1	18.7	*1.2*	18.8	*1.1*
#2	18.8	*1.3*	25.5	*8.0*

9.7 SUMMARY

In the telco loop plant, there are numerous types of digital subscriber lines deployed. The DSLs deployed can be categorized in two different types: (1) symmetric echo-canceled (EC) DSLs and (2) asymmetric (FDM) DSLs. The first class of EC DSLs includes ISDN, HDSL, and SDSL; the second class includes ADSL. The modulation technologies for the DSLs include 2B1Q for ISDN, HDSL, CAP for HDSL and MSDSL, DMT for ADSL, and CAP for RADSL.

Most echo-canceled systems, including ISDN, HDSL, and SDSL, transmit the same frequency spectra in the downstream direction as the respective system uses in the upstream direction. Assuming the same transmit power for each of these systems, the worst-case crosstalk performance occurs when the cable is filled with the same type of EC system. For a 50 pair cable, the worst-case crosstalk is 49-SNEXT. HDSL2 is an echo-canceled system that transmits a downstream PSD that differs from the upstream HDSL2 PSD.

The second type of DSL system is asymmetric FDM, which include DMT ADSL and CAP RADSL. Both the ADSL and RADSL systems use FDM to separate the upstream and downstream channels. If a cable were filled with only FDM systems, self far-end crosstalk will limit the performance. Because SFEXT is orders of magnitude less than SNEXT, the reach of FDM can be significantly greater than that where NEXT is present. So contrary to that of EC systems, the best case crosstalk environment occurs when a cable is filled with the same FDM system.

Another version of ADSL is an echo-canceled version, where the wide-band downstream channel also utilizes the narrow-band frequencies of the upstream channel. In this case, the near-end crosstalk of the downstream channel will severely limit the performance of the upstream channel because the downstream channel bandwidth completely covers that of the upstream.

Because the cable will simultaneously contain EC and FDM type systems, then performance of the DSLs in the presence of NEXT from "other" systems must be considered. The DSLs discussed in this chapter contain comparable power spectral density mask values. Therefore, the crosstalk in EC systems is dominated by SNEXT. However, the varying bandwidths of the EC systems will introduce different levels of NEXT into ADSL systems.

REFERENCES

[1] J. J. Werner, "The HDSL Environment," *IEEE JSAC* 9, no. 6, (August 1991) 785–800.

[2] ANSI T1.413-1998, Telecommunications, "Network and Customer Installation Interfaces—Asymmetric Digital Subscriber Line (ADSL) Metallic Interface."

[3] ANSI T1.601-1992, "Integrated Services Digital Network (ISDN) Basic Access Interface for Use on Metallic Loops for Application on the Network Side of the NT (Layer 1 Specification)."

[4] ETSI ETR-080, Transmission and Multiplexing (TM); "Integrated Services

Digital Network (ISDN) Basic Rate Access Digital Transmission System on Metallic Local Lines" (July 1993).

[5] Committee T1-Telecommunications Technical Report No. 28, February 1994, "A Technical Report on High-Bit-Rate Digital Subscriber Lines (HDSL)."

[6] Committee T1-Telecommunications Draft Technical Report, Issue 2 No. T1E1.4/96-006R2, April 22, 1996, "A Technical Report on High-Bit-Rate Digital Subscriber Lines (HDSL)."

[7] ETSI ETS-152 Edition 4, November 1997, "Transmission and Multiplexing (TM); High bit rate digital subscriber line (HDSL) transmission system on metallic local lines; HDSL core specification and applications for 2048 kb/s based access digital secitons."

[8] J. J. Werner, *Tutorial on Carrierless AM/PM—Part II: Performance of Bandwidth-Efficient Line Codes,* AT&T Network Systems Contribution T1E1.4/93-058, March 8, 1993.

[9] Committee T1-Telecommunications, Technical Report TR-059, 1999, "Single-Carrier Rate Adaptive Digital Subscriber Line (RADSL)."

[10] BOC Notes on the LEC Networks—1994, Bellcore Document SR-TSV-002275, Issue 2, April 1994.

[11] Committee T1-Telecommunications, Letter Ballot LB652, "Draft T1.413 Issue 2," September 1997.

[12] Bellcore Technical Advisory TA-NWT-001210 Issue 1, "Generic Requirements for High-Bit-Rate Digital Subscriber Lines," October 1991.

[13] Committee T1–T1.417-2001, "American National Standard for Telecommunications—Spectrum Management for Loop Transmission Systems."

[14] G.992.1—"ITU-T Recommendation G.991.1: High bitrate digital subscriber line (HDSL) transmission system on metallic local lines," February 1998.

[15] G.991.2—"ITU-T Recommendation G.991.2: Single-Pair High-Speed Digital Subscriber Line (SHDSL) Transceivers," February 2001.

[16] G.992.1—"ITU-T Recommendation G.992.1: Asymmetrical Digital Subscriber Line (ADSL)," March 1999.

[17] G.992.2—"ITU-T Recommendation G.992.2—Splitterless Asymmetric Digital Subscriber Line (ADSL) Transceivers," March 1999.

[18] G.996.1—"ITU-T Recommendation G.996.1: Test Procedures For Digital Subscriber Line (DSL) Transceivers," March 1999.

[19] T. Starr, J. Cioffi, & P. Silverman, *Understanding Digital Subscriber Line Technology,* Upper Saddle River, NJ: Prentice Hall, 1999.

CHAPTER 10

SPECTRUM MANAGEMENT OF DSL SYSTEMS

The telco network was originally designed for the deployment of voice grade services. In these applications the subscriber line signals utilized frequencies below 4 kHz (approximately 200 Hz to 3.3 kHz). With the development of voiceband modems, this same voice channel was used to transmit bidirectional data up to 56 kb/s per ITU-T Recommendation V.90 [1].

Clearly, the 4 kHz band of the voice channel only utilizes a small portion of the subscriber line capacity. In the 1970s, the digital data system (DDS) was developed that transmitted symmetric data on the same twisted wire pair subscriber line, but the bandwidth was not restricted to the voice band. The line code was alternate mark inversion (AMI), and bit rates supported were 2.4, 4.8, 9.6, and 56 kb/s. Later in the mid 1980s, clear channel 64 kb/s access was defined where the actual bit rate on the subscriber line was 72 kb/s, which contained 8 kb/s of signaling overhead. A key difference to voice grade services is that the access for DDS required two twisted wire pairs for duplexing; one wire pair transmits the downstream AMI signal and the other pair transmits the upstream signal [2].

Then in the 1980s, Basic Rate ISDN was developed. ISDN transports 160 kb/s on the access subscriber line using the 2B1Q (two binary one quarternary) line code, which is equivalent to 4-level pulse amplitude modulation (4-PAM). As we describe, ISDN is transmitted bidirectionally on a single twisted wire pair. Duplexing is performed using echo cancellation. Near-end crosstalk from other ISDN signals in the same cable limits the reach of ISDN [3].

In 1989, the study began on the feasibility of transmitting signals at a rate substantially higher than basic rate ISDN (160 kb/s). This study project became known as the high-rate digital subscriber line (HDSL). Although several modulation methods were considered for HDSL, the same 2B1Q line code was preferred by the industry for HDSL. The driving application for HDSL is the transmission of T1 service on loops without repeaters, and the performance objective was transmitting the1.544 Mb/s payload at carrier serving area (CSA) distances (9 kft of 26 gauge wire and 12 kft of 24 gauge wire) in the presence of near-end crosstalk from other like disturbers. To meet the performance objective, a dual duplex architecture was chosen where the 1.544 Mb/s payload was split into two parallel transmission channels, each carrying one half of the T1 payload. With the inclusion of overhead and payload, the bit rate transmitted on each wire pair of the dual-duplex HDSL is 784 kb/s. HDSL was thus targeted as a transport mechanism that would allow the operator an alternative for quickly deploying symmetric T1 grade service on two loops within CSA range without any repeaters. Provisioning of T1 service with the traditional repeatered AMI could take as long as two months if repeaters were required to be installed in the cable plant. Deploying the same T1 service using HDSL would reduce the provisioning time to as little as two to three days. T1 is a high-capacity, high-quality service targeted for business applications where HDSL is an operator option for provisioning T1 service [4], [5].

During the same time period, there was also interest from the phone companies to deploy video services to residential customers using the same subscriber line infrastructure. This is in response to the cable companies looking to deploy telephone services in their own coaxial cable infrastructure. Deployment of video service has an asymmetric transmission profile, namely, a high-capacity channel would need to be transmitted to the subscriber (downstream) whereas only a low bit–rate channel would need to transmitted back to the network (upstream) for program control. The application became known as the asymmetric digital subscriber line or simply ADSL. Video-on-demand (VOD) was the initial application driving ADSL.

Unlike HDSL, which uses echo cancellation to duplex two fully overlapped upstream and downstream channels, the asymmetric nature of ADSL made use of frequency division duplexing (FDD) for transmission of the ADSL upstream and downstream channels. With FDD, the upstream channel uses a narrow band of frequencies for transmission of the low-speed upstream channel and a separate higher band of frequencies for the transmission of the downstream channel. In addition the two ADSL bands would be placed at frequencies above those of the voice channel so ADSL would support transmission of conventional plain old telephone service (POTS), a low-speed upstream channel for the transport of signaling data, and a high-speed downstream channel for the transport of digital video and signaling. ADSL achieves downstream bit rates far beyond the preceding DSL technologies; this was made possible by asymmetric bit rates that in turn made FDD practical. ADSL performance is further enhanced by the use of Trellis coding and Reed-Solomon coding.

A key element of ADSL and it associated frequency division duplexing is that if each of the wire pairs in the cable is filled using the same upstream and downstream frequency bands, there is no self near-end crosstalk (SNEXT) present in either of the two channels because there is no overlap in the transmission of the upstream and downstream channels. With FDD only far-end crosstalk is present, which is significantly less disturbing than near-end crosstalk. Hence, if the cable plant could be managed to contain only FDD-based systems, then significantly higher bit rates (downstream or upstream) could be achieved than with echo-canceled systems. Unfortunately, it is difficult to manage or force such restrictions with competitive access in the cable plant, but limiting the amount of near-end crosstalk introduced in the cable plant is manageable.

In the initial vision, ADSL was transporting approximately 1.6 Mb/s data downstream and 24 kb/s upstream to a distance of approximately 15 kft. The definition of ADSL was later expanded to include the transport of approximately 6 Mb/s downstream and 640 kb/s upstream. In the mid-1990s the driving application shifted from VOD to Internet access. The range of bit rates was still valid, where operators can provision different bit rates for a specified quality of service.

The chosen modulation method for ADSL is discrete multi-tone (DMT), and the transmit power spectral density (PSD) for ADSL is defined in T1.413 Issues 1 and 2 and in ITU Recommendation G.992.1 [6, 7, 8]. An alternative modulation method (using carrierless amplitude and phase, CAP, modulation) was also deployed for rate adaptive ADSL (RADSL) systems prior to standards-based deployments. The corresponding PSD is defined in Committee T1 Technical Report No. 59 (TR-059) [9]. Note that RADSL uses the same upstream and downstream frequency bands, as does ADSL.

Aside from the above standards-based services, there are proprietary or nonstandard systems, both symmetric and asymmetric, deployed in the same cable plant. Examples include multirate symmetric DSL (SDSL) that use 2B1Q and single carrier modulation methods. The bandwidth and spectral shaping of each modulation approach determines the degree of disturbance into other systems in the cable plant.

There is also T1 AMI deployed in the distribution cable plant. This is known to be a very strong disturber into ADSL; so strong that it may even prohibit deployment of ADSL with any reasonable quality of service if encountered in the same cable at distances greater than 6 kft.

In addition to the well-defined DSL service described earlier, there are other nonstandard systems deployed, and new, advanced DSL systems being developed. The impact of any newly defined DSL system on already deployed systems needs to be understood prior to providing large-scale deployments. If all of the systems, old and new, are to be deployed in the same cable, we first need to understand how they interfere with each other. This quantification of disturbance is called spectral compatibility.

Spectral compatibility is the quantification of the level of disturbance that one type of DSL system has on other systems deployed in the same cable. A clear understanding of the interactions that the different types of systems have on each other in the cable is the foundation for managing the deployment of DSL systems in maintaining a specified quality of service.

In the mass deployment of DSL services, we can expect the cable to be filled with many different types of signal spectra. Given the wide variety of signals in the cable, there can be a wide range of interference between the different systems such that a set of rules or guidelines are needed to manage the deployment of these systems to assure a minimum or specified quality of service. These rules for the deployment of systems in the loop plant are called *spectrum management.* The rules of spectrum management are based on a working knowledge of the spectral compatibility of the various DSL systems and an agreement on a desired quality of service.

This chapter presents an overview of the spectrum management for the deployment of different DSL services in a single loop plant. Included are discussions on the fundamental concept of spectrum management per T1.417, along with background information leading to critical decisions in the spectrum management standard.

10.1 SPECTRUM MANAGEMENT BACKGROUND

The goal of the Telecommunications Act of 1996 was to let anyone enter any communications business and to let any communications business compete with any other communications business. This reform would affect telephone service, as we know it, namely, local and long distance service, cable programming and other video services, broadcast services, and services provided to schools.

Title I of the 1996 Telecom Act has attempted to stimulate increased competition in the provisioning of telecommunications services by opening access to the local exchange and in turn allowing the local exchange carriers (or regional Bell operating companies) to provide manufacturing and long distance service. In exchange for entering manufacturing and long distance businesses, the local exchange carriers must allow unbundled loop access [10].

Prior to the 1996 Telecom Act and after the breakup of the Bell System in 1984, the North American telephone network was divided into two separate entities, the local exchange carriers (LECs) who provide network access and local phone service, and the interexchange carriers (IXCs) who provide long distance service. Generally, the end customer would not have a choice on the network access portion of the phone service; it is the LEC that owns and provides the subscriber line access. However, customers do have a choice of long distance service providers. Generally, the LECs provided switched phone service within their local access and transport area (LATA), also known as intra-LATA service. Later the IXCs began competing with the LECs for providing intra-LATA switching service; however, it was still the LECs who provided access to the local exchange, simply because they are the ones who own the wires.

The U.S. Federal Communications Commission's (FCC) first report and order for loop unbundling was the first step in allowing competition for local access. With competition mandated for local access, two families of companies providing local access were formed: the incumbent local exchange carriers (ILECs) and the competitive local exchange carriers (CLECs). In the United States, the ILECs are the regional Bell operating companies and the independent telephone companies; these are the companies that own the wires in the loop plant. The CLECs are any other communications companies that look to provide competitive local access and communication services to end users. Because the CLECs do not have their own infrastructure of subscriber lines, the ruling of loop unbundling requires the owners of the cable (namely, the ILECs) to lease wire pairs to the competitive access providers (namely, the CLECs).

There are CLECs that provide a full range of telecommunications communications services, namely, voice and data services. There is also another group of competitive access providers that offer only data services, and they are often referred to as data local exchange carriers (DLECs). Further telecommunications service competition is provided by multiple system operators (MSOs), using cable-TV infrastructure for voice and data (cable modem) services, and by wireless data service providers, including both terrestrial- and satellite-based systems.

Digital subscriber lines are generally the means for local exchange carriers (incumbent or competitive) to offer high-speed data services. As we have seen earlier, there are a large variety of DSL-based services available. Generally, the symmetric DSL services are used for business applications and the asymmetric DSL services are for consumer and residential applications. There are exceptions to this generalization, but for the most part business applications require more symmetric transmission configurations and consumer (or residential) applications use more asymmetric transmission configurations.

Because both CLECs and ILECs will deploy service on different wire pairs in the same cable, the signals must be deployed such that the spectral integrity of the cable is maintained. The cable could potentially contain a broad mixture of symmetric and asymmetric DSL services. If not properly managed, the deployment of a broad range of signals in the cable plant can cause significant crosstalk interference into other signals in the same cable, which can cause severe degradation in the quality of service to another customer or that of a competitor. To avoid the situations of severe quality of service degradation, a set of guidelines need to be followed in the deployment of signals with varying spectral characteristics. This is the fundamental basis for having a spectrum management standard.

The spectrum management standard provides the technical specifications to help prevent service trouble resulting from uncontrolled crosstalk. In today's competitive environment, equipment vendors and service providers both desire to provide even higher bit rates to more customer (e.g., longer lines). What is the easiest way to achieve higher bit rates and operation on longer loops? Transmit a more powerful signal. Without a spectrum management standard, DSL systems would soon escalate transmit power in an effort to "shout louder than everyone else at the party."

The ILEC is both a loop provider and a service provider, because they own the cable infrastructure and provide a service (voice, data, and others) to customers via their subscriber loops. CLECs, on the other hand, generally do not own a cable infrastructure and are therefore only service providers. To provide service, they must lease lines from the loop provider, usually an ILEC. *Both the incumbent and competitive access providers must follow the same spectrum management rules when deploying services in this unbundled environment.*

In the deployment of DSL services, CLECs (or DLECs) primarily provide competitive access service to business customers. These customers are generally served with a symmetric DSL such as 2B1Q-based SDSL. The same companies will also serve the residential customers with asymmetric services, which would generally be provisioned using ADSL. ILECs also deploy both symmetric and asymmetric services. But for ILECs, a significant deployment of DSL-based services is to residential customers who are provisioned with ADSL. The business customers subscribing to high capacity service are more commonly deployed using HDSL.

With regards to policy making, the spectrum management standard does not define policy for the deployment of the DSL signals in the cable. In the United States, policy making or spectrum management rule making is the job of the FCC. The spectrum management standard would serve as a technical input to the FCC in their procedures for establishing policy for deployments of signals in a cable. Early spectrum management efforts by the FCC were based on a policy of first-in-the-cable, and a transmission was presumed acceptable when shown to have been in service somewhere with no reported troubles. The need for a sound technical basis was recognized, and the FCC requested T1E1.4 to develop a technical standard for spectrum management of local loops. Without the technical rigor of the spectrum management standard, crosstalk problems from a new type of DSL might not be recognized until many thousands of such systems had been placed in service.

10.2 THE CONCEPT OF SPECTRUM MANAGEMENT

As we have mentioned earlier, spectrum management is the administration of the loop plant in such a way that it assures spectral compatibility between the signals, services, and technologies that use the wire pairs in the same cable. In order to provide spectrum management, we need an understanding or a definition of what it means for different signals to be spectrally compatible with one another. To set this definition, we need an agreed-upon criteria or *basis* for defining the levels and conditions of spectral compatibility.

Prior to the Telecom Act of 1996 and the associated FCC rules requiring local loop unbundling and collocation of equipment, the network already had numerous types of signals deployed in the cable. The most common legacy signal is analog voice, which uses the lower 4 kHz of the frequency spectrum for provisioning POTS. Other types of signals and services deployed in the loop plant at that time include T1 AMI, DDS, ISDN, ADSL, and HDSL. These signals and others already present in the network would be the first to consider as a basis for determining spectral compatibility criteria.

The spectrum management standard (T1.417) defines a list of systems as the *basis systems* list, which forms the basis for quantifying the acceptable levels of disturbance for declaring spectral compatibility. Once these conditions are defined, any new system to be deployed in the network would be measured according to these conditions. In an unbundled loop environment, systems can only be deployed according to the guidelines for which they are declared to be spectrally compatible. In summary, any system deployed in the cable in unbundled loop environment must be spectrally compatible with the basis systems, and the spectrum management standard provides the compatibility criteria.

The purpose of the spectrum management standard is to prevent service trouble due to excessive crosstalk from other systems in the same cable. In effect the, spectrum management standard *protects* the legacy transmission systems, as well as future transmission systems. In fact, early drafts of the spectrum management standard used the term *protected systems* that was later changed to *basis systems*. The basis systems act as a surrogate for all loop transmission system. The spectrum management standard defines spectral compatibility as not interfering with any of the *basis systems* more than a specified amount.

Spectral compatibility applies to both directions of transmission, that is, the signals transmitted by equipment at both ends of the line.

T1.417 contains two methods for determining the spectral compatibility of a transmit signal. Method A defines a set of signal power limitations that specify the maximum amplitude, frequency distribution, and total power of the electrical signals at the point where the signal enters the cable. In T1.417, there are nine spectrum management classes that define such signal limitation. Any signal that falls within the specified limits of any spectrum management class is declared to be spectrally compatible within the deployment guidelines of that class. Method B defines an analytical method for determining spectral compatibility; this analytical method is an alternative method to that of signal power limitations. The driving need for defining the alternative analytical method is that future (and existing) technologies may use transmit signals whose power spectral densities do not conform to any of the signal power limitations of the spectrum management classes, but would still be spectrally compatible with the basis systems. Applying the power spectral density of any signal to the analytical method will determine the deployment guideline assuring spectral compatibility.

The first issue of spectrum management standard T1.417 has the following key assumptions: the cable plant is modeled as a 50 pair cable with all equipment at each of the loop respectively collocated. There are some DSL systems that are deployed with midspan repeaters. There are also DSL systems deployed from a remote terminal (RT) or a digital loop carrier (DLC). In general, any DSL signal that is regenerated in the mid-span of a cable such as a repeater or originates in a remote terminal is called an intermediate termination unit (or intermediate TU). The first issue of T1.417 does not define the spectral compatibility of intermediate TU systems with those deployed directly from the central office. The spectral compatibility of these systems is a subject of high priority for the second issue of T1.417. Figure 10.1 shows the different deployment scenarios mentioned: (a) direct from the central office, (b) repeatered service, and (c) RT or DLC.

Because there is about 10 dB of additional crosstalk loss between binder groups, one possible method of assuring spectral compatibility is to restrict certain DSL types from occupying selected binder groups (e.g., binder group management). Binder group integrity is the assurance of the same set of wire pairs remaining together throughout the loop plant. Binder group integrity is often not maintained when the subscriber loop is provisioned through multiple cross-connect boxes. Because binder group integrity cannot be reliably administered in a public network, the spectral compatibility guidelines in T1.417 do not consider binder group management. However, binder group separation might be a useful method of assuring spectral compatibility in situations where binder group integrity can be assured. However, with the legacy exception of T1 carrier, the FCC prohibits the restriction of services to selected binder groups because this could favor certain service providers. Furthermore, such restrictions would result in stranding wire pairs that could have provided service.

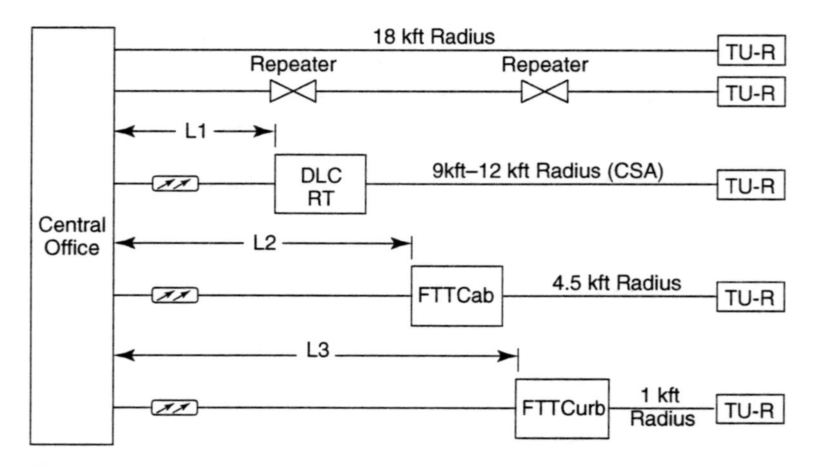

Figure 10.1 Deployment scenarios in the cable plant.

T1.417 defines spectral compatibility on the basis of allowing 49 disturbers in a 50 pair binder group. It has been suggested that special allowances be made for systems that are restricted to a small number of systems in any binder group. This method is not practical because it is not possible to reliably enforce such a restriction. Furthermore, restricting a system to a number such as 12 disturbers provides relatively little benefit compared with the 49-out-of-50 disturber assumption.

10.3 THE BASIS SYSTEMS

The basis systems are loop transmission systems deployed in the network that provide the *basis* for establishing spectral compatibility criteria for within a loop cable. Selection of the basis systems list required industry consensus in order to be adopted in the spectrum management standard.

The basis systems list per the first issue of the spectrum management standard T1.417 are as follows:

1. *Voice grade services,* which include analog voice service, network signaling, data (e.g., via voice-band modem) and tones that utilize the frequency from 0 to approximately 4 kHz.
2. *Digital data service* (DDS) based on signals defined in T1.410-1992. These signals use alternate mark inversion at a maximum line rate of 72 kb/s. The services supported are 64 kb/s and sub-rates of 2.4, 4.8, 9.6, 19.2, 38.8, and 56 kb/s.
3. *Basic rate ISDN* (BRI) based on T1.601-1999. The basic rate ISDN signal uses 2B1Q line and operates a line rate of 160 kb/s.
4. *High-bit-rate digital subscriber line* (HDSL) based on G.991.1 Annex A. HDSL was designed to transport a DS1 payload across two wire pairs, each carrying one-half of the DS1 payload plus overhead, at a distance within carrier serving area (CSA) range. Recall that CSA has a radius of 12 kft on 24-gauge wire and 9 kft of 26-gauge wire.
5. *Asymmetrical digital subscriber line* (ADSL) based on T1.413-1998 with nonoverlapped (reduced near-end crosstalk) upstream/downstream mode in G.992.1 Annex A. ADSL uses DMT modulation to transmit upstream and downstream signal in non-overlapped frequency bands.
6. *Rate-adaptive asymmetric digital subscriber line* (RADSL) based on Committee T1 TR.059-2000. RADSL uses the same frequency bands as ADSL but it uses single carrier modulation, namely carrierless AM/PM (CAP) or quadrature amplitude modulation (QAM).
7. *Splitterless ADSL* based on T1.419-2000. Splitterless ADSL uses the same DMT modulation as conventional ADSL, but the downstream channel is limited to approximately one-half the bandwidth of ADSL.
8. *HDSL2* single pair HDSL transporting a DS1(1.544 Mb/s) payload based on T1.418-2000.

9. *2B1Q-based single pair symmetric DSL* (SDSL) at bit rates of 400 kb/s, 1040 kb/s, and 1568 kb/s.
10. *Enhanced business services* (P-Phone) based on interface specification NIS S102-2, Issue 02.02 October 1998. This service utilizes the spectrum from 0 to 10 kHz providing for providing voice grade services, but the signaling is provided on a 8 kHz carrier.

The most notable criterion for systems on the above basis systems list is that each of the systems have been widely deployed and they are expected to be widely deployed, in the foreseeable future. There are some exceptions that need to be considered. Some technologies are old and are being replaced by newer technologies, where the newer technologies would have better spectral characteristics for the same quality of service than the older technologies. An example is 2B1Q SDSL will be replaced on the basis systems list with G.shdsl at an appropriate point in time. Another exception has to do with the T1 AMI signal, which is known to be a strong disturber into ADSL. Although many T1 AMI systems have been deployed in the network, it is expected that future deployments of such system would be very limited in an unbundled loop environment. Equivalent services offered with T1 AMI would be provisioned with HDSL or G.shdsl technologies, which are less disturbing to ADSL than T1 AMI.

The basis systems list above uses frequencies up to approximately 1.1 MHz, which is determined by the downstream ADSL signal spectrum. There are future DSL systems being developed that will utilize frequencies above 1.1 MHz, such as very-high bit-rate digital subscriber line (VDSL). It is expected that VDSL will become a basis system once the standard is complete and large-scale deployments occur.

For any new technology to be added to the basis systems list, the spectrum management standard T1.417 specifies the following factors in its consideration (taken from Section 4.3.1 of T1.417):

1. It is highly preferred that the system be standardized by the ITU or an ANSI accredited standards organization or that a draft standard is expected to be approved by the time the forthcoming issue of the Spectrum Management standard is expected to be published. If an effort has been made to standardize the system and there is a clear reason why the system cannot be standardized, then a physical-layer specification shall be publicly available.
2. The specification for a non-standard system shall be stable, widely accepted by most of the industry, and shall specify all aspects necessary to determine spectral compatibility (e.g., transmitted signal PSD, modulation method, coding, bit-rate, start-up process, and margin to be achieved for certain reference loops and reference noise).
3. Preferably, a new basis system should not require changes to the existing spectrum management classes (including PSD definition, total average power, transverse balance, longitudinal output voltage and deployment guidelines) to maintain spectral compatibility with the new basis system.
4. Preferably, a new basis system should not be adequately addressed by the existing systems on the basis system list.

5. New basis systems should demonstrate possible scenarios where the new system could be disturbed while other basis systems are not.

In the second issue of T1.417, G.shdsl has been added to the basis systems list, replacing 2B1Q SDSL. It is expected that VDSL will be added to the basis systems list once standardization is complete and systems are deployed.

Note that spectral compatibility is not an explicit criterion for basis systems. Alhough it is not necessary for a basis system to be spectrally compatible, in practice the basis systems generally are spectrally compatible with each other.

10.4 SPECTRAL COMPATIBILITY VIA METHOD OF SIGNAL POWER LIMITATIONS (METHOD A)

As mentioned earlier, determination of spectral compatibility per Method A is simply to show that the signal to be transmitted in the cable falls within the signal power limitations for the corresponding deployment guidelines. Following are the aspects specified for use of Method A.

- Power Spectral Density (PSD)— The PSD defines the distribution of signal power as a function frequency. The units of PSD are typically in dBm/Hz or in W/Hz.
- Transmit Signal Power—Defines the total signal power measured across the full bandwidth of the signal. The units are typically expressed in dBm.
- Deployment Guidelines—Typically defines the maximum loop reach or distance a specific signal or technology may be deployed and be declared spectrally compatible with the basis systems. As the loop length increases, the greater signal attenuation makes the victim DSL more susceptible to crosstalk. Thus, most systems will harm a basis system beyond some distance.
- Transverse balance
- Longitudinal output voltage.

Hence with method A, by knowing the signal's maximum transmit power and PSD, spectral compatibility within specific deployment guidelines may be easily determined.

10.5 FOUNDATION FOR DETERMINING ACCEPTABLE LEVELS FOR SPECTRAL COMPATIBILITY

As discussed in the previous chapter on spectral compatibility, near-end crosstalk (NEXT) is a much stronger disturber than far-end crosstalk (FEXT) because the crosstalk coupling between the wire pairs in the cable is much greater. Signals in the cable that have overlapping upstream and downstream frequency components will experience near-end crosstalk with each other. A special case of overlapping frequency spectra occurs with systems that use echo-cancellation for upstream and downstream duplexing. When the upstream and downstream signals use the same spectrum, there is full overlap of the frequency components, and the echo canceler removes the disturbance from the near-end transmitter into the near-end receiver. However, if systems with the same spectrum are deployed in a cable, then the amount of near-end crosstalk injection would be greatest due to the full spectrum overlap. Such near-end crosstalk is termed self near-end crosstalk or simply SNEXT. ISDN, HDSL, and SDSL are systems that have fully overlapping upstream and downstream signal spectra, and their reach is limited by SNEXT; hence, they are SNEXT limited systems.

ADSL (per that on the basis systems list) uses separate frequency bands to separate upstream and downstream transmission. Because there are no overlapping frequencies, ADSL has no SNEXT but it does have SFEXT. As we observed in the previous chapter on spectrum compatibility, the crosstalk coupling for FEXT is significantly lower than that for NEXT coupling. Hence, if only ADSL is in the cable, then SFEXT would limit ADSL reach. In fact, if such a situation in the cable could be assured, then ADSL could be deployed at much higher distances than would SNEXT limited systems. However, because the FCC prohibits binder group service segregation, we can expect that the cable will have a mixture of overlapping and nonoverlapping systems such that NEXT from other systems would limit the reach of ADSL to a distance significantly less than the ADSL SFEXT reach.

The question now is, What is the maximum level of NEXT acceptable into ADSL? Given that symmetric echo canceled systems are performance-limited by SNEXT and ADSL has best performance in a SFEXT environment, determination of a maximum level of NEXT for spectral compatibility with ADSL is somewhat arbitrary. Specifically, the issue is determining the spectral compatibility of 2B1Q-based SDSL with ADSL. Industry consensus is required when determining this criterion for spectral compatibility. A debate with technical and political dimensions was conducted in T1E1.4 to balance the interests of ADSL and SDSL technologies. Eventually all parties agreed to a compromise that placed some limitations on both ADSL and SDSL. Following is the agreed-upon rationale that defines the acceptable spectral compatibility between SDSL and ADSL.

The reach of SDSL is limited by SNEXT. To determine the reach of SDSL, we assume a 50 pair cable and the worst-case crosstalk of 49 SNEXT. In the computation of SDSL reach, 6 dB of SNR margin was included. A value of 6 dB SNR margin is a common practice in the industry. Figure 10.2 shows a plot of the SDSL bit rate versus the distance for the case of 49 SNEXT disturbers and 6 dB of margin. For each bit rate, the distance represents the maximum distance for which the SDSL would be deployed to guarantee 10^{-7} bit error rate with 6 dB of margin.

For each of the bit rates in Figure 10.2 we compute the maximum bit rate of ADSL at that same distance with 24 SDSL NEXT disturbers within a 50 pair binder group. The transmit signal power for SDSL is 13.5 dBm for all bit rates. In each case we assume 6 dB of margin and 3 dB of coding gain in the ADSL bit rate computations. The upstream channel PSD is –38 dBm/Hz in the passband region, and the downstream channel PSD is –40 dBm/Hz in its passband region. The re-

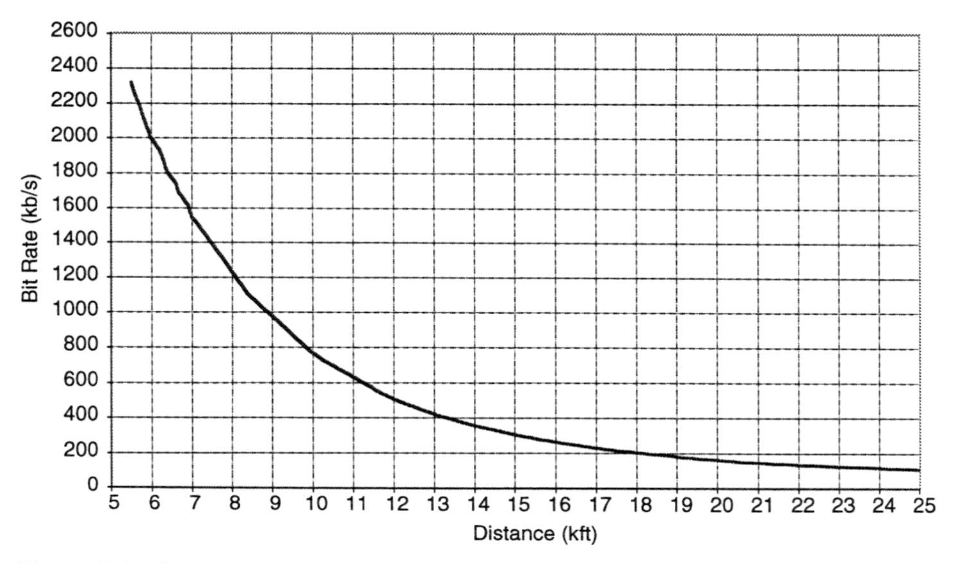

Figure 10.2 SDSL reach in the presence of 49 SNEXT.

sulting ADSL bit rate for each distance corresponds to a maximum level of near-end crosstalk from the 2B1Q SDSL system configured with its bit-rate for the same distance. Table 10.1 lists the ADSL bit rates resulting from injection of twenty-four SDSL NEXT disturbers, where the SDSL spectrum is that of the bit rate corresponding to each distance, and Figure 10.3 shows the corresponding plots.

Table 10.1 SDSL SNEXT Reach and ADSL Bit Rates

SDSL rate, kbps	SDSL reach, kft (49 disturbers)	ADSL rate in kbps (24 SDSL disturbers)		SDSL rate, kbps	SDSL reach, kft (49 disturbers)	ADSL rate in kbps (24 SDSL disturbers)	
		Up	Down			Up	Down
80	28.6	12	0	1232	8.0	862	4731
144	21.2	280	188	1296	7.8	890	4689
208	17.8	299	810	1360	7.6	918	4693
272	15.8	232	1433	1424	7.4	945	4733
336	14.4	163	2063	1488	7.2	972	4801
400	13.3	201	2583	1552	7.0	999	4890
464	12.5	274	2954	1616	6.9	1015	4826
528	11.8	395	3338	1680	6.7	1042	4942
592	11.3	446	3562	1744	6.6	1058	4896
656	10.8	502	3890	1808	6.4	1083	5021
720	10.3	560	4416	1872	6.3	1099	4977
784	9.9	609	4679	1936	6.2	1113	4928
848	9.6	649	4726	2000	6.0	1138	5036
912	9.3	688	4783	2064	5.9	1152	4958
976	9.0	727	4847	2128	5.8	1166	4849
1040	8.7	767	4904	2192	5.7	1180	4681
1104	8.4	806	4948	2256	5.6	1193	4490
1168	8.2	834	4823	2320	5.5	1207	4378

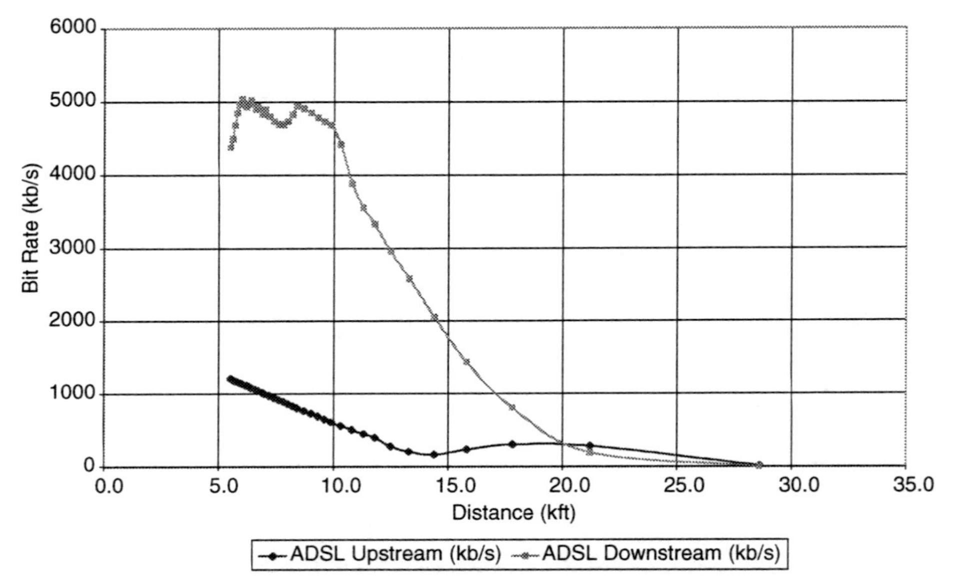

Figure 10.3 ADSL reach in presence of 24 2B1Q SDSL NEXT.

10.6 THE SPECTRUM MANAGEMENT CLASSES

Early in the development of the spectrum management standard, a proposal was considered to specify a single 0-to-infinity PSD mask that would apply to all systems. Later it was determined that no such single PSD mask would suffice because it would have to be overly restrictive to protect the basis systems. Thus, it was determined that a family of more narrow-band masks would best way to specify an easy-to-use method of determining spectral compatibility.

The spectrum management standard T1.417 defines a set of nine spectrum management classes, whereby any DSL system that meets the PSD limitations, total average power limitations, longitudinal output voltage, and transverse balance requirements for any one of the spectrum management classes is considered to be spectrally compatible with all of the basis systems if they are deployed according to the applicable deployment guidelines associated with that class. Alternatively, deployment guidelines for a system compliant with any of the spectrum management classes may be computed using the analytical method (Method B). It is worth noting that the spectrum management classes represent an overly conservative estimation of many actual subscriber line transmission systems, for example, 2B1Q and single carrier modulation such as CAP or QAM. In such situations applying the specific system's PSD to the computations of Method B may yield a better deployment guideline than that associated with the compliant spectrum management class.

- Spectrum management classes are meant to be technology independent.
- The PSD *mask* defines the absolute upper bound for spectrum management compliance.
- The PSD *template* defines a nominal signal characteristic for modeling margin of, and crosstalk due to, technologies using the PSD in reference.

10.6.1 Class 1: Very-Low-Band Symmetric

Spectrum Management Class 1 is intended to accommodate narrowband symmetric (echo canceled) transmission systems. Both the upstream and downstream channels use the same spectrum. Definition of this class is driven by the definition of ISDN per the standard T1.601. The shape of this mask was designed to accommodate numerous technologies with symmetric spectra up about 115 kHz, including passband CAP systems up to approximately 70 kBaud. Table 10.2 provides a mathematical definition of the SM Class 1 template. Plots of SM Class 1 are provided in Figure 10.4.

The total average power for a spectrum management class 1 system is ≤ 14.0 dBm measured into 135 Ω and frequencies below 115 kHz. Systems complying with SM Class 1 are considered to be spectrally compatible on any nonloaded loop facility. As long as the loop is unloaded, there are no distance limitations.

Table 2 Spectrum Management Class 1 Template Definition

Frequency Band (kHz)	PSD (dBm/Hz)
$0 < f \leq 25$	-32.5
$25 < f \leq 76$	$-32.5 - 10.35 \times log_{10}\left(\frac{f}{25}\right)$
$76 < f \leq 79$	$-37.5 - 0.5 \times \left(\frac{f-76}{3}\right)$
$79 < f \leq 85$	$-38 - 19.6 \times log_{10}\left(\frac{f-69}{10}\right)$
$85 < f \leq 100$	$-42 - 4 \times \frac{f-85}{15}$
$100 < f \leq 115$	$-46 - 7 \times \frac{f-100}{15}$
$115 < f \leq 120$	-53
$120 < f \leq 225$	$-53 - 55 \times log_{10}\left(\frac{f}{120}\right)$
$225 < f \leq 635$	$-68 - 70 \times log_{10}\left(\frac{f}{125}\right)$
$635 < f$	$-143 - 10 log_{10}\left(\frac{(1000 \times f)^{3/2}}{1134 \times 10^{1-3}}\right)$

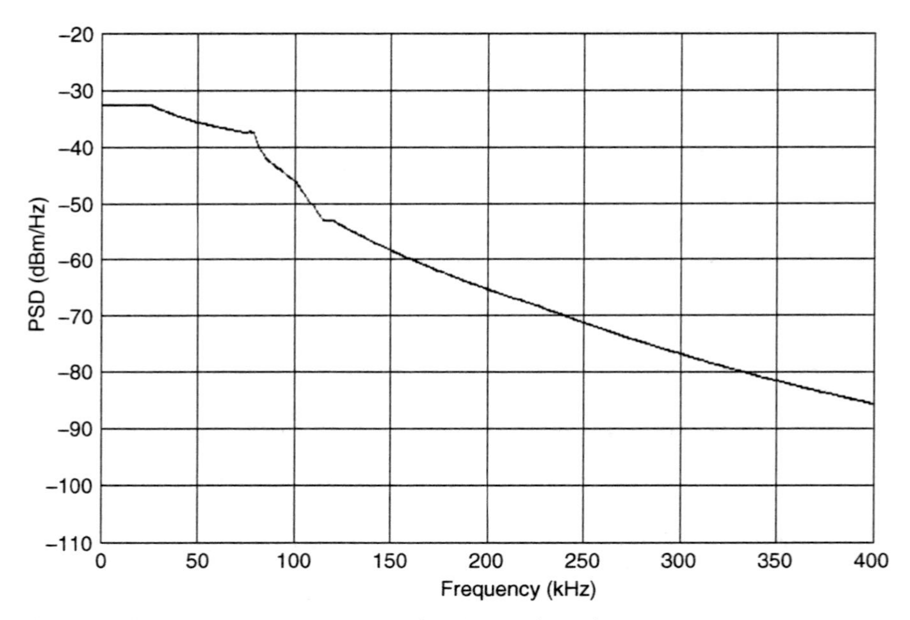

Figure 10.4 Spectrum management class 1 template plots.

10.6.2 Class 2: Low-Band Symmetric

Spectrum Management (SM) Class 2 is intended to accommodate symmetric (echo canceled) transmission systems with bandwidths slightly greater than those of SM Class 1. Both the upstream and downstream channels use the same spectrum. Definition of this class is driven by the transport of 384 kb/s symmetric data using 2B1Q technology. The shape of this mask was designed to accommodate numerous technologies with symmetric spectra up about 238 kHz. In this class passband CAP systems could support symbol rates up to approximately 120 kBaud. Table 10.3 provides a tabular definition of the SM Class 2 template. The template is constructed using linear interpolation of the frequency in kHz and the PSD in dBm/Hz. Plots of SM Class 2 are provided in Figure 10.5.

Note: The interpolation formula for computing the PSD(f) values at intermediate frequencies is provided as follows. The PSD(f) has values of PSD_1 at frequency f_1 and PSD_2 at frequency f_2, then the intermediate PSD(f) values in the frequency range of $f_1 \leq f < f_2$ is

$$PSD(f) = PSD_1 + \left(\frac{PSD_2 - PSD_1}{f_2 - f_1}\right) \cdot \left(f - f_1\right)$$

The total average power for an SM Class 2 system is ≤ 14.0 dBm measured into 135 Ω and frequencies below 238 kHz. Systems complying with SM Class 2 are considered to be spectrally compatible on nonloaded loop facilities with an *equivalent working length* (of 26-gauge wire) of 11.5 kft or less.

Table 10.3 Spectrum Management Class 2 Template Definition

Frequency (kHz)	PSD (dBm/Hz)
0	−36
25	−36
75	−36.5
100	−39
150	−45
200	−54
230	−64
245	−71
335	−72
390	−76
440	−83
475	−90
500	−98
$500 < f$	$-143-10\,log_{10}\left(\frac{(1000 \times f)^{3/2}}{1134 \times 10^{13}}\right)$

For any subscriber loop implemented with a mixture of 26-gauge and 24-gauge wire, the loop can be expressed with an equivalent working length (EWL) of 26 gauge, which approximates a 26-gauge loop having roughly the same insertion loss as the mixed gauge cable. The EWL is computed by $EWL = L_{26} + \frac{3 \cdot L_{24}}{4}$, where L_{26} is the total length of 26-gauge wire and L_{24} is the total length of 24-gauge wire.

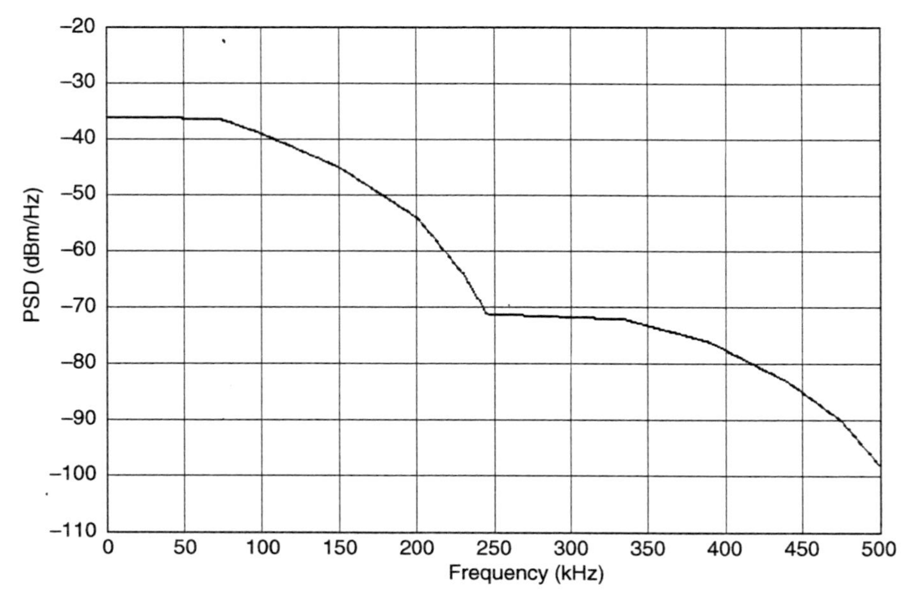

Figure 10.5 Spectrum Management Class 2 template plots.

Table 10.4 Spectrum Management Class 3 Template Definition

Frequency (kHz)	PSD (dBm/Hz)
0	−37
50	−37
125	−38
210	−41
310	−57
370	−73
550	−75
670	−85
750	−97
980	−98
1050	−102.75
$1050 < f$	$-143 - 10\log_{10}\left(\frac{(1000 \times f)^{3/2}}{1.134 \times 10^{13}}\right)$

10.6.3 Class 3: Mid-Band Symmetric

Spectrum Management Class 3 is intended to accommodate symmetric transmission systems with bandwidths greater than those of SM Classes 1 and 2. Both the upstream and downstream channels use the same spectrum. Definition of this class is driven by the definition of HDSL that transports 784 kb/s symmetric data using 2B1Q technology. The shape of this mask was designed to accommodate numerous technologies with symmetric spectra up about 370 kHz. In this class passband CAP systems could support symbol rates up to approximately 200 kBaud. Table 10.4 provides a tabular definition of the SM Class 3 template. The template is con-

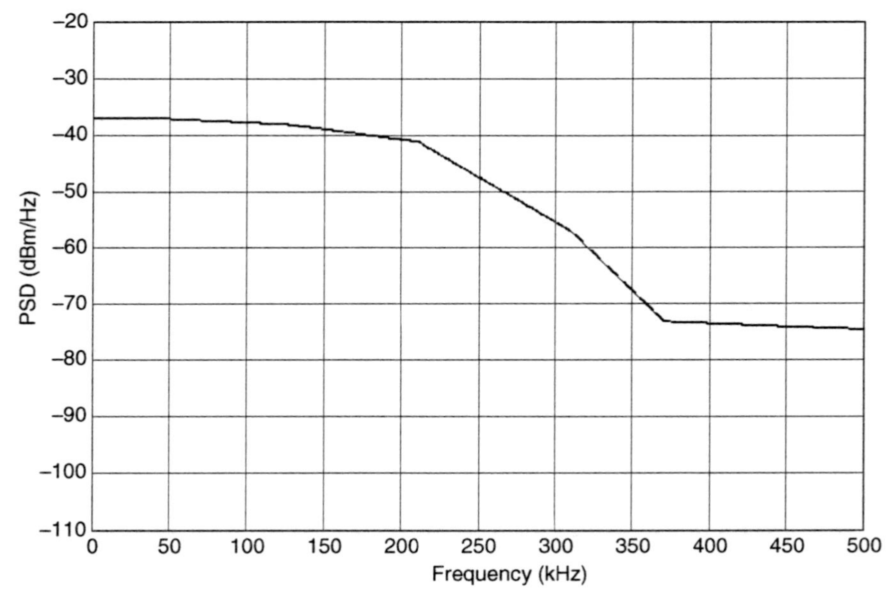

Figure 10.6 Spectrum Management Class 3 template plots.

Table 10.5 Spectrum Management Class 4 Template Downstream Definition

Frequency (kHz)	PSD (dBm/Hz)	Frequency (kHz)	PSD (dBm/Hz)	Frequency (kHz)	PSD (dBm/Hz)
≤1	−54.2	280	−35.7	1000	−89.2
2	−42.2	375	−35.7	2000	−99.7
12	−39.2	400	−40.2	>3000	−108
190	−39.2	440	−68.2		
236	−46.2	600	−76.2		

structed using linear interpolation of the frequency in kHz and the PSD in dBm/Hz. Plots of SM Class 3 are provided in Figure 10.6.

The total average power for an SM Class 3 system is ≤ 14.0 dBm measured into 135 Ω and frequencies below 370 kHz. Systems complying with SM Class 3 are considered to be spectrally compatible on nonloaded loop facilities with an EWL of 9 kft or less.

10.6.4 Class 4: HDSL2

SM Class 4 was included to accommodate HDSL2 (T1.418-2000)–based systems in the spectrum management standard. Because the SM classes are meant to be technology independent, any transmission systems whose spectra meets the Class 4 template would be considered to be spectrally compatible with the basis systems. The downstream spectrum utilizes frequencies up to about 400 kHz and the upstream spectrum utilizes frequencies up to about 300 kHz. Table 10.5 and Table 10.6 provide a tabular definition of the SM Class 4 template. Plots of SM Class 4 are provided in Figure 10.7 and Figure 10.8.

The total average power for the downstream channel of a spectrum management Class 4 system is ≤ 17.3 dBm measured into 135 Ω and frequencies below 450 kHz. The total average power for the upstream channel of a spectrum management Class 4 system is ≤ 17.0 dBm measured into 135 Ω and frequencies below 350 kHz. Systems complying with SM Class 4 are considered to be spectrally compatible on nonloaded loop facilities with an EWL of 10.5 kft or less.

10.6.5 Class 5: Asymmetric

SM Class 5 is intended to accommodate ADSL (T1.413-1998) in the SM standard. Specifically, Class 5 favors the frequency division duplex (FDD) configuration of

Table 10.6 Spectrum Management Class 4 Template Upstream Definition

Frequency (kHz)	PSD (dBm/Hz)	Frequency (kHz)	PSD (dBm/Hz)	Frequency (kHz)	PSD (dBm/Hz)
≤1	−64.2	220	−34.4	555	−102.6
2	−42.1	255	−34.4	800	−105.6
10	−37.8	276	−41.1	1400	−108
175	−37.8	300	−77.6	≥2000	−108

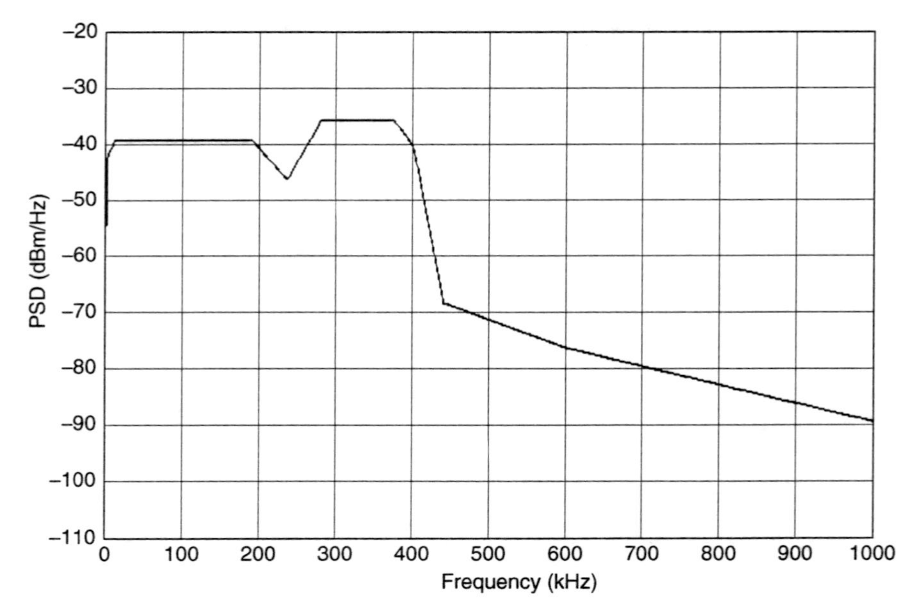

Figure 10.7 Spectrum Management Class 4 downstream template plot.

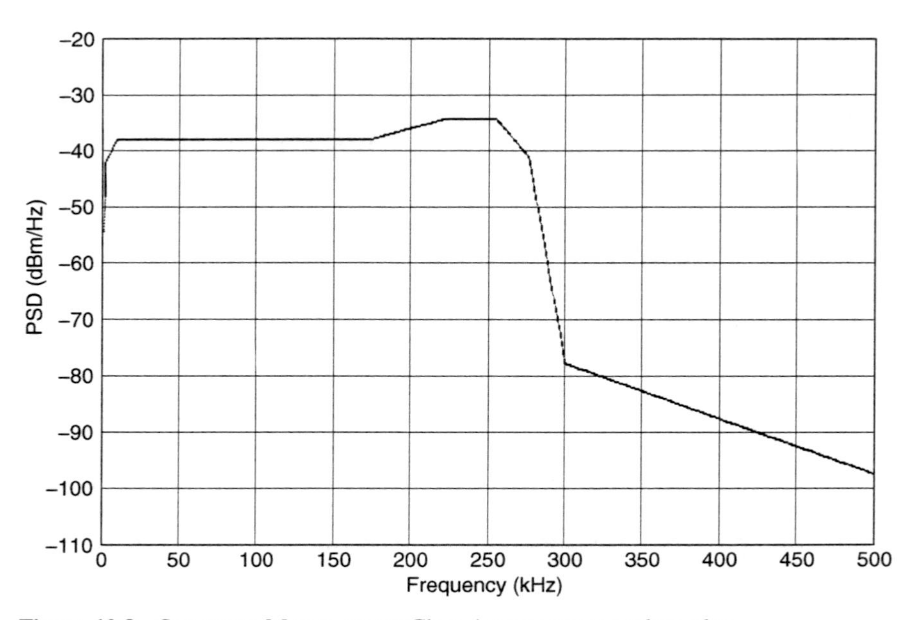

Figure 10.8 Spectrum Management Class 4 upstream template plot.

Table 10.7 Spectrum Management Class 5 Template Definition

Frequency Band (kHz)	PSD (dBm/Hz)
$0 < f \leq 4$	−101, with max power in the in 0–4 kHz band of + 15 dBm
$4 < f \leq 25.875$	$-96 + 21 \times \log_2(f/4)$
$25.875 < f \leq 81$	−40
$81 < f \leq 92.1$	$-40-70 \times \log_2(f/81)$
$92.1 < f \leq 121.4$	−53
$121.4 < f \leq 138$	$-53 + 70 \times \log_2(f/121.4)$
$138 < f \leq 1104$	−40
$1104 < f \leq 3093$	$-40-36 \times \log_2(f/1104)$
$3093 < f \leq 4545$	$\min(-36.5-36 \times \log_2(f/1104),-93.5)$
$4545 < f \leq 11040$	−110

ADSL where the upstream and downstream channels use separate frequency bands, but overlap is permitted in the low frequency bands that are common with SM Class 1. The downstream frequency spectrum uses frequencies from about 25 kHz to about 1104 kHz and the upstream frequency spectrum uses frequencies from about 25 kHz to about 138 kHz. Table 10.7 provides a mathematical definition of the SM Class 5 template. Plots of SM Class 5 are provided in Figure 10.9.

The total average power for the downstream channel of a spectrum management Class 5 system is ≤ 20.9 dBm measured into 100 Ω and frequencies between 25 kHz and 1104 kHz. The total average power for the upstream channel of an SM Class 4 system is ≤ 13 dBm measured into 100 Ω and frequencies between 25 kHz and 138 kHz. Systems complying with SM Class 5 are considered to be spectrally compatible on any nonloaded loop facilities. As long as the loop is unloaded, there are no distance limitations.

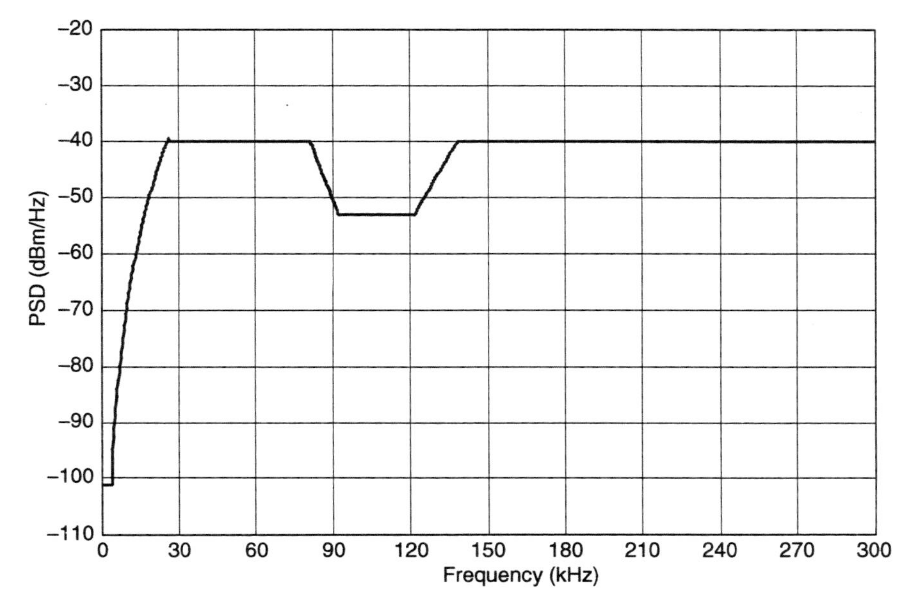

Figure 10.9 Spectrum Management Class 5 template plots.

10.6.6 Class 6:Wide-Band Asymmetric

This spectrum management class is intended to accommodate asymmetric VDSL transmission systems that utilize the frequency spectrum up to about 10–20 MHz. SM Class 6 should be based on the emerging VDSL standards, and at the time of the writing of T1.417, the VDSL standard was not completed. Therefore the template for the SM Class 6 is not yet defined and left as a placeholder for inclusion of the VDSL standard once it is complete.

10.6.7 Class 7: Very-Wide-Band Symmetric

SM Class 7 is another symmetric spectrum, which was designed to accommodate single pair 1.5 Mb/s 2B1Q systems. Both the upstream and downstream channels use the same spectrum. The shape of this mask was designed to accommodate numerous technologies with symmetric spectra up to about 776 kHz. In this class passband, CAP systems could support symbol rates up to approximately 400 kBaud. Table 10.8 provides a tabular definition of the SM Class 7 template. The template is constructed using linear interpolation of the frequency in kHz and the PSD in dBm/Hz. Plots of SM Class 7 are provided in Figure 10.10.

The total average power for an SM Class 7 system is ≤ 14.0 dBm measured into 135 Ω and frequencies below 776 kHz. Systems complying with SM Class 7 are considered to be spectrally compatible on nonloaded loop facilities with an EWL of 6.5 kft or less.

10.6.8 Class 8: Wide-Band Symmetric

SM Class 8 is yet another symmetric spectrum, which was designed to accommodate single pair 1.1 Mb/s 2B1Q systems. Both the upstream and downstream channels use the same spectrum. The shape of this mask was designed to accommodate numerous technologies with symmetric spectra up about 584 kHz. In this class passband, CAP systems could support symbol rates up to approximately 300 kBaud. Table 10.9 provides a tabular definition of the SM Class 8 template. The template is constructed using linear interpolation of the frequency in kHz and the PSD in dBm/Hz. Plots of SM Class 8 are provided in Figure 10.11.

Table 10.8 Spectrum Management Class 7 Template Definition

Frequency (kHz)	PSD (dBm/Hz)	Frequency (kHz)	PSD (dBm/Hz)
0	−40	775	−77
100	−40	1000	−77
150	−40.5	1100	−80
200	−41.5	1300	−86
300	−42	1500	−102
390	−42	1900	−104
420	−43	2000	−107
500	−51	>2000	$-143-10\,log_{10}\left(\frac{(1000 \times f)^{3/2}}{1.134 \times 10^{13}}\right)$

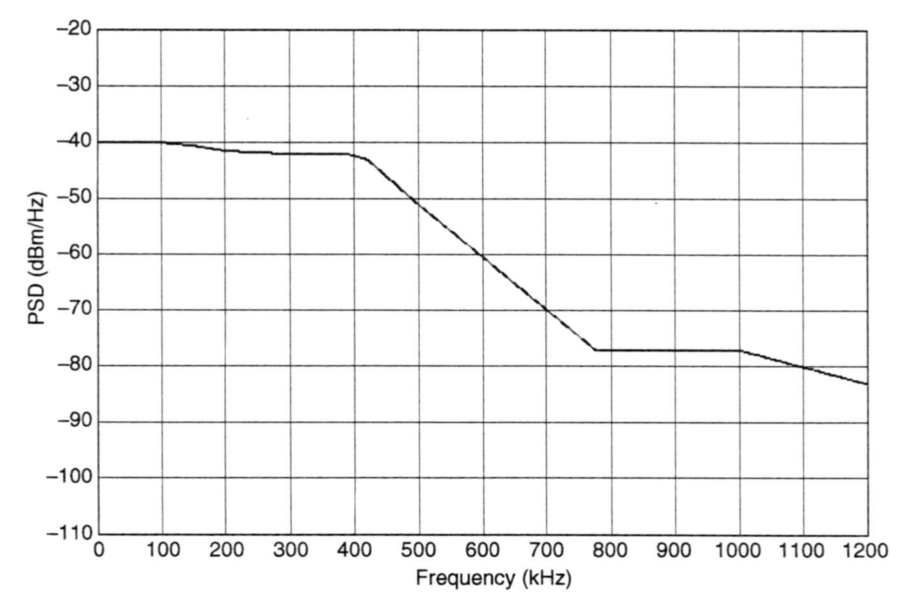

Figure 10.10 Spectrum Management Class 7 template plots.

The total average power for an SM Class 8 system is ≤ 14.0 dBm measured into 135 Ω and frequencies below 584 kHz. Systems complying with SM Class 8 are considered to be spectrally compatible on nonloaded loop facilities with an EWL of 7.5 kft or less.

10.6.9 Class 9: Overlapping Asymmetric

SM Class 9 is intended to accommodate ADSL (T1.413-1998) with overlapping upstream and downstream spectra in the SM standard. The downstream frequency spectrum uses frequencies from about 25 kHz to about 1104 kHz, and the upstream frequency spectrum uses frequencies from about 25 kHz to about 138 kHz. Table 10.10 provides a mathematical definition of the SM Class 9 template. Plots of SM Class 9 are provided in Figure 10.12.

The total average power for the downstream channel of an SM Class 9 system is ≤ 20.9 dBm measured into 100 Ω and frequencies between 25 kHz and 1104 kHz. The

Table 10.9 Spectrum Management Class 8 Template Definition

Frequency (kHz)	PSD (dBm/Hz)	Frequency (kHz)	PSD (dBm/Hz)	Frequency (kHz)	PSD (dBm/Hz)
0	−39	400	−53	1120	−95
60	−39	500	−66	1500	−95
200	−40	550	−75	2000	−107
250	−40.5	750	−76	>2000	$-143-10\,log_{10}\left(\frac{(1000 \times f)^{3/2}}{1.134 \times 10^{13}}\right)$
315	−41	950	−84		

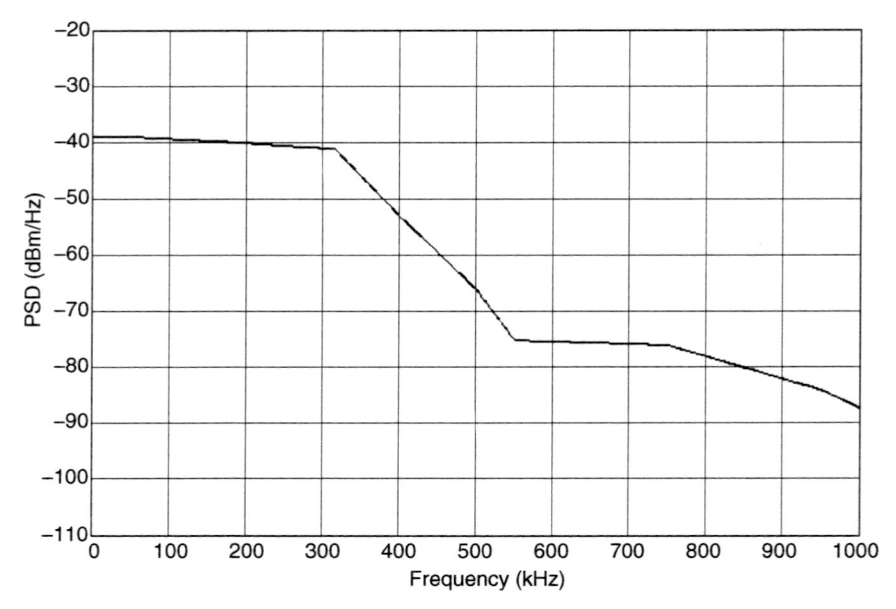

Figure 10.11 Spectrum Management Class 8 template plots.

total average power for the upstream channel of an SM Class 4 system is ≤ 13 dBm measured into 100 Ω and frequencies between 25 kHz and 138 kHz. Systems complying with SM Class 9 are considered to be spectrally compatible on any nonloaded loop facilities. As long as the loop is unloaded, there are no distance limitations.

10.7 TECHNOLOGY-SPECIFIC GUIDELINES

In addition to the technology-independent SM classes described, the SM standard T1.417 also contains technology-specific guidelines. In the first issue of T1.417, there is one specific underlying technology, namely, pulse amplitude modulation; however, there are three distinct families of technology specific guidelines. The

Table 10.10 Spectrum Management Class 9 Upstream Template Definition

Frequency Band (kHz)	PSD (dBm/Hz)
$0 < f < 4$	−101, with max power in the in 0–4 kHz bands of + 15 dBm
$4 < f < 25.875$	$-96 + 21.5 \times \log_2(f/4)$
$25.875 < f < 138$	−38
$138 < f < 307$	$-38 - 48 \times \log_2(f/138)$
$307 < f < 1221$	−93.5
$1221 < f < 1630$	$\min(-90 - 48 \times \log_2(f/1221), -93.5)$
$1630 < f < 11040$	−110

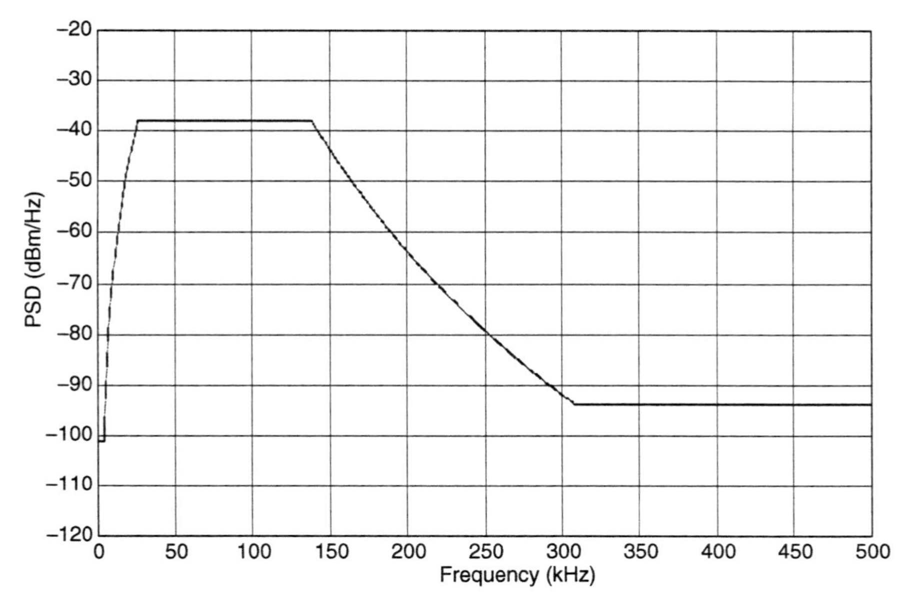

Figure 10.12 Spectrum Management Class 5 or Class 9 upstream template plots.

technology specific guidelines include 2B1Q-based SDSL, PAM per G.991.2 Annex A, and PAM per HDSL4 using 776/784 kb/s asymmetric PSDS.

10.7.1 2B1Q SDSL

Although this is a nonstandard technology in North America, there have been many systems deployed using multirate 2B1Q SDSL in the loop plant. 2B1Q is four-level pulse amplitude technology whose bandwidth is directly proportional to the bit rate of the system. The spectral compatibility of 2B1Q SDSL with the basis systems is computed using the following mathematical expression for the PSD:

$$PSD_{SDSL}(f) = \frac{2.7 \times 2.7}{135 \times f_{sym}} \cdot \left(\frac{\sin\left(\frac{\pi f}{f_{sym}}\right)}{\left(\frac{\pi f}{f_{sym}}\right)} \right)^2 \cdot \frac{1}{1 + \left(\frac{f}{\frac{240}{392} \cdot f_{sym}} \right)^8} Watts/Hz,$$

where f_{sym} is the symbol rate in units of symbols per second, the symbol rate is one-half the bit rate, and f is the frequency in Hz. Because this is a variable bit rate, hence variable bandwidth transmission system, the spectral compatibility with the basis systems will vary with the symbol rate of SDSL. We therefore need to define a deployment guideline for each of the symbol rates supported by SDSL so that there is no degradation to the services provided by the basis systems. Complying with the 2B1Q SDSL template at a specific symbol rate assures spectral compatibility with the basis systems if deployed according to the deployment guidelines associated with that symbol rate. A complete list of deployment guidelines is provided in Table 10.11.

10.7.2 G.shdsl

G.shdsl (i.e., G.991.2) defines a multirate transmission system based on trellis-coded pulse amplitude modulation (TC-PAM). Annex A of G.shdsl defines a family of symmetric PSDs that is a function of the symbol rate. As with 2B1Q SDSL, the spectral compatibility of G.shdsl with the basis systems will be vary with the symbol rate of G.shdsl-based systems. To compute the spectral compatibility of the G.shdsl systems with the basis systems, the following mathematical expression for the G.shdsl PSD is used:

$$PSD_{SHDSL}(f) = \begin{cases} \dfrac{K_{SHDSL}}{135} \cdot \dfrac{1}{f_{sym}} \cdot \left[\dfrac{\sin\left(\dfrac{\pi f}{f_{sym}}\right)}{\dfrac{\pi f}{f_{sym}}} \right]^2 \cdot \dfrac{1}{1 + \left(\dfrac{f}{f_{3dB}}\right)^{12}} \cdot \dfrac{f^2}{f^2 + f_c^2}, & f < f_{int} \\ 0.5683 \times 10^{-4} \times f^{-15}, & f_{\text{int}} \le f \le 1.1 \text{ MHz} \end{cases}$$

Table 10.11 SDSL Deployment Guidelines

Designation	PSD	Maximum 2B1Q SDSL line bit rate (kbps)	2B1Q SDSL deployment guideline, EWL (kft)
TS101	SM1 PSD template	300	all non-loaded loops
TS102	$SDSL_u(f)$ with $f_{sym,} = 160{,}000$	320	15.5
TS103	$SDSL_u(f)$ with $f_{sym,} = 168{,}000$	336	14.5
TS104	$SDSL_u(f)$ with $f_{sym,} = 208{,}000$	416	13.5
TS105	$SDSL_u(f)$ with $f_{sym,} = 232{,}000$	464	12.5
TS106	$SDSL_u(f)$ with $f_{sym,} = 264{,}000$	528	12
TS107	$SDSL_u(f)$ with $f_{sym,} = 296{,}000$	592	11.5
TS108	$SDSL_u(f)$ with $f_{sym,} = 328{,}000$	656	11
TS109	$SDSL_u(f)$ with $f_{sym,} = 360{,}000$	720	10.5
TS110	$SDSL_u(f)$ with $f_{sym,} = 392{,}000$	784	10
TS111	$SDSL_u(f)$ with $f_{sym,} = 456{,}000$	912	9.5
TS112	$SDSL_u(f)$ with $f_{sym,} = 520{,}000$	1040	9
TS113	$SDSL_u(f)$ with $f_{sym,} = 552{,}000$	1104	8.5
TS114	$SDSL_u(f)$ with $f_{sym,} = 616{,}000$	1232	8
TS115	$SDSL_u(f)$ with $f_{sym,} = 712{,}000$	1424	7.5
TS116	$SDSL_u(f)$ with $f_{sym,} = 840{,}000$	1680	7
TS117	$SDSL_u(f)$ with $f_{sym,} = 936{,}000$	1872	6.5
TS118	$SDSL_u(f)$ with $f_{sym,} = 1{,}064{,}000$	2128	6
TS119	$SDSL_u(f)$ with $f_{sym,} = 1{,}128{,}000$	2256	5.5
TS120	$SDSL_u(f)$ with $f_{sym,} = 1{,}160{,}000$	2320	5

Table 10.12 Values of the G.shdsl PSD Parameters

Line Bit Rate *LBR* **(kbps)**	K_{SHDSL}	f_{sym} **(ksymbols)**	f_{3dB}
$LBR \neq 1544$ or 1552	7.86	*LBR* 13	$1.0 \times f_{sym}12$
$LBR = 1544$ or 1522	8.32	*LBR* 13	$0.9 \times f_{sym}12$

where f_{int} is the intersection frequency in which the two functions in $PSD_{SHDSL}(f)$ intersect in the range of 10 kHz to f_{sym}, K_{SHDSL} is a scaling coefficient, f_{sym} is the symbol rate, f_{3dB} is the low-pass shaping filter cut-off frequency, and f_c is the high-pass transformer cut-off frequency. The values for the G.shdsl PSD parameters are given in Table 10.12. Because G.shdsl is a variable bit rate resulting in a variable bandwidth transmission system, the spectral compatibility with the basis systems will vary with its symbol rate. As with SDSL, the deployment guidelines for G.shdsl need to be defined as a function of symbol rate. Complying with the G.shdsl template at a specific symbol rate assures spectral compatibility with the basis systems if deployed according to the deployment guidelines associated with that symbol rate. A complete list of deployment guidelines is provided in Table 10.13.

10.7.3 776/784 HDSL4 Asymmetric Spectra Using TC-PAM

HDSL4 technology is designed to transport a 1.544 Mb/s (DS1) payload on two twisted wire pairs using TC-PAM. This is the same core technology as that for

Table 10.13 Deployment Guidelines for G.shdsl

Designation	**G.shdsl Line Bit Rate (kbps)**	**G.shdsl deployment guideline, EWL (kft)**
TS201	LBR ≤ 592	All non-loaded loops
TS202	600 ≤ LBR ≤ 616	15.0
TS203	624 ≤ LBR ≤ 648	14.5
TS204	656 ≤ LBR ≤ 688	14.0
TS205	696 ≤ LBR ≤ 800	13.5
TS206	808 ≤ LBR ≤ 832	12.5
TS207	840 ≤ LBR ≤ 896	12.0
TS208	904 ≤ LBR ≤ 952	13.0
TS209	960 ≤ LBR ≤ 1000	12.5
TS210	1008 ≤ LBR ≤ 1088	12.0
TS211	1096 ≤ LBR ≤ 1160	11.5
TS212	1168 ≤ LBR ≤ 1320	11.0
TS213	1328 ≤ LBR ≤ 1472	10.5
TS214	1480 ≤ LBR ≤ 1536	10.0
TS215	1544 ≤ LBR ≤ 1552	10.5
TS216	1560 ≤ LBR ≤ 1664	10.0
TS217	1672 ≤ LBR ≤ 1880	9.5
TS218	1888 ≤ LBR ≤ 2008	9.0
TS219	2016 ≤ LBR ≤ 2320	8.5

Table 10.14 PSD Template for HDSL4 Downstream Channel

Frequency (kHz)	PSD (dBm/Hz)	Frequency (kHz)	PSD (dBm/Hz)	Frequency (kHz)	PSD (dBm/Hz)
≤0.2	−51	110	−58	250	−51.5
2	−41	135	−46.5	400	− 46.5
5	−37.5	145	−40.5	600	−70
50	−37.5	150	−38.5	1000	−89.2
80	−40.5	155	−37.5	2000	−99.7
90	−45	200	−40.25	3000	−108
105	−58	210	−43	≥3100	−110

G.shdsl, where the bit rate and transmission spectrum are fixed to transport 784 kb/s or 776 kb/s on each wire pair end-to-end.

To obtain optimum reach and spectral compatibility with other services in the cable, HDSL4 uses a set of asymmetric PSDs for transmission of the upstream and downstream channels. The downstream channel template is listed in Table 10.14 and the corresponding PSD plot is shown in Figure 10.13. In the construction of the PSD, linear interpolation of the frequency and PSD points is used. The upstream channel template is listed in Table 10.15 and the corresponding plot is given in Figure 10.14.

Given the optimized spectral shaping for the given bit rates, signals of HDSL4 technology are considered to be spectrally compatible with the basis system when deployed on any nonloaded loop facility. The only restriction is that an HDSL4 transceiver is not located near the customer end with the downstream spectrum traveling in the downstream direction. The far-end crosstalk may cause a spectral incompatibility with other systems served directly from the central office.

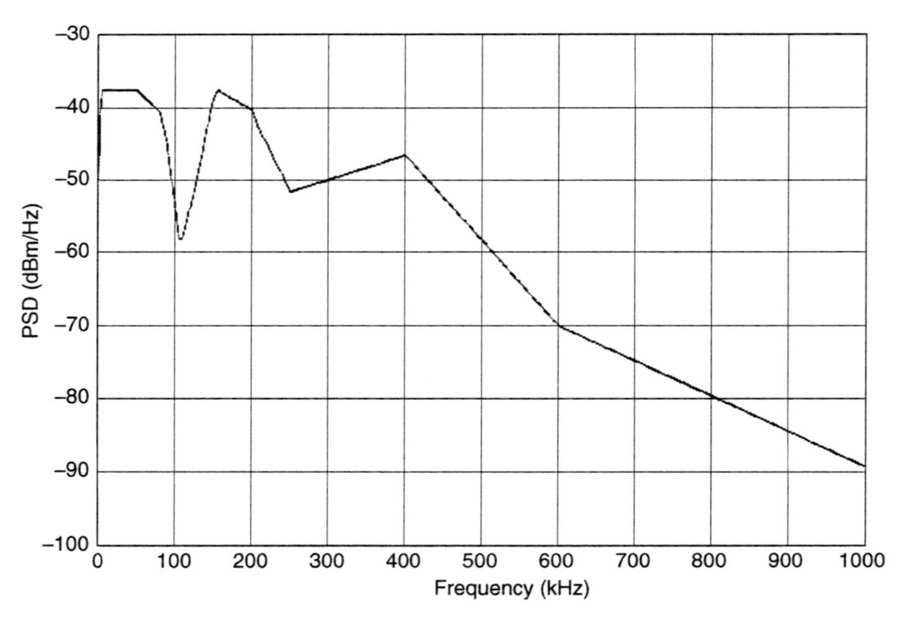

Figure 10.13 Plot of PSD template for HDSL4 downstream channel.

Table 10.15 PSD Template for HDSL4 Upstream Channel

Frequency Band (kHz)	PSD (dBm/Hz)
$0 < f \leq 200$	-51
$200 < f \leq 2000$	$-41 + 10(f-2000)/1800$
$2000 < f \leq 5000$	$-37 + 4(f-5000)/3000$
$5000 < f \leq 50000$	-37
$50000 < f \leq 125000$	$-37-((f-50000)/75000)$
$125000 < f \leq 130000$	-38
$130000 < f \leq 307000$	$-38 - 142 \log_{10}(f/130000)$
$307000 < f \leq 1221000$	-93.5
$1221000 < f \leq 1630000$	$\min(-90 - 48 \times \log_2(f/1221000), -93.5)$
$f < 1630000$	-110

10.8 ANALYTICAL METHOD (METHOD B)

The goal of the analytical method, also referred to as Method B, is to provide an alternative technique to demonstrate or quantify the spectral compatibility of a technology with the basis systems. The result of this activity will be the determination of a deployment guideline of a loop transmission technology that would be considered to be spectrally compatible with the basis systems.

There could be numerous reasons for wanting to use the analytical method to determine spectral compatibility as opposed to using the method of signal power limitations, that is, via the SM classes. One reason could be that the technology to be deployed does not meet any of the nine SM classes defined earlier;

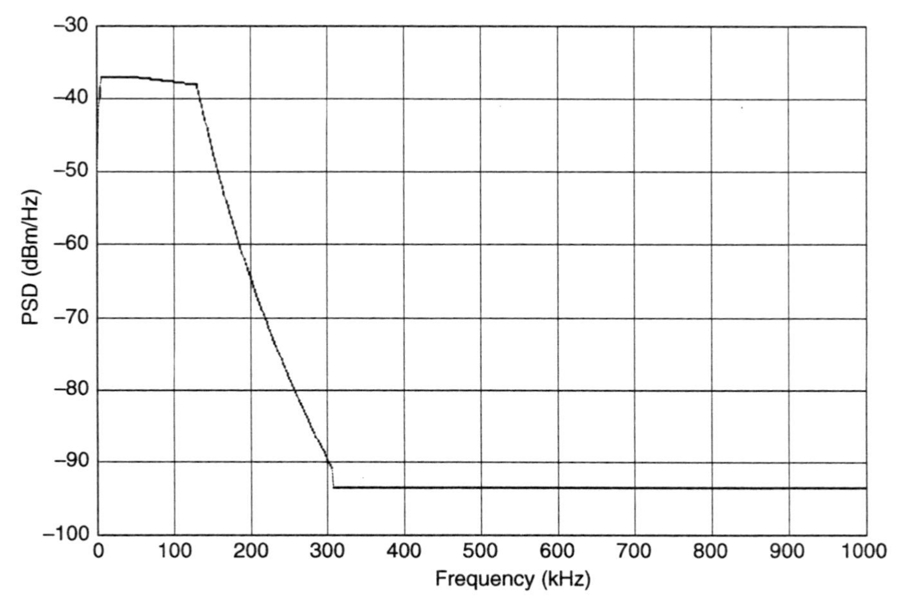

Figure 10.14 Plot of PSD template for HDSL4 upstream channel.

the analytical method could be used to determine a deployment guideline for which this technology under test could be deployed and be spectrally compatible with the basis systems. Another reason could be that although the technology under test meets one or more of the spectrum management classes, the spectral characteristics of the technology could be such that the analytical method would allow a greater deployment guideline than that provided by the spectrum management classes.

One point to note is that although the analytical method does quantify the spectral compatibility of the technology under test with the basis systems, it does not quantify the spectral compatibility with other technologies in the cable that are not on the basis systems list. To assure performance of the new technology in the cable, crosstalk evaluations should be performed using crosstalk from each of the SM classes.

10.8.1 Determination of Spectral Compatibility with Basis Systems Using Method B

The analytical method requires that the spectral compatibility of the new technology (or technology under test) be evaluated for each of the basis systems. In other words, a spectral compatibility evaluation of some form needs to be done relative to each of the following basis systems:

- Voice-grade services and technologies
- Enhanced business services
- DDS per T1.410
- ISDN BRI per T1.601
- HDSL
- ADSL and RADSL technologies
- HDSL2
- 2B1Q SDSL
- In the second issue of T1.417, G.shdsl replaces 2B1Q SDSL on the basis systems list.

The general process for determining the spectral compatibility with each basis system is an iterative process that requires determining the performance target of the basis system based on a reference disturber, computing the performance of the basis system with disturbance from the new technology, and then comparing the results. In some cases the determination of spectral compatibility is done by evaluating relative margin and in other cases by evaluating absolute margin. Figure 10.15 shows a high-level crosstalk model for performance evaluation. In general, the following basic steps are performed in determining the spectral compatibility of a "new" technology with each basis system.

1. First, determine the target margin of the basis system at the target equivalent working length Z kft of 26-gauge wire. This target margin of the basis system

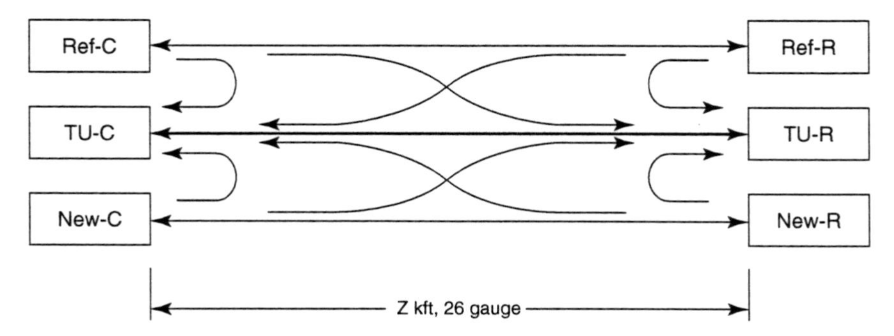

Figure 10.15 System model for evaluating crosstalk disturbance into basis system.

is computed by applying a reference disturbance to the input of a basis system receiver operating on a loop with an EWL of Z.

2. Second, calculate the margin of the basis system operating on the same loop of EWL Z using crosstalk disturbance with the "new" technology and a mixture of disturbers, including the "new" technology.
3. If the new margin computed in step 2 is greater than the target margin from step 1 less some delta (i.e., target margin—delta) specified for each basis system, then this new technology is spectrally compatible with the specific basis system at the working length Z. Depending on how much greater the margin is, it may be desirable to repeat steps 1–3 using a longer EWL of Z′. As long as the computed "new" margin is greater than the target margin less the specified delta, then the new technology is spectrally compatible with the basis system at EWL of Z′.
4. If the new margin is less than the target margin less a specified delta, then the "new" technology is NOT spectrally compatible with the specific basis system at EWL of length Z. We must then reduce the EWL to a shorter length of Z′ and repeat the computations of step 2. The loop lengths in the evaluations have a resolution of 500 ft. Keep reducing the EWL in 500 ft increments until a value of Z′ meets the target margin less the specified delta determined in step (1).
5. Steps (1) through (4) are repeated for each of the basis systems and the resulting EWLs are recorded. Select the shortest Z′ from all of the tests, and this is the resulting deployment guideline that will assure spectral compatibility with all of the basis systems.

NEXT and FEXT Models

As described in Chapter 9 on spectral compatibility, the near-end crosstalk model used in the analytical method is the simplified NEXT model of

$$NEXT(f, n) = S(f) \cdot |H_{NEXT}(f, n)|^2 = S(f) \cdot X_{49} \cdot \left(\frac{n}{49}\right)^{0.6} \cdot f^{3/2}$$

where $S(f)$ is the PSD of the interfering system, $X_{49} = \dfrac{1}{1.13 \times 10^{13}}$ is coupling coefficient for 49-NEXT disturbers, n is the number of NEXT disturbers, and f is the frequency in Hz. The far-end crosstalk model is defined as

$$\begin{aligned} FEXT(f,n) &= S(f) \cdot |H_{channel}(f)|^2 \cdot |H_{FEXT}(f,n)|^2 \\ &= S(f) \cdot |H_{channel}(f)|^2 \cdot \left(\frac{n}{49}\right)^{0.6} \cdot k \cdot l \cdot f^2 \end{aligned}$$

where $S(f)$ is the PSD of the interfering system, n is the number of FEXT disturbers, $k = 8 \times 10^{-20}$ is the FEXT coupling coefficient for 49-FEXT disturbers, l is the channel length, and f is the frequency in Hz.

In some cases, the two-piece model is used for modeling next. The two-piece NEXT model is defined as

$$NEXT(f,n) = \begin{cases} S(f) \cdot \alpha_1 \cdot f^{B_1}, f < 20\,kHz \\ S(f) \cdot \alpha_2 \cdot f^{B_2}, f \geq 20\,kHz \end{cases}$$

where f is frequency in Hz, $S(f)$ is the power spectral density of the interfering system, and α_1 β_1 α_2 β_2 are constants defined in Table 10.16 corresponding to 1, 10, 24, and 49 disturbers. The two piece model is plotted in Figure 10.16.

Mixing Crosstalk

When a mixture of crosstalk is required in basis system compatibility evaluation, the FSAN (Full Service Access Network) method of combining crosstalk is used. Instead of adding directly the crosstalk power terms, each component term is raised to power of 1/0.6 and the resulting terms are added together. The final sum is then raised to the power of 0.6. This summing operation may be expressed as

$$\text{FSAN Cross Talk Sum} = \left(\sum_{i=1}^{N} Xtalk(f, n_i)^{\frac{1}{0.6}}\right)^{0.6}$$

where $Xtalk(f, n_i)$ is either the NEXT or FEXT noise component, n_i is the number of disturbers of the ith noise component, and N is the total number of unlike disturber types.

Table 10.16 Constants for the Two-Piece NEXT Model

Number of Disturbers	α_1	β_1	α_2	β_2
1	6.598×10^{-11}	0.6	8.881×10^{-15}	1.5
10	7.071×10^{-10}	0.5	9.518×10^{-14}	1.4
24	1.925×10^{-9}	0.45	1.580×10^{-13}	1.4
49	4.782×10^{-9}	0.4	2.391×10^{-13}	1.4

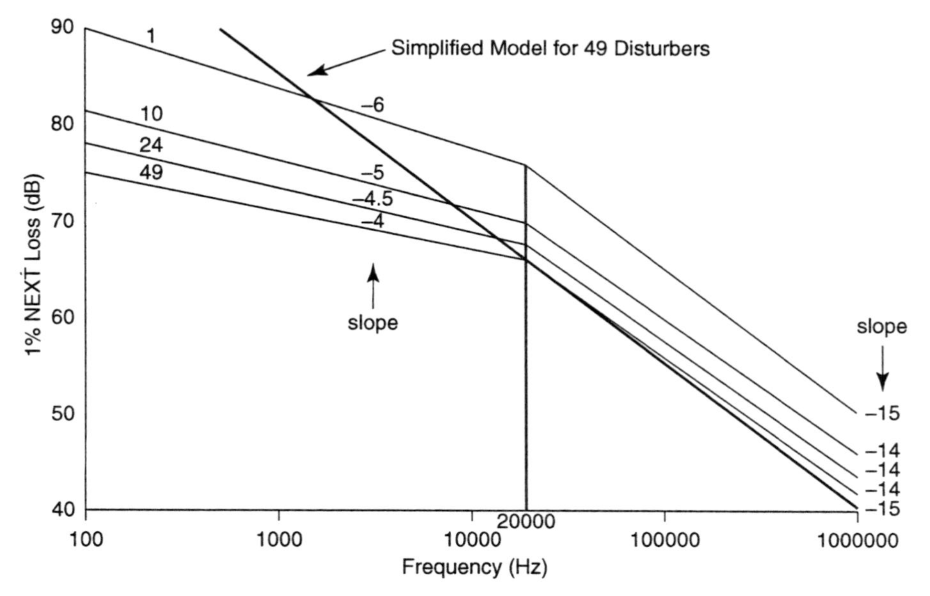

Figure 10.16 Two-Piece NEXT.

Ideal DFE Equation for PAM-Based Systems

Computation of margin for DFE-based PAM systems, for which 2B1Q is a 4-level PAM system, is computed using the ideal DFE equation for PAM-based systems. Expressed in dB form, the ideal DFE equation for PAM-based systems is

$$\text{Margin in dB} = \frac{1}{f_B} \cdot \int_0^{f_B} 10 \cdot \log_{10}(1 + f_{SNR}(f)) \cdot df - SNR_{req}$$

where f_B is the symbol rate (or Baud rate) of the PAM-based system, SNR_{req} is the required SNR for the basis system, and $f_{SNR}(f)$ is the folded receiver signal-to-noise ratio. The folded SNR for PAM-based systems is expressed as

$$f_{SNR}(f) = \sum_{n=-2}^{1} \frac{S(f + f_B \cdot n) \cdot |H(f + f_B \cdot n)|^2}{N(f + f_B \cdot n)}$$

where S(f) is the desired signal's transmit PSD, $|H(f)|^2$ is the squared magnitude channel insertion gain, and N(f) is the total noise PSD seen at the receiver input.

Ideal DFE Equation for Single-carrier Based Systems

Computation of margin for DFE-based single-carrier systems, such as CAP or QAM, is computed using the ideal DFE equation for QAM-based systems. Expressed in dB form, the ideal DFE equation for QAM-based systems is

$$\text{Margin in dB} = \frac{1}{f_B} \cdot \int_0^{f_B} 10 \cdot \log_{10}(1 + f_{SNR}(f)) \cdot df - SNR_{req}$$

where f_B is the symbol rate (or Baud rate) of the single-carrier system, SNR_{req} is the required SNR for the basis system, and $f_{SNR}(f)$ is the folded receiver signal-to-noise ratio. The folded SNR for QAM-based systems is expressed as

$$f_{SNR}(f) = \sum_{n=0}^{3} \frac{S(f + f_B \cdot n) \cdot |H(f + f_B \cdot n)|^2}{N(f + f_B \cdot n)}$$

where $S(f)$ is the desired signal's transmit PSD, $|H(f)|^2$ is the squared magnitude channel insertion gain, and $N(f)$ is the total noise PSD seen at the receiver input.

Computations for Multicarrier (DMT) Systems

Computation of margin for DMT systems may be done as follows. DMT is a multicarrier system that effectively divides the transmit spectrum into N evenly spaced tones with a carrier spacing of Df = 4.3125 kHz. Bits are allocated to each tone based on the Shannon capacity of the individual tone compensated for by the SNR gap. The capacity in bits/Hz at an individual carrier frequency f_i is given by

$$C(f_i) = \log_2\left(1 + \frac{SNR(f_i)}{10^{\Gamma/10}}\right)$$

where $SNR(f_i)$ is the signal-to-noise ratio at center tone frequency f_i and the SNR is expressed as

$$SNR(f_i) = \frac{S(f_i) \cdot |H(f_i)|^2}{N(f_i)},$$

$S(f_i)$ is the desired signal PSD at center tone frequency f_i, $|H(f)|^2$ is the squared magnitude of the loop insertion gain, $N(f)$ is the total noise that includes both crosstalk and background noise, and G is the signal-to-noise ratio gap expressed in dB and it is equal to 9.75 dB – (coding gain in dB) + (margin in dB).

The total capacity of the multicarrier system is computed by integrating the capacity across all of the tones. Because we are computing capacities of discrete tones, we simply add all of tone across the carrier frequencies. The reciprocal of the tone spacing is the DMT symbol interval T. The total bit capacity in the symbol interval is expressed as

$$C = \sum_{i=1}^{N} C(f_i) = \sum_{i=1}^{N} \log_2\left(1 + \frac{SNR(f_i)}{10^{\Gamma/10}}\right)$$

and the total bit rate R (bit/sec) is computed by $R = C/T$, where T is the DMT symbol interval in seconds, which is also equal to the reciprocal of spacing between the center tone frequencies.

To compute the margin of a DMT system, we first need to know the line bit rate that the DMT system will be operating. By knowing the bit rate, we can determine the capacity C (in bits/Hz) by $C = R \cdot T$. A simple way to compute the margin of a DMT system is to average the signal-to-noise ratio across all of the active tones N by

$$SNR_{Avg} \approx \left[\prod_{i=1}^{N} (SNR(f_i)) \right]^{1/N}$$

and the margin expressed in dB (γ_m) is computed by

$$\gamma_m = 10 \cdot \log_{10}\left(\frac{SNR_{Avg}}{2^{C/N} - 1} \right) + \gamma_c - 9.75 \text{ dB},$$

where γ_c is the coding gain expressed in dB.

The above computation methods are used for computing the spectral compatibility of a new technology with the basis systems. The following sections describe the conditions for spectral compatibility with each of the basis systems.

10.8.2 Voice Grade Services Spectral Compatibility

In general, voice grade services contain a family of technologies that utilize the frequency spectrum from 0 to approximately 4 kHz. Such services and technologies include speech signals, dual and single tone signaling, low frequency signaling, and digital and analog data. Performance of speech data is generally subjective in nature, and the SM standard does not provide any guidance for evaluating the subjective effects of crosstalk into speech signals. However, the performance of data services can be quantified. Hence, spectral compatibility of new technologies with voice grade services assume that a V.90 modem is the victim technology, since V.90 modems have the greatest sensitivity to noise than any other voice grade data signal. Spectral compatibility into V.90 systems is evaluated simply by computing the total crosstalk power in the 0–4 kHz voice band and comparing it to set thresholds.

Evaluations into the voice band are done for both the upstream and downstream channels. In each case we assume 49-NEXT disturbers into the 0–4 kHz frequency band. The 49-disturber NEXT coupling model used for this frequency band is

$$NEXT_{49-disturbers} = 10 \cdot \log_{10}\left(\frac{f^{2/5}}{2.11 \times 10^8} \right)$$

where f is frequency in Hz and is in the range from 200–20,000 Hz. For voice grade services, there are two evaluations that need to be performed: (1) absolute values of the PSD in the range from 200 to 4000 Hz and (2) total NEXT noise in the 200 to 4000 Hz band.

First, the near-end crosstalk PSD resulting from the new technology disturbing signal must be less than or equal to –97.5 dBm/Hz at any frequency in the band from 200 to 4000 Hz. In other words, when we pass the new technology PSD through the above 49-NEXT coupling equation, all PSD values in the frequency band from 200 to 4000 Hz cannot exceed –97.5 dBm/Hz. Mathematically, this is expressed as

$$PSD_{Disturber} + 10 \cdot \log_{20}\left(\frac{f^{2/5}}{2.11 \times 10^8}\right) \leq -97.5 \text{ dBm/Hz.}$$

This criteria requires that the disturbing signal PSD be less than –29 dBm/Hz across the frequency band from 200 to 4000 Hz.

Second, the total NEXT noise in the 200–4000 Hz band coming from 49-disturbers must be less than or equal to –75 dBm/Hz. Mathematically, this test is described as

$$10 \cdot \log_{10}\left[\int_{200}^{4000} PSD_{Disturber}(f) \cdot \left(\frac{f^{2/5}}{2.11 \times 10^8}\right)\right] \cdot df \leq -75 \text{ dBm}$$

where the PSD is expressed in linear units such as mW/Hz.

The above two criteria together defines a restriction in the total crosstalk power in the voice band while allowing for some variation in the PSD value at different frequencies in the band.

10.8.3 Enhanced Business Services Spectral Compatibility

Enhanced business services utilize the frequency band from 0 to 10 kHz, but the voice services supported under this umbrella utilize the same 0–4 kHz band as those of conventional voice grade services. The primary difference with conventional voice grade services is that the enhanced business services support signaling functions that are modulated by an 8 kHz carrier. Therefore, spectral compatibility with enhanced business services must cover the signaling frequency band from 6 to 10 kHz while the lower frequencies are covered by the criteria for voice grade services.

As with conventional voice grade services, we use the techniques of evaluating absolute values of the PSD in the range from 6000 to 10,000 Hz for determining the spectral compatibility with enhanced business services.

The near-end crosstalk PSD resulting from the new technology disturbing signal must be less than or equal to –96 dBm/Hz at any frequency in the band from 6000 to 10,000 Hz. In other words, when we pass the new technology PSD through the above 49-NEXT coupling equation, all PSD values in the frequency band from 6000 to 10,000 Hz cannot exceed –96.0 dBm/Hz. Mathematically, this is expressed as

$$PSD_{Disturber} + 10 \cdot \log_{10}\left(\frac{f^{2/5}}{2.11 \times 10^8}\right) \leq -96.0 \text{ dBm/Hz.}$$

This criteria requires that the disturbing signal PSD be less than –29 dBm/Hz across the frequency band from 6000 to 10,000 Hz.

10.8.4 DDS per T1.410 Spectral Compatibility

The digital data system defined in T1.410 uses AMI technology for transmission of data signals from bit rates of 2.4 kb/s to 64 kb/s. The AMI signal uses a 50 percent duty cycle, which is similar to that of T1 service. The computation of the spectral compatibility with DDS is provided with the following method.

DDS Margin Computation

The transmit signal spectrum of AMI signal for DDS is defined as

$$S(f) = K \cdot \left[\frac{\sin\left(\frac{\pi \cdot f}{2 \cdot f_0}\right)}{\left(\frac{\pi \cdot f}{2 \cdot f_0}\right)} \right] \cdot \frac{1}{\left[1 + \left(\frac{f}{1.3 \cdot f_0}\right)^2\right]} \text{W/Hz}$$

where f_o is the signaling rate (or line bit rate) and K is a constant that is a function of the bit rate. Two bit rates are used for determining the spectral compatibility with DDS, namely, 9.6 kb/s and 64 kb/s. Note that for the 64 kb/s payload service, the line rate contains an additional 8 kb/s of overhead so the bit rate for evaluating spectral compatibility is 72 kb/s. For 9.6 kb/s, $K = 1/(3{,}634{,}868)$ and for 64 kb/s (i.e., 72 kb/s line rate) $K = 1/(6{,}848{,}000)$. The margin in dB of a DDS receiver is computed by

$$\text{Margin (in dB)} = -10 \cdot \log_{10}\left[\frac{1}{f_B}\int_{-f_B/2}^{f_B/2} \frac{1}{SNR_{folded}(f) + 1} df\right] - SNR_{req}\, dB$$

where SNR_{folded} is the folded SNR expressed as

$$SNR_{folded}(f) = \sum_{n=-\infty}^{\infty} \frac{S(f - n \cdot f_B) \cdot |H(f - n \cdot f_B)|^2}{N(f - n \cdot f_B)}$$

and SNR_{req} is the required SNR of 17.3 dB for a 10^{-7} bit error rate.

DDS Reference Crosstalk

Each bit rate of DDS has a specified evaluation loop for computing spectral compatibility. For 72 kb/s line rate, the evaluation loop is 13 kft of 26-gauge wire. For 9.6 kb/s, the evaluation loop is 27 kft of 26-gauge wire. To assess the affect of crosstalk onto DDS from a "new" technology, a relative comparison with 49-disturbers of ISDN basic rate crosstalk is made; in this case, the Spectrum Management Class 1 template represents the ISDN PSD. Hence, the reference crosstalk disturber into DDS is ISDN basic rate. If the new technology produces the same or higher margins than from ISDN, then the new technology is considered to be spectrally compatible with DDS.

Spectral Compatibility Criteria

To be considered spectrally compatible with DDS systems, the new technology must not produce any greater disturbance than that from 49-disturbers from SM Class 1 into DDS when computed under the following test conditions:

1. 49 "new" technology NEXT and FEXT
2. 24 spectrum management class 1 NEXT/FEXT + 24 new technology NEXT/FEXT

A background noise of –140 dBm/Hz should also be included. Evaluation must be done for both 9.6 kb/s and 64 kb/s payload conditions. As mentioned earlier, the required SNR for DDS is 17.3 dB, which is the operating point for a 10^{-7} bit error rate. There is no delta value defined when evaluating spectral compatibility with DDS.

10.8.5 ISDN per T1.601 Spectral Compatibility

For ISDN, the spectral compatibility is computed using the ideal DFE equation for PAM-based systems. The evaluation loop should be either 17.5 kft of 26-gauge wire or the maximum length of 26-gauge wire that the technology is capable of operating, whichever is shorter. The reference crosstalk environment for ISDN is 49-NEXT from the SM Class 1 template. For this case, the two-piece NEXT model as defined above is used.

To be considered spectrally compatible with ISDN systems, the new technology must not produce any greater disturbance than that from 49-disturbers from SM Class 1 into ISDN when computed under the following test conditions:

1. 49 "new" technology NEXT and FEXT
2. 24 spectrum management class 1 NEXT/FEXT + 24 new technology NEXT/FEXT

A background noise of –140 dBm/Hz should also be included. This test basically says that no new technology shall introduce more crosstalk into ISDN than that from 49-self-NEXT ISDN disturbers.

10.8.6 HDSL Spectral Compatibility

For HDSL, the spectral compatibility is computed using the ideal DFE equation for PAM-based systems. The evaluation loop should be either 9 kft of 26-gauge wire or the maximum length of 26-gauge wire that the technology is capable of operating, whichever is shorter. The reference crosstalk environment for HDSL is 49-NEXT from the SM Class-3 template. This is effectively the same as 49-SNEXT HDSL disturbers. For this case, the simplified NEXT model is used.

To be considered spectrally compatible with HDSL systems, the new technology must not produce any greater disturbance than that from 49-disturbers from SM Class 3 into ISDN when computed under the following test conditions:

1. 49 new technology NEXT and FEXT
2. 24 SM Class 3 NEXT/FEXT + 24 new technology NEXT/FEXT

A background noise of –140 dBm/Hz should also be included. This test basically says that no new technology shall introduce more crosstalk into HDSL than that from 49-self-NEXT HDSL disturbers.

10.8.7 ADSL and RADSL Spectral Compatibility

ADSL used DMT technology; RADSL uses single-carrier (CAP) technology. Both systems use the same spectra for upstream and downstream transmission. Showing spectral compatibility with ADSL also assures spectral compatibility with RADSL. Spectral compatibility with ADSL can be evaluated using the method described above for DMT-based systems. For ADSL and RADSL, the simplified NEXT coupling model is used.

There are three performance levels required in demonstrating spectral compatibility with ADSL, namely:

1. 4850 kb/s downstream and 645 kb/s upstream on a test loop of 9 kft of 26-gauge wire
2. 3095 kb/s downstream and 415 kb/s upstream on a test loop of 11.5 kft of 26-gauge wire
3. 425 kb/s downstream and 105 kb/s upstream on a test loop of 15.5 kgt of 26-gauge wire

The respective reference crosstalk talk environments for each of the above performance levels are defined as follows:

1. 24 SM Class 3 template NEXT/FEXT disturbers
2. 24 SM Class 2 template NEXT/FEXT disturbers
3. 24 SM Class 1 template NEXT/FEXT disturbers

Margin is computed first for performance level A. If the tests for performance level A pass, then margin is computed for performance level B and so on. Spectral compatibility with ADSL is declared only when the new technology crosstalk gives greater than or equal to 6 dB of margin at the specified bit rate and loop length. If the performance level A tests fail, then the test is redone a new reference bit rate and a shorter loop length. The crosstalk and loop length environments for each of the upstream and downstream performance levels are provided as follows.

Downstream Performance Level A

Loop length L is 9kft of 26-gauge wire or less for a data rate of 4850 kb/s. The crosstalk conditions are

1. 24 new technology NEXT and FEXT
2. 12 spectrum management class 3 NEXT/FEXT + 12 new technology NEXT/FEXT

Downstream Performance Level B

The following table provides the loop lengths and corresponding data rates:

Loop Length (kft of 26-gauge wire)	Data Rate (kb/s)
$9 < L \leq 10$	$4595 + 255 \times (10 - L)$
$10 < L \leq 11.5$	$3095 + 1000 \times (11.5 - L)$

The corresponding crosstalk conditions are

1. 24 new technology NEXT and FEXT
2. 12 SM Class 2 NEXT/FEXT + 12 new technology NEXT/FEXT

Downstream Performance Level C

The following table provides the loop lengths and corresponding data rates:

Loop Length (kft of 26-gauge wire)	Data Rate (kb/s)
$11.5 < L \leq 11.8$	3095
$11.8 < L \leq 13.0$	$2045 + 875 \times (13.0 - L)$
$13.0 < L \leq 14.0$	$1265 + 780 \times (14.0 - L)$
$14.0 < L \leq 15.5$	$425 + 560 \times (15.5 - L)$

The corresponding crosstalk conditions are

1. 24 new technology NEXT and FEXT
2. 12 SM Class 1 NEXT/FEXT + 12 new technology NEXT/FEXT

Upstream Performance Level A

The following table provides the loop lengths and corresponding data rates:

Loop Length (kft of 26-gauge wire)	Data Rate (kb/s)
$L \leq 6$	950
$6 < L \leq 9$	$644 + 102 \times (9 - L)$

The corresponding crosstalk conditions are

1. 24 new technology NEXT and FEXT
2. 12 SM Class 3 NEXT/FEXT + 12 new technology NEXT/FEXT

Upstream Performance Level B

The following table provides the loop lengths and corresponding data rates:

Loop Length (kft of 26-gauge wire)	Data Rate (kb/s)
$9 < L \leq 11.5$	$415 + 92 \times (11.5 - L)$

The corresponding crosstalk conditions are

1. 24 new technology NEXT and FEXT
2. 12 SM Class 2 NEXT/FEXT + 12 new technology NEXT/FEXT

Upstream Performance Level C

The following table provides the loop lengths and corresponding data rates:

Loop Length (kft of 26-gauge wire)	Data Rate (kb/s)
$11.5 < L \leq 13.0$	$202 + 142 \times (13.0 - L)$
$13.0 < L \leq 15.5$	$104.5 + 39 \times (15.5 - L)$

The corresponding crosstalk conditions are

1. 24 new technology NEXT and FEXT
2. 12 SM Class 1 NEXT/FEXT + 12 new technology NEXT/FEXT

10.8.8 HDSL2 Spectral Compatibility

For HDSL2, the spectral compatibility is computed using the ideal DFE equation for PAM-based systems. The evaluation loop should be either 9 kft of 26-gauge wire or the maximum length of 26-gauge wire that the technology is capable of operating, whichever is shorter. For HDSL2, the simplified NEXT coupling model is used.

Because HDSL2 uses asymmetric spectra, that is, the upstream transmit spectrum is different from the downstream spectrum, the reference crosstalk environment is defined separately for the upstream and downstream channels.

1. For the downstream channel, the HDSL2 reference crosstalk environment is 24-T1 template disturbers and 24 SM Class 4 template disturbers. This models the worst-case disturbance into HDSL2 downstream channel.
2. For the upstream channel, the reference crosstalk environment is 24 SM Class 3 disturbers and 24 SM Class 4 disturbers. For this case, the simplified NEXT model is used.

The following crosstalk combinations are used to determine the spectral compatibility of new technologies with HDSL2. For the downstream channel:

1. 49 new technology NEXT/FEXT
2. 24 G.shdsl at a line bit rate of 1848 kb/s + 24 new technology NEXT/FEXT
3. 24 new technology NEXT/FEXT + 24 SM Class 4 template NEXT/FEXT
4. 12 G.shdsl at a line bit rate of 1848 + 12 SM class 4 template NEXT/FEXT + 24 new technology NEXT/FEXT

For the upstream channel:

1. 49 new technology NEXT/FEXT
2. 24 SM class 3 template NEXT/FEXT + 24 new technology NEXT/FEXT
3. 24 new technology NEXT/FEXT + 24 SM class 5 template NEXT/FEXT
4. 12 SM class 3 template NEXT/FEXT + 12 SM class 5 template NEXT/FEXT + 24 new technology NEXT/FEXT.

In all of the above cases, a background noise of –140 dBm/Hz should be included.

The required SNR for HDSL2 is 27.7 dB – 5.1 dB for the trellis coding gain, which results in net required SNR of 22.6 dB. The resulting margin from the DFE computations with the new technology should be no more than an HDSL2_delta lower than that for the reference case. The value of HDSL2_delta is zero for all the downstream test cases. For the upstream channel, the values of HDSL2_delta are provided in the following table:

Upstream Channel Test Case	**HDSL2_delta (dB)**
49 new technology	0.3
24 SM1 + 24 new	0.3
24 SM5 + 24 new	0.0
12 SM3 + 12 SM5 + 24 new	0.4

10.8.9 2B1Q SDSL Spectral Compatibility

For 2B1Q SDSL, the spectral compatibility is computed using the ideal DFE equation for PAM-based systems. Like ADSL, 2B1Q SDSL is a multirate system. To determine spectral compatibility with 2B1Q SDSL, there are three performance classes required for spectral compatibility evaluation. The three performance classes are 2B1Q SDSL at:

1. 400 kb/s on 13.5 kft of 26-gauge wire
2. 1040 kb/s on 8.5 kft of 26-gauge wire
3. 1568 kb/s on 7.0 kft of 26-gauge wire

All three of these performance classes need to be considered in the evaluation of the spectral compatibility with SDSL upstream and downstream channels.

The reference crosstalk noise for 2B1Q SDSL is 49-SNEXT disturbers. For 2B1Q SDSL, the simplified NEXT coupling model is used.

A reference crosstalk environment is defined for each of the SDSL performance classes. The reference crosstalk environments for each performance class are as follows:

1. 400 kb/s SDSL: 49 SM Class2 template disturbers
2. 1040 kb/s SDSL: 49 SM Class 8 template disturbers
3. 1568 kb/s SDSL: 49 SM Class 7 template disturbers

The SDSL margin computation in the presence of new technology disturbance is done with the following crosstalk configurations:

1. 49 new technology NEXT
2. 24 reference disturbers + 24 new technology NEXT

In all cases, a background noise of –140 dBm/Hz should be included.

The required SNR for 2B1Q SDSL is 21.4 dB. Spectral compatibility with 2B1Q SDSL requires that the SDSL margins computed with the new technology disturbers is no more than an SDSL_delta lower than the margins computed with the reference disturbers. The values of SDSL_delta are as follows:

SDSL Bit Rate (kb/s)	SDSL_delta (dB)
400	0.5
1040	0.7
1568	0.6

REFERENCES

[1] ITU-T Recommendation V.90, "A Digital Modem and Analogue Modem Pair for Use on the Public Switched Telephone Network (PSTN) at Data Signalling Rates of Up to 56,000 Bit/s Downstream and Up to 33,600 Bit/s Upstream," September 1998.

[2] Bell System Technical Journal, "Special Issue on DDS," 1982.

[3] ANSI T1.601-1999 for Telecommunications—"Integrated Services Digital Network (ISDN)—Basic Access Interface for Use on Metallic Loops for Application on the Network Side of the NT (Layer 1 Specification)," 1999.

[4] Committee T1, "Technical Report No. 28—A Technical Report on High-Rate Digital Subscriber Line (HDSL)," February 1994.

[5] T1E1.4/96-006R2—Draft Second Issue HDSL Technical Report (TR-028).

[6] T1.413 Issue 1: ANSI T1.413-1995 for Telecommunications—"Network and Customer Installation Interfaces—Asymmetric Digital Subscriber Line (ADSL) Metallic Interface," 1995.

[7] T1.413 Issue 2: ANSI T1.413-1998 for Telecommunications— "Network and Customer Installation Interfaces—Asymmetric Digital Subscriber Line (ADSL) Metallic Interface," 1998.

[8] ITU-T Recommendation G.992.1, "Asymmetrical digital subscriber line (ADSL) transceivers," July 1999.

[9] Committee T1, Technical Report No. 59, "Single-Carrier Rate Adaptive Digital Subscriber Line (RADSL)," September 1999.

[10] FCC, " First Report and Order and Further Notice Of Proposed Rulemaking," FCC 99-48, CC Docket No. 98-147, March 31, 1999.

CHAPTER 11

DYNAMIC SPECTRUM MANAGEMENT (DSM)

A 500-meter cable of 50 twisted pairs has a capacity of 5 Gbps in each direction to be shared among the customers of the telephone plant, greater than HFC and, yes, even greater than a single fiber shared in the most popular projected fiber-to-the-home architecture, known as a passive optical network (PON). Today's DSL operates at less than 1 percent of this capacity. Dynamic spectrum management offers the promise of eventually realizing the goal of broadband connection at 100BT-like speeds to every customer of every phone line in the world, thus enabling the broadband age.

Dynamic spectrum management (DSM) is the newest DSL method that enables an array of highly desirable improvements to DSL service:

1. Automatic detection and/or prevention of service faults caused by crosstalk
2. Greater mixture of symmetric and asymmetric services
3. Higher and more reliable data rates

DSM can be thought of as an adaptive form of the earlier spectrum management (see Chapter 10).

Figure 11.1 shows a reference diagram for DSM. The DSL interface is presumed to be one of the standardized DSLs in common use. As depicted in Figure 11.1 and Figure 11.2, a single service-provider's *DSL maintenance center* performing DSM-Data coordination may optionally accept line and crosstalk information in a specified format across an interface *DSM-D*. The DSL spectrum maintenance center (SMC) can process the information received and provide such processed data results for the use of DSL service providers according to specified criteria and format at interface *DSM-S*. The SMC performing DSM-Control coordination may also dynamically specify downstream and upstream DSL line spectra and signals of any modems to which it is attached when those modems are fully DSM capable. Thus, the SMC may be implemented within a DSLAM (especially when used in a fiber-fed remote DSL-LT). DSM is adaptive in that the DSL-LT and DSL-NT can

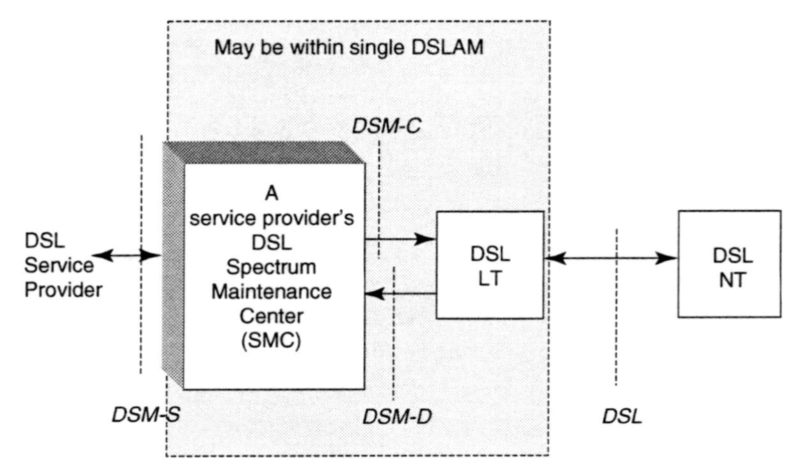

Figure 11.1 DSM reference diagram.

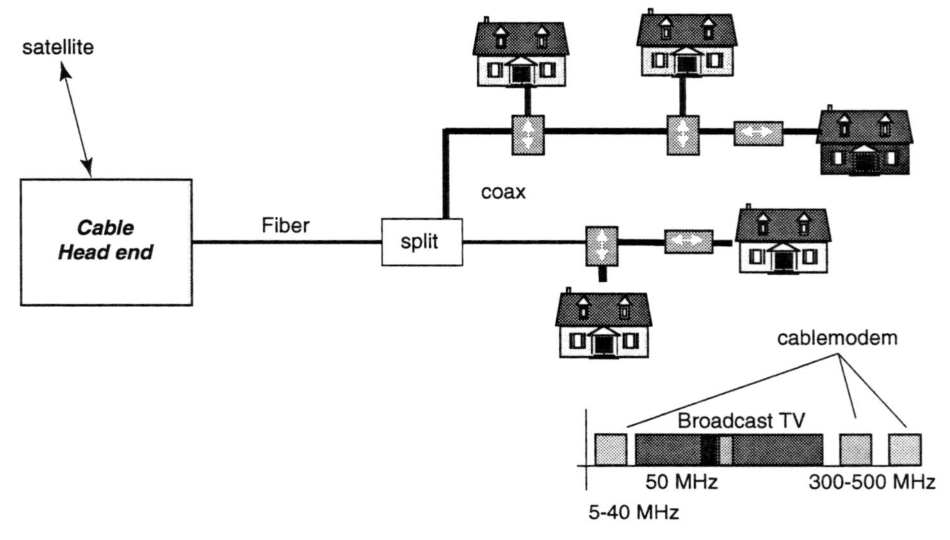

Figure 11.2 Basic cable modem architecture, HFC (hybrid-fiber coax). (All links shared among the common users to that link.)

react to recommendations provided automatically by the SMC. The DSL service provider may also adaptively react to reported situations in the loop plant by altering loops to which specific DSL modems are attached or by taking other corrective actions. It is also possible for significant DSM improvement when the lines operate autonomously (without any directed control from the service provider or SMC other than perhaps the data rate desired). Especially in the autonomous case, DSM standards will enumerate training procedures that cause a DSM modem to use no more power nor bandwidth than it needs for reliable operation. This chapter explores the possible improvements as DSL progresses from a highly autonomous state to one in which ultimately signals are co-generated at remote DSL-LT's of the future.

DSL standards such as ADSL (Chapter 3) and VDSL (Chapter 7) have capably enabled individual transmission links to play near their individual point-to-point fundamental theoretical limits, igniting a worldwide interest in broadband access. To date, these standards normally operate autonomously without active concern for other lines, whether or not their own performance exceeds or misses the capability demanded by the individual customer associated with that line. High performance of one line could generate large crosstalk interference into other lines, perhaps preventing the desired reliable transmission of a good data rate on some of those other lines. Spectrum management in Chapter 10 is an early attempt to mitigate such "DSL binder hogging" by any one DSL line through enforced imposition of a fixed set of DSL spectrum masks on each type of DSL modem. These masks compromise different customer's interests based on presumed fixed channel conditions. The management thus enforced by early spectrum management only best balances the customers' varied performance in the one fixed situation presumed in the design of the masks. DSM attempts to balance customer interests adaptively as a function of line situation and of binder and loop topology and of crosstalk activity. Dynamic behavior may occur at several periodic intervals, as well as with several degrees of coordination from autonomous to co-signal generation. The time interval may be per-service order, per-day, per-session, or continuously adaptive.

At time of writing, DSM was in early drafting stages within the T1E1.4 DSL standardization group. This chapter attempts to illustrate the basic concepts and issues in DSM. Section 11.1 address the evolution of DSL unbundling and services, showing that unbundling is inevitably migrating to a packet or "wholesale" level, especially as fiber is increasingly deployed in the feeder portion of the telephone loop plant. Section 11.1 also outlines the fundamentals of multiuser communication theory that apply to DSL. Section 11.2 then addresses a cable characterization that can be used to enable DSM, given the evolution described in Section 11.1. Section 11.3 provides technical description of the improvements of early autonomous DSM that can dramatically improve existing DSL deployments without need for the SMC. Section 11.3 illustrates the significant benefits of a method known as "iterative water-filling" that may be implemented autonomously and that always improves DSL binder performance if used correctly no matter what the mixture of DSM-practicing or nonpracticing modems occurs. Section 11.4 then proceeds to the more sophisticated vectored DSL systems that do require DSLAM-side coordination and enable extremely high data rates and variable mixtures of symmetry in downstream/upstream data rates. Section 11.5 investigates methods for on-line in-service measurement of the critical channel and crosstalk parameters. Appendix 11A describes some mutiuser detection methods that can be used in the absence of DSM to limited benefit in all DSL systems.

This chapter should be viewed as an early snapshot of DSM activities and capabilities. The area is likely to evolve and improve considerably after the publication of this book.

11.1 DSL UNBUNDLING EVOLUTION

Unbundling is the incumbent local exchange carrier's (ILEC's) lease of a telephone line or some part of its bandwidth to a competitive local exchange carrier (CLEC). Unbundling first began in the United States as a consequence of the 1996 Federal Telecommunications Act. Current unbundling practice with DSL service usually allows the CLEC to place modulated signals directly on their leased physical copper-pair phone line, sometimes referred to as the lease of "dark copper." Such unbundled signals may have services, and consequently spectra, that differ among the various service providers. The difference in spectra can magnify crosstalking incompatibilities caused by electromagnetic leakage between lines existing in close proximity within the same cable. ILECs and CLECs then try to ensure mutual spectral compatibility by standardizing the frequency bands that can be used by various DSL services. This standardization is known generally as *spectrum management* (see Chapter 10). However, there are many DSL types and bandwidths, and service providers that are often competitors, which complicate the process of such spectrum-management standardization. Further, the cooperation and connection between spectrum regulators and DSL standards groups is still in early evolution, so that regulators may allow practices different than those presumed in the process of spectrum-management standardization. For instance, the Tauzin-Dingell bill passed by the U.S. House and before the U.S. Senate at time of writing would (unintentionally) create a structure that allows the highest possible data rates and services technically, although its encouragement of competition and the consequent rate of DSL service installations is debated.

In advanced DSL service the location of the line terminal (LT or "central-office side"), as well as network termination (NT or "customer premises side"), can vary. That is, not all LT modems are in the same physical location. Often the location may be an optical network unit or cabinet, where placement and attachment of CLEC equipment may be technically difficult if not economically infeasible. The difficulty arises because there may be no spare fibers for the CLEC to access the optical network unit (ONU), and/or the ONU may not be large enough to accommodate a shelf/rack for each new CLEC. Placement of such CLEC equipment for dark copper is often called "colocation" when it is in the central office (CO). Space and facilitation of such CO colocation for unbundling of the dark copper is mandated by law in the United States. Presently, many ILECs are finding regulator acceptance of their control of all physical-layer signals that emanate from remote terminals (as long as they can prove they provide wholesale packet-level unbundled service fairly), as opposed to emanating from the CO where the physical-layer unbundling is forced by law. This represents a change in architecture with respect to what many standards groups have presumed in SM studies. The control of all the physical-layer signals by a single service provider allows potential coordination of the transmitted signals in ways that can be beneficial to the achievable data rates, reliability, and complexity of DSL service, such as is now studied in DSM. Even without such coordination, there is much that DSM can offer, as we shall see in Section 11.3.

11.1.1 Cable Modem Architecture—DSL's Competition

Figure 11.2 illustrates a general architecture common to cable-TV service providers. Of particular interest is that the cable system is operated by a single service provider, and the coaxial-cable bandwidth used is shared by all users. Cable modem technology basically uses time-division multiple access of the different users on a shared coaxial segment in a common up or down frequency band. The up band can be located below 40 MHz, but additional up bands may be appropriated from the existing TV channels for upstream transmission at higher frequencies. At least one downstream TV channel is also appropriated for the downstream cable modem, and shared again in the time domain. Cable systems today are operated by a single operator, and this operator controls all content (e.g., which Internet service[s] or voice service[s] may be offered, as well as what TV channels are offered). However, the FCC [2] in the United States has opened discussion on whether cable operators will be forced to provide other content. Even if not forced, many cable operators are investigating allowing competitive-service suppliers on the same cable, effectively implementing wholesale "unbundling." Also, as illustrated in Figure 11.2, as fiber moves into the HFC network, a single fiber will attach to many homes, replacing the coax and providing higher bandwidth for all services. That fiber would consequently be shared in the time-domain according to the same conventions as with the coax (just with more channels available for all services)—this is similar to PONs (passive optical networks), sometimes also studied as an alternative by telephone service providers for fiber migration to the customer. A single service provider, perhaps eventually restricted from controlling content, would control that fiber. Various mixes of time-, frequency-, or code-division access will not change this aspect of a single common carrier, with likely multiple services/contents provided on the system. Such a system is now, and will be in the future, a competitor to DSL.

To emphasize and draw a comparison later with DSL, in the cable system the issue of competition and unbundling is forced to a higher level, which we call here "packet unbundling" by the physical coordinated nature of the shared media. A single common carrier, the cable operator, maintains the physical layer and the consequent bits that flow over that layer. In DSL systems with current unbundling practice, the bits are managed independently by each DSL service provider, and indeed the physical-layer signals may be different, so different that they cause harmful interference to each other. Spectrum management attempts to contain this harmful interference in DSL, whereas in cable, there is no such spectrum management because all signals provided by a single service provider are necessarily compatible.

Because the physical medium is shared, a MAC (medium access control) is *required* in the cable system to coordinate data to different users. An aspect of having a MAC is that it moves the unbundling problem to a higher layer in the protocol stack. The MAC doesn't care who's providing the incoming data—it just routes it to the right customer. A similarity of the DSL and cable system is that the DSL crosstalking has the same electrical interference problem as the shared common media in cable, which is increasingly important at the higher frequencies used by DSLs on shorter lines from LTs, and causes a performance dependence between lines. However, DSL has yet no MAC to accommodate this problem, and DSM can be construed as a first step towards enabling such a MAC. This no-MAC observation provides a market-competition argument to support the eventual DSL-regulatory migration to packet unbundling, which is inevitable anyway if multiple fibers to each home are to be avoided as DSL evolves (such arguments appear in Section 11.1.2). As the cable system evolves to greater bandwidths and more fiber, a single fiber eventually reaches all customers, and its bandwidth is shared among any common customers and content providers.

11.1.2 DSL Evolution

An often-presumed DSL evolution appears in Figure 11.3(a) with remote-terminal-based DSL. Individual uncoordinated twisted pairs run to each customer. The content of a pair is controlled by the service provider whose modem attaches to that pair in the line terminal (LT). If that LT modem is in a central office, several service providers may compete for the privilege to supply DSL service to that customer as mandated by law. However, the issue is yet formally undecided at the LT outside the central office (be it for ADSL, VDSL or any other DSL), although at

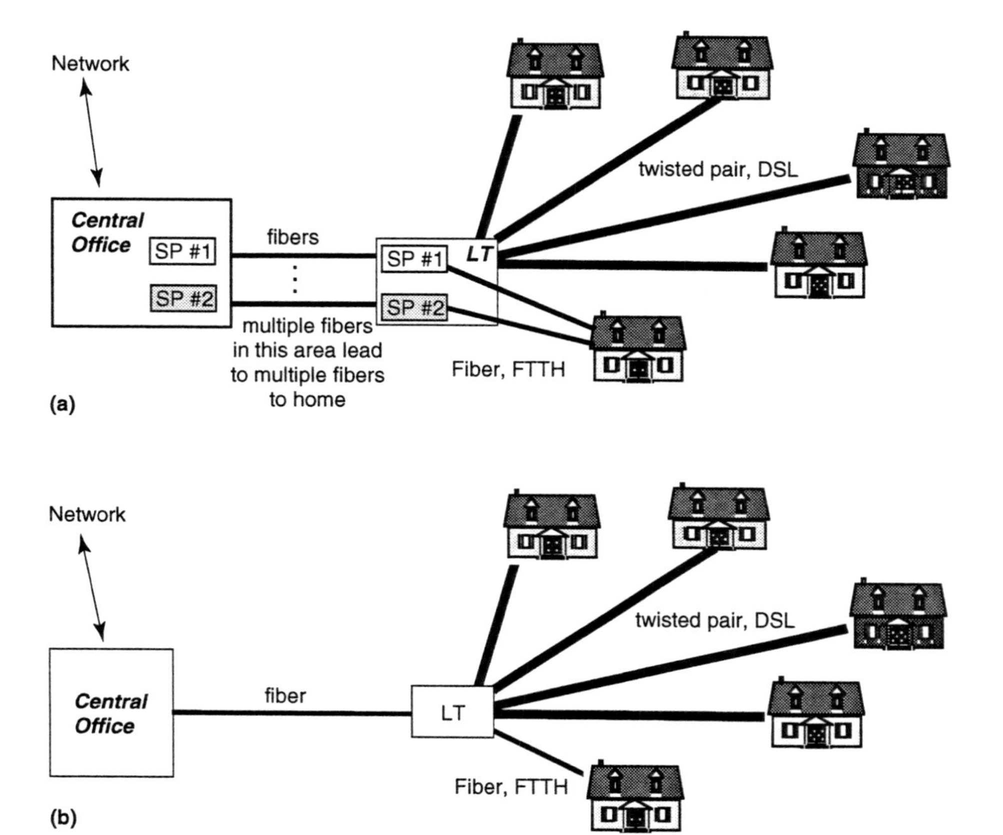

Figure 11.3 (a) Mutliple service provider, line-unbundled LT-based DSL; (b) Packet-unbundled DSL evolution.

least one ILEC in the United States (SBC) has permission to instead "packet unbundle" at a higher digital layer in the protocol stack at the LT. Figure 11.3(a) illustrates how a second service provider would connect, with their own fiber from the central office to the DSLAM presumed if physical-layer unbundling were continued.[1] A third service provider would have their own fiber, and so on, resulting in many fibers to the LT. As this system evolves, the loop plant eventually has many fibers to each customer in FTTH to maintain the present form of physical-layer unbundling. Although a multiplicity of fibers connecting to DSLAMs colocated in a CO is perhaps a common expectation, the purpose of the use of fiber is to avoid many parallel wires/paths to a customer. Thus, clearly present physical-layer unbundling leads to a ludicrous technical evolution of multiple fibers to every customer; nonetheless existing DSL spectrum-management decisions (see [28]) address only this evolution path. Note this evolution is different that the cable sys-

[1]If this separate fiber is not the case, then the ILEC controls a crucial link, which is then packet unbundling.

tem's evolution, which has one fiber shared among many customers, as discussed in the previous section.

A clear alternative is to maintain one fiber as in Figure 11.3(b), but carry the different service providers' signals on that same fiber; that is, packet unbundling or wholesaling. Technically, when one common fiber carrier carries all the signals, necessarily there is a demultiplexer in the LT for all the individual digital signals. If the common carrier must implement this demultiplexer, that carrier might also implement the modem. This allows coordination of the lines at the LT, which can lead to enormous gains in data rate as in Sections 10.3–10.5, as well as a nearly arbitrary mixture of asymmetric and symmetric services. SM then becomes more of a multiplexing problem, than one of just fixed worst-case minimization of crosstalk between lines that may be operated by different service providers.

Figure 11.4 predicts the timeline of argued evolution of DSL from its present configuration to a likely packet unbundled future. The ILEC could actually be any service provider, and the twisted-pair network in the final step might actually be a private network. Early DSM could have the network maintenance center for DSL collecting performance/line information from the lines and possibly making recommendations to the lines/DSLAMs as to maximum binder-benefical data rates to attempt. Largely in early DSM, each DSL line operates autonomously without guidance from an SMC—even in this case, considerable improvement is possible, as in Section 11.3. Clearly the DSL maintenance center can also provide information to the ILEC service personnel as to potential or identified problems, resulting in either manual or automatic repair/prevention. As DSM progresses and DSL evolves to fiber-fed re-

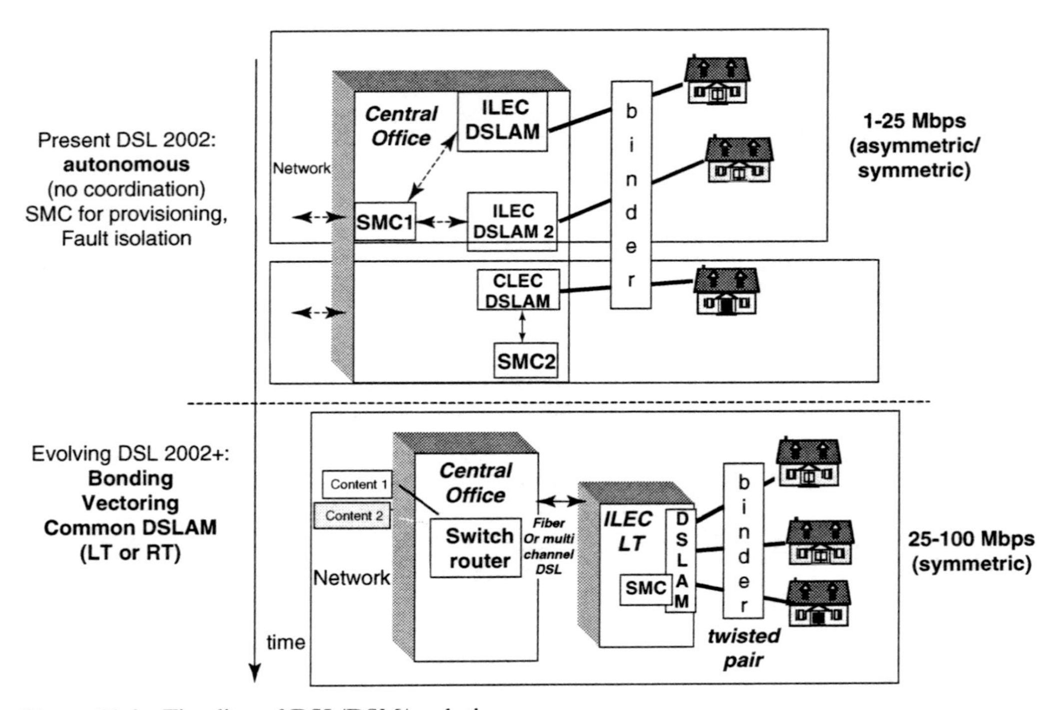

Figure 11.4 Timeline of DSL/DSM/evolution.

mote terminals or line terminals, then highly advanced maintenance can be used remotely or placed directly at the common copper interface to customers in the LT. Signals with such later coordination could ultimately be cogenerated to avert crosstalking problems between lines that otherwise dramatically reduce data rates. The data rates listed are representative of what is possible at each stage of evolution with existing modem technologies, but with increasing management and coordination.

Ultimately coordinated drops lead to very high data rates, easily 100 Mbps symmetric service (effectively allowing 100BT connections that customers today often find in their internal office network) proliferating to all ends of the copper network. Such speed actually matches or exceeds the best envisioned by fiber networks that would require wide-scale replacement of facility that may not be practical (all the way to customer) over the same time-frame of evolution, and at a lot less cost than the all-fiber system.

11.1.3 The Essential Steps of DSM

Having no spectrum management is much like driving in a country with no traffic laws, no traffic signals, and no police. Static spectrum management is like driving with traffic laws, signals, and police control under an ultra-conservative dictatorship. Dynamic spectrum management is like having more mutually beneficial laws with traffic-controlled stoplights, car pool lanes, and drivers trained to at least somewhat respect their lanes and use their good common sense in avoiding collisions.

Figure 11.5 illustrates the essential steps of any DSM system. In autonomous DSM operation, the service provider (DSL operator) sets data rates for the DSL services they will provide. Each modem is configured to run in one of three modes:

RA—rate adaptive (maximize data rate)

MA—margin adaptive (use all available power at given fixed data rate)

FM—fixed margin (use only power needed to guarantee high-quality service)

The last mode is most conducive to best overall use of the binder and is finding increasing use with enlightened service providers who want to see all services perform as best as is possible. The first two steps can lead to dramatic improvement in overall binder performance if the service provider sets the modes and data rates according to DSM-standard guidelines, and require absolutely no coordination among DSL lines or service providers. Such improvements are the subject of Section 11.3.

For yet greater performance as a function of individual situations, information about DSL lines can be collected, perhaps many times. Information acquired can be as simple as the DSL modem reporting it is operational at the supplied data rate (or can operate at some other data rate, lower or higher). Such information in the SMC can lead to setting yet better (revenue improving) rates and modes of the DSL modems, or perhaps suggest to the service provider other operations (like putting in a fiber, replacing DSL modems with better/newer ones, etc.). With in-

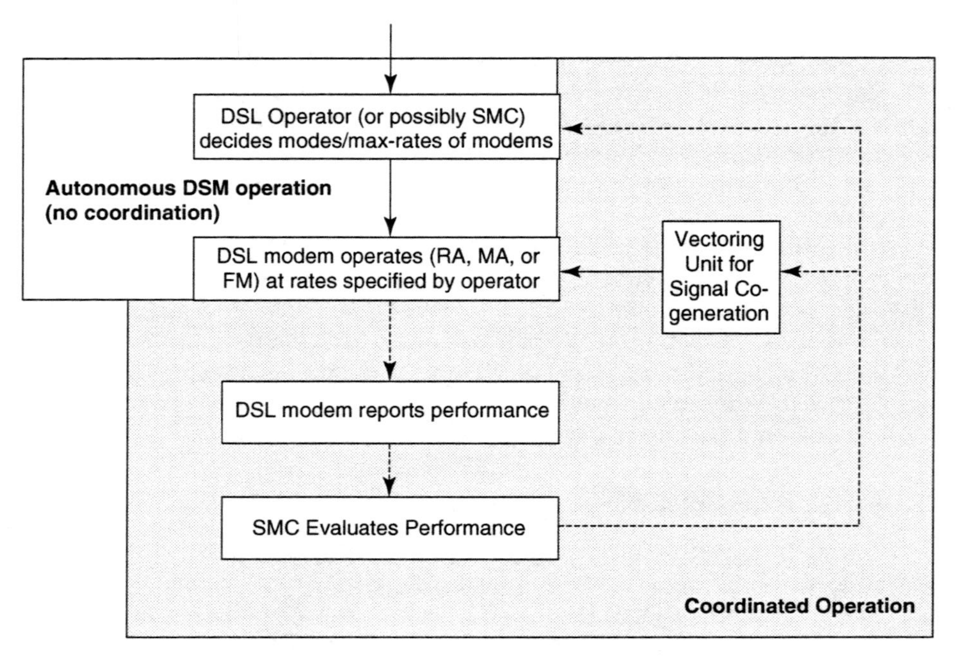

Figure 11.5 Essential steps of DSM.

creasing sophistication of the DSM system, information acquired about the network can include loop characteristics (e.g., topology, channel and crosstalk description), transmission parameters (e.g., data rate, power, bandwidth, energy/bit allocation in frequency/time), and traffic information (e.g., statistics, user traffic characterization, real-time measurements). Ultimately, vectored generation of DSL modem signals can lead to the highest performance and (because of shared processing) the highest port densities, lowest power consumed, and lowest costs of the modems. Thus, DSM provides an enormous opportunity for service providers to improve DSL service and increase rates, enabling a migration to hundreds of millions of customers connected at high speeds, while simultaneously providing equipment vendors and component suppliers an opportunity to differentiate their products.

11.2 MULTIUSER BASICS

Dynamic spectrum management embodies multiple-user communication concepts. All the situations introduced in Section 11.1 contemplate the potential of multiple DSL users who occupy lines in the same binder. These lines may interfere with one another. The resultant crosstalk often is the limiting factor in DSL deployments, determining speeds, ranges, and, indirectly, costs of installation. The possibility of coordinated control of some or all lines for global benefit of all customers is a goal of DSM.

This chapter primarily addresses frequency domain dynamic spectrum management. It is also possible to coordinate the transmission of many lines in the time domain, for burst-mode systems. Such packet level network use is an additional improvement that can occur independently with proper DSM and system design.

DSM's consequent shared use of the communications facility falls into a mathematical science that is called "multiuser information theory" (see [3],[4]). Thus, although DSL initially appears to be a single user on each line, crosstalk effects actually create a shared or matrix channel. The best transmission strategy for a multiuser situation is sometimes not the best transmission strategy for an isolated channel. Figure 11.6 illustrates this multiuser channel for the DSL case. Each of the users on the right can correspond to a customer premises in DSL where one or more phone lines enter the premises for the access to a particular user. On the left, typically the users are in a common building or facility, like a CO or a remote/line terminal. Any coordination of users' signals may thus be easier on the left or service provider side. In the most general case, the lines may have no means of coordination whatsoever. Static spectrum management standards such as T1.417 [5] are an early attempt to address the uncoordinated situation. DSM standards contemplate the judicious use of the self-optimizing spectra of various DSL modems, particularly ADSL and DMT VDSL, but are not limited to those most heavily deployed DSLs. DSM additionally contemplates optional use of the controller, also shown in Figure 11.6. This controller may combine the processes of information acquisition, optimization, and/or control. The controller may specify only the data rates of the various downstream and upstream transmissions or may specify more complete signal-generating information such as spectra or even vectored transmissions. Clearly in the upstream direction, the maximum possible specification is the spectra of physically separated modems. The upstream spectra would be controlled via commands sent over the transmission link, and are indicated by the dashed lines in Figure 11.6. Downstream signals may have their signals cogenerated. Clearly channel information may also flow along these dotted lines to the controller to assist its determination of best spectrum policy for the binder group illustrated by the vertical dashed line. Problems can be isolated and/or averted by avoiding difficult crosstalk coupling paths, thus simplifying deployment and reduc-

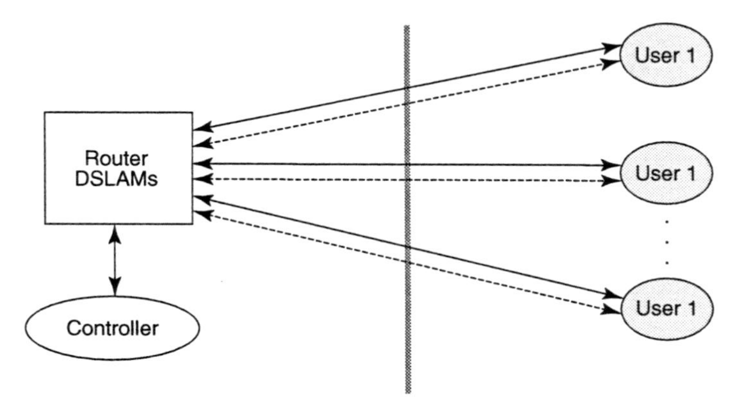

Figure 11.6 Basic multiuser DSL configuration.

ing costs of service. Also, a well-designed controller/DSLAM configuration might well lead to lower complexities and power consumption on the DSL lines (e.g., better performance could be exploited by sending less power, and thus reducing costs of critical line-driver circuits. Additionally, signal processing may be shared among the lines, further reducing complexity and power for the DSLAM.

This section develops the basic terminology for the use of multiuser information theory concepts in DSL in Section 11.2.1 on multiuser coordination levels and Section 11.2.2 on the matrix channel. A theory known as generalized decision feedback equalization (GDFE) is introduced in Section 11.2.3 and can be used to develop nearly every concept and structure in multiuser information theory's handling of crosstalking channels. Sections 11.3–11.5 will follow with specific implementations of various mutiuser structures with increasing levels of permitted coordination from none to substantial.

11.2.1 Multiuser Channels and Rate Regions

Multiuser channels as in Figure 11.6 are necessarily characterized by several variables that this subsection summarizes.

Number of lines: L is the number of lines or twisted pair in the cable (or group of lines to be investigated). Typical values for L are 50 or 25 for binder groups, or possibly 2 for a "quad"—4 wires twisted as a 4-wire unit. L may also be any number and could correspond to only those lines of a specific service provider within a binder. Associated with L is a line index $l = 1, ..., L$.

Number of users: U is the number of users. Typically $U = L$, but with advanced systems and/or DSL evolution, it is possible that the number of users is less than the number of lines when lines are combined (such as with perhaps bonded applications) to increase available data rates to a single customer. The number of users could also be greater than the number of lines if one line is being shared by a customer to accommodate several services/users. Associated with U is a user index $u = 1, ..., U$.

Packet Size: N is the packet size. N specifies the size of a block of transmitted symbols that may be coordinated on any line (N is often the number of tones in DMT-based DSL systems). There is a packet or tone index $n = 1, ..., N$.

Transmit Group Size: M is the transmit group size, which is the number of DSL transmitters that may be coordinated, so that $M \leq L$. Sometimes a subscript is used, M_u, to indicate that different users may coordinate different numbers of lines. When $M = L$, which could happen in an LT for VDSL downstream, the signals generated by the DSLAM are coordinated simultaneously for all lines.

Receiver Group Size: K is the transmit group size, which is the number of DSL receivers that may be coordinated, so that $K \leq L$. Sometimes a subscript is used, K_u, to indicate that different users may coordinate different numbers of lines.

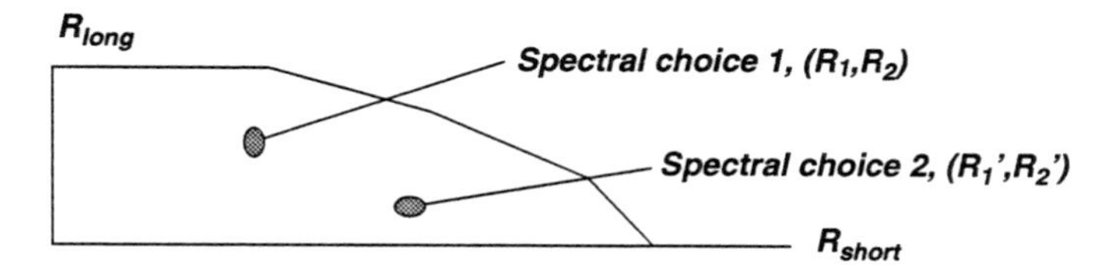

Figure 11.7 Illustration of rate regions for two DSL lines.

Rate Regions: Figure 11.7 illustrates a rate region for two users on long and short loops in the same binder. Each point in the shaded area indicates a combination of data rates that can be achieved. Each point may thus correspond to different spectrum use on the two lines. Increasing the data rate on one line may cause the possible data rates on the other line to decrease. As the number of lines increases, the rate region becomes a multidimensional version of Figure 11.7—hard to draw, but certainly a computer-maintainable description of the combinations of L-rate tuples that can be achieved within a binder.

Coordination Levels: Level 0 coordination means there is no means of coordination between different DSL lines, that is, the earliest practice in DSL as in Figure 11.8. Information acquisition may or may not be implemented in Level 0 coordination, but there is no coordination of signals or spectral use. There may, however, be autonomous control procedures that implement some form of DSM. Level 1 and Level 2 coordination mean that the optimization of spectra on one or more lines can follow the recommendations provided by an SMC. Level 3 coordination further means that the signals on two or more lines can be generated simultaneously in a vector of transmit signals.

Multiuser Detection (MUD)—Appendix 11A: MUD can be used with all levels of coordination by all types of DSL receivers to estimate all the transmit data of all crosstalking lines simultaneously. The number of users is equal to the number of lines $U = L$ and each user transmits only on one line $M = 1$,

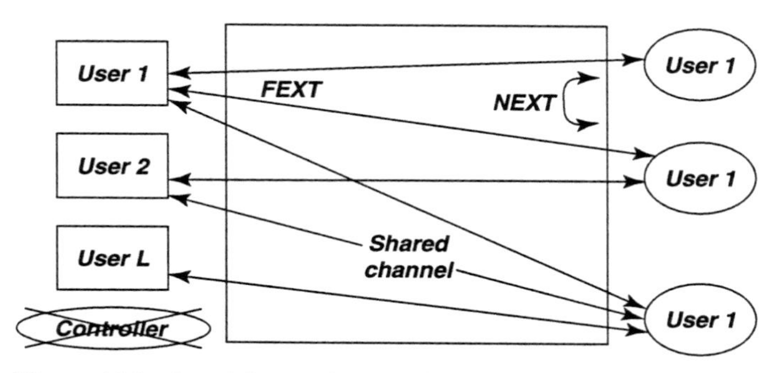

Figure 11.8 Level 0 coordination (no coordination)—MUD problem.

but its signal may leak into other lines, for which there is only the $K = 1$ receiver that may try to estimate all or some of the transmit data signals. Even though the estimate of another line's data may be poor, the use of the poor estimate to reduce crosstalk can still provide a performance improvement in many situations. Most MUD systems without spectrum balancing or vectored transmission provide some modest gain only with only one or two significant crosstalkers and high complexity.

Spectral Balancing—Section 11.3: Spectral balancing may be used with or without coordination. Figure 11.9 illustrates the multiuser situation with spectral balancing. This situation is known in information theory as the "interference channel." The spectra of one or more lines can be autonomously specified (or specified under control) to improve performance for the aggregate group of lines. In autonomous operation, the points selected in the rate region may be more conservatively selected than when dynamic rate assignment is possible if the rate-assignment control unit knows the rates of all or some of the lines involved in DSM. The number of users is $U = L$ in the most common situation and $M = K = 1$.

Vectored DSL—Section 11.4: Vectored DSL occurs only with Level 3 coordination. Figure 11.10 illustrates the situation where more than one of the line's inputs are coordinated for downstream transmission in the DSLAM. The same configuration allows use of two or more lines' outputs simultaneously to estimate any of the upstream user's data. The downstream channel has $U = M = L$, and $K = 1$ in the most common situation and is known as a **vector broadcast channel** in information theory. The upstream channel has $U = K = L$, and $M = 1$ in the most common situation and is known as a **vector multiple-access channel** in information theory. **Full vectoring** in Figure 11.11 occurs when $U = M = L = K$ and inputs and outputs at both ends of the channel (i.e., think of a coordinated set of four twisted-pair in Ethernet) are coordinated.

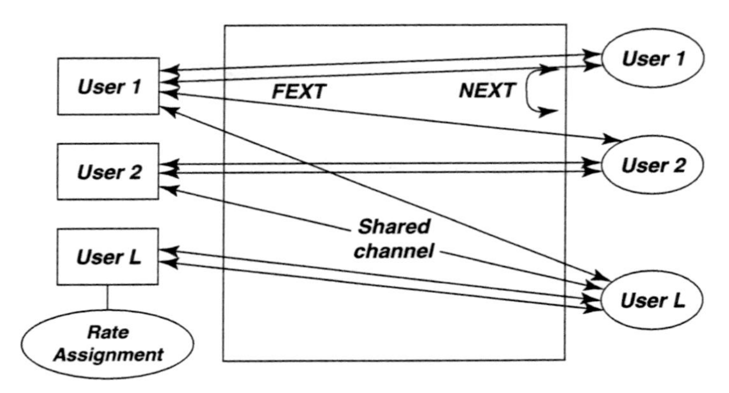

Figure 11.9 Controller used to negotiate and control spectra in spectrum balancing.

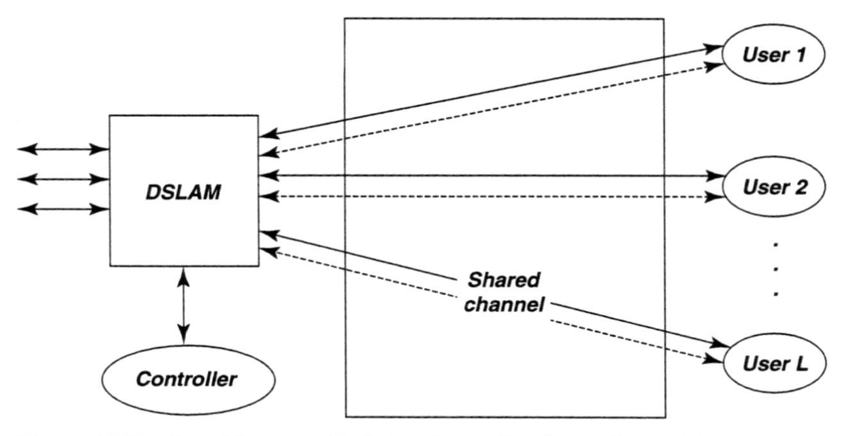

Figure 11.10 Level 2—one-sided signal coordination at DSLAM.

Some of the types and levels of coordination are not mutually exclusive. For instance, not all the lines may report or allow control of spectra or signals. It is possible to have a few older DSL lines that are uncooperative and used fixed spectrum masks, whereas others may report their channel characteristics, and still yet others may have the additional ability to allow control of their spectra. There may be groups of lines that allow vectoring, but may or may not be able to be coordinated with other lines or groups of lines. Cogent DSL evolution may need to encompass DSM strategies and solutions that at least always do as well as a system without the additional incremental allowance of cooperation.

11.2.2 The Matrix Channel

Channel Transfer Function

Any of the $L \times L$ shared channels in Figures 11.8 to 11.11 can be represented by a matrix equation

$$\mathbf{Y}(f) = \mathbf{H}(f) \cdot \mathbf{X}(f)$$

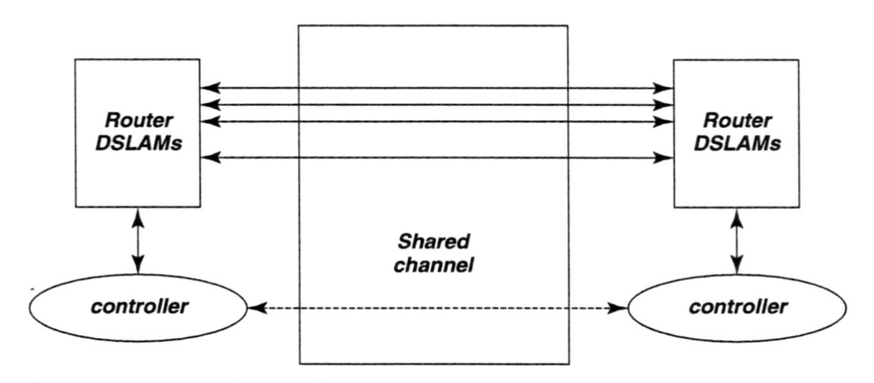

Figure 11.11 Level 2 two-sided coordination.

where $\mathbf{H}(f)$ is an $L \times L$ matrix of channel transfer functions, $\mathbf{X}(f)$ is an $L \times 1$ vector of channel inputs, and $\mathbf{Y}(f)$ is a $L \times 1$ vector of channel outputs. This relation generalizes in a simple manner the traditional scalar relation that is used in point-to-point DSL (often with any coupled crosstalk signal from another line instead modeled simply as a component of noise added to the channel output).

The matrix channel transfer function is

$$\mathbf{H}(f) = \left[H_{km}(f) \right]_{k=1,\ldots,L\ m=1,\ldots,L}$$

where $H_{km}(f)$ is the channel transfer function from line m to line k.[2] A continuous transform is indicated here to avoid the difficulty associated with possibly asynchronous clocking of DSL circuits. Many of the transfer functions between lines may be zero or practically negligible, but there are a few dominant crosstalkers into each line (and which is dominant may depend strongly on frequency). A method for determining $\boldsymbol{H}(f)$ appears in [6].

There are of course two transfer functions for downstream and upstream; however, basic reciprocity in electromagnetic theory forces these two functions to essentially be the same mathematically

$$H_{km}(f) = H^*_{mk}(f),$$

although real cables can exhibit deviation from reciprocity. For line signals traveling in opposite directions, NEXT coupling is also potentially of interest even for frequency-division multiplexed systems (as in DSL). In this case, the NEXT insertion-loss transfer function from line m in one direction to line k at the same side of the binder is $G_{km}(f)$. The $L \times L$ matrix is $\mathbf{G}(f)$, and $G_{kk}(f)$ measures echo on line k. The above does not address the additional complexities of repeatered lines where reciprocity of course does not apply, and each line between repeaters should be individually considered.

Also of interest and easier to compute sometimes are the channel magnitude profiles

$$|H_{km}(f)|^2_{k=1,\ldots,L\ m=1,\ldots,L} \text{ or } |G_{km}(f)|^2_{k=1,\ldots,L\ m=1,\ldots,L}$$

The virtual binder group is that set of couplings or pairs of lines M_k, such that the maximum coupling from m into k exceeds a threshold $\max_f |H_{km}(f)| > \sigma_k$ or $\max_f |G_{km}(f)| > \sigma_k$. These are the only lines' crosstalk that will concern line k. The measurement of $\boldsymbol{H}(f)$ and determination of thresholds and procedures for locating virtual binder groups need discussion within DSM. The number of significant crosstalkers of a given type has been assumed in earlier static spectrum management—clearly the virtual binder group and the matrix channel transfer function

[2]The transfer function is closely related to and 6 dB larger than the insertion loss of the line, the latter of which is measured by taking the ratio of the line input voltage with the line present to the voltage at the load when the line is removed. This relation holds exactly when the binder matrix load and binder matrix source are matched to the matrix characteristic impedance of the binder.

contain all the same information on an individual line basis, plus more information that is pertinent to each situation/DSL deployment. The measurement of binders and their matrix characterization is now just beginning in standards groups.

The channel is often replaced by a vector packet time–domain matrix model that is derived by inverse transforming $\mathbf{H}(f)$ to the time domain to get $\mathbf{h}(t)$ and then sampling $\mathbf{h}(t)$ at some desirable sampling rate $1/T'$, and rectangularly windowing to some convenient sufficiently long packet interval in time, call it NT' to get $\mathbf{h} = \mathbf{h}(t)|_{t=0:(N-1)T'}$. Then a packet time–domain vector equivalent of the channel can be formed, and with a slight reuse and abuse of notation, a packet channel model can be formed from the samples in the matrix $\mathbf{h}$ by reordering corresponding to inputs, outputs, and convolution (perhaps with guard intervals to make sure there is no overlap between packets) to get $\mathbf{Y} = \mathbf{HX}$. This latter model is actually the one used most heavily in signal processing and receiver design. The sampling rate needs to be chosen sufficiently high so that all signals of interest are represented. Further, the inputs of the different DSL channels represented in $\boldsymbol{X}$ may be at different symbol rates than the sampling rate. Polyphase interpolation of the input-channel convolution to the correct sampling rate may be necessary. For more on such models, see [7]. Clearly, the fundamental model is easy. The interpolation may take some effort to understand, but that is not of consequence to basic principles being developed in this chapter although it will eventually be very important to a designer of detailed receiving equipment.

Noises

Noise sometimes includes crosstalk, and always did in earlier SM studies. Any crosstalking signal not in M_k is still a contributor to noise. Other noise sources are the inevitable thermal-line, resistive-termination, and radio-frequency noises. Impulse noise is also of great concern, and was addressed in Chapter 3.

Noise will be identified by a power spectrum $N(f)$. This residual noise has been modeled in all types of spectrum management as Gaussian. Dynamic spectrum management in this chapter will presume that the DSL modems autonomously measure this spectrum and report it, rather than assume it to be some fixed model.

The measurement procedure and accuracy for the noise power spectrum needs definition and specification. For example procedures, see [8] and [9].

Source Information

The power level transmitted by any DSL line is known to that line only and cannot be measured absolutely on another line. There is always an ambiguity without specific knowledge of the transmit power spectrum of whether a signal measured at significant levels has absolute scaling determined by transmit power or by coupling between lines. Fortunately, this does not matter unless power levels are to be mutually controlled across many lines. In this case, the power spectral density of the transmitter $S_m(f)$ can be reported by a modem.

Today, g.dmt.bis (g.992.3) modems already provide all the source, noise, and $|H_{km}(f)|$ information necessary.

Clocking information is important; fortunately, signal processing methods exist at any point in a system to determine appropriate clock offset between any two signals in a virtual binder group [8], [10].

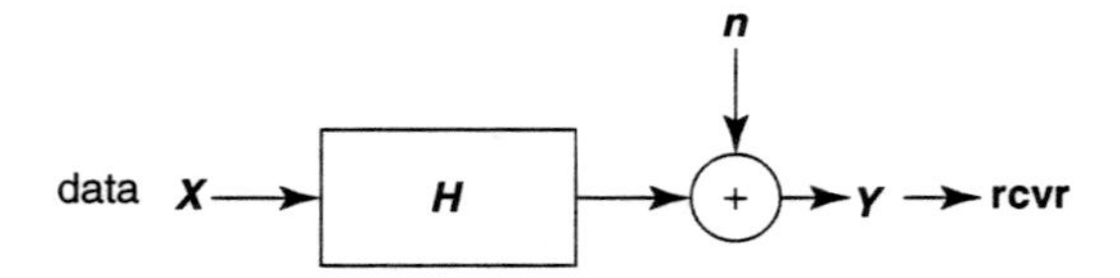

Figure 11.12 Packet modeling of matrix channel.

A Packet-Channel Model

The packet-channel model appears in Figure 11.12 and is applicable to all levels of coordination. The additive noise includes any crosstalk not already in **X.** The input vector corresponds to U users, but may actually be LN dimensional where N again is the packet length for the set of users (which may be a maximum over the set). Basically N is an observation interval in the most general case over which the receiver will elect to sample and store successive outputs from each of the DSL lines. The matrix thus can be very large if N is sufficiently large to model the inherent ISI in each of the DSL channels. Any consequent matrix operations on the received vector $\boldsymbol{Y}$ that occur even in suboptimal receivers would have enormous complexity. Fortunately a simplification exists for DMT, which is the most widely used modulation format for DSL. When the cyclic prefix is used (and especially when the digital duplexing of Chapter 7 allows desirable alignment of DMT transmit symbols on all coordinated lines), each of the individual tones in a DMT system has an independent model of the type in Figure 11.12. The size of the matrix is then just the size of the virtual binder group at each frequency, which is typically a 2×2 matrix, or sometimes a 3×3 matrix and in no case more than a 4×4 matrix. Matrix operations on each tone can then be independently executed with very small matrices, or essentially scalar complexity. Multiuser systems with nonfrequency-domain processing (i.e., PAM, QAM, CAD) have inherently much larger matrices and consequently an enormously higher complexity. Time-domain approaches have received less attention because they are suboptimal and complicated with multiple users and intersymbol interference. Furthermore, almost the entire wide literature on multiuser spectrum allocation today is predicated on the use of frequency-domain modulation methods. Ordering of the tones for simplest processing is important and presently remains an area of proprietary technology.

In any case, whether for each tone or for a large packet, the channel model mathematically is

$$\mathbf{Y} = \mathbf{HX} + \mathbf{N}$$

where the dimensions of $\mathbf{Y}$ and $\mathbf{X}$ depend on the number of lines L and on the packet size N and thus $\boldsymbol{H}$ could be as large as an $LN \times LN$ matrix. For DMT systems, this is never at any frequency more than a 4×4 matrix because the number of significant crosstalkers at any one frequency rarely exceeds 4.

11.2.3 GDFE Theory

Generalized decision feedback equalization (GDFE) is described in several other references [11–14] and resembles the simple DFE often known to transmission theorists for removing intersymbol interference from a single-channel transmission. In its most general form, the GDFE structure views interuser interference and intersymbol interference essentially on the same level and works to eliminate all crosstalk or intersymbol interference from a transmitted signal. Even DMT systems are a special case of the GDFE theory—indeed a very special case when the structure has best performance guaranteed. The equivalent of the DMT single-user channel partitioning [1] across the user dimensions is known as **vectoring** and is determined by **singular value decomposition** of the channel matrix $\boldsymbol{H}$. Vectoring leads to best multiuser performance when it can be implemented.

The GDFE applies to any channel of the form $\mathbf{Y} = \mathbf{HX} + \mathbf{N}$ and hence to the packet DSL model in Figure 11.12. The GDFE attempts to estimate the vector of all the users' inputs $\boldsymbol{X}$ by the structure shown in Figure 11.13. A feedforward matrix $\boldsymbol{W}$ causes the matrix-filtered channel output $\boldsymbol{Z} = \boldsymbol{WY}$ to appear triangular in structure. The triangular matrix created is such that it is monic, that is, it has ones on all diagonal elements, $\mathbf{B}$. The result is

$$\mathbf{Z} = \mathbf{WY} = \mathbf{BX} + \mathbf{E}$$

where $\boldsymbol{E}$ is an error matrix that is essentially "white" noise, that is, $R_{ee} = E[\mathbf{EE}^*]$ is diagonal. The first output of the packet $\boldsymbol{Z}$ is free of crosstalk and can be estimated by a simple slicer or memoryless decoder. This first decision, presumed correct, can then be used to subtract the crosstalk or ISI from the second packet output in $\boldsymbol{Z}$. Then simple decoding is applied to that second output, and the process proceeds through the entire packet, each time subtracting crosstalk or ISI and then making a simple decision. A GDFE structure of one sort or another will appear in all multiuser situations of interest in DSL. This section introduces the general concept, and then later sections use the GDFE later as a tool for the solution of each particular level of allowed coordination in DSL. The GDFE decision process is prone to errors in multiuser transmission and thus the simple slicer and feedback subtraction may be replaced by an iterative decoder as described in the appendix at the end of this chapter.

With the correct choice of $\boldsymbol{W}$ and $\boldsymbol{B}$, it is possible to show that the GDFE-processed system is equivalent to an AWGN noise channel that has the maximum possible SNR that theory admits for the multiuser channel. Good codes (like turbo codes) applied outside the system will drive the performance to the highest possible sum of user rates on the channel. A GDFE structure exists for all possible input choices, and clearly some inputs will be better than others, so an additional step is to optimize the input. Best inputs often separate the spectra of users, as well as determine several independent best bands for each user, in which case the GDFE often decomposes into a set of N independent structures of the type in Figure 11.13. In no case is this more evident than when all lines use the digitally duplexed DMT system that synchronizes frames of transmission as described in Chapter 7. In this case, there is an independent GDFE for each tone, and the only interference left is crosstalk on the same tone between different DSL lines. This later residual independent-per-tone crosstalk can be handled easily and well in several ways, as will become evident.

With correct decisions, the GDFE essentially creates NL (the dimensionality of $\boldsymbol{X}$) parallel AWGNs, one for each tone of each user. Many of these parallel channels may have zeroed inputs. The rest of the AWGNs pass energy and may be

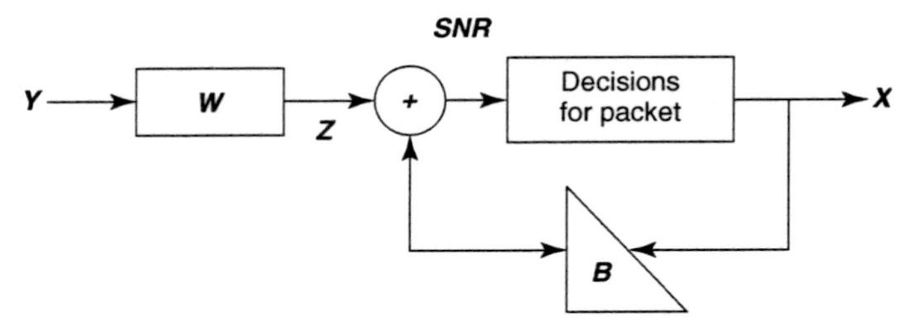

Figure 11.13 Generalized decision feedback equalization.

loaded with a number of bits proportional to the log of the gap-reduced-SNR (see [1]) on that particular channel. For any given choice of input spectra on all the users, the GDFE will have a product of such SNRs that is maximum (and equal to $2^{2I(X;\mathbf{Y})}$ where $I(\mathbf{X};\mathbf{Y})$ is the mutual information for the vectored channel), and thus the data rate for that choice of input spectra is maximum. The best overall SNR requires input optimization and maximizes the best sum of rates of all the users, which is $C = \max I(\mathbf{X};\mathbf{Y})$. The best sum is only one possible solution for the multiuser situation, and the rate region is actually often more of interest, of which the rate-sum maximum will just be one of the boundary points.

In formulating the input $\boldsymbol{X}$, this higher level development ignores the differences in clocks or sampling rates that may exist between different DSLs, especially different types of xDSLs. Nonetheless, these offsets can always be determined by signal processing and used to interpolate and construct an appropriate matrix $\boldsymbol{H}$ for each packet of transmitted signals from all the users. This chapter pursues this interpolation in Section 11.5.

DSL's Noteworthy Simplification—Ginis's QR GDFE for Square Channels

A special case of the GDFE occurs when a zero-forcing criteria is imposed on the nominally involved computation of the $\boldsymbol{W}$ and $\boldsymbol{B}$ matrices and the channel is square (that is, coordination occurs at signal level). This ZF-GDFE is found by setting the residual non-FEXT crosstalk noise to zero. Nominally, this is a horrible thing to do in GDFE theory and can result in catastrophic misconclusions; however, for DSL, FEXT matrices $\boldsymbol{H}$ the off-diagonal terms are at least 100 times smaller than the on-diagonal terms in practice when DMT is used to simplify the problem into a set of independent GDFEs for each tone. This is essentially equivalent to noting that if all DSL modems on one side of the link, that is, the DSLAM side, are colocated, then it must be true at each receiver that the coupling of FEXT over the common length between transmitter m and receiver m must necessarily have less gain from any other transmitter $i \neq m$. Of course an upstream transmitter close to the DSLAM may insert energy at a level much larger than the DSL signals traveling upstream on adjacent lines. This does not affect the $\boldsymbol{H}$ matrix, but rather it is a function of transmit power levels. Thus, although FEXT from adjacent lines could be larger than the on-line signal itself, the relative scaling of FEXT specified by the $\boldsymbol{H}$ transfer function is still much smaller from adjacent lines than from the on-line being investigated (or the diagonal terms in $\boldsymbol{H}$ must be the largest). This observation was first made by George Ginis.

The ZF-GDFE is always essentially equivalent in performance to the GDFE under these situations, and furthermore the ZF-GDFE can be much more easily specified as Ginis notes from the "QR" decomposition of the matrix ***H:***

$$\mathbf{H} = \mathbf{QG}$$

where $\mathbf{Q}$ is an orthogonal matrix satisfying $\mathbf{QQ}' = \mathbf{I} = \mathbf{Q}'\mathbf{Q}$ and $\mathbf{G}$ is a triangular matrix. The factorization $\mathbf{QG}$ exists for any square matrix (even singular ones), and is not necessarily unique. It is convenient to write $\mathbf{G}$ in terms of a monic matrix $\mathbf{G} = \mathbf{DG}_{monic}$ where $\mathbf{D}$ is a diagonal matrix but is unique if all users are present (nonsingular) and the order of the users is specified.

In this case, the GDFE becomes

$$\mathbf{W} = \mathbf{D}^{-1}\mathbf{Q}'$$

and

$$\mathbf{B} = \mathbf{G}_{monic}.$$

The signal-to-noise ratio for each user on each tone with correct decisions becomes

$$SNR(m,n) = \frac{D_m^2 \cdot P_x(m,n)}{\sigma^2}$$

where D_m is the mth diagonal (user) element of $\mathbf{D}$, $P_x(m,n)$ is the power of the mth user on the nth DMT tone, and σ^2 is the background AWGN (e.g., –140 dBm/Hz). The product of the SNRs is the highest possible for any receiver and achieves the best possible data rate (with good codes). Water-filling optimization can be applied to all the user's tones as a set to maximize the sum of the data rates that can be achieved with such a channel. Clearly, perhaps by design, it is possible for some of the diagonal elements to be zero (corresponding to perhaps intentional silencing of a particular user on a particular frequency). These users on any tone should be deleted from further consideration, which simply means that the corresponding row of $\mathbf{Q}'$ in the product $\mathbf{Q}'\mathbf{D}^+$ is deleted. Note the use of the + sign in the exponent to indicate "pseudoinverse," which means 1/0 is defined to be 0 in this case. In practice, presumably this user's zero contribution would be known in advance, and the corresponding entry eliminated in advance from the $\mathbf{H}$ matrix (which has the same effect as using the psuedoinverse).

Furthermore, with the use of such a structure, the coordination need only be on the receiver side. H is whatever it is for the possible use of transmitters that cannot coordinate their signals. Thus the QR-DFE applies directly to upstream DSL transmission where the DSLAM can possibly coordinate all the received signals (especially in situations like a remote terminal where all signals emanating from the user terminate on a common DSLAM). Downstream, the channel matrix is factored in a left-handed QR factorization $\mathbf{H} = \mathbf{DG}_{monic}\mathbf{Q}$ (not the same ***Q, G*** and ***D*** as above), and the Ginis precoded structure of Figure 11.14 applies instead. Each user channel needs to scale itself prior to a modulo (wraparound) device that is similar to the Tomlinson precoder of [1].

For some levels of lesser coordination between lines, the **H** matrix is nonsquare and necessarily then has zero singular values. **QR** factorization is still possible but is nonunique, and some of the diagonal elements of the resulting **R** matrix will be thus be zero. Unfortunately, the GDFE will estimate only certain nonsingular "pass-space" dimensions of the input that correspond to linear combinations of the multiple users in general. Thus, simple slicing decisions become somewhat ludicrous unless the inputs have been carefully chosen to lie in this space with simply implemented decisions. Thus instead, a maximum likelihood decision on the signal $\mathbf{Z} = \mathbf{WY} = \mathbf{WHX} + \mathbf{WN}$, which means either trying all possible vector values for $\mathbf{X}$ or somehow approximating this (as in the appendix) with iterative decoding. Thus, the feedback section in Figure 11.13 is replaced by a ML decoder, or an approximation to it like an interative decoder.

More generally, the MMSE GDFE settings can be found in any case via the following notational simplifications. The GDFE error vector is

$$\mathbf{E} = \mathbf{B}(\mathbf{X}-\mathbf{WY})$$

with mean-square matrix $\mathbf{R_{EE}} = E[\mathbf{EE}^*]$ thus determined as

$$\mathbf{R_{EE}} = \mathbf{B}[\mathbf{R_{XX}} - \mathbf{R_{XX}}\mathbf{H}^*(\mathbf{H}\mathbf{R_{XX}}\mathbf{H}^* + \sigma^2\mathbf{I})^{-1}\mathbf{H}\mathbf{R_{XX}}]\mathbf{B}^*$$

for any triangular monic matrix value of $\boldsymbol{B}$ and optimum value of $\mathbf{W} = \mathbf{B}\mathbf{R_{XX}}\mathbf{H}^*(\mathbf{H}\mathbf{R_{XX}}\mathbf{H}^* + \sigma^2\mathbf{I})^{-1}$ The minimizing $\boldsymbol{B}$ can be determined by generalized Cholesky factorization [13]–[14] of the inverse of the internal matrix in the equation above for $\mathbf{R_{EE}}$, and regular Cholesky factorization if this internal matrix is nonsingular. In either case, then

$$[\mathbf{R_{XX}} - \mathbf{R_{XX}}\mathbf{H}^*(\mathbf{H}\mathbf{R_{XX}}\mathbf{H}^* + \sigma^2\mathbf{I})^{-1}\mathbf{H}\mathbf{R_{XX}}] = \mathbf{G}^{-1}\mathbf{S}^{+}\mathbf{G}^{-*},$$

where **S** is diagonal, **G** is monic lower triangular (same structure as B), and the optimum feedback section is $\mathbf{B} = \mathbf{G}$. A zeroed entry on the diagonal of **S** is indicative of an input that cannot be estimated because it has some component along a direc-

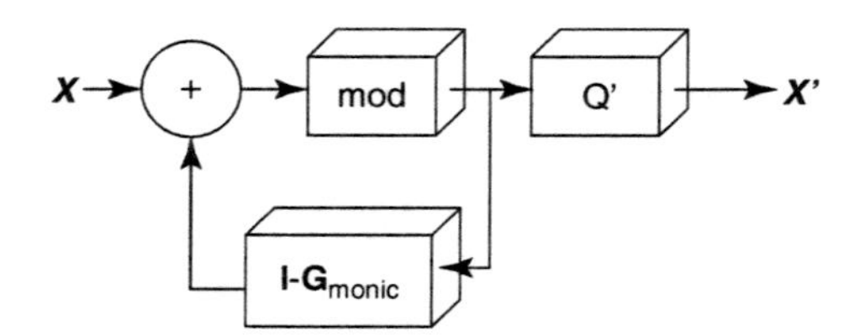

Figure 11.14 Ginis precoder—the channel output is HX′ = DX where D is diagonal, so simply scaling by D^{-1} can be executed separately at each independent receiver.

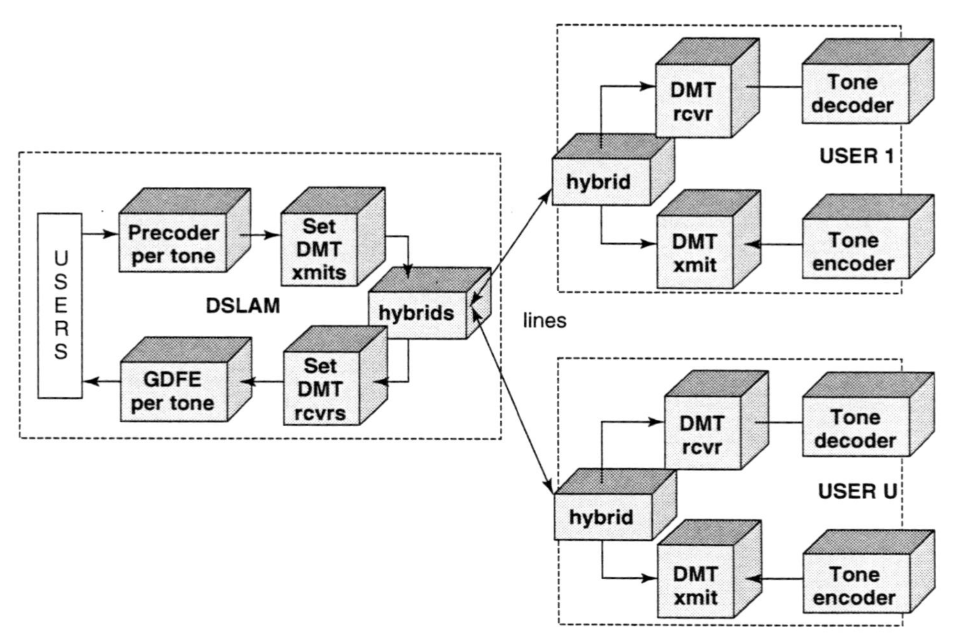

Figure 11.15 One-sided vectoring for DMT VDSL.

tion that the channel does not pass (which can occur often in multiuser situations). The MMSE is infinite, and the system fails in this case to estimate all input symbol values. However, individual components may not have infinite MMSEs themselves and are estimated by the GDFE with the error corresponding to the diagonal element of the matrix $\mathbf{R}_{EE}$ corresponding to that component. Simply stated, the input components that cannot be estimated cannot help in any of the other symbol estimates, and their respective decision outputs should just be zero as far as the feedback to other components is concerned. The corresponding elements of $\mathbf{B}$ should also be zeroed in the MMSE equation. Generalized Cholesky factorization proceeds according to a Gram-Schmidt decomposition of the matrix in the above equation (see [13, 14]) and essentially leaves zeroed values of the diagonal matrix $\mathbf{S}$ when prediction errors associated with linear combinations of previously processed elements are zero. The corresponding values of a row or column of the triangular matrices are set to force the zero prediction (and the choices for $\mathbf{G}$ become nonunique). An adaptive implementation would need to know those input symbols that cannot be estimated because the channel does not pass them. The implementation should delete them to avoid numerical "blow-up" of the GDFE. This effect will dominate multiuser situations while it only occurs with ludicrous design choices in single-user DFEs. A better development that suggests aversion of transmission on singular modes of the multidimensional channel when possible appears in [11–14] but is beyond the mathematical scope of this chapter. Fortunately, the case with DMT and independent GDFEs per tone does not exhibit any of these problems and can also use Ginis's simplifications.

11.3 SPECTRAL BALANCING (ITERATIVE WATER-FILLING)

Spectral balancing can be defined as a wide class of methods where systems may alter their power spectral densities for the common good of all DSL systems. This may be performed autonomously, and can be used with or without coordination. Generally, autonomous DSM improves all DSL systems' performance with respect to their current performance level (even if some have a fixed, nondynamic spectra). The more systems that execute DSM, the larger the improvements for all.

The situation for spectral balancing refers to the "interference channel" in information theory that appeared earlier in Figure 11.9. Each DSL transmitter and receiver are of the usual type for the line, but the spectra used may be varied dynamically in response to the crosstalking situation in the binder.

The interference channel best spectra and rate regions remains an unsolved problem in information theory today. However, Section 11.3.1 introduces a very good method known as **iterative water-filling,** developed by Yu and Rhee [24] and investigated for DSL/DSM by Yu. Iterative water-filling may often be close to optimum, can be used to generate rate regions, and leads to enormous performance improvements with respect to existing practice of SSM in DSL. It also admits a distributed implementation with no need for coordination (which may not be admitted by absolutely optimum solutions whenever they are found). Lastly, a solution found by iterative water-filling will do no worse than those chosen by existing standards in static SM, so there is never any degradation with respect to existing practice. Section 11.3.2 then proceeds with examples of the large improvements possible using iterative water-filling.

11.3.1 Yu's Iterative Water-filling

Water-filling is an algorithm well known in DSL because most DMT-based modems use it (or approximations to it) to decide the energy and information distribution for the modem adaptively as a function of line conditions. Wei Yu, assisted by colleague G. Ginis, noticed that a certain optimization procedure used in multiple-access spectrum optimization that he was studying has an interpretation of being equivalent to several iterative and successive uses of the water-filling procedure that is optimum for a single-user channel. Water-filling is illustrated for a single point-to-point channel in [1]. The spectrum is found by plotting the inverse of the SNR(f) for the channel and filling that inverse curve with water/power until all power available has been used. The water power will lay at a constant level as shown in Figure 11.16. Mathematically, the sum of the noise-to-signal ratio at each frequency plus the power spectral density at that same frequency must equal a constant (the water level):

$$\text{constant} = \lambda = S_x(f) + \frac{S_n(f)}{|H(f)|^2}$$

where f can represent a discrete set of tones in a DMT system, $F = \frac{n}{T}, n = 1, ..., N$, or a continuous variable of frequency. QAM systems may implement the spectrum derived from a DMT system to approximate the same performance, as long as that spectrum is continuous. When not continuous, QAM cannot perform at the same

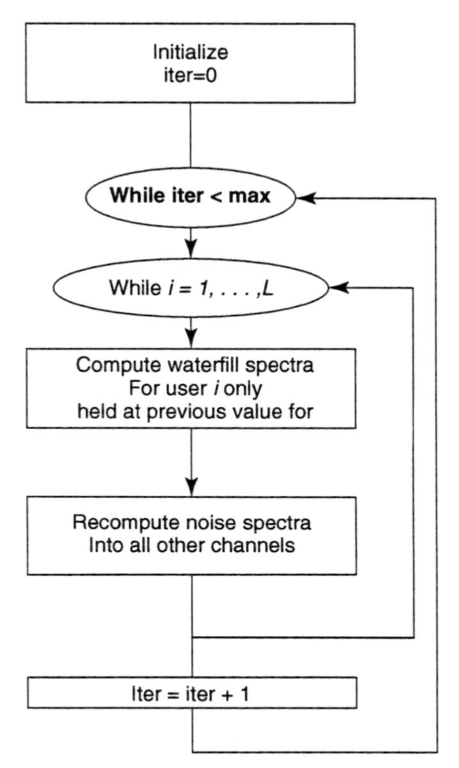

Figure 11.16 Iterative water-filling concept.

levels as DMT. This section assumes all systems are multicarrier and have a mechanism for implementing a water-filling spectrum once derived. Systems with fixed spectra have crosstalk that must be either considered as part of noise or handled with the more complex and less effective MUD methods of the appendix.

The highest spectral efficiency achievable at each tone or spectral point is computed as

$$b(f) = log\left(\frac{\lambda|H(f)|^2}{S_n(f)}\right),$$

and practical systems usually use a code with some constant gap to capacity (including any needed margin) and so follow a similar equation

$$b_n = \log_2\left(1 + \frac{SNR_n}{\Gamma}\right),$$

where b_n is the number of bits transmitted in the nth subchannel's QAM symbol, and the data rate for a single user is thus

$$R = \frac{1}{T}\sum_{n=1}^{N} b_n.$$

A set of $U = L$ users, each with data rate R_l on the lth line (combined perhaps into a vector of data rates $\mathbf{R} = [R_1\ R_2 \ldots R_L]$), has an average rate provided by[3]

$$\overline{R} = \frac{1}{L}\sum_{l=1}^{L} R_l = \sum_{l=1}^{L}\sum_{n=1}^{N}\frac{b_{n,l}}{L} = \frac{\mathbf{R}\cdot\mathbf{1}}{L},$$

for which an objective might be maximization, and is equivalent to maximizing the sum of the data rates of the lines. In general, one user's line rate might be more important than that of another and so a weighting could be assigned to each channel μ_l such that $\sum_{l=1}^{L}\mu_l = 1$, and the function

$$f(\mathbf{R}) = \mathbf{R}\cdot\boldsymbol{\mu} = \sum_{l=1}^{L}\mu_l\cdot R_l,$$

which will equal $\overline{R}$ when $\mu_l = l/L$. Maximization of $f(\mathbf{R})$ over $\mathbf{R}$ for all nonnegative choices of $\boldsymbol{\mu}$ produces the set of achievable data rates in the rate region for the capacity. The gap may be different on different users (and even on different tones for different users) but is often set constant and equal to some reasonable value based on desired margin of service (say 6 dB) and coding gain (say 4 dB) and perhaps then left at around 10 dB in practice. Aggressive designs with turbo coding and more confidence on reliability might reduce this to 3 dB. All procedures will follow the same with any gap.

Iterative water-filling can be optimum in some cases and leverages the observation that if all spectra except one users' spectrum are held constant, then the resulting solution is again a water-fill solution with other users channel-filtered energy at any frequency being viewed as a component of the noise spectra at that frequency. Clearly then, a necessary (but not always sufficient) condition for an optimum solution would be that each and every user's spectra individually satisfied a water-filling constraint with the sum contributions of the other water-filling spectra included. Both Chung [37] and Leshem [38] have shown that such a point, often called a "Nash equilibrium," always exists. Chung [39] has also shown that the IW procedure will converge to it for DSL. For the simple rate-sum maximization, the iterative water-filling process is simple—just cycle through water-filling problems for all users (starting initially with one user and all others zeroed or perhaps initialized to using flat spectra) several times. The process will converge to a solution that satisfies the water-filling condition simultaneously for all users necessarily. This procedure applies directly when all the users have equal rate, so that the average rate (or sum rate) is desired to be as large as possible.

A method to approximating, but lower bounding, the rate region for the interference channel was also provided by Yu. The iterative water-filling procedure in Figure 11.16 could be augmented by an outer set of loops for each of the users

---

[3] $\mathbf{1}$ is a column vector of L ones.

that would reduce the transmit power constraint individually for one of the users for each execution of iterative water-filling. Each user will individually lower power, then transmit the necessarily lower (or the same) data rate, while other users may see an increase. Running through possibilities of reduced power on all users in all combinations over some grid of possible power sets should also trace a region of implementable and achievable rates for spectral balancing on the interference channel. This region lower bounds the rate region. It is the rate-region achievable with iterative water-filling.

Distributed Water-filling

Yu's power loop suggests a mechanism for avoiding essentially all coordination in spectrum balancing. If each DSL system independently executes water-filling at a presumably slow speed of spectrum change, the entire system implements iterative water-filling in a distributed fashion, assuring the maximum sum of total data rates for the given power levels. Thus, no coordination is necessary. Today, ADSL modems are a good example of an interative water-filling system, except that they often rate-adapt, or worse yet, margin-adapt instead of minimizing transmitted power. However, they are well known to always converge in the RA, MA, or FM mode of operation. A DSL system having excessive margin, or equivalently, a higher data rate, should have its power decreased (which can be done automatically by that system until margin at the desired data rate is the minimum acceptable). This is a FM (fixed margin) operation when the power is minimized at a fixed margin sufficient to ensure excellent performance. The use of FM mode improves other systems that otherwise would have insufficient margin. Coordination is unnecessary as long as the attempted data-rate-tuple is in the achievable rate region of iterative water-filling. Coordination in the form of the SMC knowing the rate region more accurately allows a DSM-coordinated iterative water-filling system to more aggressively chose points near the boundary of the achievable rate region. Clearly, just a small amount of coordination in terms of knowing which systems have excessive margin, and permitting or not permitting them to exploit it based on other system's margins and data rates, allows further distributed water-filling improvement. Discussions with several service providers reveal that their SMCs do indeed already store data rate (and line length also) for all DSL lines.

11.3.2 Examples of Improvements

This section provides four examples of the benefit of iterative water-filling: (1) RT-located-ADSL FEXT into co-located ADSL; (2) 10MDSL to/from ADSL (and VDSL), (3) ADSL to/from VDSL; and (4) mixture of asymmetric and symmetric VDSL service in the same binder.

RT-Located-ADSL FEXT into CO-located ADSL

Figure 11.17 comes from [29] and illustrates a basic problem in ADSL deployment that is solved by iterative water-filling in DMT ADSL modems (when the FM mode option is selected). Several authors and at least four companies have independently verified the improvements in [28], [30], [31], [32].

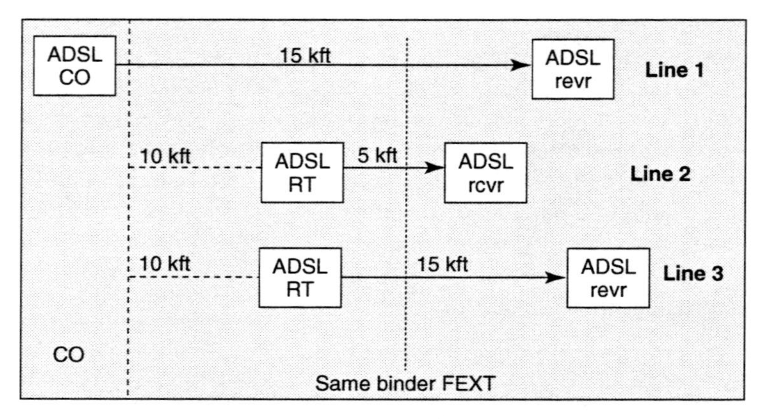

Figure 11.17 Illustration of RT-located-ADSL FEXT into CO-located ADSL (for Verizon experiment in [31], no line 3, and line 1 is 20 kft, while line 2 is 14 kft fiber, 6 kft copper).

The ADSL receiver on line 1 in Figure 11.17 will sense large FEXT from the RT-located ADSL downstream transmitters on lines 2 and 3, both of which are only 5 kft away from the ADSL receiver. Under nominal static spectrum management (T1.417,[5]) assumptions, line 1 has very low data rate—that is, the achievable data rate is 300 kbps downstream. Verizon [31] reports this problem also for the case where line 1 is 20,000 feet long, and the RT on line 2 (there is no line 3 in Verizon results) is 6000 feet from the line 1 receiver (or 14,000 fiber feet from central office). In the Verizon case, the line 1 data rate is only 100 kbps, whereas line 2 happily trained to over 9 Mbps. Both were operated in rate-adaptive (RA) mode. Clearly, line 2 is "hogging" line 1's possible data rate with strong FEXT. Instead, if both lines 1 and 2 are operated with FM training, so the iterative water-filling now works to minimize power at 6 dB of margin, and the rate region in Figure 11.18 is obtained. A good operating point might be to hold the short line (2) to 2 Mbps, in which case the long line (1) gets nearly 1.8 Mbps also—*a factor of nearly 20 increase in data rate on the long line.* Table 11.1 illustrates the improvements/choices for the Verizon experiment. Note the sum of data rates is highest in RA mode, but line 1 does not get much data rate as line 2 takes all.

Telcordia [32], Voyan [28], and Stanford [30] all report also on the situation exactly shown in Figure 11.17 with line 3 present and absent. The long line 1 data rate is limited to 300 kbps again because of ADSL FEXT. Figure 11.19 illustrates the rate regions in this case for lines 1 and 2 with line 3 silent. Clearly, the data rate of line 1 can be improved by a factor of nearly 10, to almost 3 Mbps.

The rate region on the right in Figure 11.19 shows the data rates when line 2 is held at 1.6 Mbps—clearly here, 1.5 Mbps is possible on each of the other two long lines. This represents a factor of 4 increase in data rate with respect to the case where line 2 can "hog" the capacity of the binder. Absolutely no coordination is used in Figures 11.18 or 11.19. The operator simply sets the data rate of the short-line modem less aggressively and operates all modems in FM mode.

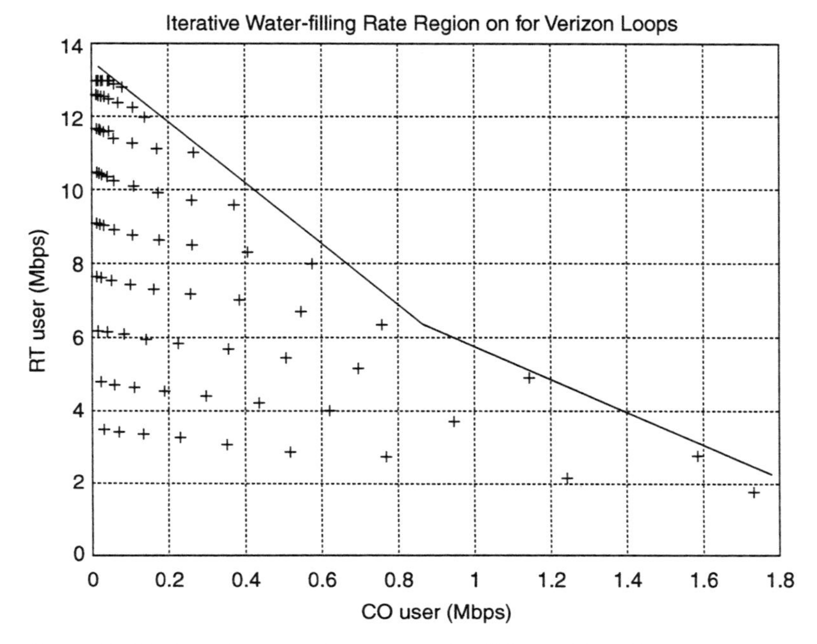

Figure 11.18 Fixed-margin iterative water-filling achievable rate region for Verizon experiment in [31].

The question of spectra is also of interest corresponding to the rate region points. For instance, in the case of the right-hand plot in Figure 11.19, the spectra were nearly flat and had the following levels:

Line 1 used tones 35–130 at about –43 dBm/Hz.

Line 2 used tones 35–220 at –90 dBm/Hz.

Line 3 used tones 35–100 at –51 dBm/Hz.

Note that line 2 did not turn off at low frequencies, so this is not in this case an FDM-like solution, but rather simply reduced its transmit spectrum by a large amount. In more extreme cases, like the Verizon experiment of Figure 11.18 and [31], the line 1 spectra was concentrated at low frequencies 35–100 at high power-spectral-density level, whereas line 2 was above tone 100 at a very low power-

Table 11.1 Illustration of Gains on Long Line for Verizon Expt.

Mode of DMT ADSL	Line 1 data rate	Line 2 data rate
RA (static-spectra) mode	0.1 Mbps	9 Mbps +
FM (DSM) mode	1.8 Mbps	2 Mbps
FM (DSM) mode	1.4 Mbps	4 Mbps
FM (DSM) mode	0.9 Mbps	6 Mbps
FM (DSM) mode	0.6 Mbps	8 Mbps

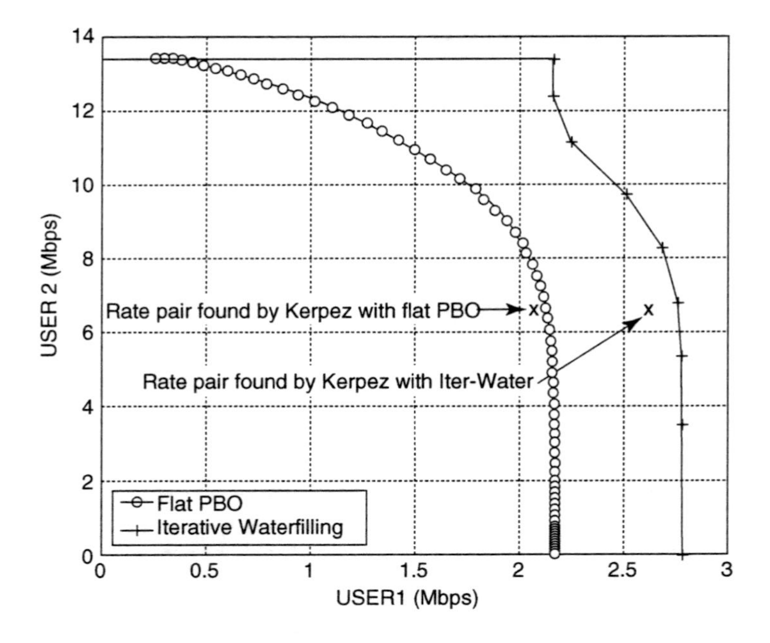

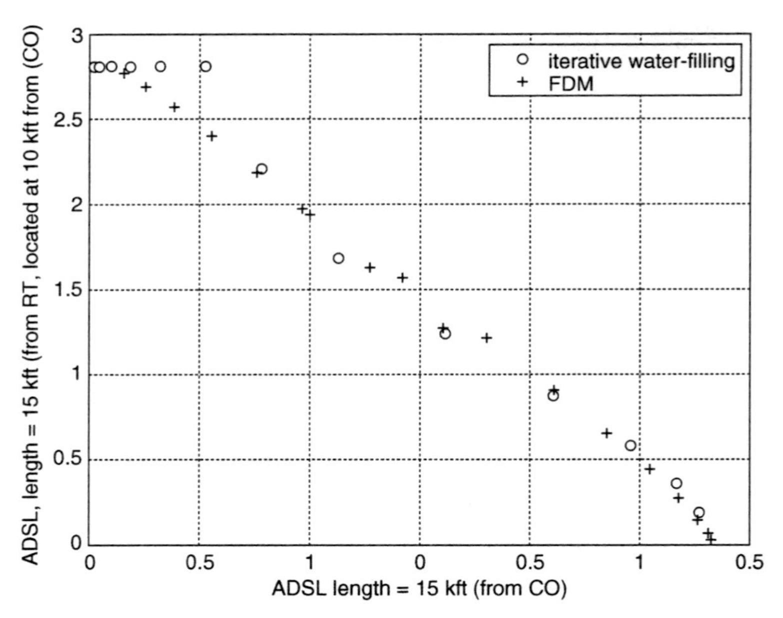

Figure 11.19 Rate regions for lines 1 and 2 with line 3 off (top); for lines 1 and 3 with 2 held at 1.6 Mbps (bottom).

spectral-density level. In each case, the spectra will adapt to the correct point in the rate region with the corresponding spectra, no matter which line starts first. 6 dB of fixed margin was used in all plots.

In practice, the FM operation has such a wide range of signals, it is implemented in two parts:

1. An initial gross politeness back-off (0 to 50 dB is allowed in g.dmt ADSL modems) that is set according to an appropriate table for FM mode.
2. Bit-swapping in steady state, which can provide from +2.5 dB to –13.5 dB gain alteration and of course allows tones to be zeroed so that no energy is carried.

The initial politeness back-off allows conversion devices (e.g., ADCs and DACs) to be operated with full precision when large amounts of power reduction are present on short lines. Doing the entire back-off with swapping could lead to precision loss in the converters, so politeness is used to avoid this implementation problem.

10MDSL, self, and into ADSL and VDSL

10MDSL is, at the time of this writing, a new standards project in the United States. The objective is basically 10 Mbps symmetric transmission on 1 or more DSL lines. Clearly with static SM, the upstream transmission of 10 Mbps creates a strong crosstalking NEXT into downstream ADSL (or VDSL). Basically—without the iterative water-filling DMT method in FM mode illustrated here—this 10MDSL is not feasible technically as the upstream from static spectra will annihilate downstream ADSL. To solve the problem, both directions of transmission in a DMT modem (this time using up to 1024 tones) are allowed to adapt in FM mode, thus

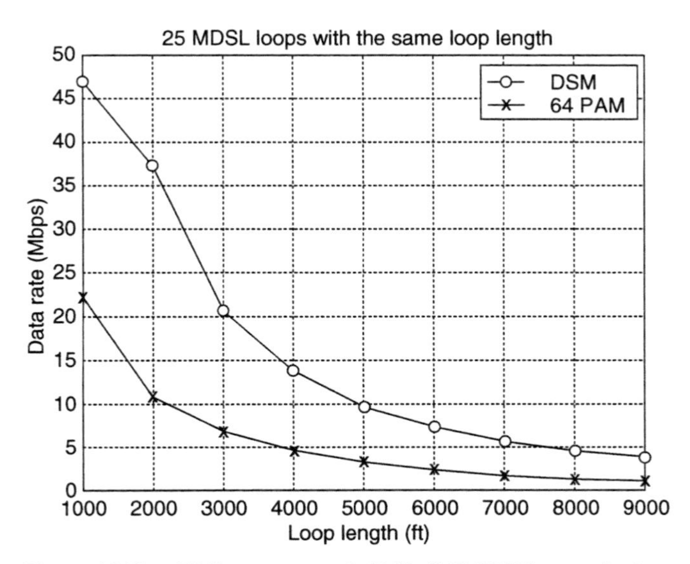

Figure 11.20 10Mbps symmetric DSL (10MDSL) on a single-line with 25 lines all of same length. No coordination used.

determining NEXT and FEXT on a per-binder best basis for the selected data rates of all the DSL modems. The first concern will be 10MDSL upstream into 10MDSL downstream, and then the combinations with ADSL and VDSL will be illustrated.

Figure 11.20 illustrates that static-spectra such as a 64 PAM ("SH"DSL) modem has very poor range on a single line, slightly more than 2000 feet. This number has been verified by a number of groups [33] who report from 2000 feet to 2500 feet, depending on noise and coding-gain assumptions. For the same coding

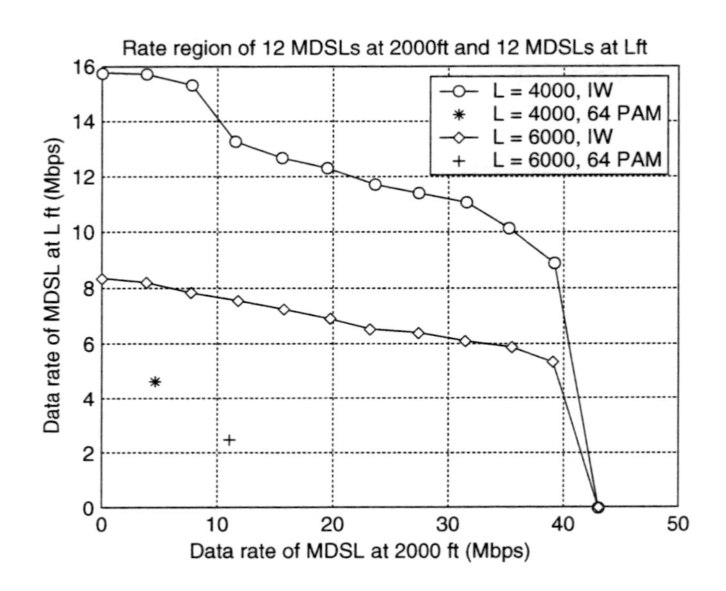

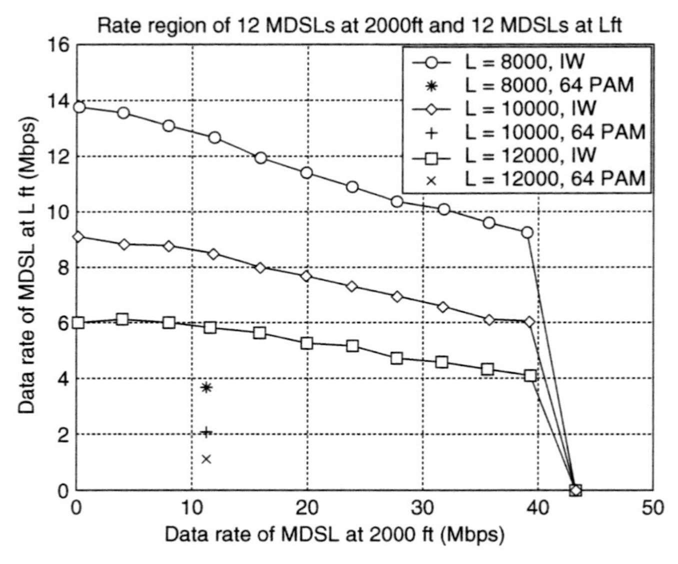

Figure 11.21 Illustration of self-10 MDSL rate regions with with DSM's iterative water-filling.

Table 11.2 Augmentation of 10 MDSL results with FM-IW (DMT) Results

Aggregate Bit Rate (symmetric payload)	# of Twisted Pairs	Average Bit Rate Per Pair	Objective Loop Length	Desired Loop Length	FM Mode IW Result
10 Mb/s	1	10 Mb/s	2.5 kft	>3.5 kft	5 kft
10 Mb/s	2	5 Mb/s	4 kft	>5 kft	8 kft
10 Mb/s	3	3.33 Mb/s	5.5 kft	>6.5 kft	9 kft
10 Mb/s	4	2.5 Mb/s	7 kft	>8 kft	10 kft
10 Mb/s	5	2 Mb/s	8 kft	>9 kft	Achieved
4 Mb/s	4	1 Mb/s	12 kft	>12 kft	Achieved
2 Mb/s	4	512 kb/s	15 kft	>15 kft	Achieved

gain of 3.8 dB, the FM-mode DMT system achieves 5000 feet (or equivalently gets 4x the data rate at 2000 feet). Voyan has verified the 5000-foot range point in [34]. Figure 11.21 shows the more realistic situation of mixture of various loop lengths of 10MDSL, and instead now plots rate regions. Figure 11.21 holds 1/2 the lines (12 MDSLs) at 2000 feet and varies the length of the another 12 between 2000 feet and 12,000 feet to see the achievable rates. Again the FM-mode interative water-filling maintains the 5000 foot range of 10 Mbps, but also allows very high data rates on the short loop while the long loops are doing quite well. Table 11.2 summarizes the goals of the 10 MDSL project, along with what was achieved in the FM mode, which leaves a considerable margin for error with respected to the project goals.

Figure 11.22 also provides some rate regions for ADSL with 10 MDSL. When deployed from the same point, the rate regions are very rectangular, indicating little effect (because of the iterative water-filling) of 10 MDSL on ADSL range. Figure 11.23 illustrates the interesting and highly likely possibility of an existing CO-deployed ADSL in a binder with a new 10 MDSL at an RT—the situation of Figure 11.17 is repeated with the 10MDSL on line 2, which is 4000 feet beyond the fiber terminal at 10,000 feet and various lengths of line 1.

Figure 11.23 clearly illustrates the annihilation of the ADSL circuit with static spectra–like SHDSL systems, while the DSM-based iterative water-filling is showing some degradation to ADSL, but data rates of 1.5–2 Mbps are preserved even in the presence of 10 MDSL. Clearly, static spectra (in this case particularly SHDSL PAM solutions) fail horribly, while the DSM provides a very clear advantage that is overwhelming.

Figure 11.23 on the bottom illustrates the VDSL and 10MDSL rate regions. Again, static spectra (see the points on left in Figure 11.23 close to horizontal axis) are failing to meet requirements while DSM is clearly functioning at very high rates. For more information, see [30].

ADSL to/from VDSL Examples

A problem feared by service providers in potential deployment of VDSL is the downstream FEXT that VDSL ONU emits into the ADSL downstream receivers attached to lines in the same binder. For instance, the situation investigated in Figure 11.24 has a VDSL ONU at 3000 feet from its customer, and ADSL cen-

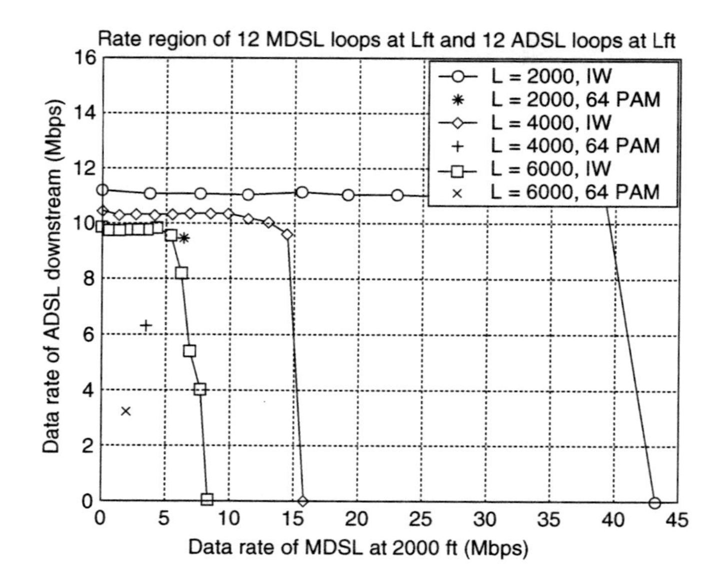

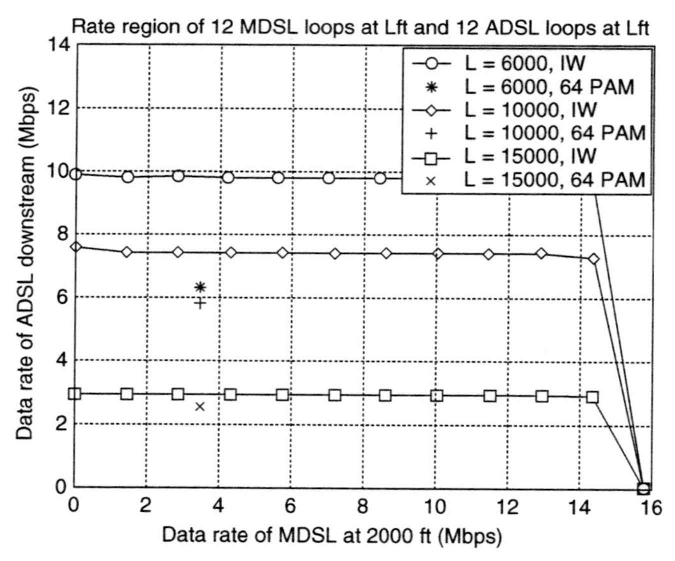

Figure 11.22 Illustration of rate regions of 10 MDSL with ADSL (deployed from same point).

tral office at 9000 feet from its customer. Both customers are at approximately the same location. VDSL downstream signals below 1.1 MHz create a large FEXT for ADSL, severely reducing its performance under static spectrum management. Some studies leading to [28] have found the interaction to be so detrimental to the performance of both ADSL and VDSL, leaving static SM with a difficult trade-off that discouraged VDSL deployment.

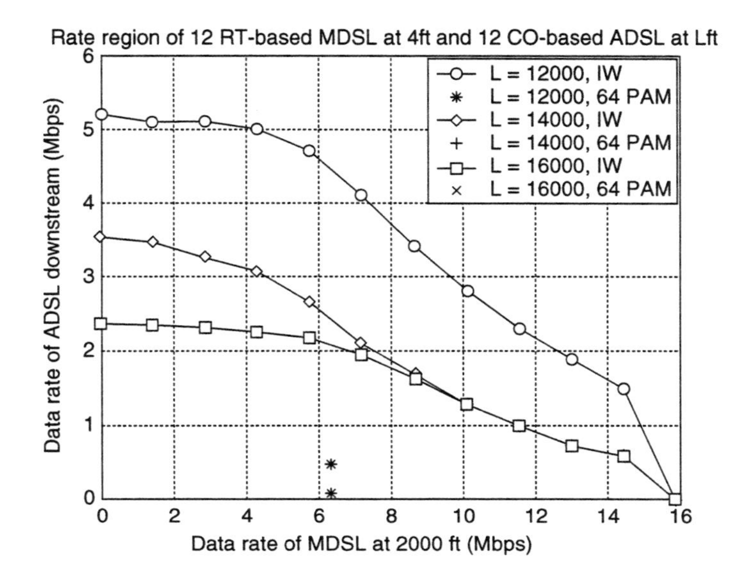

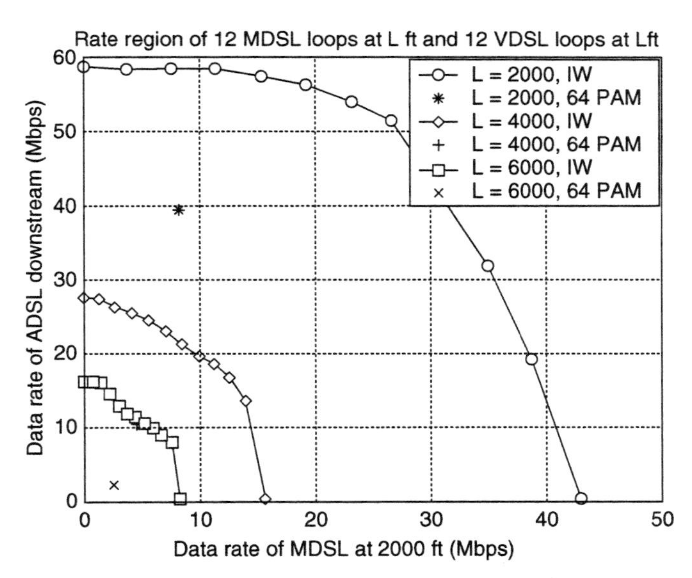

Figure 11.23 Illustration of remote-10 MDSL (10 kft from CO) and ADSL located at CO, and also MDSL/VDSL rate regions (bottom).

Iterative water-filling again solves the interference problem, providing good rates for both ADSL and VDSL beyond requirements. Under this new presumption of both modems converging to the iterative water-filling solution creates the rate region of Figure 11.24. Figure 11.24 illustrates downstream rate region for 4 VDSL modems at 3000 feet and 4 ADSL modem at 9000 feet (both on 26 gauge). All four ADSL modems have the same spectrum and so do all four VDSL modems

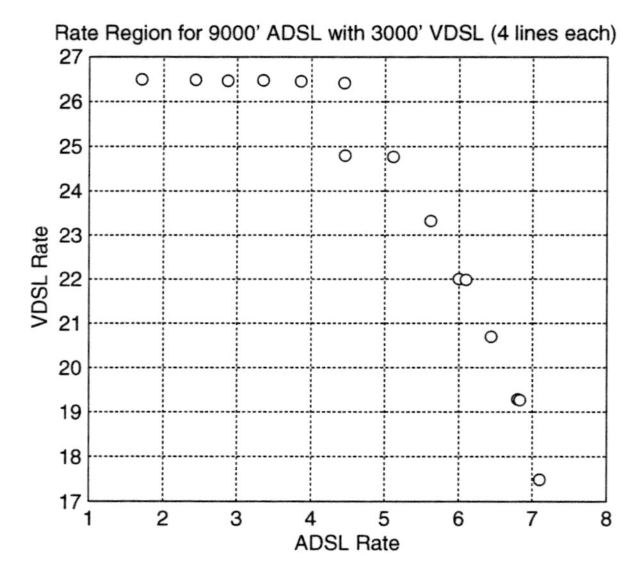

Figure 11.24 Rate region for ADSL and VDSL example.

in this case because the line lengths and impairments are the same (it would be hard to draw the 8-dimensional region, but we also wanted to ensure more than just one crosstalker for each of ADSL and VDSL as that would lead to data rates larger than those in Figure 11.24). The upstream data rates are not shown, but are 3 Mbps for VDSL and 500 kbps for ADSL. The nominal situation of T1.417, using the 998-standardized spectrum for VDSL of –60 dBm/Hz, causes the downstream rate for ADSL to be the 1.5 Mbps best case, while VDSL hopes to achieve 18 Mbps downstream in recent standards. Clearly, the rate regions here indicate that the VDSL systems can operate at 25–26 Mbps, while the ADSL system operates somewhere between 4 and 5 Mbps. The spectra of the two signals is shown in Figures 11.25–11.27 for various power levels of the ADSL and VDSL. Note that the ADSL signals tend to operate a full power, but over very different bandwidths as the iterative water-filling determines a cut-off frequency that is a function of the data rate attempted on the ADSL line. The VDSL signal does not vary so much in used band, but does vary power level automatically with margin-adaptive water-filling used (which can be implemented to implement the lowest power to achieve a given margin of 6 dB at a given data rate as in our examples). The VDSL PSD level varies over a wide range, taking only as much power as is necessary to implement a target bit rate (say 26.6 Mbps) while then reducing crosstalk into the ADSL so its performance then increases.

Figures 11.28(a) and (b) show the rate regions as the length of the VDSL or ADSL lines are varied, respectively.

VDSL Examples

As an alternative, this subsection investigates some trade-offs in VDSL when iterative water-filling is used for a 1000-foot loop and a 3000-foot loop. Our investi-

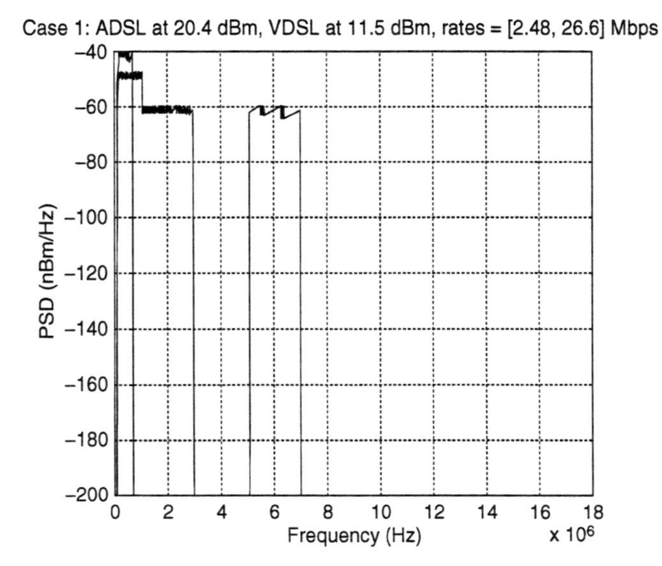

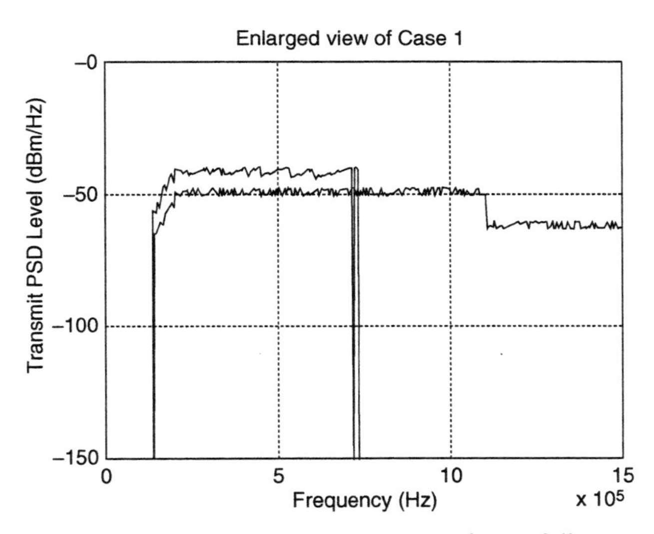

Figure 11.25 ADSL and VDSL spectra when at full power (2.48 Mbps, 26.6 Mbps).

gations in this section are actually done for 4 loops at 3000 feet and 4 loops at 1000 feet. The objective here is to determine what symmetric services might be offered on the 1000-foot loops while the 3000-foot loops maintain a 26/3 Mbps asymmetric VDSL service.

Any point in the upstream rate region in Figures 11.29 and 11.30 can be paired with any point in the downstream rate region of Figures 11.29 and 11.30. In 1999, various service providers argued as to the relative merits of asymmetric

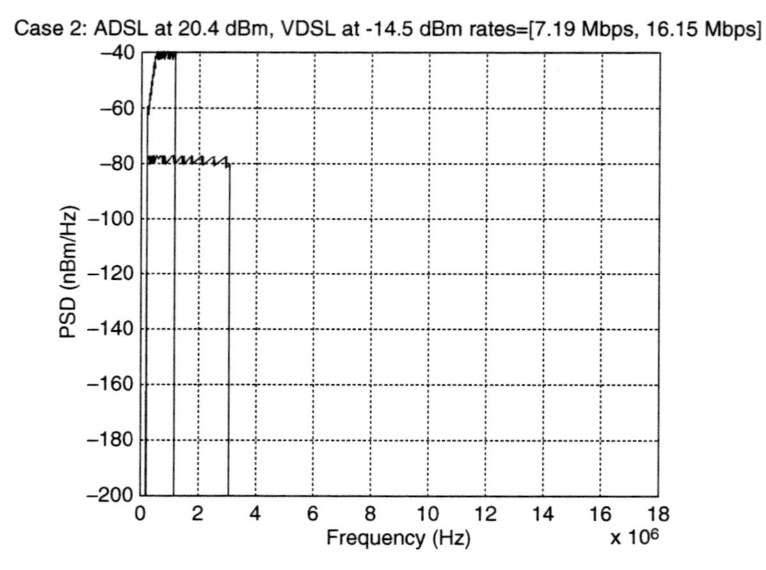

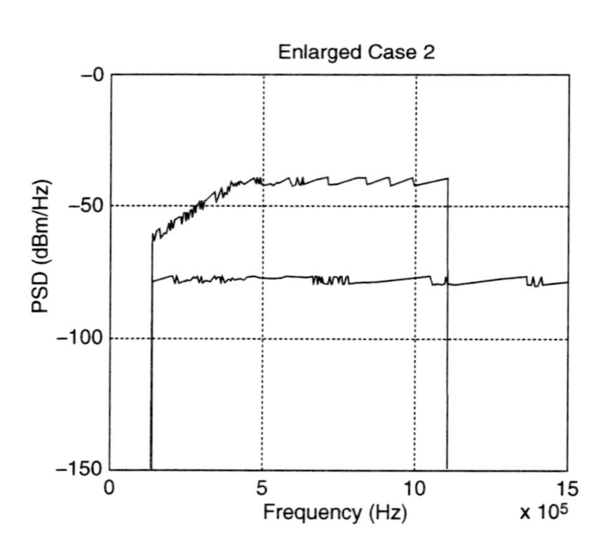

Figure 11.26 ADSL and VDSL spectra with full ADSL power, but low VDSL power (7.19, 16.2) Mbps.

versus symmetric service, eventually reaching a compromise where no group saw a data rate it liked with 18 Mbps/1.5 Mbps downstream service being argued by one group of service providers as perhaps the bare minimum for video service, whereas 6 Mbps symmetric was argued by another group to be below what they wanted but perhaps interesting. Figure 11.29 shows that even within the 998 plan of VDSL, DMT water-filling VDSL modems avoid each other within the common band in a

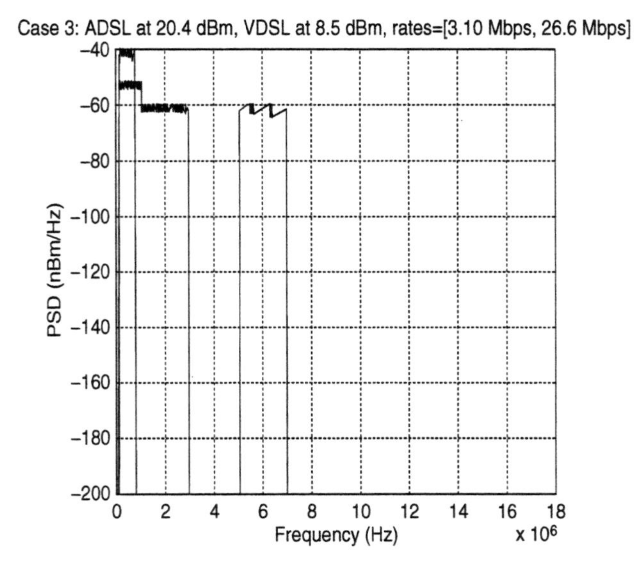

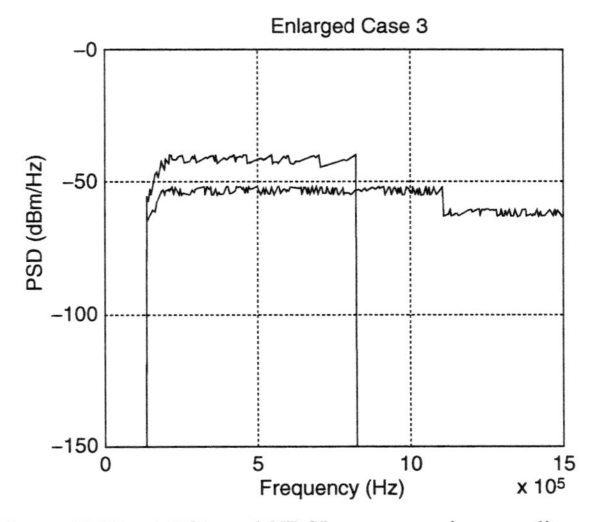

Figure 11.27 ADSL and VDSL spectra at intermediate condition [3.1, 26.6] Mbps.

mutually beneficial way autonomously, leading to 26/6 Mbps on a worst-case 3000-foot VDSL loop and at least 20 Mbps symmetric on the 1000-ft loop. Such a situation might well occur in practice with residential loops being further than small business loops. In any case, the data rates significantly exceed those that were projected by static spectrum management. Figure 11.36 alters the cut-off frequencies of up and down, given that with DSM, the 998 plan may not have provided the best

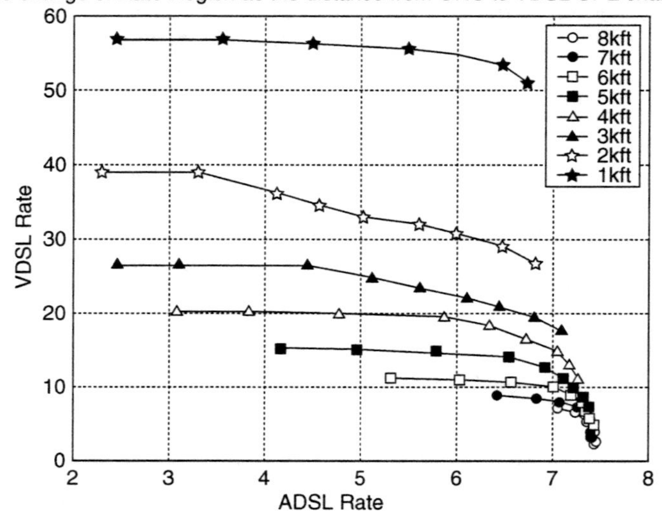

Figure 11.28(a) VDSL/ADSL with varying VDSL line length.

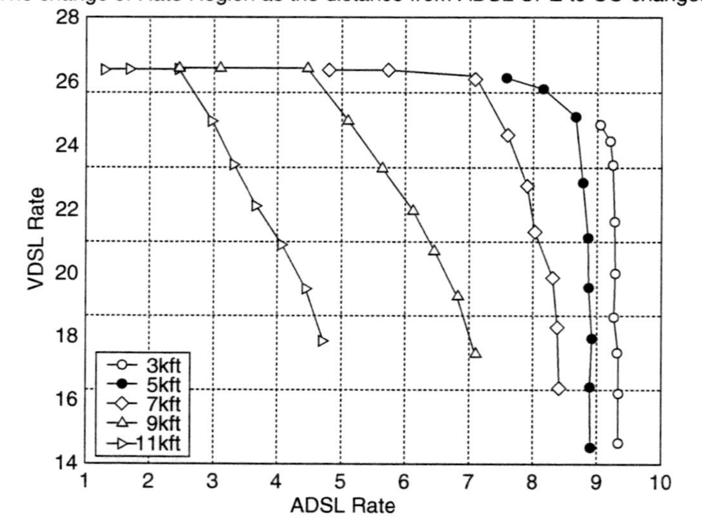

Figure 11.28(b) VDSL/ADSL with varying ADSL line length.

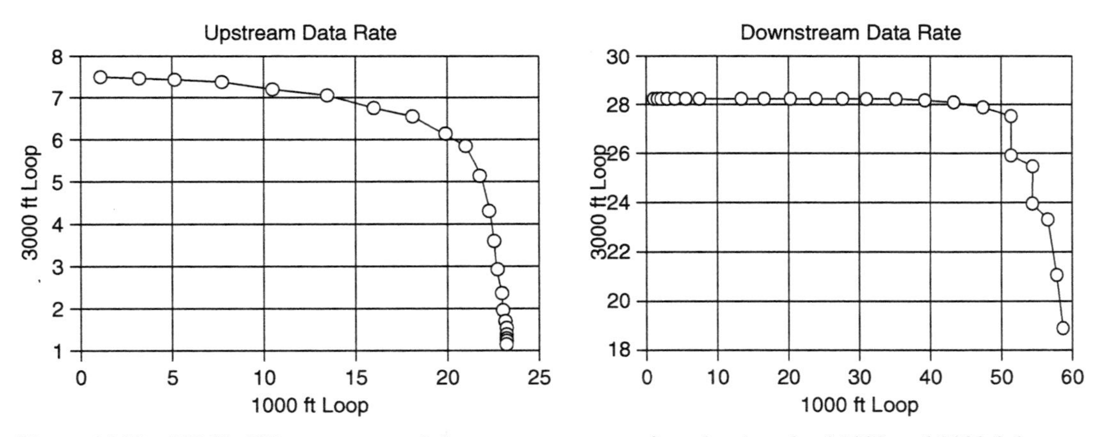

Figure 11.29 VDSL 998 upstream and downstream rate regions for 4 each of 1000 and 3000 ft loops in same binder with distributed Level 0 iterative water-filling.

trade-off. Indeed, that leads to 26/13 Mbps on the asymmetric VDSL loop at 3000 ft while the 1000 ft loop has at least 30 Mbps symmetric service. These cut-off frequencies for up/down need not be fixed if DSM is used.

Figure 11.31 illustrates the trade-off versus line length as the loop length of VDSL instead varies with respect to the best-known (and not used because it cannot be implemented in the static-spectra-only QAM VDSL) reference-noise, spectrally shaped, power back-off methods of VDSL and the corresponding ADSL data rates.

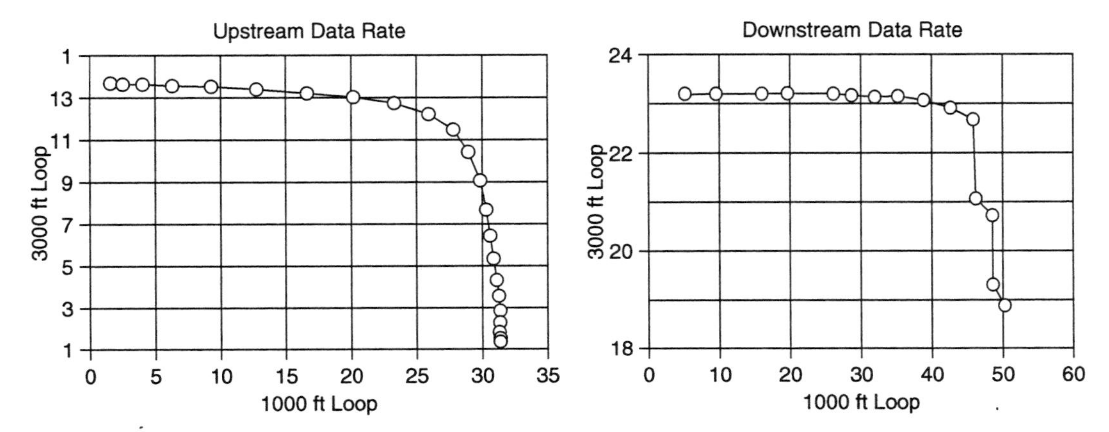

Figure 11.30 VDSL flexible upstream and downstream rate regions for 4 each of 1000 and 3000 ft loops in same binder with Level 0 distributed water-filling, but Level 1 control of up/down cut-off frequencies.

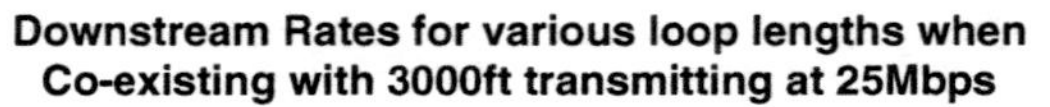

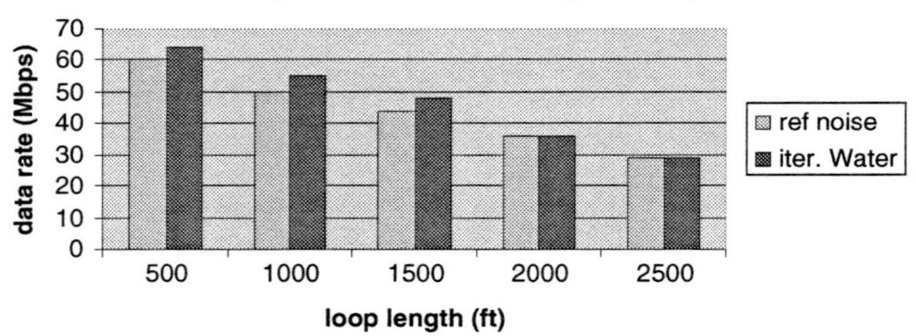

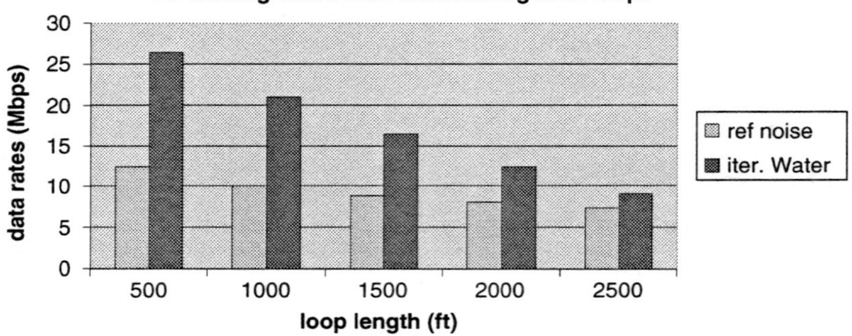

Figure 11.31 Downstream data rate comparisons with 4 lines at 3000′ at 25 Mbps; Upstream data rate comparisons with 4 lines at 3000′ at 6.7 Mbps.

11.4 VECTORING

Coordination within a common DSLAM or modem unit allows co-generation (or co-reception) of multiple (preferably all) lines at the DSLAM. Such situations occur naturally as DSL evolves to remote/line terminals fed by fiber. In such DSLAMs at remote terminals, it is possible and feasible to coordinate signals because all or most of the lines are terminated on common equipment. In downstream transmission, joint generation of a vector of transmit signals over a common packet period is feasible (and possibly more cost-efficient than individual generation). This is the downstream broadcast vectored system of Figure 11.32. Vectored multiple access for upstream signals occurs at the receiver as in Figure 11.33.

For DSL, both vectored situations meet the condition for Ginis's QR zero-forcing GDFE when the upstream and downstream signals travel in different frequency bands when only FEXT is of concern. This condition is simply that the off-diagonal entries of $\mathbf{H}$ are much smaller than the diagonal entries. Section 11.4.1

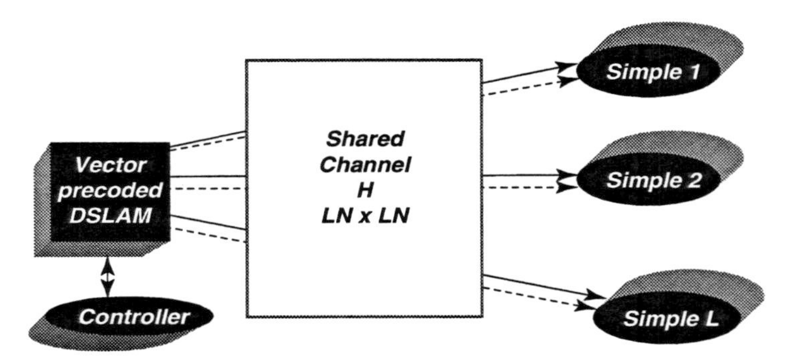

Figure 11.32 Vector broadcast system for downstream DSL with cogeneration of transmit signals (but separate simple receivers).

discusses the multiple-access upstream problem briefly, whereas Section 11.4.2 discusses the broadcast downstream problem. When either end of the link can be coordinated at that same end, then NEXT can be canceled with a simple multidimensional echo canceler. Section 11.4.3 illustrates and discusses such Ethernet-in-the-first-mile (EFM)–like results, as well as DSL results.

11.4.1 Upstream Multiple Access

The vector upstream multiple-access DSL system of Figure 11.33 has typically $M = 1$ transmit lines per user but typically has $K = L$ coordinated line receivers for reception. It is possible with some DSL systems (such as EFM) to have $M = 2,4$ coordinated upstream transmitters also. It is also possible that only some of the lines in the DSLAM can be coordinated, or there are groups of lines that can be coordinated within each group, but not between groups. However, this subsection focuses on the case where $K = L$ and $M = 1$ for purposes of illustration. Extensions

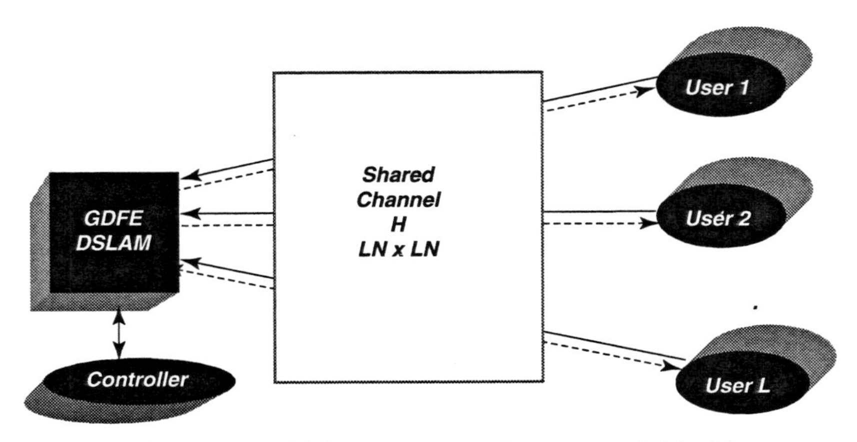

Figure 11.33 Vector multiple-access system for upstream DSL with coordinated reception but separated transmitters.

then follow naturally by combining the concepts in Section 11.3 with those in this section where coordination is used among the various groups.

In this vectored upstream multiple-access DSL system, iterative water-filling generates the optimum vectored spectrum choice for all the upstream users, as first noted by Yu and Rhee [24].[4] The vectored DSLAM thus determines the spectra for each of the individual separated users and communicates these spectra to individual upstream transmitters through some kind of control or back channel. The receiver is a GDFE with L sampled inputs from the lines. When DMT is used and all lines are synchronized (which is easy with the common DSLAM and less costly as well), the GDFE becomes a simple 2×2 or at most 4×4 GDFE on each tone where only the significant crosstalkers at that frequency are processed. Indeed with this small number of users, ML detection is easy, and the feedback section need not be implemented (and thus error propagation is avoided). Indeed, soft cancellation can be used to simplify the ML if desired. Furthermore, Ginis's QR simplification can be applied to derive the feedforward and channel(ML)/feedback coefficients directly from the channel matrix with essentially no loss in performance with respect to optimal. Ordering of the tones is not illustrated here, as it is often considered to be a proprietary secret of the manufacture—here we just presume such an order exists. Only complexity of implementation depends on ordering.

Figure 11.34 illustrates the exact receiver for the upstream case for each tone of a DMT system. The quantities $G_{i,n}$ are the gains for the nth tone of the ith user as determined by iterative water-filling. The structure for non-DMT systems is the same except that the various matrices become much larger. Rate regions for the different users can be derived from the iterative water-filling procedure by lowering the power for individual users and recomputing the rates of the other users, thus sketching an approximation to the rate regions. Individual rate requests of customers can then be evaluated for feasibility as a function of the rate region. Section 11.4.3 has some sample rate region plots.

11.4.2 Downstream Broadcast

The vector broadcast DSL system of Figure 11.35 has typically $M = L$ coordinated transmitters per user but typically has $K = 1$ receiver for each line. It is possible with some DSL systems (such as EFM) to have $K = 2,4$ coordinated receivers also. It is also possible that only some of the lines in the DSLAM can be coordinated, or there are groups of lines that can be coordinated within each group, but not between groups. However, this subsection focuses on the case where $K = 1$ and $M = L$.

The optimum spectra for each broadcast user has been determined by Yu and Cioffi [25] and requires a two-step iteration consisting of

1. Determining a worst-case noise autocorrelation matrix that has the same diagonal noises as the original downstream lines in terms of minimum mutual

---

[4]Cheng and Verdu [3] studied a non-vectored situation where all of the upstream users effectively were added together, in which case iterative water-filling reduces to their solution. However, Cheng and Verdu's solution does not apply to the vectored case (and their result of FDM being optimal no longer holds).

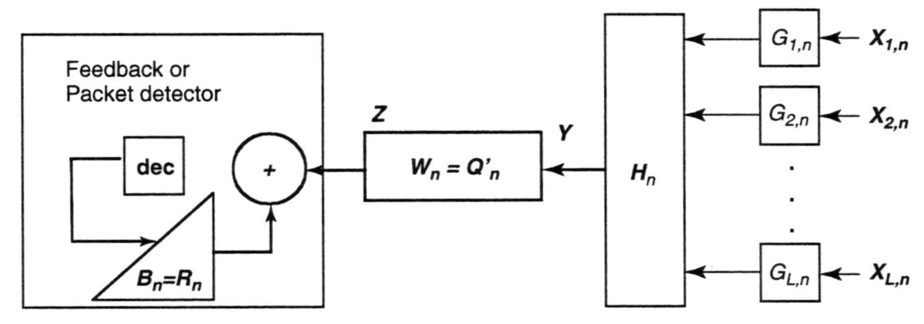

Figure 11.34 Exact illustration for one tone (n) of GDFE upstream vector receiver.

information (this involves solution of a generalized Lyapunov equation as noted in [25])

2. Determining a GDFE for that noise autocorrelation that when implemented as a precoder maximizes rate sum and simultaneously diagonalizes the GDFE feedforward matrix—this is a water-filling step

This solution will be essentially equal to that provided by Ginis Precoder, which is easier to compute when iterative water-filling is instead used to decide the spectra of each of the downstream signals. Thus the optimum algorithm, while slightly better, may require at least conceptually considerably more complexity for design for little gain with respect to the easily understood iterative water-filling.

Figure 11.35 illustrates each tone of the Ginis Precoded implementation of the downstream broadcast vectored DSL system. The receiver gains correspond to the diagonal inverses that occur in the QR factorization of the matrix H. Receiver implementation is particularly simple for each tone in this case. The precoding is such that simple slicing (in the absence of trellis or turbo codes) is sufficient in each receiver. Rate regions can be computed in the same way previously described of executing several instances of iterative water-filling with individual users' powers reduced.

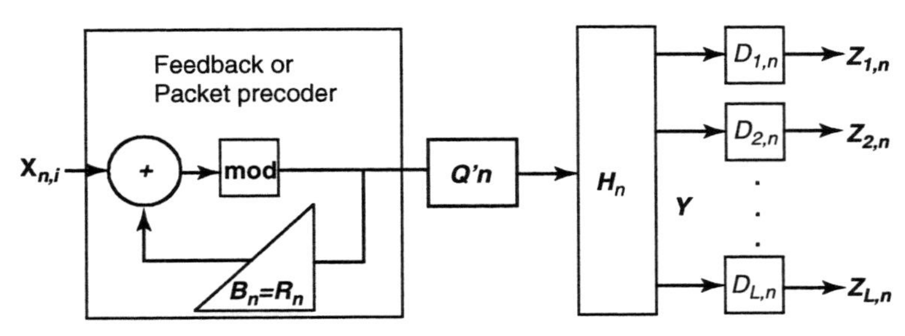

Figure 11.35 Exact illustration for one tone (n) of GDFE downstream vector receiver.

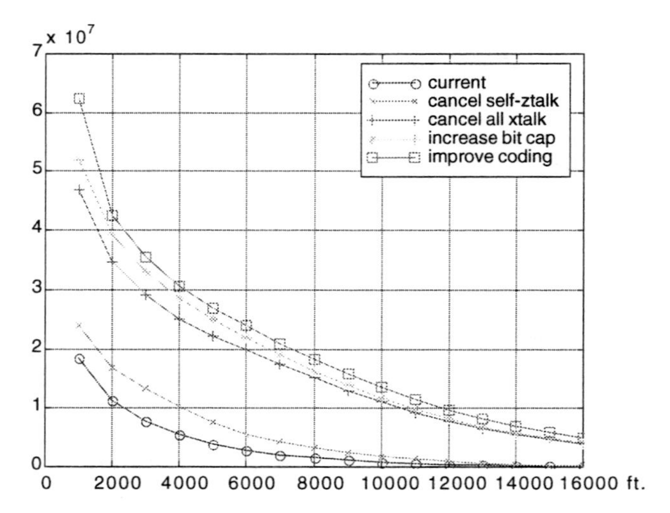

Figure 11.36 Vectored ADSL data rates versus range.

11.4.3 Examples

It is interesting to compute the performance of vectored systems in a few cases. Figure 11.36 illustrates the type of improvement possible in ADSL if each ADSL line in the binder used Level 2 coordination or vectoring with the existing ADSL frequency assignments. The lowest data rate versus range curve in the figure illustrates a theoretical upper bound on the performance of the best ADSL receivers today without vectoring nor coordination. The middle curve represents vectoring, whereas the upper two curves recognize that continued improvement after vectoring could occur by increasing the bit cap and using better (turbo or LDPC) codes. Note the level of improvement is quite large, basically a factor of 3 or more at all data rates.

Figure 11.37 repeats the vectoring exercise for VDSL and compares against theoretically best transmission when the current noncoordinated VDSL is used

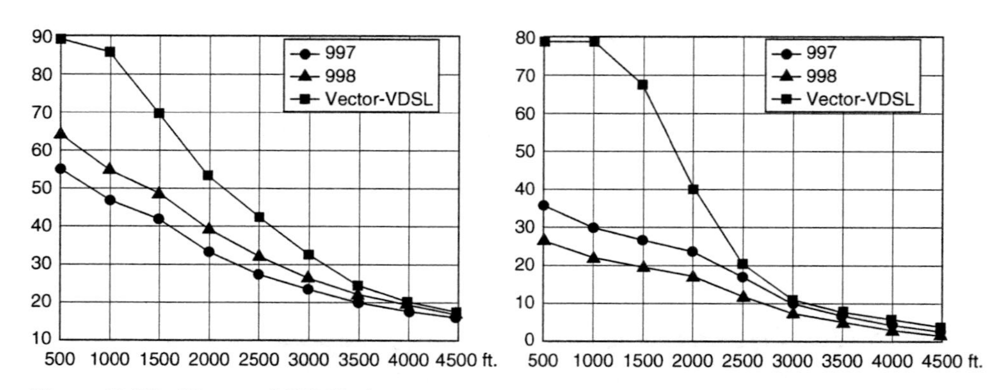

Figure 11.37 Vectored VDSL data rates versus range.

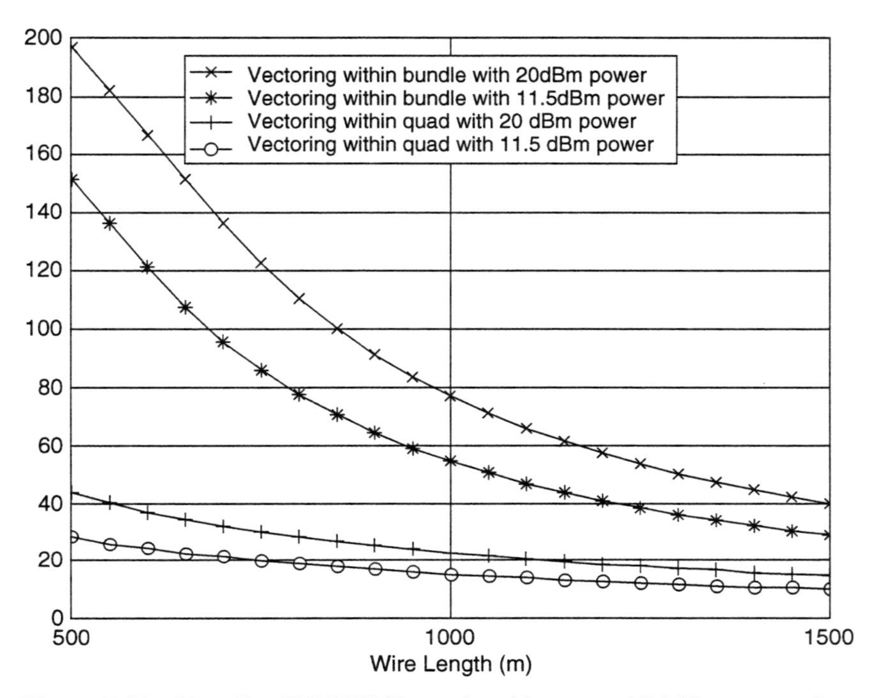

Figure 11.38 Short line EFM/VDSL results with vectored DMT—symmetric transmission over full band.

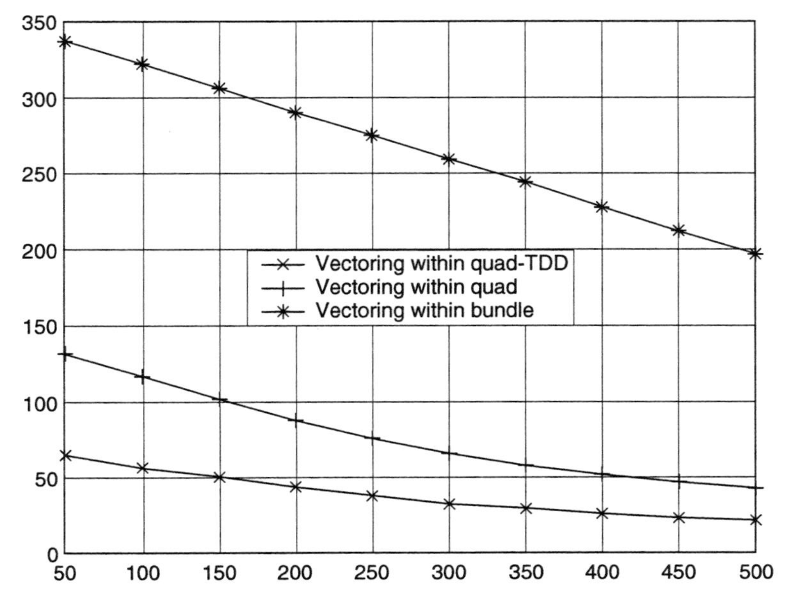

Figure 11.39 Longer line EFM/VDSL results with vectored DMT—symmetric transmission over full band.

with both American (997) and European (998) band allocations. Again the data rate is symmetric and considerably higher than with no coordination.

Figures 11.38 and 11.39 show the results of full vectoring (with NEXT cancellation) for potential short and longer line use of DMT VDSL with EFM. Also shown are results when vectoring is only used within groups of 4 lines.

11.5 MIMO CHANNEL IDENTIFICATION

All the methods for multiuser coordination need information about the multiple-input multiple-output (MIMO) channel, $\mathbf{H}(f)$ or its magnitude $|\mathbf{H}(f)|$ and/or the the noise spectra on each line's output. This information may be derived by each line itself without coordination. However, coordination may allow more swift and easy determination of the characteristics. With no coordination, the identification of crosstalking coupling into a specific channel is called "blind" training as the data sequences on other lines are unknown. When these sequences are known either through coordination or because a regularly inserted synchronization pattern on another line has been identified, the DSM system "uses training" or simply is "not blind."

Level 3 coordination is the easiest for channel identification and noise spectral estimation and is discussed in Subsection 11.5.1. Essentially special training sequences can be used to allow easy and accurate identification of every possible cross coupling between the lines, with any remaining noise typically modeled as Gaussian and identified easily by removing crosstalk and measuring the spectrum of the residual error. Some systems may instead only allow input/output packets of a channel to be available at certain points in time to the network maintenance center (data acquisition unit in Figure 11.6). Then the methods in Subsection 11.5.1 can be used with some additional processing as described in Subsection 11.5.2. Absence of coordination or information acquisition forces the blind methods of Subsection 11.5.3, which are effective but can take much longer to train.

11.5.1 Method of Least Squares

Level 2 coordination likely will occur only with synchronized DMT systems in the binder group, which fortunately then means that each tone sees or provides crosstalk from other users only on exactly that same tone. This leads to enormous simplification in channel identification. Each modem can use a known 4-pt QAM training sequence that has been randomly generated for robustness and low PAR. The current patterns used in ADSL and VDSL are sufficient. Different lines will need to have a different phase (achieved by bit-level staggering of that sequence by one DMT symbol from each line to the next). The adaptive combiner shown in Figure 11.40 is implemented for each tone, with no more than 4 coefficients (as other crosstalkers are much smaller) that should provide nonzero crosstalk coefficients H_{kmn} on the nth tone. Training may take longer than for a single modem and can be implemented by a least-squares fit over S training symbols as

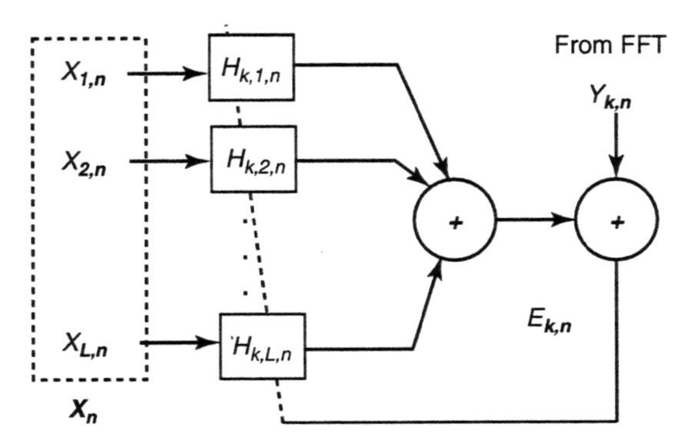

Figure 11.40 Adaptivechannel (per tone) for L users.

$$\mathbf{H}^*_{k,n} = \left[\sum_{s=1}^{S} Y_{k,n}(s) \cdot \mathbf{X}^*_n(s)\right]\left[\sum_{s=1}^{S} \mathbf{X}_n(s)\mathbf{X}^*_n(s)\right]^{-1}$$

Values in $\mathbf{H}_{k,n}$ that have magnitude below some threshold (or relative threshold) can be ignored in subsequent signal processing (with Level 2 coordination). A simple LMS gradient algorithm (see [1]) can instead be used if the above matrix aversion is to be avoided, and is illustrated in Figure 11.40 according to

$$\mathbf{H}^*_{k,n}(s+1) = \mathbf{H}^*_{k,n}(s) + \mu \cdot E_{k,n}(s) \cdot \mathbf{X}^*_n(s).$$

The procedure is greatly facilitated in the case of Level 3 coordination, which is also the one of the few cases in which the full (gain and phase) $\mathbf{H}_{k,n}$ is desirable with Level 3 coordination, and $\mathbf{H}_{k,n}$ is needed to set either the GDFE or precoder coefficients. It is sometimes also necessary with MUD methods as discussed in the appendix. In this case, coordinated known training is highly unlikely to occur and so identification becomes more complicated (at least conceptually) and is discussed in Section 11.5.2.

Noise spectrum estimation proceeds for tone n of user k by forming the error sequence

$$E_{k,n}(s) = Y_{k,n}(s) - \mathbf{H}^*_{k,n}\mathbf{X}_n(s)$$

and, for example, computing an average spectrum value at each frequency n and each line output k according to

$$S_{k,n} = \frac{1}{S} \cdot \sum_{s=1}^{S} |E_{k,n}(s)|^2.$$

11.5.2 Packetization

Levels 0–2 coordination does not admit a known training pattern on all lines, thus complicating the procedure. Sometimes a multiuser system might benefit from knowing the full **H** also, in particular, some of the MUD methods in the appendix.

DSM standardization efforts [26] describe the methods for collecting either type of data. Level 2 data acquisition simply implies that a DSL modem has the ability to release $SNR(f = n/T)$ for that line in DSM standards, whereas Level 3 data acquisition implies that input/output packet characterizations of the channel (with either live[5] or training data) are available. Level 3 data acquisition can occur even when Level 3 coordination is not possible because the input/output packets can be released over short periods of time to the acquisition unit with need for line-signal coordination.

Virtual Binder Group Identification and Alignment

Level 3 data collection provides I/O DSL packet data at specified times of operation of a DSL line. The first issue then is the time stamping and collection. An information-acquisition unit specifies a time of operation approximately around which the DSL modem collects S successive packets of DSL I/O information and provides to the acquisition unit for later use in processing. The approximate time is established through the network clock (or any reasonably accurate clock). Presumably, the information-acquisition unit requests data from several modems in the same binder at about the same time. Knowledge of the input and output packets then enables isolation of the important crosstalkers (i.e., those that are large) for any given line through cross-correlation with the various other lines' input sequences. Such cross-correlation is computed by

$$R_{yx,ij}(l) = \frac{1}{S \cdot N} \sum_{s=1}^{S} \sum_{k=1}^{N} y_j((s-1) \cdot S + k + l) \cdot x_i((s-1) \cdot S + k)$$

which will have a maximum over the "lag" index $l = 0, \ldots, \pm \overline{L}$, where $\overline{L}$ is a maximum possible offset based on common clock tolerances between systems. The designer presumes a relatively small frequency offset between clocks for this calculation so that the position of the maximum is not significantly altered, basically meaning that not too many symbols should be used ($S \leq 10$). If this maximum exceeds a threshold, typically set as a function of the received signal y_j's energy level and a minimum noise floor (like −140 dBm/Hz) as significant, then both a loop in the virtual binder and its rough offset in time $\overline{l}_{j,\max}(i)$ are determined. The sequences of inputs and outputs on different channels are presumed interpolated through standard signal processing to the same sampling rate (roughly) for purposes of the lag and virtual-binder-group determinations. Once a crosstalker has been determined as significant, then the input can be exactly interpolated to

[5]One need not be concerned over the release of user data because the data released would be DMT symbol values, from which it is not possible to know the corresponding user messages because the data scrambling would be in an arbitrary phase not known to the acquisition unit. Clearly, highly secure DSL transmissions independently need encryption in any case, and would also remain just as highly secure if only symbol values $X_{i,n}(s)$ (and corresponding channel outputs) are released because the encrypting code in that case is unknown along with the scrambler phase.

the same clock as the line j's clock by phase locking. Such phase locking can be recomputed by postulating that the exact offset is

$$l_{j,\max}(i) = \bar{l}_{j,\max}(i) + \Delta \cdot s + \varepsilon$$

and reinterpolating the input of the oppositive channel to this offset for a few exact values in the vicinity of the original average value $\bar{l}_{j,\max}(i)$ and recomputing the cross-correlation. A new maximum over Δ, ε values is found, and then that interpolated value defines the offset between the lines. (This offset may need to be maintained for subsequent MUD or other multiuser operations, which can be achieved by a number of traditional phase-locking methods.) If all lines are synchronized as in Section 11.4, there is no need for such clock recovery.

Interpolation

Interpolation among non-level-3–coordinated DMT systems leads to the structure of Figure 11.40 and the phase-locking and virtual-binder-group identification of the previous subsection. Typically these systems all have 4.3125 kHz tone spacing (whether VDSL, ADSL, or SDSL) and thus interpolation is not necessary and corresponding tones are aligned. For mixes of any other combination of DSL, interpolation between sampling rates is required. The vector $\mathbf{H}_k$ represents the contribution of every DSL system input (over several time samples with a packet, or over several tones when a DMT system) from an input packet into the corresponding packet at the output of the kth line. The structure of Figure 11.40 is still applicable, except that many input samples from each of the lines and many output samples for the kth line appear. Thus, no special interpolation procedure is required other than knowing which samples on each input (and how many) contribute to each of the outputs.

The time-domain signal in the cross-correlation $x(t)$ must be constructed at approximately the sampling rate of the receiver. There are many methods for such interpolation from the same data supplied at another sampling rate. Typically, this is facilitated if input/output samples are supplied at sampling rates two or more times higher than the actual symbol rate of time-domain (PAM and QAM) signals, allowing linear interpolation between samples to often be sufficient. (For DMT systems, interpolation is easy in the frequency domain—see [1] and the data need not be oversampled.) This book does not contribute further to the considerable literature on interpolation via digital signal processing.

Zeng has provided an example in Figures 11.41–11.43 where Level 3 data acquisition has been used (presumably for Level 0–2 coordination, not Level 3 coordination, otherwise the methods of Section 11.A would be used directly). The line of interest is one of the ADSL signals. Here, 4 basic-rate ISDN, 4 HDSLs, and 5 ADSLs all share the same binder with significant mutual crosstalking. The spectra of the crosstalkers appear in Figure 11.41, while the cross-correlation with one of the HDSLs appears in Figure 11.42. Note the shape is the same as that of the crosstalk coupling channel response and has a definite peak identifying the timing offset between the two signals. Finally Figure 11.43 shows the resulting error magnitude as the number of training symbols increases. After a few hundred packets of

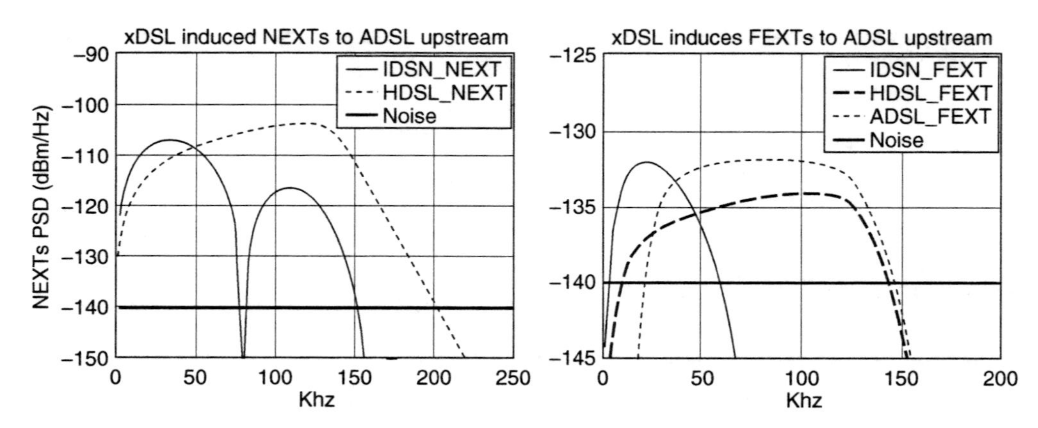

Figure 11.41 Crosstalker spectra into an upstream ADSL.

training, the residual error is acceptably small. For more on these methods, see Zeng and Cioffi [27].

Channel state machines as in the appendix and Figure 11.45 can be used to monitor state changes between different signals for the situation where different DSLs are turning on and off. A coupling function can only be identified when the corresponding DSL is energized.

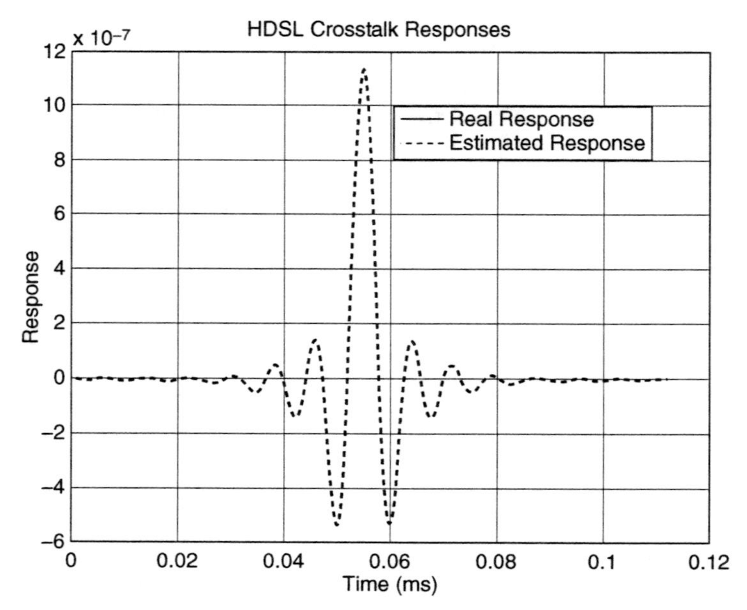

Figure 11.42 Cross-correlation for identification of offset between one of the HDSLs and the ADSL.

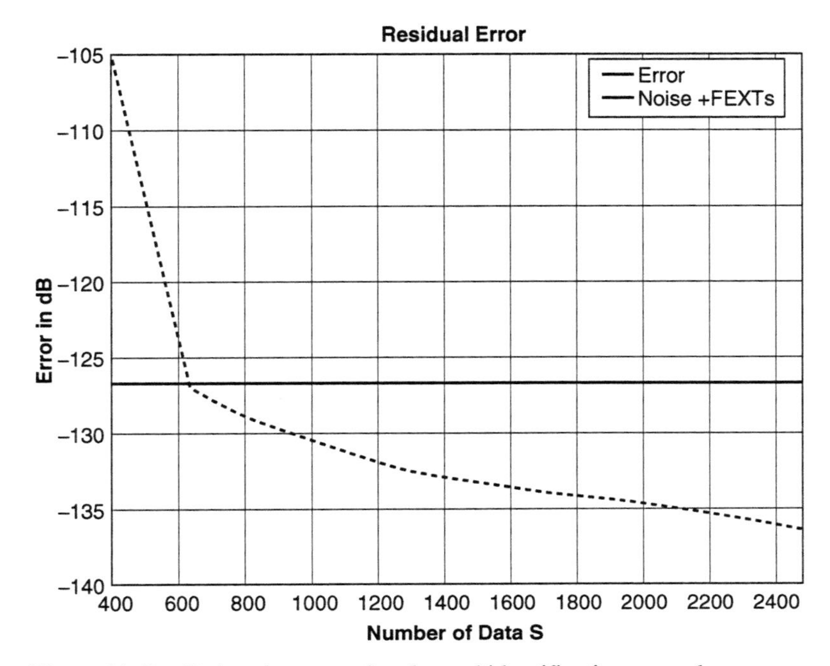

Figure 11.43 Estimation error for channel identification procedure.

11.5.3 Blind Training

Without Level 3 data acquisition, the DSL system on each line must identify crosstalkers "blindly" if MUD methods are to be used by that line (more advanced methods than MUD are necessarily not available if there is no Level 3 coordination nor data acquisition). The lack of coordination complicates channel identification. This area has been studied by Aldana and Cioffi [9].

Initially, the conception of blind training proceeds with the least-squares concept of Section 11.5.1. The least-squares solution could theoretically be recomputed for every possible transmitted data sequence and then that sum of squared errors that is smallest over all such calculations corresponds to a value of the matrix **H** that is the maximum-likelihood estimate of the matrix channel. It is not possible to do better than this estimate. However, clearly the number of possible sequences even with the per-tone reductions of DMT could simply be too numerous to compute. A possibility is to iteratively compute this estimate by construction of the probability density of the joint input sequence and channel matrix. This is an area of active research for many in wireless transmission and could prove fruitful when applied to DSL, but has not been fully investigated at this time. However, Aldana has studied one version of such iterative estimation/decoding which is the well-known **expectation maximization** (EM) method [9]. In this case, the average value over all possible transmitted sequences is precomputed or estimated in the least-squares equation. This essentially maximizes the likelihood function on average. It is still somewhat complex, but possible to do most of the difficult computation offline and store it for many uses.

The interested reader is referred to [9] for a detailed analysis and description of Aldana and Cioffi's EM method for DSL where it is shown to be unusually effective in identifying the matrix channel blindly. Work to simplify calculations further is likely to occur by groups interested in transmission systems that do not have Level 2 coordination of signals and cannot implement sharing methods in a distributed manner. Fortunately, distributed-iterative water-filling eliminates the need to know the channel in all places so it is only those systems that must know it for multiuser detection that would need to use the EM algorithm.

11.6 PREDICTIONS FOR THE DSL AGE

The possibilities for DSL are seemingly unlimited when augmented by dynamic spectrum management. Results in this chapter have shown that 100 Mbps symmetric transmission on 500 meters of cable is possible, are 100 Mbps at 800 meters, 50 Mbps at 1.5 km, and 25 Mbps in excess of 2 km. As fiber enters the loop plant and reduces loop lengths, DSM-charged DSL can enable 100 Mbps 100BT speeds to every customer in the future, and with less power and cost than existing DSL systems. The opportunity for the telecommunications industry is just unprecedented in terms of applications, service, equipment and chip revenue. A 500-meter binder of 50 pairs could carry 5 Gbps total capacity symmetrically and reliably, more than is dreamed of for even larger numbers of customers in HFC and even in PON all-fiber networks. The bandwidth in the billions of copper twisted pairs in service in this decade can and will overwhelm the existing network capabilities, creating additional opportunities for new service and equipment revenue. Customers already know how to use and are acclimated to 100 BT speeds in their offices and internal networks, and thus an enormous market effectively already exists for DSL connection.

DSM-based DSL welcomes, enables, and implements the information age. Let DSL/DSM pave the way for the message of the future.

APPENDIX 11A—MULTIUSER DETECTION

Multiuser detection (MUD) is a technique that can be used without any need for standardization or coordination of any type, and simply enhances receiver performance by trying to detect and eliminate all large crosstalking signals. MUD has limited benefit as it typically only works for a few large crosstalkers and has high complexity. Sections 11.3 and 11.4 show DMT systems can migrate naturally to mutually beneficial spectra that simplify the design so much that effectively MUD is not necessary and would add nothing. However, not all DSLs are DMT, thus leaving MUD as the only solution sometimes. Static SM standards attempt to impose some regulation on future DSL systems that do not coordinate in any way. Debate in creation of SM standards, and compromise in setting this debate, led to each type of DSL having considerable crosstalk into and from many other DSL services. Fortunately, multiuser detection can often provide some significant improvement even when there is no coordination and many different types of DSLs share a binder.

The rate regions of Section 11.1 are of less interest in MUD because there is no ability to change the data rate of any particular DSL system. Clearly if the designer of any system did have access to rate regions, then that designer would at least know whether his design was feasible or not. Figure 11.44 shows a revised GDFE structure where for each (downstream or upstream) receiver. The equalizer **W** accepts one channel input sample stream of N samples in a packet, and outputs up to L signal streams that can each be (unfortunately) linear combinations of all $U = L$ users. The GDFE is also not unique in this case, and there are actually an infinite number of possibilities with the same performance. The ensuing ML detector or approximations thereof may be complicated or simplified, depending on the choice. A special case of all users being synchronized and DMT leads to a linear combination of the L user's tone-n signals only on tone n of the output. The ML detector can be easily implemented with an L-dimensional user-feedback structure augmented by ML detection separately on each tone. Otherwise, any DSL input symbol from any user within the packet time slot can and often will affect all of every other user's samples in the time slot. Nonetheless, given the proliferation of DSL types in use, the latter situation is the most likely at least in existing or older DSL deployments.

Section 11.A.1 discusses some basic prerequisites for the multiuser detection problem, and then defines the problem mathematically. Complexity (except possibly in some cases with DMT as noted above) is then shown to be very large and not feasible. Section 11.A.2 then proceeds with a feasible approximation to ML detection derived by Wu-Fan-Cheong-Choi [15],[16],[17] and abbreviated the *soft canceler.* Some limitations to this canceller are observed as the number of crosstalkers increases, motivating Zeng's Canceler in Section 11.A.3. All the MUD methods tend to peform increasingly less well as the number of significant crosstalkers exceeds two, motivating simpler and higher-performance methods that require some level of coordination in Sections 11.3 and 11.4.

11.A.1 Basic Receiver Prerequisites for MUD

This subsection introduces some topics that will appear as difficult practical problems for the designer in the use of MUD *only.* Both synchronization issues and channel-identification issues are discussed. *It is important to note up front that these issues ONLY arise when there is NO COORDINATION.*

Figure 11.43 Use of W, B for MUD with GDFE processing.

Different DSLs may be asynchronous, even if of the same type. The actual DAC sampling clock may not be synchronized to a central office clock (even if the data are) and stuffing/robbing may be used to align data with such an asynchronous crystal. Fortunately, if a signal is strong enough to be a significant crosstalker, then at the very least its clock sampling phase can be derived by simple phase lock loops. The block model **H** will need to interpolate for each packet a model of the type:

$$y_k(t)_{t=iT'} = \sum_j h_{km}(t-jT_m) \cdot x_{m,j}$$

for each packet, so that **H** is a time-varying matrix with its values in each successive packet being different if there are clock differences between DSL users' sampling/symbol rates. Assuming that clock differences from packet to packet on any user are relatively small, a frequency domain linear-phase interpolation (as in [1]) or rotor is one mechanism by which to simply interpolate all the individual terms in the model **H** to the correct values for each block if the relative clock offset is known from basic phase-lock loops (PLLs) for all possible inputs. Two clocks that are relatively close in both frequency and/or phase may interpolate by the same factor until they have accumulated a significant phase difference that can be discerned. This text will not deal further with phase differentiation other than to say this is a problem well understood in the signal processing community, and there are wealth of methods for deriving two clocks that are close in frequency (see [18]). If the two DSLs are different types, clocks will be more easily discerned at widely disparate frequencies (but crosstalk reduction then otherwise becomes more complex).

Crosstalk types can be identified through spectral estimation of the noise/error signals on the line. Both the types of and numbers each of crosstalkers that are significant can be estimated through profiling. Profiling is discussed in [10], [19] at length. Figure 11.45 illustrates the basic concept. A state machine is

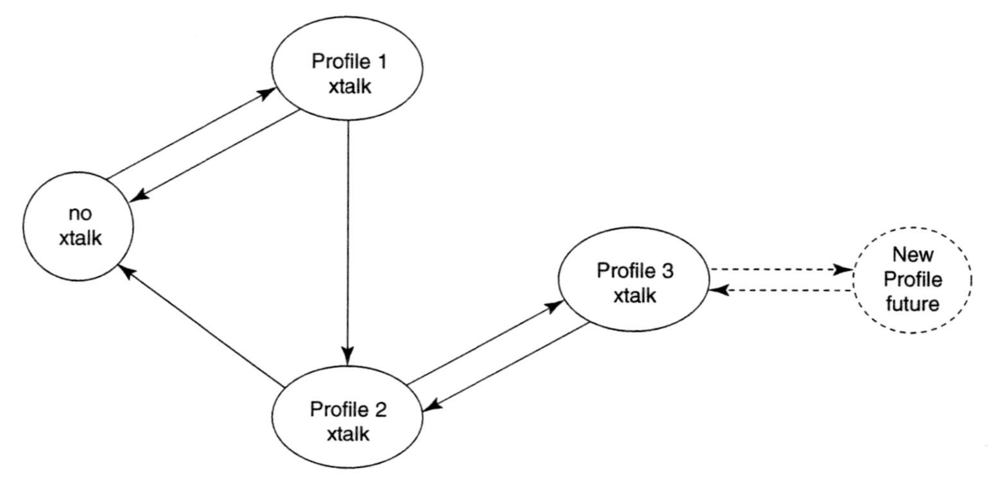

Figure 11.45 Profiling state machine.

constructed adaptively for each of the possible line states ever encountered by any DSL receiver. There is always one state that corresponds to no other DSLs. Other states may correspond to situations like 1 ADSL and 1 HDSL of significance to this line/receiver, 2 VDSLs of significance, 1 ISDN and 2 T1's of significance, and so on. Each of these services has a distinct bandwidth use that is defined by earlier spectrum management standards such as T1.417 [5]. Any time a new spectrum is measured for noise (see [1]) that significantly differs by more than some threshold related to desired data rate and margin on the loop of interest, a new state is added to Figure 11.45. The state of the channel is always either the newly defined state or some earlier determined state, depending on essentially the difference between the identified spectrum and that associated with each state. If the difference metric for any existing state is the smallest and below the threshold, the channel is defined to be in that state; then the number, type, and crosstalk coupling for the crosstalkers in that state has been previously defined (or must be determined for a new state, see Section 11.5), and PLLs are used to find timing differences.

PLLs and profiling are used to establish the channel matrix **H** for each packet. Once **H** is known, the maximum likelihood estimate of **X** is simply satisfies.

$$\min_{\mathbf{X}} \|\mathbf{Y} - \mathbf{H}\mathbf{X}\|^2,$$

a very easily written and understood concept that may take an eternity to compute if the packet length is large and the number of inputs crosstalking is significant. Maximum likelihood detection is the best possible MUD (for equally likely inputs) and often is very impressive in terms of its ability to reconstruct a given user in the presence of severe crosstalk. Figure 11.46(a) illustrates an example of a G.pnt home network signal and a VDSL signal that are on adjacent lines in a binder group. Both overlap in the 5–10 MHz frequency range. If the normal DMT VDSL detector is used, the margin versus length plot in the lower curve of Figure 11.44(b) occurs. Thus, G.pnt dramatically impairs normal VDSL performance, and no effort was made to ensure spectral compatibility, for instance, of these two signals. Thus,

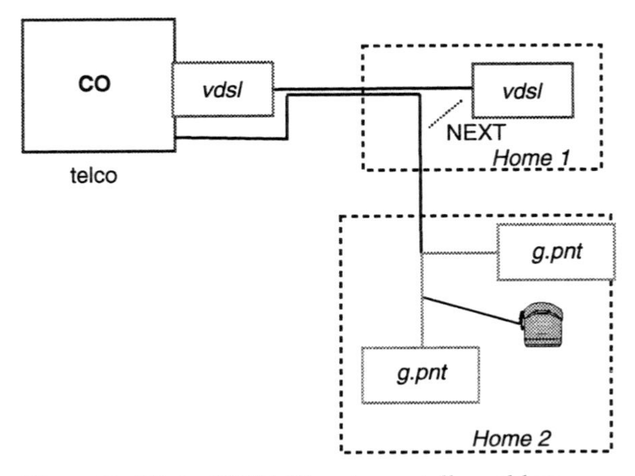

Figure 11.46(a) VDSL/G.pnt crosstalk problem.

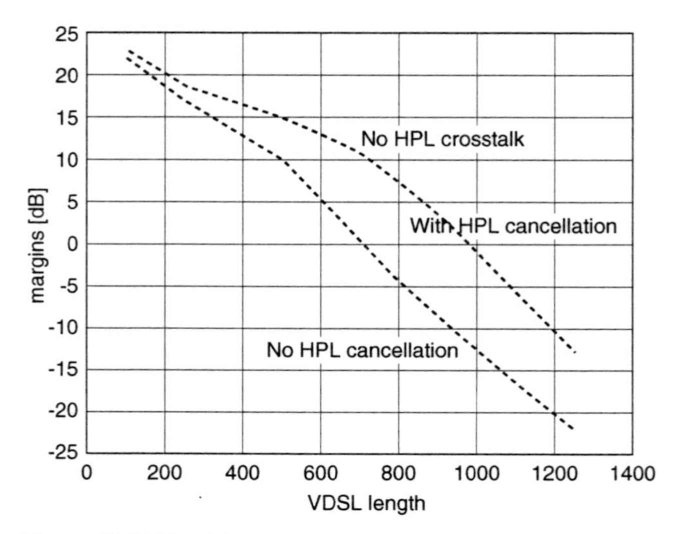

Figure 11.46(b) MUD ML detector for VDSL at 26 Mbps.

they are a good example for multiuser methods. The upper curves in Figure 11.46(b) correspond to no G.pnt crosstalk (obviously an upper bound on performance) and the ML detector. They are indiscernible in terms of VDSL performance, so the ML detector has a very large improvement in range. For more details, see [17].

11.A.2 Soft Cancellation

Unfortunately, multiuser ML detection becomes extremely complex for most practical situations of interest and provides only a motivation to search for approximate methods that do nearly as well. An aspect of ML detection often not well appreciated is that it minimizes the probability of a packet error as illustrated in Section 11.A.1. In practice, one might argue that minimizing the probability of error separately for each symbol in a packet might be better, and indeed such minimization leads to iterative construction of the best data symbol for each and every user sample within a packet. **Iterative decoding** recursively constructs the probability distribution of each and every element of the vector of all the users' input symbols in **X.** The channel matrix **H** and estimates of the other elements of the vector **X** are used to update an estimate of the probability distribution of any element in **X.** The iterative-decoding procedure proceeds sequentially through the elements of **X,** one by one, each time updating its probability distribution (which amounts to computing its average value and variance about that value in various approximations rather than computing the distribution itself). When each element has been processed once, the cycle continues through each element again. After about 5–10 passes, the distributions converge to a stationary setting, and then the **X** estimates themselves (not probabilities) are detected by finding those that maximize each probability distribution. For crosstalking channels, this process will closely approximate ML detection. Figure 11.47 illustrates the basic process.

Iterative decoding for crosstalk and ISI is somewhat more complicated than that for codes (turbo or LDPC) in that the actual probability distribution of symbol values (rather than for bits) is used. The process can be rederived for individual bits as well, but that is a natural extension of the algorithm here, and therefore is not presented here. The basic criterion for each symbol value is

$$\max_{\xi i}\left[p_{x_i/\mathbf{Y}}(\xi_i)\right]$$

where ξ_i is a place holding variable for the discrete values in the distribution of the ith symbol x_i in **X.** Such a detector is called a MAP detector (see [1]) and minimizes probability of symbol error. Realizing that the observed channel output **Y** is common to the criterion for all possible values of ξ_i, the criterion can be written instead as

$$\max_{\xi i}\{p_{\mathbf{Y}/x_i}(\xi_i)\cdot p_{x_i}(\xi_i)\} = \max_{\xi i}\{p_{ext}(\xi_i)\cdot p_{priori}(\xi_i)\}$$

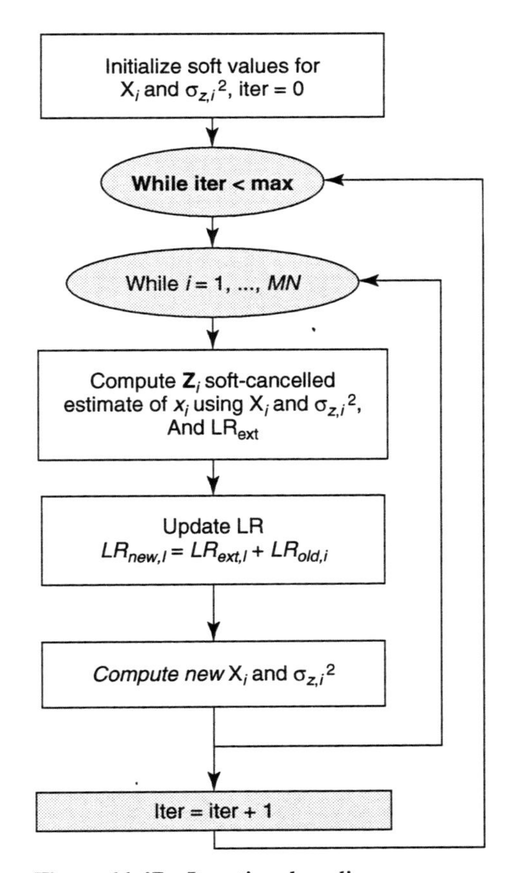

Figure 11.47 Iterative decoding procedure for soft cancellation MUD.

or in its preferred log likelihood form where $LR = \log(p)$

$$\max_{\xi i} \{LR_{\mathbf{Y}/x_i}(\xi_i) \cdot LR_{x_i}(\xi_i)\} = \max_{\xi i} \{L_{ext}(\xi_i) \cdot LR_{priori}(\xi_i)\}.$$

The log-likelihood follows a form that is iteratively computed in interative decoding as

$$LR_{new,i} = LR_{ext,i} + LR_{old,i}.$$

The idea is to get a new estimate of the LR by taking its old estimate and adding an update or "extrinsic" term that is based on the current probability distributions of all the other symbols.

The initializing old value can be the log of a uniform distribution. The extrinsic values can be computed at any iteration by soft cancellation. Soft cancellation observes that the matrix $\mathbf{G} = \mathbf{WH}$ may be rewritten as $\mathbf{G} = [\mathbf{g}_1\, \mathbf{g}_2 \ldots \mathbf{g}_{MN}]$ so that $\mathbf{WHX} = \sum_{i=1}^{MN} \mathbf{g_i} x_i$. If all the other symbols were known exactly, then a "hard" canceler would produce

$$\mathbf{Z}_i = \mathbf{WY} - \sum_{j \neq i} \mathbf{g}_j x_j = \mathbf{g}_i x_i + \mathbf{N}.$$

MAP detection would be simply finding the value of $x_i = \xi_i$ that minimizes $\|\mathbf{Z}_i - \mathbf{g}_i \xi_i\|^2$, essentially a handful of computations, or perhaps a simple slicing. However, the other values of $x_{j \neq i}$ are not known, so hard cancellation is not possible—any kind of ordering and hoping for correction decisions is prone to severe error propagation should any of the previously estimated values be incorrect. Thus, the objective becomes computing the distribution or LR of symbol x_i from the distributions (LRs) of all the other symbols and the old distribution (LR) of this symbol. This can actually be exactly computed, but with enormous effort so soft-cancellation makes an approximation where the soft or average value of each $x_{j \neq i}$ is used in the cancellation rather than a poor hard value. Only the $LR_{new,i}(\xi_i)$ is computed, and no decision is made (yet). Given $LR_{old,i}$ from the initial uniform distribution assumption, or from a previous interation of the algorithm once started, the decoder then needs to compute only $LR_{ext,i} = LR_{\mathbf{Y}/x_i}(\xi_i)$. The decoder computes the average or soft value of any symbol as $\chi_i = E[x_i]$ from the current distribution estimate. Then this value is used to compute the soft value for

$$\overline{\mathbf{Z}}_i = \mathbf{WY} - \sum_{j \neq i} \mathbf{g}_j \chi_j$$

and also the variance about the soft value

$$\sigma_{z,i}^2 = \sum_{j \neq i} \|\mathbf{g}_j\|^2 \sigma_{x,j}^2 + \sigma^2$$

where $\sigma_{x,i}^2$ is the variance of symbol x_i as computed from its most recent distribution (LR). Consequently the extrinsic LR is then approximated by

$$LR_{ext,i}(\xi_i) \approx -\frac{1}{2}\log\left(2\pi\sigma_{z,i}^2\right) - \frac{1}{2\sigma_{z,i}^2}\|Z_i - \mathbf{g}_i \cdot \xi_i\|^2 .$$

Thus, the GDFE structure is used for soft cancellation with the average values (based on iterated distributions) used instead of previous decisions. The new LR is computed by adding the extrinsic to the old LR, and then this new LR can be inverted to the probability distribution, from which the new soft or average value $\mathcal{X}_i$ is computed for use in subsequent iterations. This process converges in 5–10 passes, despite the approximations.

Some Practical Cautions

If the soft value $\mathcal{X}_i$ exceeds the maximum value for the constellation significantly, it should be discarded and not used as clearly there is a "bad" value in the distribution that would cause this. Thus, for the pass of all the other symbols that follows, this value is not used in the soft cancellation, and a value of zero contribution to the variance is used. This has been found to speed convergence of the algorithm significantly.

In designs that may not coordinate, it may still be of value to leave some parts of different user's bands nonoverlapping or equivalently to have a symbol or two from one or more of the users in any packet otherwise free of interference. This helps "get the iterative decoder going" in the right direction. Usually the amount necessary is a tiny fraction of the total user bandwidth.

With many, many crosstalkers, iterative decoding and even ML tend to perform poorly. This is not a problem with the soft canceler, but rather the basic information theory fundamentals of the system being such that multiuser detection can not be expected to work. The total transmitted information exceeds the capacity region of the system, or at least is structured so poorly that low-probability-of-error detection of all signals is not possible. Thus, soft cancellation can be expected to work when fundamental information and communication theory limits for the ML detector suggest that near-error-free performance is possible.

Initialization

The very first soft values for cancellation would be based on uniform distributions for the input and therefore somewhat useless. Clearly, more conventional receiver design could be used to obtain estimates for the soft values on the first iteration. This can be achieved by processing the channel output according to

$$\mathcal{X}_0 = \mathbf{W}\mathbf{G}^{-1}\mathbf{Y}$$

and then limiting the resulting values for any element of $\mathcal{X}_0$ to the largest point in the constellation if it is large and then (and otherwise) using the values as average values for the soft canceler. The initial estimate of $\sigma_{z,i}^2$ can be found from the diagonal elements of $\sigma^2\mathbf{W}\mathbf{G}^{-2}\mathbf{W}^*$.

11.A.3 Simplied Cancellation by Grouping of Excess Dimensions

Although a large simplification with respect to ML detection, soft-cancellation's iterative decoding can still be excessively complex for reasonable implementation. Further complexity reduction profits from specific structure of the specific channel that may be obvious in certain situations. Examples include HDSL or SDSL crosstalk into ADSL and also ADSL into VDSL. This concept of this subsection was motivated by Dr. Steve Zeng, but represents a generalization of the concepts in his Stanford dissertation. When viewed in a general perspective, one also finds the work of Pederson and Falconer [20], Pal et al. [21], Im and Werner [22], and Voyan Technology [23] to also be examples of the same general principle, although the general previous view of this work was that each individually was different.

Returning to the GDFE, the settings are a function of ordering of the various elements of the vector $\boldsymbol{X}$. One possible grouping that can lead to simplification in many cases is to segregate all the values of the channel of interest into a vector $\mathbf{X}_1$ and all the rest into another vector $\mathbf{X}_2$. The basic idea is to try to estimate without iterative methods the net effect of the elements $\mathbf{X}_2$ in terms of their interference into $\mathbf{Y}$ and cancel that interference from $\mathbf{Y}$. Then $\mathbf{X}_1$ is estimated alone. Clearly, a higher-level iterative decoding could be used once $\mathbf{X}_1$ is estimated to produce a second better estimate of $\mathbf{X}_2$ and so on. This kind of method is effective when the second signal has components that are relatively free of crosstalk. Examples occur in DSL when a system such as SDSL or HDSL introduces NEXT that has upperband sidelobes that are captured by the higher sampling rates of ADSL. These bands may be significantly larger than the ADSL signal being transmitted over a long line. Thus, a reasonable estimate of the entire HDSL or SDSL signal may be constructed from just viewing the upper band. If that estimate is accurate, it can be used to reconstruct the NEXT over the entire band of the ADSL signal and eliminate it, thus allowing considerably higher performance.

The channel is modeled specifically as

$$\mathbf{Y} = [\mathbf{H}_1 \, \mathbf{H}_2] \begin{bmatrix} \mathbf{X}_1 \\ \mathbf{X}_2 \end{bmatrix} + \mathbf{N}$$

where H_1 is the channel from X_1 to output, and H_2 is the channel from X_2 to output.

Figure 11.48 redraws the GDFE with the specific separation into the two components. A GDFE estimates $\mathbf{X}_2$ and that estimate is then filtered by $\mathbf{H}_2$ prior to its subtraction from $\mathbf{Y}$ before a second GDFE based on the model $\mathbf{Y} = \mathbf{H}_1\mathbf{X}_1 + \mathbf{N}$ refines the estimate of $\mathbf{X}_1$. If the first signal is a DMT signal, then the second GDFE simplifies to essentially just a "pass-through" if the IFFT and FFT are absorbed into the channel model for $\mathbf{H}$. There are various choices for the feedback section of the first GDFE, and $\mathbf{B}_2 = \mathbf{I}$ is the typical choice for this first GDFE and equivalent to those made by Falconer and Pederson, Werner, and Zeng. Zeng investigated for this particular choice to determine whether using decisions, or simply making no decision before passing to the matrix $\mathbf{H}_2$, works better. Neither Falconer and Pederson, Pal, nor Werner addressed this concept and in fact did not show that

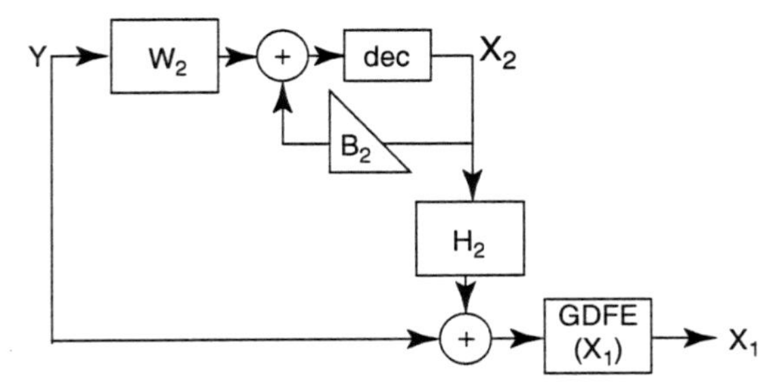

Figure 11.48 Simplified canceler configuration.

decisions might be made at this point, choosing instead to find a matrix corresponding to $\mathbf{W}_2\mathbf{H}_2$ without the possibility of an intermediate decision. Voyan Technology [23] appears to have been so cognizant, and released performance results that are similar to Zeng's without a detailed description of the processing. Zeng found in some cases making decisions works better whereas in other cases it does not. The approach taken here with the GDFE notes that $\mathbf{B}_2$ can be any lower triangular (monic) matrix and the performance will then be better. However, as the number of users increases beyond two, that first GDFE has increasing problems with error propagation; and even with no decisions, the system often breaks down. Some level of higher-level interative decoding may be used, but generally it also breaks down as does ML detection with large numbers of users.

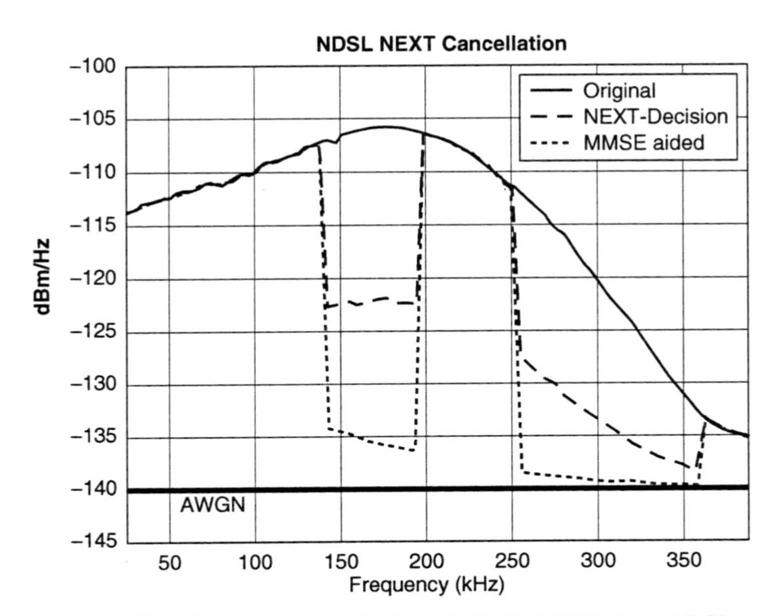

Figure 11.49 Illustration of reduction of HDSL NEXT into ADSL in the 150 to 200 kHz band that accrues to choosing X_2 to correspond to signals in the 250 to 350 kHz band. Effect of both decisions and no decisions with $B_2 = 0$ is shown.

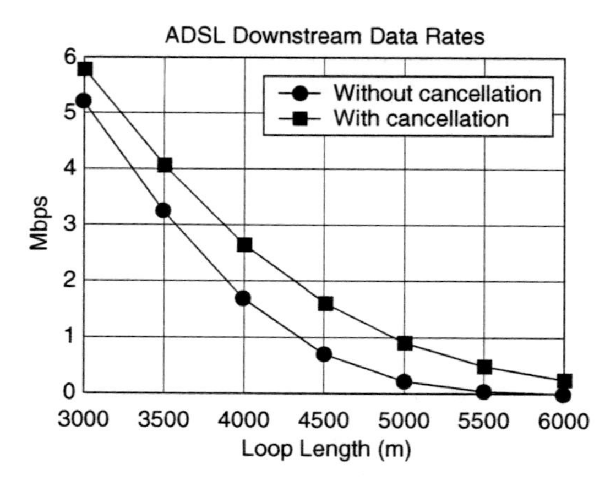

Figure 11.50 Illustration or rate/range improvement for situation illustrated in Figure 11.24.

Zeng's final conclusion was that the system works, perhaps with the need of some higher-level iterative decoding as in the soft canceler, as long as fundamental ML would also work. However, this is usually just for one additional crosstalk signal, thus limiting iterative decoding cases. To perform better, some level of coordination is necessary with many users and the gains may be larger.

Previous investigators failed to realize the full potential of the general viewpoint in Figure 11.13. For instance, Figure 11.49 from Zeng [23] illustrates the effect of using part of an HDSL signal out of band to cancel its in-band content. In this case, $\mathbf{X}_2$ corresponds to signals only in one of the two nulled bands shown and not the entire HDSL signal. Although the method has a very good elimination of signal in that particular band, there are other choices of $\mathbf{X}_2$ that would lead to yet more complete cancellation of the HDSL signal. Similar observations would occur for all previous work, but nonetheless Figure 11.49 illustrates the concept works very well for one dominant crosstalker. One would find that as the number of crosstalkers increases beyond two, it is exceedingly difficult to recover $\mathbf{X}_2$ because even ML detection will not do well in this case. Figure 11.50 for this same situation illustrates the range/rate improvement for the situation shown in Figure 11.49. Again, the improvement is very significant and similar improvements have been introduced by Voyan Technology [23].

REFERENCES

[1] T. Starr, J. M. Cioffi, and P. Silverman, *Understanding Digital Subscriber Lines* (Upper Saddle River, NJ: Prentice-Hall, 1999).

[2] Deployment of Advanced Telecommunications Report. Federal Communications Commission (USA) 00-290, August 2000.

[3] S. Verdu, *Multiuser Detection* (UK: Cambridge University Press, 1998).
[4] T. Cover and J. Thomas, *Elements of Information Theory* (New York: Wiley, 1991).
[5] American National Standard, "Spectrum Management for Loop Transmission Systems," T1.417-2001, New York: ANSI.
[6] C. Zeng, C. Aldana, A. Salvekar, and J. M. Cioffi, "Crosstalk Identification in xDSL Systems," *IEEE Journal on Selected Areas in Communications* 19, no. 8, (August 2001), 1488–1496.
[7] P.P. Vaidyanathan, *Multirate Systems and Filter Banks* (Englewood Cliffs, NJ: Prentice Hall, 1993).
[8] Carlos Aldana, Atul Salvekar, Jose Tellado, and John Cioffi, "MAP Crosstalk Profile Matching for Multicarrier Systems," IEEE International Conference on Telecommunications, 2001, Helsinki, Finland, June 2001.
[9] Carlos Aldana, and John Cioffi, "Channel Tracking for Multiple Input, Single Output Systems Using EM Algorithm," IEEE International Conference on Communications, 2001 (pdf format).
[10] Atul A. Salvekar, Carlos Aldana, Elisabeth de Carvalho, and John Cioffi, Crosstalk Profile Detection for use in Multiuser Detection," ICC 2001, Helsinki, Finland, June 2001.
[11] J. M. Cioffi and G. D. Forney, Jr., "Generalized Decision-Feedback Equalization for Packet Transmission with ISI and Gaussian Noise," in *Communication, Computation, Control, and Signal Processing* (a tribute to Thomas Kailath) Ed. A. Paulraj, V. Roychowdhury, and C. Schaper, Ch. 4. (Boston: Kluwer, 1997).
[12] G. Ginis and J. M. Cioffi, "Vectored-DMT: A FEXT Canceling Modulation Scheme for Coordinating Users," Proceedings of IEEE International Conference on Communications 2001, Vol. 1, Helsinki, Finland, pp. 305–309, June 2001.
[13] G. Ginis and J. M. Cioffi, "A Multi-user Precoding Scheme Achieving Crosstalk Cancellation with Application to DSL Systems," Proceedings of the 34th Asilomar Conference, Pacific Grove, CA, pp. 1627–1631, October 2000.
[14] J. Cioffi, EE 379c textbook, "Digital Transmission Theory, Volume I," http://www.stanford.edu/class/ee379c/.
[15] Zining Wu, *Coding and Iterative Detection for Magnetic Recording Channels* (Boston: Kluwer, 2000).
[16] K. W. Cheong, "Multiuser Detection for DSL Applications," Ph.D. dissertation, Stanford University, December 2000.
[17] K. W. Cheong, W. J. Choi, and J. M. Cioffi, "Multiuser Soft Interference Canceler via Iterative Decoding for DSL Applications," *IEEE Journal on Selected Areas in Communications*, 19, no. 2 (February 2002).
[18] S. L. Marple, Jr., *Digital Spectral Analysis with Applications* (Upper Saddle River, NJ: Prentice Hall, 1987).
[19] A. Salvekar, "State Detection Techniques for Digital Subscriber Line Systems," Ph.D. dissertation, Stanford University, January 2002.
[20] B. R. Pederson and D. D. Falconer, "Minimum Mean Square Error Equalization in Cyclostationary and Stationary Interference-Analysis and Subscriber-Line Calculations," *IEEE Journal on Selected Areas in Communications* 9, no. 6 (August 1991), 931–940.

[21] D. Pal, G. N. Iyengar, and J. M. Cioffi, "A New Method of Channel Shortening with Applications to Discrete Multi-Tone (DMT) Systems," Proc. 1998 IEEE International Conference on Communications, pp. 763–768, May 1998.

[22] G. H. Im, and J. J. Werner, "Bandwidth-Efficient Digital Transmission over Unshielded Twisted-Pair Wiring," *IEEE Journal on Selected Areas in Communication (JSAC)* 12, no. 9 (December 1995), 1643–1655.

[23] Voyan Technology, "Crosstalk Compensation for ADSL" *ANSI Contribution T1E1.4/2000-251R1,* Dallas, TX, January 2001.

[24] W. Yu, W. Rhee, J. Cioffi, and S. Boyd, "Iterative Water-filling for the Vector Multiple Access Channel," submitted, *IEEE Transactions on Information Theory.* See also *ANSI Contribution T1E1.4/2001-200R4,* Greensboro, NC, (November 2001).

[25] W. Yu and J. Cioffi, "Sum Capacity of Gaussian Vector Broadcast Channels," submitted, *IEEE Transactions on Information Theory.*

[26] J. Cioffi et al., "Scope and Mission for Dynamic Spectrum Management," ANSI *T1E1.4 Contribution T1E1.4/2001-188R5,* Greensboro, NC, November 8, 2001.

[27] C. Zeng and J. Cioffi, "Crosstalk Cancellation in xDSL Systems," *IEEE Journal on Selected Areas in Communications* 19, no. 2 (February 2002).

[28] ITU G.993.1 Standard, "Very-High-Speed Digital Subscriber Line Foundation," 2002.

[29] M. Tsatsanis and I. Kanellakopoulis, "Identification of Crosstalk Using MIB-reported Data," *ANSI Contribution T1E1.4/2001-278,* Greensboro, NC, November 5, 2001.

[30] J. Cioffi, J. Lee, and S. T. Chung, "10MDSL beyond All Goals, and Spectrally Compatible with ADSL and VDSL, from CO or RT," *ANSI Contribution T1E1.4/2002-129,* Atlanta, GA, April 8, 2002.

[31] G. Sherrill, J. Cioffi, S. T. Chung, J. Fang, and W. Yu, "Response to 2001-273R1 Using Measured Verizon DSL SNRs," *ANSI Contribution T1E1.4/2002-069,* Vancouver, BC, February 18, 2001.

[32] K. Kerpez et al., "Response to 2001-273R1 using Telcordia DSL Analysis," *ANSI Contribution T1E1.4/2002-063,* Vancouver, BC, February 2002.

[33] K. Kerpez, "Composite PSD Template for Upstream VDSL that Accommodates 10 MDSL ," *ANSI Contribution T1E1.4/2002-100,* Atlanta, GA, April 8, 2002.

[34] M. Tstatsanis, "Efficient Use of Asymmetric Masks for 10MDSL Symmetric Services," *ANSI Contribution T1E1.4/2002-118,* Atlanta, GA, April 8, 2002.

[35] J. Cioffi and J. Fang, "Proposed Channel Modeling Test for DSM," *ANSI Contribution T1E1.4/2002-172R3,* Atlanta, GA, November 18, 2002.

[36] T1E1.4 Draft DSM Report, T1E1.4/2002-018Rx.

[37] S. T. Chung, S. J. Kim, and J. Cioffi, "On the Existence of a Nash Equilibrium in the Frequency Selective Interference Channel," *ISIT,* 2003.

[38] A. Leshem, "On the Existence of Nash Equilibrium Points for the Gaussian Interference Game," submitted for publication.

[39] S. T. Chung, J. Lee, et al., "On the convergence of Iterative Water-filling in the Frequency Selective Interference Channel," *ISIT,* 2003.

CHAPTER 12

CUSTOMER PREMISES NETWORKING

One end of each DSL connects to the customer premises network. The premises may be a business office in a commercial building, a business office in a home, a residential consumer's home, a unit in an apartment building, or even a public kiosk located in a shopping mall. In some cases, the customer premises network is simply the phone line connected directly to a PCI card installed within a personal computer (PC), or a phone line connected to a DSL modem with a 10BASE-T Ethernet interface to a single PC.

Large business installations may consist of one or more DSLs connecting to a router with firewall functions that then connect to an Ethernet local area network (LAN). IEEE 802.3 (CSMA/CD 10BASE-T), Fast Ethernet (100bT), Gigabit Ethernet, and 10 Gigabit Ethernet LANs provide access for many PCs and servers to access the Internet or an intranet via the router. Ethernet LANs are thoroughly addressed in other books,[1] so this book focuses on customer premises networks for residential consumers, small business offices, and home offices.

The home provides the opportunity of a huge potential market with the challenges of demanding very-low-cost, self-installation by a novice home owner, and serving many of the following applications:

- PCs, information appliances, printers, scanners, digital camera
- Internet access: DSL, cable modem, satellite dish, wireless
- Video: TVs, cable, satellite dish, VCR, DVD player, video camera
- Audio: stereo, CD recorder/player, MP3 recorder/player, radio, AC3
- Game machines
- Intercom, baby monitor, doorbell
- Surveillance camera
- Telephones: wired and wireless, facsimile, answering machine, caller ID[2]
- Voice-band modems
- Alarm: fire, intrusion, water, freeze, carbon monoxide, power loss, medical
- HVAC (heat, ventilation, air conditioning) control
- Energy management (utility peak load control), vacation setting
- Telemetry: gas, water, power meter reading
- Control: lights, window covering, garage door, yard-watering, pool, door locks
- Appliances (e.g., refrigerator door open, remote diagnosis of dishwasher trouble)
- Exercise machines
- Clocks (synchronize all clocks)
- Community alert: weather, pollution, flood, fire, water restriction
- Home weather station
- Vehicles and lawn mower

George Jetson (the futuristic cartoon character) might have accessed his personal weather station from his lawn mower, but in the real world there is one leading application for mass-market home data networking: *high-speed Internet access* by multiple devices within the home. A market study conducted by Intel in 1998 found that 86 percent of homes with two or more PCs are connected to the Inter-

---

[1]Also see www.ieee802.org.

[2]Note that voice may include voice-over IP that is networked as data, in addition to traditional voice conveyed as an analog signal.

net. There are about 20 million multi-PC homes in North America. Connecting only one PC to the Internet soon leads to the desire to connect the remaining PCs or information appliances to the Internet. Once the home data network is present, other applications, such as house-wide stereo and lighting control, will follow. Other leading applications for data networking in the home are printer and file sharing. An example of file sharing is access of an address list stored on one PC from another information appliance in the home. File sharing is also useful for audio and video: a family collection of CDs, MP3s, and videos accessible by every stereo and TV in the house. With a wireless LAN, information appliances may be located anywhere within the home and even outside near the home (such as a pool-side laptop computer to visit a baseball team's Web site while listening to a game on radio).

In the ultimate scenario, multimode information appliances could use the home network to access multiple wide area networks (WANs): a satellite dish for video, DSL for the Internet, and the public telephone network for phone calls. The term *information appliance* applies to any device that has intelligence (an embedded microcomputer/controller) and is connected to the home network. Examples of information appliances are PCs, a kitchen display/keyboard terminal, servers, printers, home alarm system, and a garage door opener (if connected to the home network). A device called a *residential gateway* or an IAD (integrated access device) interfaces to the DSL, cable modem, or satellite dish on the WAN side, and the home LAN (e.g., HomePNA [Home Phoneline Network Alliance], 10baseT, 802.11b wireless LAN) on the premises side.

There have been many attempts by various companies and consortia to create a mass-market, all-purpose home network that does everything from controlling light dimmers to distributing video throughout the home. Thus far, the universal home network has not replaced the status quo using separate ad hoc home networks dedicated for each application: video (coax), telephone (twisted pair), intercom (separate twisted pair), security alarm (often yet another separate set of wires), light control (signals modulated over inside power wires, such as X10), and data (dedicated wires for10baseT). It is not clear if there ever will be an ideal unified home network for all applications, but some degree of integration is likely. The recent development of the Konnex standard in Europe is an encouraging attempt at a unified home networking standard.

12.1 HOME NETWORK MEDIA

The general rule for home networks is that it must be self-installed by a novice home owner, and this means no new wires run between rooms. There are some persons who have the technical aptitude to install new Category 5 wiring themselves or the willingness to hire a professional installer. In the past, these persons have often installed an Ethernet LAN using twisted wire pairs between rooms, as is common in business offices. However, with the advent of low-cost and high-performance wireless LANs and other new home network techniques, there is no need for even the most sophisticated homeowner to undertake the rewiring of an existing house.

The following alternative home networking media are discussed:

- Inside telephone wires
- Inside coax cable
- Inside AC power wiring
- Dedicated data wiring
- Radio LAN
- Infrared

12.2 INSIDE TELEPHONE WIRING AND ADSL

As shown in Figure 12.1, the telephone distribution cable connects to the customer premises via a *drop wire* that typically consists of two to four pairs of wire. Starting in the year 2000, the industry widely adopted self-installation by customers using the splitterless customer premises configuration shown in Figure 12.2. The splitterless configuration uses the same pair of inside wires for POTS and ADSL data; this pair of wires is known as *line-1* and connects to pins three and four of the telephone wall jacks. Despite being called "splitterless," this configuration typically includes a line-sharing splitter at the central office end of the line.

The splitterless customer premises configuration was introduced with ITU Recommendation G.991.2 (g.lite), which contained a fast-retrain function to permit

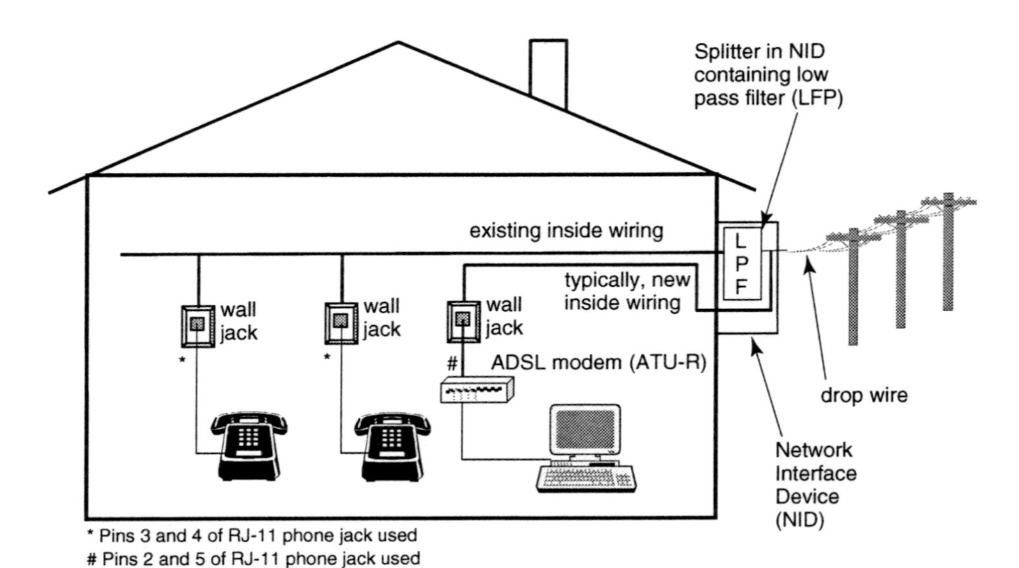

Figure 12.1 Customer premises configuration with shared splitter.

the ADSL modem to quickly adapt to changes in the transmission channel resulting from telephones on the same line going on-hook or off-hook. During the development of G.991.2, it was discovered that the fast-retrain was not adequate to assure good voice-band and data performance when some types of phones were directly connected to the same line used by the ADSL. Thus, it is necessary to place a low-pass in-line filter at the connecting cord to every phone and answering machine. Subsequently, it was also discovered that the provision of in-line isolating filters not only assured good performance for G.991.2, but also the full-rate G.991.1 (g.dmt) and T1.413 standard equipment. So, what was originally intended as a g.lite feature, has evolved to a general practice for splitterless configurations using G.991.1 and T1.413 full-rate equipment to facilitate potential future upgrade to service rates above 1.5 Mb/s.

Prior to the year 2000, the predominant configuration for the customer premises required a splitter located at the point where the telephone wires entered the premises. For single family homes, the splitter is often placed within a wiring junction box called the network interface device (NID), located on the outside wall of the premises. As shown in Figure 12.1, the splitter consists of a low-pass filter (LPF) in series with the existing inside telephone wiring connected to the *line 1* phones within the premises. A separate pair of inside wires connect the ADSL modem to the telephone line; often a new CAT-5 wire is installed. New wire is often installed because a spare pair of inside wires is not available. Even if a spare wire pair is available, the wire quality may be inadequate.

The splitterless configuration is now practiced for approximately 90 percent of ADSL installations because it avoids the cost and inconvenience of a technician visiting the customer premises. The customer simply plugs the ADSL modem into any telephone wall jack, inserts a small in-line filter by every telephone and answering machine, and runs the setup program on their computer. The customer

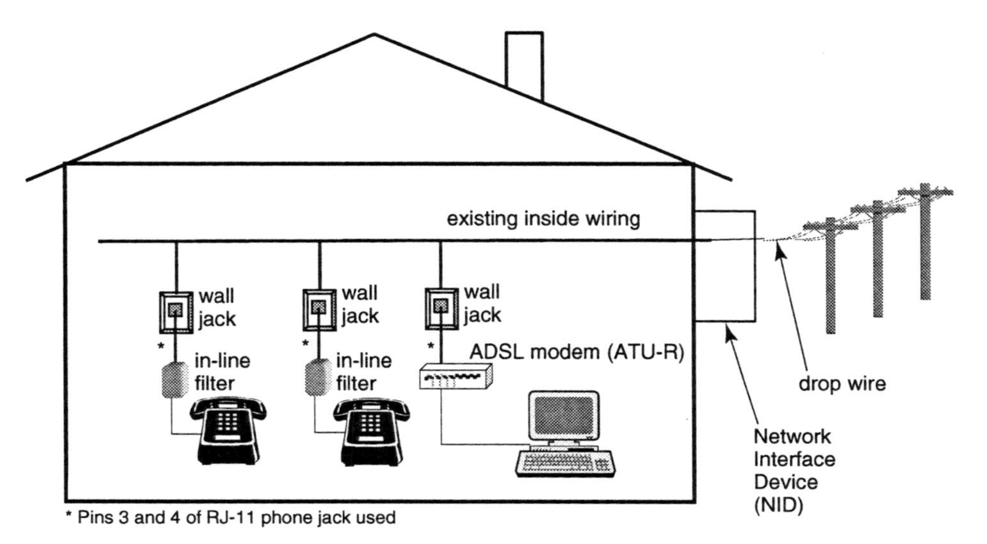

Figure 12.2 Splitterless customer premises configuration.

self-installation success rate is approximately 95 percent. A portion of the cases where self-installation fails is due to poor quality inside wiring that is susceptible to noise ingress and also presents many short bridged taps at the customer-end of the line. Thus, the shared-splitter configuration continues to be applied in a few cases to overcome the impairments presented by the existing inside wiring. The splitter at the NID isolates (at higher frequencies) most of the inside wiring from the newly installed CAT5 wire connecting to the ADSL modem. The configuration with the splitter at the NID is also preferred for customer installations having a reporting security alarm service. There have been reports of splitterless ADSL installation causing malfunction of alarm systems that automatically dial a voice-band modem call to the alarm service station. Further information on customer premises configurations is provided in DSL Forum TR-007.

12.2.1 In-line Filter (Microfilter)

T1E1.4 developed U.S. standard T1.421-2001 for customer premises in-line filters (also known as microfilters) to isolate telephone instruments from DSL equipment sharing the same inside wire pair. The in-line filter consists of a low-pass filter (LPF) with a RJ11 jack connected to one side of the filter and a RJ11 plug connected to the other side (Figure 12.3). The in-line filter plugs into a wall jack, and the telephone connecting cord plugs into the RJ11 jack within the in-line filter. Typically, a separate in-line filter is used for each telephone, answering machine, voice-band modem, or other telephone-type equipment. However, an Y-adapter would allow more than one telephone instrument to share the same in-line filter. Every telephone instrument connected to the DSL line must be connected via an in-line filter. The in-line filters permit only voice-band signals to and from the telephone instruments. This solves two problems resulting from the telephones and the DSL sharing the same line: (1) noise audible in the telephones due to DSL signals being cross-modulated by the telephone, and (2) impaired DSL transmission due to noise from the telephone equipment and changes in line impedance when the telephone goes on-hook or off-hook.

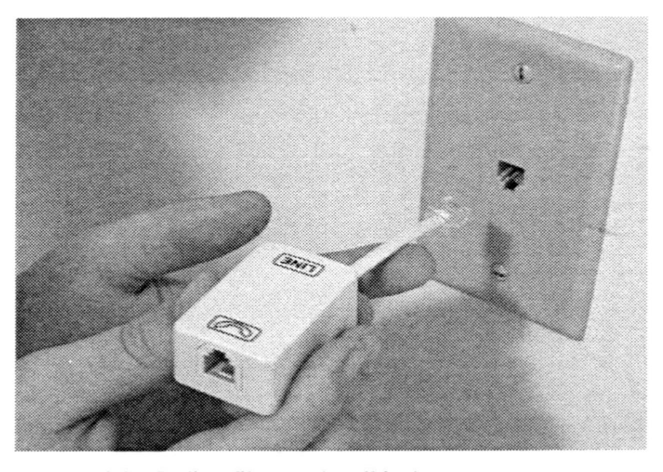

Figure 12.3 In-line filter and wall jack.

Without a filter, approximately one-third of telephones will exhibit objectionable hissing in the ear piece when connected to an ADSL line. The impedance characteristics of telephone models vary widely, with many telephones presenting a high impedance during on-hook, and as low as 10 ohms at ADSL frequencies when off-hook. An ADSL transceiver that has adapted to the line characteristics is suddenly presented with radically different channel characteristics when a phone goes on- or off-hook, and thus is forced to cease transmission while it readapts to the new line characteristics. This is particularly troublesome if pulse-type dialing is used where the line repeatedly changes between on-hook and off-hook. Furthermore, a very large noise burst can occur when a telephone goes off-hook while the ringing signal is near its maximum voltage; this event is called a *ring-trip*.

In-line filters were originally created for use with ITU G.992.2 (G.lite), but the in-line filters are useful for G.992.1 (G.dmt), T1.413, and certain proprietary DSLs. HomePNA usually does not require in-line filters, but the filters can help in cases where the telephone equipment causes excessive noise at high frequencies. Several 900 MHz and 2.4 GHz cordless phones emit noise that interfere with DSL. In some cases, power supplies or electric motors (for an answering machine tape drive) in the telephone equipment can generate high frequency noise.

The simplest in-line filter design is non*symmetrical*, meaning that one side of the filter must face the telephone and the other must face the wall jack. Nonsymmetrical filters (also called polarized filters) are clearly marked "line" and "phone" to indicate the correct orientation. The in-line filters are designed to plug directly into the wall jack to isolate the connecting cord from the house wiring. This improves the high frequency characteristics of the premises wiring by effectively shortening the length of the wiring stub. Because telephone connecting cords have a plug at both ends, it is possible to connect the in-line filter at phone-end of the connecting cord. This would result in the polarized filter facing the wrong direction, and thus diminishing the quality of service. In the case of a phone cord already connected to a wall jack behind a bookcase, gaining the necessary access to the wall jack could be very inconvenient. As a result, some in-line filters are designed to be *symmetrical*, meaning that they work as well regardless of which side of the filter faces the phone. Symmetrical filters are an advantage for people with jacks behind large heavy objects such as bookcases. This design is also helpful for others because there is no way to connect the filter backward. In-line filters are also packaged within a shim-plate adapter that fits between a wall-phone and the standard wall plate.

In-line filter design must simultaneously assure:

- High voice-band return loss
- Low voice-band insertion loss
- Low resistance (25 ohms or less) to prevent "current hogging" when different types of phones are in use
- High stop-band attenuation above 25 kHz
- Nearly constant high-frequency impedance presented to the line when phone is on-hook and off-hook
- Maintain a high impedance from 25 kHz through 10 MHz presented to the line

- Capable of withstanding high voltage surges
- Very low cost (typically, less than $1 in volume purchase)

Detailed Discussion of T1.421—The In-Line Filter Standard

All these attributes must be maintained when connected to any type of telephone, and when several in-line filters are connected to the premises wiring. The T1.421-2001 in-line filter standard requires that the specifications be met when up to five in-line filters are connected in parallel. In practice, tolerable service will often be achieved when more than five in-line filters are connected. In-line filters are designed to let Caller ID signals pass through. The in-line filter characteristics are specified up to 10 MHz to help provide for future DSL technologies and local phone-line networking (e.g., HomePNA). One in-line filter design employs a third-order Chebychev II symmetrical design.

T1.421 includes the terms, purpose, and the scope of the testing, on-hook testing, off-hook testing, and high-band frequency testing. The physical characteristics and packaging are also discussed. We concentrate on the portions of the standard where the filter testing takes place.

The purpose of the standard is to present the electrical and physical characteristics of an in-line DSL filter. The filter is then used to protect the voice-band equipment from the high-frequencies generated by the digital data over voice in the 25 kHz–10 MHz range. Filters are also used to protect data over voice equipment from impedance changes and other impairments caused by the off-hook and on-hook transitions of voice-band equipment.

The in-line DSL filter standards benefit users and service providers of voice/data services by simplifying interpretation of customer provided equipment (CPE) and by defining the parameters of a single filter up to a total of five filters. In installation of more than five filters, it is necessary to add dynamic filters that exhibit all of their filtering only when the associated telephone set is off-hook. The shared use of the telephone line by voice and DSL equipment can cause some serious operating problems. Some of these problems include:

1. Reduced DSL amplitude
2. Nonlinearity of POTS devices causing intermodulation distortion
3. DSL signals causing audible POTS interference on POTS lines
4. DSL impedance mismatches caused by POTS equipment
5. Unterminated wire stubs (bridged tap) acting like a filter and degrading performance

Attenuation distortion is the profile of loss versus frequency across the channel bandwidth. The loss is measured at any frequency in the 200 Hz–2.8 kHz band, as compared with a 1004 Hz reference.

Another very common term found in telecommunications is the decibel. The decibel (dB) is a unit of signal strength that expresses the relationship of the received signal to the signal generator (test set) or source. It is a unit of measurement that expresses a reference point and another point, above or below, the reference. A dB is equal to 20 times the log of the ratio between two numbers. A ratio of 10 is 20 dB, a ratio of 100 is 40 dB, and a ratio of 1,000 is 60 dB, and so on. If a filter has a –40 dB noise reduction, the ratio between the circuit output noise and the input noise is 40 dB or 100. You may also see the terms dBm, and dBmV used. A dBm is the power, in dB, related to a one *milliwatt* (thousandth of a watt) reference. Often seen in the measurement of signal levels is dBmV. This is the reference to one *millivolt* (thousandth of a volt)

The in-line DSL filters must comply with the caveats regarding Caller ID and Voice Band Data compatibility. This compatibility ensures that you can transmit data, including V.90 modems and fax machines in both the on- and off-hook states. Now let's take a look at the T1.421 standard requirements. We begin with on-hook testing.

On-Hook Voice Band (200 Hz–2.8 kHz) Requirements

A phone is *on-hook* when it is *not* being used to make a call. These requirements are for on-hook transmission of such services as calling number (Caller ID) delivery, and Message Waiting Indicator. The standard calls for a single filter and then four bridged filters, for a total of five filters.

On-Hook Voice Band Insertion Loss and Distortion. The tests performed in Section 5 of T1.421 are intended to ensure that the information sent, while the phone is on-hook, will pass through the filter to the Caller ID box without error. On-hook transmission includes such information as calling number (Caller ID) and Message Waiting Indicator. Filter testing involves both a single filter and five filters to simulate a typical home with multiple extension phones or other devices. Testing is conducted within the POTS telephone band (200 Hz to 2.8 kHz)

These tests are concerned with insertion loss and insertion loss distortion. **Insertion loss** is the difference in the level of power before and after inserting a device (filter) in the circuit. Some of the original signal is lost to resistance and reactance (component resistance to AC or pulsating DC) in a circuit. It is the difference in readings with the filter in and out of the circuit. Distortion in a circuit is any change in the waveform from the original, except volume. A volume increase is called amplification, and a decrease is called attenuation. The distortion is measured by subtracting the insertion loss at 1 kHz from that measured at a frequency between 200 Hz and 2.8 kHz.

On-Hook Envelope Delay Distortion. In this test the measured frequency ranges from 200 Hz to 2.8 kHz. Envelope delay is the measure of the phase versus frequency characteristic of the circuit. Different frequencies travel at different (unequal) rates in a given medium. Envelope delay distortion measures the distortion that occurs when the phase (unequal rate) shifts with the frequency over the bandwidth of interest. It is an important measurement because some data modems use phase to distinguish the bits (phase linearity). If 1004 Hz is the reference frequency, or zero point value, then the phase change difference will be the envelope delay distortion. It is measured in microseconds, and the impairment may not exceed 250 microseconds.

Off-Hook Voice Band (200 Hz–2.8 kHz) Requirements

The phone is *off-hook* while it is in use, for example, during a call or while dialing.

Off-Hook Voice Band Insertion Loss. The main difference with these tests and the previous tests is that they are done in the off-hook mode. Insertion loss is the loss experienced by the signal as it passes through the componentry. The signal power can be degraded by impairments and the energy extracted from it by the componentry. It is measured before and after adding from one to five filters to the circuit. The idea is to see how much loss the filter(s) add to the circuit.

Off-Hook Insertion Loss Distortion. There are two sides to an in-line filter: the network or LINE side, and the PHONE side. These tests are conducted in the off-hook mode. The tests are a measure of the signal power before and after adding a filter to the circuit. Measured at both sides of the filter, network and phone, we are within specified limits. Insertion loss distortion is computed by subtracting the insertion loss (1004 Hz) from the loss at any other frequency between 200 Hz and 2.8 kHz.

Off-Hook Impedance Distortion Tests. POTS performance can be compromised by both variations in the local loop and an imbalance in the front end of a hybrid transformer in a modem. One unique property of a two-wire circuit is an impairment called echo or return loss. It is caused when part of the signal energy sent from the source (test set) is reflected back toward the transmitter. This condition is much like the energy feedback heard in a public address system when a speaker is too close to a microphone. Variations in the local loop can change the sidetone balance of a telephone instrument. This test measures three specific return-losses: low-band singing return loss (200 Hz–3.4 kHz), echo return loss, and high-band singing return loss (3.4 kHz–4 kHz).

The nature of the distortion we are looking to test is called sidetone balance. **Sidetone balance** is the difference between the voice signal that is spoken and the signal that is heard at the same telephone set. Improper sidetone balance can cause a very distorted signal, loud as heard by the calling party, and soft as heard by the called party. A good filter should not cause any noticeable difference in the received signal at either end of the conversation.

Some low-cost phones are so poorly designed that they are subject to acoustical feedback if the handset is placed too near the cradle or on a flat surface. Placing a good filter in a circuit with a such a phone may not prevent feedback. This test is performed at the network and phone sides of the filter. Four filters are added in parallel to accomplish this test.

Envelope Delay Distortion

Voice channels that utilize filters often exhibit an impairment known as envelope delay distortion. This discussion refers us back to the point where we defined the problem. Simply stated, Envelope Delay Distortion tracks the phase characteristics of a channel with different frequencies arriving at the end of a circuit at different times. The references are from the central (telephone) office to the CPE and the customer's equipment back to the CO. Our band of interest is 200 Hz–3.4 kHz.

Transverse Balance

When we speak of a balanced line, we are talking about a transmission path in which two wires (a pair) are used to carry the desired signal to the customer from the sender. A balanced line is one with the ability to prevent signals from being introduced into another pair (crosstalk) by its proximity to that pair. The undesirable signal is known as "noise." Careful design of a phone keeps the balance high by keeping the impedance (total resistance in an AC circuit) to ground high. Twisting the wires (a tighter twist is usually better) is another way that the balance of a circuit is kept high in the wiring that feeds the phone as it becomes part of the overall circuit. Noise on a phone line is often due to an imperfect balance condition.

Electromagnetic noise is present when the wire in a cable radiates into neighboring pairs causing a condition known as crosstalk. This is commonly the greatest noise impairment on twisted pair cable and can interfere with DSL performance.

Transverse balance is defined as the ratio of the disturbing metallic voltage (V_m) and the resulting longitudinal voltage (V_l) of the filter under test, expressed in decibels. The equation is: $20 \log V_m/V_l$ (dB). What this means is that a metallic voltage (reference) is applied to the filter in both an on-hook and an off-hook condition. The filter must maintain a high degree of balance and not convert the metallic voltage into a longitudinal voltage. Longitudinal signals tend to couple to nearby wiring pairs causing crosstalk. Most people have heard crosstalk in the form of another faint conversation, sometimes half of a conversation, when they called someone on the telephone.

In the testing setup, a ground is applied to the filter under test with the phone in an on-hook then an off-hook condition. The balance of the filter is measured with a current applied in the off-hook condition.

High Frequency Band (HB) (25 kHz–10 MHz) Performance

On-Hook HB Stop Band Attenuation. The following tests are performed with the line in both the on-hook and off-hook conditions. An electrical network of inductors and capacitors which is designed to pass either low-band (LB) or high-band (HB) frequencies is called a filter. The filter takes advantage of inductors, which pass LB frequencies easily, but offers increasing resistance to higher frequencies. The reverse is true of capacitors that pass HB frequencies easily, but severely attenuate LB frequencies.

By combining these two properties in a filter, we have a "stop-band" filter that effectively blocks certain bands of frequencies. Stop-band filters are often required and cause the signal to lose energy as it deviates from the frequencies allowed by the pass band. This is called the cut-off frequency. Filter design is difficult because you must make many assumptions based on the average phone line. Any given phone line can have a multitude of impairments present. Lower frequencies are especially difficult and cause imprecise impedances caused by bridged tap, gauge changes, and remote loads. The characteristic impedance of the line itself can change when these conditions are encountered.

Filters can separate high frequency data from voice and yet allow the Caller ID and Message Waiting Information to pass through to your set. The on-hook test is performed with no current flowing in the test loop and an input signal of 10 dBm measured at a balun (balanced/unbalanced) in the circuit. The balun is an impedance matching transformer that matches the 50 Ohm test set signal to the 100 Ohm impedance of the circuit under test.

Off-hook HB (25 kHz–10 MHz) Bridging Loss

This test measures the HB attenuation due to the loading effect of up to five filters in the circuit. Bridging loss is the effect of adding filters and is measured by comparing the readings with the filters(s) in and out of the circuit.

ADSL Band Intermodulation Distortion. This group of tests (Figure 12.4) is difficult to produce because you must generate several frequencies at the same time, and then inject them into the circuit under test to measure the noise components they cause.

The purpose of this testing is to measure the noise components produced by the insertion of a set of continuous tones into a simulated ADSL channel. Multiple tones are then introduced into the upstream direction of the ADSL channel and monitored by the spectrum analyzer in the downstream direction.

As a filter is added, a nonlinear channel condition is created by the loading effect of the filter(s) terminated on the POTS port. These noise components are

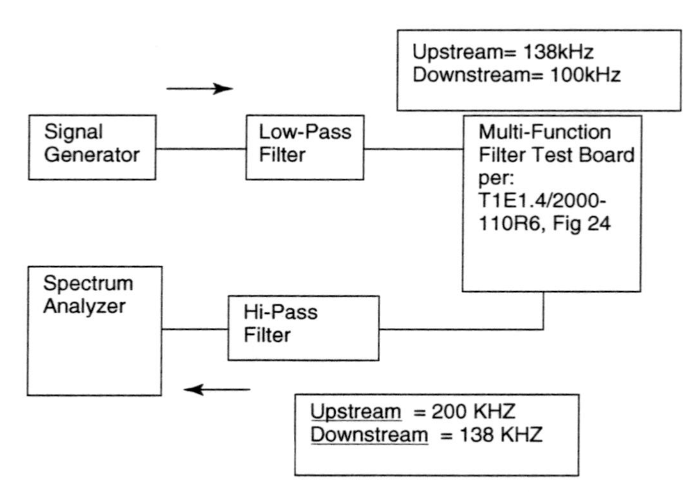

Figure 12.4 ADSL band intermodulation distortion testing diagram.

measured in both the upstream and downstream directions as the frequency of the applied signals are changed.

12.2.2 Inside Wire

After its long and difficult journey from the CO to the customer premises, the DSL signal may find its greatest hurdle: the wiring within the customer's premises. Extensive investigation by the authors into the nature of existing in-home telephone wiring has reached the conclusion that very little is known about inside wiring. There is a wide variety of practices. Inside wiring in homes consists of:

- One to eight wire pairs, with two or three pairs most common
- Twisted pairs, flat wire (no twist), and quad wire (four conductors twisted together)
- CAT5, CAT3, D-type inside wire, shielded wire, and just about every other type of wire
- AWG 22 and 24 unshielded, flat or quad type wire is most common
- Topologies: star, bus, tree, and ring (see Figure 12.8)
- In the United States typically have four to ten wall jacks; in Europe, one to three wall jacks
- Approximately 75 percent of single family homes in the United States have an NID
- Approximately 20 percent of U.S. homes subscribe to two or more phone lines
- Approximately 40 percent of U.S. homes connect more than three phones to a line

Standards for high-quality *structured* inside wiring have been in place since 1994. Regrettably, to save a few dollars, new home construction often does not follow these standards. The ISO/IEC 11801, CENELEC EN 50173, and Canadian T525-94 standards specify inside wire types and installation configurations. TIA/EIA standards 568A and 570A specify 24 AWG unshielded twisted pair (UTP) category 3 or 5 wire placed in a star topology (also known as *home run* wiring). Less than 2 percent of U.S. homes currently follow these standards. The pair-to-pair crosstalk and loss for UTP CAT3 is worse than typical outside plant telephone cables[3], whereas UTP CAT5 is as good or better than the outside plant telephone cable characteristics. Type D inside wire (DIW), which exists in older office buildings, has poor crosstalk characteristics above 1 MHz.

In addition to being a test access point, the NID contains primary overvoltage protection that shunts excessive voltage due to lightning or power-cross to ground. There are three types of primary protectors. Carbon block and gas tube protectors generally do not impair DSL operation, but they can age and may eventually need replacement. Some types of semiconductor primary protectors have poor high frequency characteristics that can impair DSL operation; fortunately they are not common. Some home owners have placed a secondary lightning protectors inside

[3]Outside plant cables employ a variable twist rate to reduce crosstalk.

their house. For example, the secondary protector might be plugged into the telephone wall jack where a fax machine or voice-band modem is connected. This secondary protector is distinct from an AC power surge protector that is not connected to the phone line. The secondary protectors that connect to the telephone line often impair DSL operation and HomePNA operation. These types of secondary protectors are frequently found in "deluxe" power strips. Even though they help analog modems, they usually impair DSL transmission.

An additional item found in the NID is some areas is a metallic termination unit (MTU), which aids telephone line testing from the CO by presenting a known impedance at the entry point to the house. Unfortunately, older types of MTUs have poor high frequency characteristics that impair DSL operation.

There has been much debate about the reliability of dry (no DC current flow) ADSL and HomePNA connections in the outside plant and house wiring. In normal telephone service, an off-hook DC current of 20–50 mA flows from the CO–48V line circuit through all of the connections and jacks to the phone. When the various connectors and wiring splices in the total phone circuit age and corrode, they can produce an open circuit, but the –48V feed voltage from the CO will break down this oxide and reseal the open connections, and the traditional phones circuit continue to operate satisfactorily. Basic rate ISDN and certain other types of DSL have provided for *sealing current* (also known as wetting current), a current up to 20 mA provided by the line termination at the CO for the purpose of breaking down the oxide layer in wire splices. Field and laboratory studies performed by Telcordia (formerly Bellcore) have produced contradictory conclusions regarding the effectiveness of sealing current. Some experts doubt the value of sealing current. Many DSL lines without sealing current are providing reliable service today. Most telephone companies have used *gas-tight* splice methods since the early 1980s; these splices should not require sealing current. Sealing current is not needed for types of DSL that have DC current flowing for line powering the transceiver at the customer-end of the line. Also, ADSL operating over POTS (line sharing) has no need for sealing current because DC current flows whenever the phone is used.

The wall jack and plugs used to connect ADSL and HomePNA may cause trouble because there is no DC loop current flowing in this portion of the circuit to clean the jack and plug contacts. In addition, the existing wall jacks may already be corroded and continuing to age. Plugging in a new RJ11 cord for the ADSL or HomePNA connection may temporarily wipe off the oxide on the existing wall jack contacts from the wiping action (but not always, it may take a few insertions to clean the connection). However time and environments will eventually reoxide the connection especially in certain high humidity areas, such as jacks located in outside walls with air drafts and/or glued paneling which can accelerate the aging. As the connections age, the transmission will become noisy and intermittent and eventually fail completely.

There is, however, an easy customer solution to wall-jack and plug problems that is neither costly nor difficult, and no new jack wiring is required:

1. Colocate the ADSL (or HomePNA) modem connection with a telephone device connection such as a phone, answering or fax machine, using a dual line

adapter (or an in-line filter with a convenience ADSL/HomePNA jack) in the existing wall jack. One should be sure the dual line adapter and filter have at least 50 micro-inches gold plating over 100 micro-inches nickel to ensure long life. The adapters from some distributors may only have a gold flash of 5 micro-inches of gold which will allow early failures.

2. Use a line cord with RJ11 plugs with a minimum 50 micro-inch gold plating. For other crosstalk and RFI reasons, this cord should also be a single pair twisted. The traditional 4 conductor flat cords may be suitable for voice band frequencies and telephones, but a 14 foot 4 conductor flat cord has only 3 dB of crosstalk loss in the upper HomePNA band between the Line 1 to Line 2 part of the cable.
3. The manufacturer of the ADSL or HomePNA modem should also specify the minimum 50 micro-inches gold plating on the RJ11 jacks in their modems.

12.2.3 Dial-up Alarm Systems and ADSL

Many security alarm systems contain a relay that connects a voice-band modem in the alarm panel to the telephone line when an alarm is to be reported. Normally, the relay connects the telephone line to the various telephone devices within the home (see Figure 12.5).

When an alarm is reported, the relay in the alarm panel "seizes the line" and a voice-band modem in the alarm panel dials the phone number for the Alarm

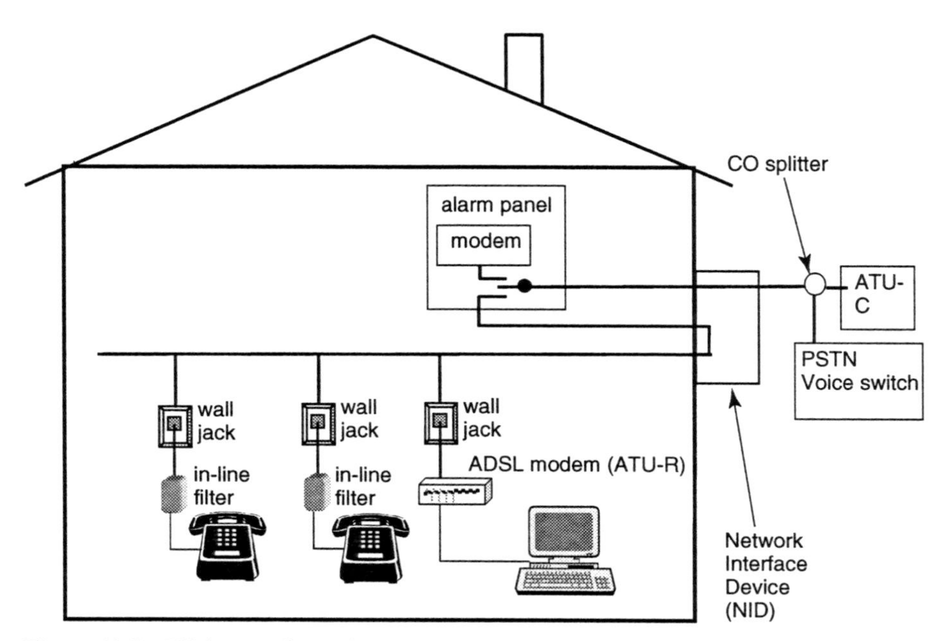

Figure 12.5 Wiring configuration for home with splitterless ADSL and an alarm system.

Company's monitoring office. Alarm system operators have reported communications problems with home alarm panels after the installation of ADSL service. It has been suggested that this trouble was caused by the ADSL signals transmitted by the ATU-C overloading the receiver in the voice-band modem in the alarm panel. The alarm panel modem front end has relatively little rejection of signals above the voice-band. Some parties debate the validity of this scenario.

To avoid the potential for such trouble, most ADSL service providers ask prospective ADSL customers if they have a reporting alarm service. Customer self-installation using in-line filters (Figure 12.2) is not attempted for customers with reporting alarm service. Instead, a technician installs a shared splitter at the NID (Figure 12.1) so that only the ADSL modem receives the wideband signals. Thus, the NID splitter blocks the out-of-band energy to the voice-band modem in the alarm panel. An alternative solution, is to place a suitable low-pass-filter at the front end of the voice-band modem in the alarm panel. The filter is not easy to retrofit into existing alarm panels, but is a desirable design improvement for future alarm panels. A less likely solution would be to report alarms via IP messages through the data network.

12.3 INSIDE TELEPHONE WIRE-BASED HOME NETWORKS

Several methods use existing inside telephone wiring as the physical medium for home networking. The most popular method was developed by the HomePNA, and is discussed in the next paragraph. Alternatively, some or all of the telephone wall jacks in a home may be replaced by new wall jacks containing active data repeaters connecting to the existing inside wiring. With this method, each active wall jack supports one point-to-point physical-layer connection to the wiring hub (in a star-wired topology) or two point-to-point physical-layer connections to each neighboring wall jacks (in a chain-wired topology). The active-jack method can attain high-data performance, but the installation of new wall jacks is too demanding for self-installation by most home owners. A third method is a LAN using the inside telephone wire pairs not used for telephone service. The spare wire pairs rarely have adequate balance, impedance, and low enough noise for use by traditional LAN technology. Furthermore, many homes do not have spare inside telephone wire pairs that are actually connected to most wall jacks.

With HomePNA, information appliances (such as PCs) are easily interconnected via a LAN that uses the existing inside telephone wiring. Each information appliance connects to a low-cost HomePNA adapter which then plugs into a standard telephone wall jack (see Figure 12.6).

The very same inside wire pair that carries traditional telephone service throughout the home also carries the HomePNA data modulated into a different frequency band (4 to 10 MHz for version 2.0 of HomePNA) (see Figure 12.7). HomePNA devices connect to pins three and four of the telephone wall jack; this is *line one* for multiline wired homes. More important, this is the *only* telephone pair for homes wired for only one line. Thus, HomePNA connects to the one pair that is

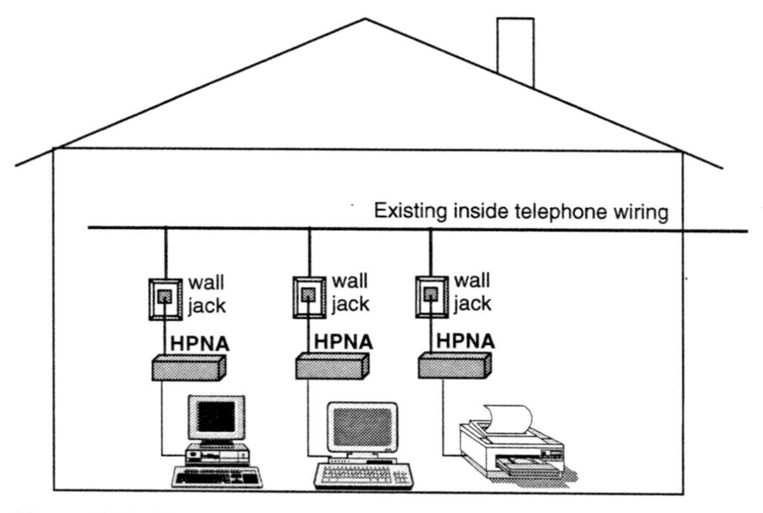

Figure 12.6 HomePNA configuration.

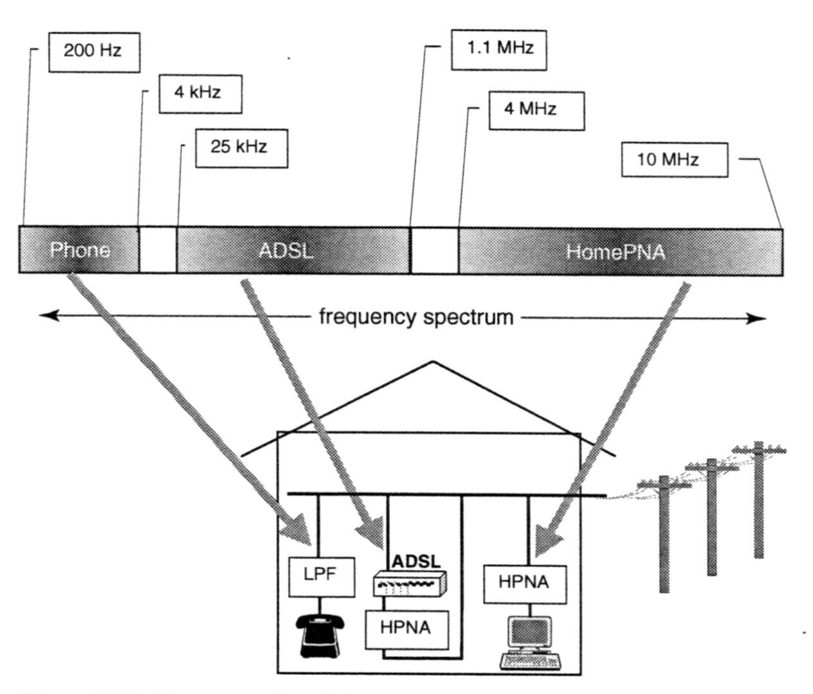

Figure 12.7 Frequency bands for phone, ADSL, and HomePNA.

surely present and working in virtually all homes. In a remarkable triple-use of the inside telephone wires, the same pair of wires can also carry ADSL signals within the 25 kHz to 1.1 MHz frequency band.

Up to twenty-five information appliances within a building are easily IP (Internet protocol) networked simply by connecting the USB (universal serial bus) or 10baseT Ethernet interface of each information appliance to a low cost HomePNA adapter that plugs into a nearby telephone wall jack. HomePNA version 1.0 provides an aggregate LAN bit rate of up to 1 Mb/s, and HomePNA version 2.0 provides up to 10 Mb/s. The actual maximum LAN throughput may be less for a home with very long wiring runs, poor quality wire, or very noisy environments (such as near a radio transmitter).

HomePNA versions 1.0 and 2.0 are designed to interwork with each other. Version 1.0 uses pulse position modulation in the 5.5 MHz to 9.5 MHz frequency band. Version 2.0 uses a more complex method: trellis-coded frequency diversity quadrature amplitude modulation (FDQAM), which modulates the same data into two separate frequency bands within the 4.0 to 10 MHz band. It is expected that most transmission impairments would not severely impair both frequency bands. Adaptive equalization is also used to reduce the impairments from the home wiring. HomePNA version 2.0 has largely superceded version 1.0 in the marketplace. For typical inside applications, the packet throughput of HomePNA 2.0 (up to 7 Mb/s) is up to three times the packet throughput realized for wireless LANs such as IEEE 802.11b (often 2 to 3 Mb/s).

HomePNA was designed to overcome the challenges of transmitting high speed data over the various conditions found in home telephone wiring:

- Wire types ranging from CAT5 to quad wire and flat wire (not twisted)
- Various wiring topologies: star, tree, ring (see Figure 12.8)
- Bridged taps within the home
- Noise from many sources: answering machines, AM radio, electric motors, light dimmers, fluorescent lights
- Line impedance changes due to phones being on-hook or off-hook

Based upon HomePNA field studies, it is estimated that HomePNA will operate reliably at the rated maximum bit rate upon initial installation in more than 90 percent of U.S. homes, at a reduced bit rate for about 5 percent of homes, and not work reliably for about 5% of homes. Many of the cases of trouble with HomePNA are due to noise from telephony equipment (fax, modem, caller-ID, and answering machine), which can be resolved by placing an in-line low-pass filter in series with the telephony device. On rare occasions, HomePNA causes audible noise in telephone receivers; this too can be resolved by placing an in-line low-pass filter in series with the telephone. Another source of trouble is the home telephone wiring being either too long or having too many long bridged taps. As a rule, the maximum wire distance between HomePNA devices is 500 ft. Inside telephone wiring often takes indirect routes so that two outlets that appear to be only 50 ft apart could be hundreds of feet apart in the wiring.

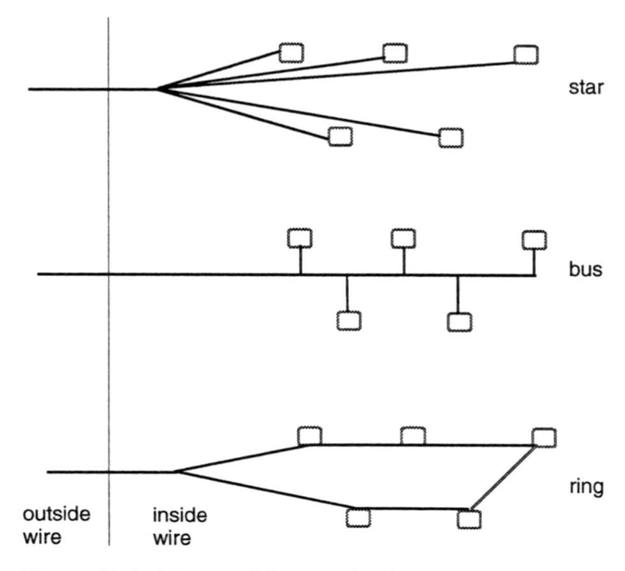

Figure 12.8 Home wiring topologies.

HomePNA offers the lowest cost home LAN solution using existing wiring ($38 as of the time of this printing). However, it does not work well in all homes, and there may not be a phone wall jack accessible at some of the locations where an information appliance resides.

The frequency bands used by HomePNA and VDSL overlap to a large degree. Analysis of HomePNA and VDSL located at the same premises indicate that both systems would be greatly impaired unless the two systems use separate inside wiring and a low-pass filter was placed at the entrance to the premises to prevent the HomePNA signals from exiting to the outside wiring. This would require installation by a skilled technician. There is a possibility for adverse interference between nearby customers sharing the same distribution cable. One customer with VDSL and a neighbor with HomePNA could experience interference problems. This could be especially acute for an MDU (multiple dwelling unit, such as an apartment building). Radio frequency egress from HomePNA may be a concern in some countries due to radiation from poorly balanced inside wiring. Security is another concern for HomePNA; in rare instances a neighbor in the same telephone distribution cable could pick up enough signal crosstalk to monitor or possibly send unauthorized messages.

HomePNA is supported by a large number of equipment and silicon vendors, including many PC-related companies. Further information may be found at *www.homepna.org.* The ITU SG15 is developing an international standard based on HomePNA version 2.0 under the name G.pnt (phone network transceivers). A framework ITU Recommendation based on HomePNA 2.0, G.989.1 was approved in 2001. ITU G.989.1 provides a specification for the frequency band and other basic aspects; approval of a more complete specification is expected in 2002.

12.4 COAX CABLE-BASED HOME NETWORKS

The Cablehome™ project by CableLabs® is developing technology for a coaxial cable-based home network to support both data and the quality-of-service capabilities needed for voice networking. Telephones and information appliances could be attached where cable outlets exist within the home. Because many homes have cable wall jacks on only one or two locations within the home, cable-based networking may be limited to those homes prewired for cable to many rooms. Information on CableHome may be found at www.cablelabs.com.

12.5 AC POWER WIRING-BASED HOME NETWORKS

Local communication using the power wiring within a premises has seen widespread use in homes and business since 1978 for low-speed signals to control lights, HVAC (heating, ventilation, air conditioning), and appliances. New systems are now addressing power line LANs at much higher data rates. Premises LANs using inside power wires are distinct from WANs using the public power line grid to communicate data between homes and a public data network. Whereas power line communications with a premises have proved to be reliable (at least a lower bit rates), the feasibility of WANs using outside power lines to communicate more than a few kb/s are in doubt due to electromagnetic interference (EMI), and network topology problems. There have been many bold announcements about plans for power-line WANs with data rates up to 2 Mb/s, but many of the projects have been quietly shut down. More information is available at www.plcforum.org.

Power-line LANs permit information appliances to be anywhere there is a power outlet. In many houses, telephone jacks are found in only a few places, whereas power outlets are found in every room including the garage, and sometimes even outside the house. However, power-line LANs do not provide as much mobility as wireless LANs.

The X10 power line control system consists of a control station that plugs in to any power outlet, and remote modules that plugged in series with the power cord to each controlled appliance. The control station can control up to 256 stations by sending addressed binary commands at about 1 kb/s. The appliance modules can turn the power on, off, or perform a light dimming function. X10 can also poll remote modules for their status; this is used for security alarm systems. At the time of this printing, X10 is priced under U.S. $15 per appliance module. The manufacturer of X10 equipment claims that more than 100 million X10 units have been shipped.

Sending data faster than a few kilobits per second must overcome a challenging transmission environment. Inside power wiring has poor balance at high frequencies; this presents the dual challenge of dealing with high levels of radio frequency ingress noise and restricting radio frequency emissions so as to not interfere with radio systems. Power lines have high levels of conducted noise caused by motors, light dimmers, and switching-type power supplies. The line impedance changes greatly when appliances are turned on or off. Furthermore, inside power wiring has many branches that cause signal reflections (e.g., bridged taps).

CEBus (consumer electronics bus) is specified in an open standard (EIA 600) and uses spread spectrum transmission over home power wiring at approximately 10 kb/s. CEBus specifications also address networking via coaxial cable, wireless, CAT5 wire, and infrared. CEBus is primarily targeted at interconnecting home entertainment audio/video systems, and control of lights and appliances. The CEBus MAC is similar to CSMA/CD, and common application language (CAL) is specified to define messages between nodes. Further information may be found at www.cebus.org. CEBus is supported by HomePlug Powerline Alliance, and further information may be found at www.intellon.com.

The LonWorks protocol was developed by Echelon Corporation and adopted by the IEEE and EIA (electronics industry alliance) as standard EIA 709.1. LonWorks is widely used in business and commercial building for control of lights, security, and HVAC. Twenty kb/s is transmitted with a 16-bit CRC check for reliability. The LonWorks MAC uses the LonTalk protocol, and is based on CSMA/CD.

Intellon Inc, Enikia Inc., and Itran Communications Inc. have a variety of home power-line networking schemes based on OFDM (orthogonal frequency division multiplexing, similar to DMT), claiming to achieve data rates of 1.5 Mb/s or higher. These companies and about seventy-five others have joined together to form the HomePlug Powerline Alliance (www.homeplug.com) to develop an industry specification for low-cost, interoperable, high-speed Ethernet-based networking via power wiring inside homes. A specification is available to members of the HomePlug association, and products built to this specification came to the market during 2002, priced at about $100/node and able to support data rates of up to 11 Mb/s. Initial field experience shows that rates near 11 Mb/s are feasible for most homes. However, a data rate of about 1 Mb/s may be more suitable for successful operation in virtually all homes. It is not yet known if these systems can achieve performance, reliability, and cost that is competitive with other home networking solutions. The HomePlug system specifies OFDM for the physical layer (PHY) and CSMA/CA for the MAC. The frequency band used is approximately 4 MHz to 21 MHz. The noise in the band is generally lower than for frequencies below 4 MHz. Both point-to-point and broadcast modes are supported. The HomePlug specification provides for optional 56-bit DES encryption for security. To function correctly, HomePlug equipment must be connected to the same AC wiring phase; many homes have two wiring phases and it is not obvious which plugs are on the same phase. Surge protectors and power back-up units (UPS) must not be in-line with the HomePlug equipment. Further information is available at www. homeplug.com.

Power-line based LANs must deal with many of the same issues as other LAN and DSL technologies. To be competitive, data rates far above one megabit per second should be achieved even in large homes. Radio frequency emissions must not interfere with radio systems. Special care must be taken to avoid excessive latency while providing the necessary robustness against noise. Also, adequate security must be provided as eavesdropping could be possible, especially in multiple dwelling units.

12.6 DEDICATED DATA HOME NETWORKS

For those who are able to install their own dedicated LAN wiring or pay others to do it, and are willing to be confined to the few sites where the LAN connectors are located, the ultimate in performance and reliability is available.

The principal LAN specifications are listed in the following table:

Standard Name	Gross Bit Rate	Wire Distance
IEEE 802.3	10 Mb/s	10 BASE-T, 100 meters of CAT3 wire
IEEE 802.3v	100 Mb/s	100 BASE-T, 100 meters of CAT5 wire
IEEE 802.3z	1 Gb/s	1000 BASE-T, 100 meters of CAT5 wire
IEEE 802.ae	10 Gb/s	Fiber only

Further information on LANs may be found as www.uts.utexas.edu/ethernet/ethernet or www.10gea.org.

The following are not considered to be LANs, but the are used to interconnect equipment via short connectorized jumper cables.

Specification Name	Gross Bit Rate	Wire Distance
IEEE 1394[4]	400 Mb/s	4.5 meters per hop
USB 1.1	12 Mb/s	5 meters
USB 2.0	480 Mb/s	5 meters
Fast SCSI	160 Mb/s	about 2 meters
RS 232	115 kb/s	4 meters
Parallel bus	920 kb/s	about 3 meters

Further information on USB is available at www.usb.org.

12.7 RADIO LAN HOME NETWORKS

Home radio LANs may connect one or more information appliances to a home gateway and can also enable direct communication between the information appliances. Radio LANs are distinct from public wireless WANs, such as those using microwave multipoint distribution service (MMDS), local multipoint distribution service (LMDS), and satellite technology. In comparison to wireless WANs, the radio LANs provide higher throughput performance and lower cost. Wireless LANs are primarily targeted at computer-related nodes, although audio and video applications may also be supported (see Figure 12.9). Although home radio LANS

[4]IEEE 1394 is also known as FireWire; 1.6 Gb/s expected for a future version of IEEE 1394.

are slightly more expensive than other home LAN solutions such as HomePNA, radio LANs provide attractive benefits:

- No wiring to install; LAN installation is quick and easy, and additional terminals are easily added later.
- Information appliances may be located anywhere in any room.
- Information appliances may be located in the nearby out-of-doors (e.g., poolside).
- Local mobility with no tether is especially important for the laptop and handheld computers.
- Standards permit the same wireless LAN modem to be used at other sites such as the business office, hotel, and home.

Wireless LANs have their limits due to several impairments:

- Signal loss increases with distance much more rapidly than the $1/R^2$ rate for free-space radiation. Walls, furniture, plants, and even people attenuate the radio signals. Metal and reinforced concrete walls and floors are particularly limiting. Glass, wood, and plaster cause little attenuation.
- Radio frequency noise from sources in the same frequency band such as microwave ovens (2.4 to 2.45 GHz), intrusion detection alarm systems, some types of electrical motors, cordless phones, and other types of wireless LANs. For example, operating a Bluetooth wireless LAN nearby an 802.11 wireless LAN can reduce 802.11 throughput by up to 33 percent depending on the traffic patterns. Effects of the noise may be spatially localized; thus moving the radio transceiver a few feet may avoid the noise.
- Multipath distortion caused by delayed echoes of a signal being received. This effect is similar to the effect of bridged tap on DSL transmission and is illus-

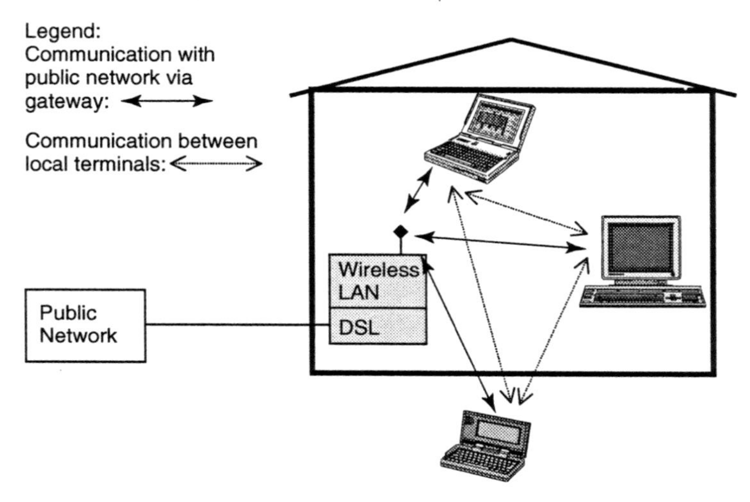

Figure 12.9 Wireless home LAN.

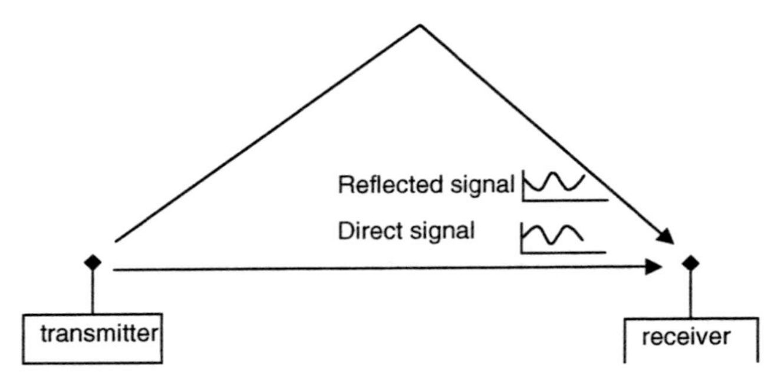

Figure 12.10 Multipath distortion.

trated in Figure 12.10. The effect of multipath distortion is often most severe in narrow portions of the frequency band.

- Intersymbol interference (ISI): differences in propagation delay at different frequencies causes the received symbols to spread out in the time domain. The effects of ISI increase with symbol rate.

Another problem for wireless LANs is called a *hidden node*; this occurs when node A can receive the transmissions from nodes B and C, but nodes B and C are too far from each other to receive the other's transmissions. As a result, transmissions from nodes B and C can collide without detecting the collision. To prevent this, some wireless LAN specifications, such as 802.11, provide for request-to-send (RTS) and clear-to-send (CTS) messages that are relayed through the access point. Although node B does not receive the RTS message from node C, it will receive the CTS message from the access point (node A in the example above), and thus node B will know to wait before transmitting.

Wireless LANs employ several countermeasures to overcome impairments:

- Spread spectrum transmission of a lower signal power across a wider frequency band than traditional single-carrier radio transmission. This permits reception despite strong narrowband impairments such as noise and multipath distortion. Spread spectrum transmitted signals have the additional benefit of causing less interference into other radio systems.
- Antenna diversity: since the effects of multipath distortion change greatly with position, receiving signals from multiple antennas separated by a few inches often will overcome the effects of multipath distortion.
- Forward error control (FEC) coding can correct for the effects of short term noise.
- Frequency equalization can reduce the effects of noise, signal loss, and multipath distortion.

Regional regulatory agencies, such as the FCC in the United States, set requirements for each portion of the radio frequency band. Rules are defined for each range of frequencies, including the characteristics of radio transmissions (e.g., signal power), the appropriate applications, and whether the user must be licensed. Radio LANs use RF bands that do not require user licenses. The use of an unlicensed band allows anyone to quickly and easily set-up a radio LAN, but does not provide statutory protection against RF interference from other unlicensed systems operating in the same RF band.

Figure 12.11 shows two wireless LAN configurations. The nodes in the independent basic service set (IBSS, aka an ad hoc network) connect only to each other with no wired infrastructure used other than the optional addition of a DSL or other external network connection via one of the nodes. The infrastructure

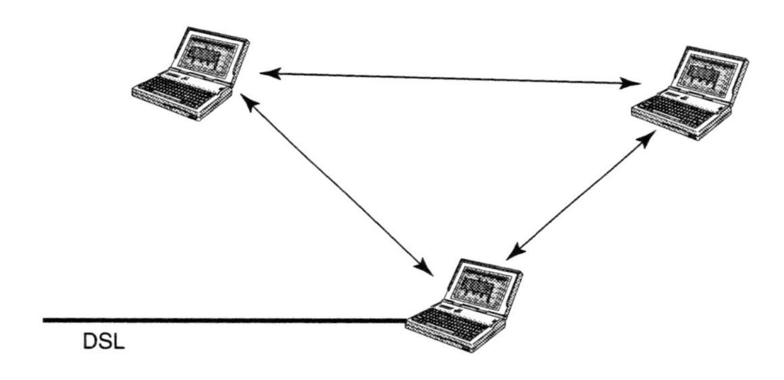

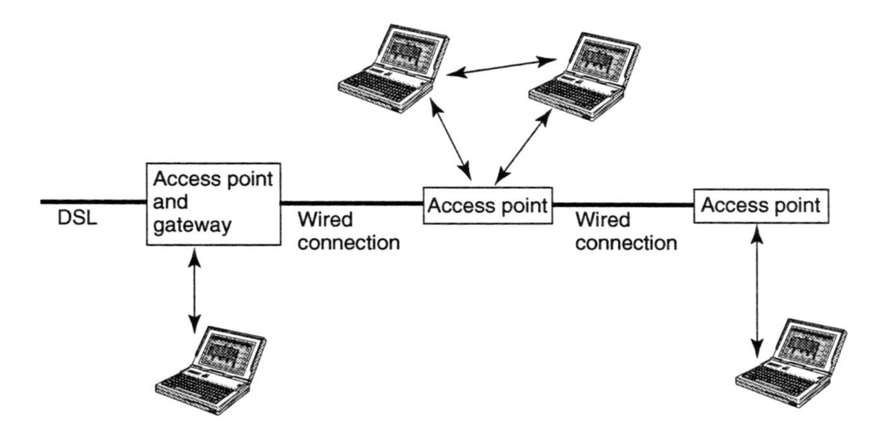

Figure 12.11 Wireless LAN configurations.

based wireless LAN (also known as an extended service set, ESS) adds *access points* that are connected to each other via a wired infrastructure. The access point functions as a LAN hub and a gateway between the LAN and the WAN. The distributed access points permit the wireless LAN to cover a greater area, and logical segmentation of access points assigned to different radio frequencies also permits greater aggregate network capacity. As shown in the figure, one of the access points may also contain a gateway function with a link to an external network. Wireless LAN specifications such as 802.11 permit nodes to roam among many access points.

12.7.1 IEEE 802.11, 802.11b, and Other 802.11 Wireless LANs

The 802.11 wireless LAN standard for operation at 1 and 2 Mb/s was approved in 1997 by the IEEE and ISO/IEC. It was followed in 1999 by the IEEE approval of the 11 Mb/s 802.11b standard that has quickly become one of the most attractive business and home LAN solutions. Many equipment vendors adopted 802.11b, and its success is making 802.11 obsolete. IEEE 802.11b products are sometimes marketed under the *Wi-Fi* name, and typical 2002 prices were $90 per "access point" wireless hub plus $63 per PCMCIA adapter. Prices are expected to drop lower as the market grows toward the forecasted 3 million units per year by 2003. The original 802.11 standard consisted of three PHY specifications: infrared operating at wavelengths of 850 to 950 nm with up to 2 Watts of peak transmitted power, DSSS radio, and FHSS radio. The DSSS is the most often used PHY for 802.11 products. The radio transmissions for North America and most of Europe are in the 2.412 to 2.472 GHz frequency range; this is known as the ISM band. Japan, Spain, and France use different frequency ranges.

The 802.11 direct sequence spread spectrum (DSSS) transmission at 1 Mb/s uses differential binary phase shift keying (DBPSK) modulation and at 2 Mb/s uses differential quadrature phase shift keying (DQPSK) modulation. DSSS spreads the signal evenly across the transmission band by using an eleven-chip pseudonoise code similar to that used by code division multiple access (CDMA) mobile cellular telephone systems and global positioning system satellites. However, unlike cellular telephone systems, all 802.11 stations use the same pseudonoise code, and it operates with less processing gain (10.4 dB) to permit higher bit rates. In the United States, 802.11 transmits up to 1 Watt of RF power.

The 802.11 frequency-hopping spread spectrum (FHSS) transmission at 1 Mb/s uses 2-level Gaussian frequency shift keying (2-GFSK) modulation, and at 2 Mb/s uses 4-GFSK. FHSS transmits a relatively high amplitude signal in a small portion of the usable frequency band for 400 ms (or less), and then the transmission moves to a different sub-band. All stations hop simultaneously using the same sequence of sub-bands.

The 802.11 and 802.11b media access control (MAC) uses CSMA/CA that is similar to the CSMA/CD algorithm specified for 802.3 Ethernet. The maximum 802.11 packet size is 1,500 bytes. To improve throughput, the 802.11 and 802.11b MAC assigns highest priority to message acknowledgments (ACKs). The 802.11 standard specifies optional security features called *wired equivalent privacy* (WEP), including encryption and authentication. The encryption encodes messages using the RC4 algorithm with a 40-bit key plus a 24-bit node identifier (hence WEP is often said to use 64-bit encryption). Authentication assures that messages are from known nodes. This provides security appropriate for applications needing protection against casual interception, but WEP can be quickly circumvented by widely available methods. IEEE 802 is developing further security measures for applications where privacy and authentication is critical.

Furthermore, the administration of security may become difficult for installations with a very large number of stations. WEP's security is reduced by the use of a modest-sized key and the use of the same encryption key for all nodes connected to an access point. Some vendors provide proprietary security enhancements, including 128-bit encryption. Because 802.11 and 802.11b reach at most a few hundred feet, an eavesdropper must be in close proximity.

The 802.11 MAC specifies CSMA/CA for asynchronous packet-oriented distributed coordination function (DCF), and a point coordination function (PCF) for isochronous information such as voice and video. The PCF function is not suitable for LANs with a large number of stations. The 802.11's provisions for voice applications are not as complete as HomeRF.

The IEEE 802.11b standard specifies DQPSK modulation with an 8-chip DSSS code known as complementary code keying (CCK) in the 2.4 GHz ISM frequency band. 802.11b provides for backward compatibility with 802.11 DSSS systems, but the FHSS and infrared PHYs are not included in 802.11b. The 802.11b marketing makes claims of 11 Mb/s at a distance of 500 ft, and 1 Mb/s at a distance of 1700 ft. Actual user data throughput and real-world distances are far less than the marketing claims. Due to protocol overhead, the effective data throughput of 802.11b is up to 7 Mb/s for large packets and distances up to 80 ft with ideal conditions; for smaller packets the effective data rate can be less than 4 Mb/s. In typical indoor conditions beyond about 40 ft, the bit rate reduces in several steps (11, 5.5, 2, and 1 Mb/s) toward a cut-off distance of 80 to 250 ft depending on the environment. Throughput can be greatly reduced while a microwave oven is operating within 15 ft of the 802.11b device. Despite the misleading marketing claims, 802.11b serves most wireless LAN applications well and sales of 802.11b products exceed all other types of wireless LANs. Proponents of 802.11b claim that the prices for 802.11b products will become one of the lowest cost home LAN solutions.

The IEEE also developed the 802.11a standard, which operates at a gross bit rate of 54 Mb/s and has effective throughout of 24 Mb/s in the single channel mode. Bit rates of 72 to 108 Mb/s are possible in the multi-channel "turbo" mode. 802.11a uses orthogonal frequency division modulation (OFDM) in the 5 GHz band and operates over distances of up to 100 feet. 802.11a products became available during 2002, and future versions are expected to provide backward compatibility with 802.11 and 802.11b. The IEEE 802.11 committee is also considering the development of a 22 Mb/s wireless LAN standard.

The Wireless Ethernet Compatibility Alliance (WECA) is an industry association dedicated to interoperability between all 802.11 products. Information regarding wireless ethernet and interoperability programs may be found at www.standards.ieee, www.wirelessethernet.org, and www.wi-fi.net.

12.7.2 HIPERLAN

ETSI has standardized HIPERLAN that operates at gross bit rates up to 23.5 Mb/s using Gaussian minimum shift keying (GMSK) in the 5.15 to 5.3 GHz band. HIPERLAN was first implemented in 1999.

12.7.3 HomeRF

HomeRF operates at a gross bit rate of 1.6 Mb/s using a frequency shift keyed FHSS transmission defined in the shared wireless access protocol (SWAP) specification. As the name implies, HomeRF is primarily targeted at networking within the home. The key distinctions of HomeRF is its low cost: about $90 per PC and its support of both packet mode data (using a CSMA/CA MAC) and isochronous transport of up to four voice channels using 32 kb/s ADPCM coding. A portion of the capacity is reserved for time division multiple access (TDMA) for high-quality digital enhanced cordless telephony (DECT). The maximum effective data throughput is about 800 kb/s. HomeRF operates in the 2.4 GHz ISM frequency band with a nominal 100 mW of transmitted RF power, and has an effective range up to 150 ft indoors. HomeRF includes a low-complexity encryption algorithm and optional data compression to improve data throughput.

The first HomeRF products were available late in 1999, shortly after the SWAP 1.1 specification was published. The planned introduction of a new 10 Mb/s HomeRF-2 is expected. Both HomeRF-1 and HomeRF-2 products are expected to be lower cost than 802.11b products. Advocates claim that HomeRF's support of high-quality voice, video, and data is an advantage over 802.11b; however, it is likely that there will be ways for 802.11b do support these applications as well. Further information may be found at www.homerf.org.

12.7.4 Bluetooth

Named for a tenth-century Danish king, Bluetooth is designed to provide a very low-cost one Mb/s *personal area network* (PAN) within a radius of 30 ft. Bluetooth can easily cover a room, but not a typical house. With a transmitted RF power of 2.5 mW, Class 2 Bluetooth devices can be very compact. For example, Bluetooth devices fit within cellular phones, personal digital assistants (PDAs), and even a wristwatch. Somewhat larger Class 1 Bluetooth devices can reach about 300 ft by transmitting up to 100 mW. Bluetooth is primarily targeted at interconnecting cellular phones, PDAs, PCs, and PC peripherals without the inconvenience of wires. This permits collaboration among the electronic devices within the vicinity of a person.

Bluetooth uses Gaussian binary frequency shift keying with FHSS at 1600 hops per second within the 2.4 GHz to 2.4835 GHz ISM frequency band. The effective data throughput is 721 kb/s in one direction with 57.6 kb/s in the opposite direction, or 432.6 kb/s simultaneously in both directions. Less throughput may be realized in noisy environments (e.g., noise from 802.11b or microwave ovens nearby). Bluetooth's frequency hopping permits it to avoid narrowband noise, but the low-transmitted power makes Bluetooth highly susceptible to wideband noise.

Both asynchronous (packet) and isochronous (circuit) type information transport is supported. Encryption and authentication are included for secure communications.

Bit rates up to 20 Mb/s are being considered for a future version of Bluetooth. Further information regarding Bluetooth may be found at www.bluetooth.com.

12.8 INFRARED HOME NETWORKS

Infrared LANs operate at gross bit rates up to 4 Mb/s or 16 Mb/s depending on the type of system. Due to the line-of-sight propagation with limited reflections, infrared LANs are generally confined within a room. Further information may be found at www.irda.org.

CHAPTER 13

DSL CPE AUTOCONFIGURATION

13.1 AN OVERVIEW OF THE PROBLEM OF DSL CUSTOMER PREMISES EQUIPMENT (CPE) MANAGEMENT AND CONFIGURATION

Ensuring that the DSL customer can configure and manage the DSL equipment at their premises with a minimum of difficulty is critical to the success of any DSL service that is offered to the public. Broadband access in general and DSL in particular bring network and management complexities that formerly were the concern only of the larger enterprise to the home and small business. DSL promises a wide range of new and useful services to these customers; however, if they cannot install, configure, and manage these services easily and reliably, the services will not be utilized.

Among the management issues raised by use of DSL for the customer include:

- The variety of DSL services that can be offered to the customer by a carrier or service provider. Each of these services may require different configurations and different configuration tools. For example, a home that supports both an Internet access service and a voice-over DSL service will require each of these services to be configured.
- Different providers of DSL service use different end to end architectures on their networks when they provide a DSL service. For example, some carriers use PPP over Ethernet for providing IP services to their customers, others use IP directly on AAL5. A vendor of CPE can design their equipment to work with either protocol stack. However, in order for the CPE to work with the particular carrier environment the CPE must know both the stack being used and particular parameters used by the network for that stack. It is unreasonable to expect the customer to know these technical details required to configure their devices to communicate over the DSL network.
- The DSL environment is complex and typically has divided realms of responsibility for both the network and services offered over the network. Figure 13.1 illustrates the three management realms that are typical for a DSL service.

  A DSL network may support multiple service providers, each of whom is responsible for the particular services that they offer. The configuration of equipment at the customer's site to support these services may be the responsibility of the service provider, the customer, or it may be shared. A user may simultaneously access services from several service providers.

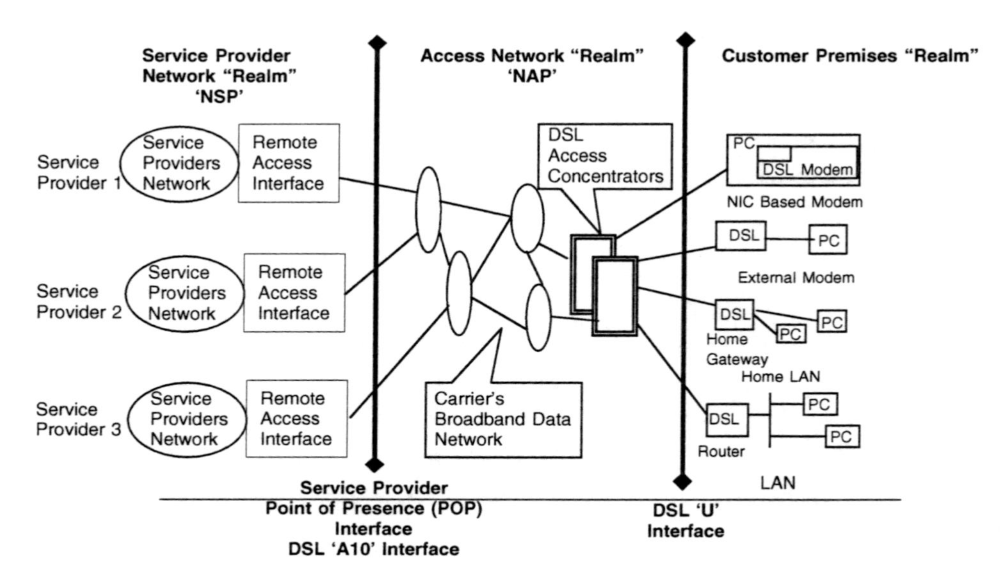

Figure 13.1 Division of realms of management control in a DSL network.

The carrier is responsible for configuring their network to support the end users' access to the service providers. Each service accessed may require its own path through the carrier's network and the specific parameters configured into the CPE and into equipment controlled by the service provider to use the path. However, ideally the amount of information that the three "realms" must exchange should be kept to a minimum.

- Most of the customers utilizing DSL for broadband access have little knowledge of network technology and little time for configuring or managing the network on their premises. In addition to the novelties of using a broadband access network, the customer may be installing a LAN for the first time in their home and small business. Neither the carrier nor the service provider can determine in advance what a customer will install in their home or business, yet the customer may expect assistance when they have difficulties connecting their new network to the DSL service. Both the carrier and the service provider have limited resources to provide this assistance.

The following features, which address the issues raised above, are required of the configuration tools for DSL CPE:

1. CPE installation by the user should not require a user manual or other specialized knowledge about the carrier's network or network configuration.
2. CPE installation and setup should require a minimum of manual configuration on the part of the user or the carrier's technicians. Complete autoconfiguration without human intervention is the ultimate goal.
3. CPE installation and setup should be consistent across multiple DSL access networks. Variations should be isolated from the end user. CPE from various vendors should work with many different DSL access architectures.
4. CPE must work regardless of the relationships between subscriber, network, and service providers. A carrier may support multiple service providers.
5. The CPE configuration methods must allow for the evolution of the services that are provided by the carriers and service providers.
6. The network operator should not be required to contact the end user for information about their installed CPE. That is, the CPE should be able to automatically notify the network of its capabilities over the configuration interfaces provided by the network. This is in contrast to the situation for ISDN where the user must provide the carrier with the SPID (service profile ID) of their CPE at the time service is ordered.
7. CPE and the DSL access network configurations should always be able to automatically maintain compatible settings. Reconfiguration of the ADSL access network should be allowed independently of the end user CPE. The user should never have to take action when the network is reconfigured. Conversely, rearrangements at the customer premise must never require the manual reconfiguration of the DSL access network by the carrier.

8. Communications—either manual or automated (bonding)—between the service providers and carriers should be kept to a minimum. The DSL industry is highly "horizontally stratified" such that several separate providers may be cooperating to provide the service to the customer. In many cases, the carrier that provides the DSL service, sometimes known as the network access provider (NAP), provides only the lower portions of the protocol stack as seen by the customer. The service provider, often known as the network service provider (NSP), is responsible for the remainder of the stack and access to the remote applications. Both the NAP and NSP require the ability to interact independently with the CPE to configure their parts of the service.
9. Widely deployed and standard configuration tools (e.g., PPP, SNMP, DHCP, etc.) should be used wherever possible.
10. The function of existing (or future) user applications (e.g., Web browsers or e-mail programs) should not be affected by the DSL configuration methods.
11. The tools to allow autoconfiguration of the CPE must be designed for various regulatory environments.

13.2 RELATIONSHIP BETWEEN THE STACK, THE NETWORK, AND THE CPE CONFIGURATION

A management architecture for DSL CPE, which meets the requirements mentioned above, can be built based on the observation that communications between the customer's equipment and the services available over a DSL network progress deeper into the network as each layer of the communication stack initializes. As the CPE initializes its communications protocol stack, it is able to receive configuration information from the part of the network that it is currently able to communicate with. The particular place in the network that the CPE has "reached" at any time during the initialization has the information that is required to get to the next further place in the network. The configuration tools associated with the protocol being initialized at that particular stage of the initialization process can be used to initialize the CPE. The network elements that initialize the CPE at that stage can be expected to have the specific protocol parameters required to initialize that particular layer of the protocol stack. Figure 13.2 illustrates this principle.

At time "T0" the connection between the customer premises and the network is completely uninitialized. There are no paths over the network established where any configuration information can be transported to the CPE from over the network.

At time "T1" the DSL Physical Layer (PHY) is initializing. The CPE is communicating with the edge of the carrier's network (to the DSLAM). The initialization features of the PHY interface protocol allow for the exchange of PHY configuration information. At the end of this stage, the CPE can communicate with the edge of the carrier's core network (the DSLAM).

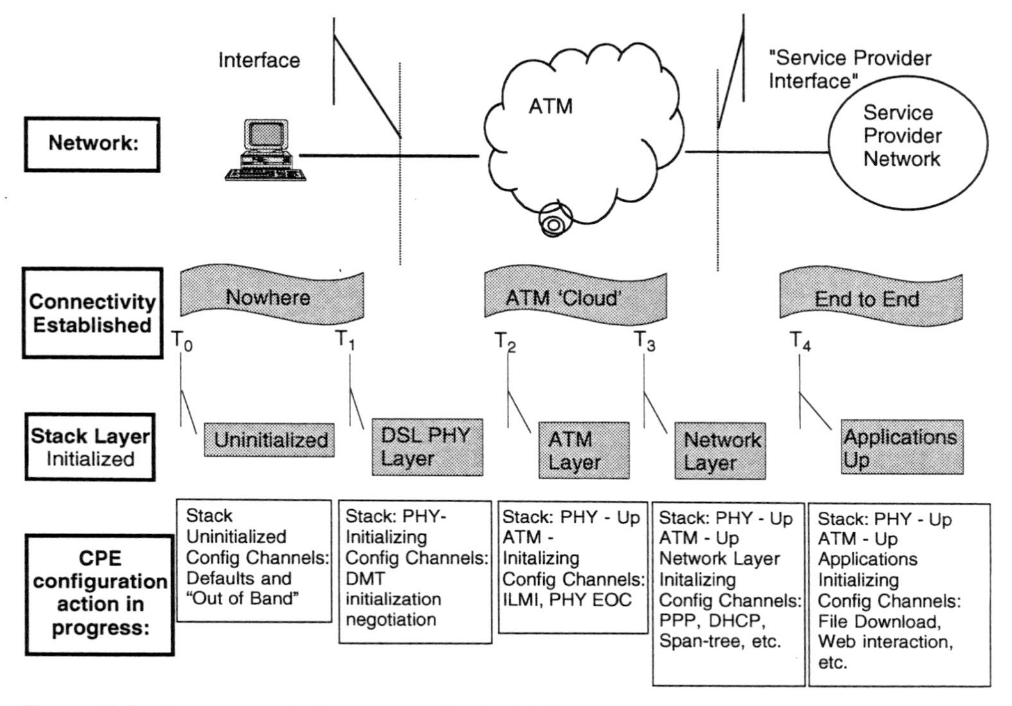

Figure 13.2 Progressive initialization of DSL CPE.

Assuming that the core network is ATM based, at "T2" the CPE is able to begin bringing up its ATM layer. Because communication is being established to the ATM elements in the carrier's network (of which the DSLAM is typically the one closest to the subscriber), information can be passed to the CPE about the configuration of the ATM layer from these elements via the ATM ILMI (integrated local management interface) [1] the standard configuration interface for an ATM link. In the case of the use of other link level protocols, such as frame relay, the initialization tools provided with those protocols can provide the CPE with the appropriate parameters from the network elements.

When time "T3" is reached, the CPE has established communications with the far side of the carrier's core network. The network protocols (such as IP) can now be initialized over the ATM connections that have been established.

At "T4" the CPE has been configured with the information required to utilize its protocol stack to communicate with the service provider at the far end of the network. Specific applications can now be initialized over this end-to-end service.

This philosophy of configuration supports most of the requirements named earlier. There are several primary advantages of this view of initialization:

- Existing and standard tools of network management can be used to allow the DSL network to initialize the CPE.

- Elements in the network provide information to the CPE about the parameters that they themselves control. This limits the amount of bonding between "management realms" required to support the autoconfiguration.
- Evolution of the network services is supported. An element supporting the lower parts of the stack does not have to be cognizant of the configuration required for the higher layer protocols. This allows the higher layer protocols to evolve without involving the elements that are only concerned with establishing the lower layers of the stack.

13.3 STANDARDS FOR AUTOCONFIGURATION OF DSL CPE

The DSL Forum is in the process of developing a series of recommendations for the autoconfiguration of DSL CPE which incorporates this philosophy of progressive initialization. An overall architecture is described in TR-46 [10]. The configuration of CPE to support the lower layers of an ATM-based stack is defined in TR-37 [8]. TR-44 [9] describes the configuration of the network parameters in CPE for those architectures that support an IP stack. There is work in progress to describe the tools for the autoconfiguration of CPE supporting interfaces more complex than those described in TR-44. Additionally, in the future, the DSL Forum expects to develop test plans to validate interoperability of equipment using their configuration framework.

13.4 THE DSL FORUM FRAMEWORK FOR AUTOCONFIGURATION

DSL Forum TR-46 [10] describes a progressive configuration architecture for DSL CPE. Figure 13.3, adapted from TR-46, provides an overview of the Forum's framework for autoconfiguration.

As with the scheme illustrated in Figure 13.2, the DSL Forum architecture assumes that the autoconfiguration progresses both up the stack and into the network. The Forum's autoconfiguration scheme then divides the space of CPE into three categories:

1. *Autoconfiguration of Basic IP Services.* Basic IP services are those IP Services that a user would use for best-effort fast Internet access. The higher layers of these services can be configured using PPP and DHCP.
2. *Autoconfiguration of Complex IP Services.* Complex IP services are those services for which CPE cannot be autoconfigured using only PPP and DHCP. These services include the configuration of firewalls, VPNs, and complex routing functions. The framework defines an ACS (autoconfiguration service) as the repository and protocol used for providing the appropriate parameters for these functions to the CPE.

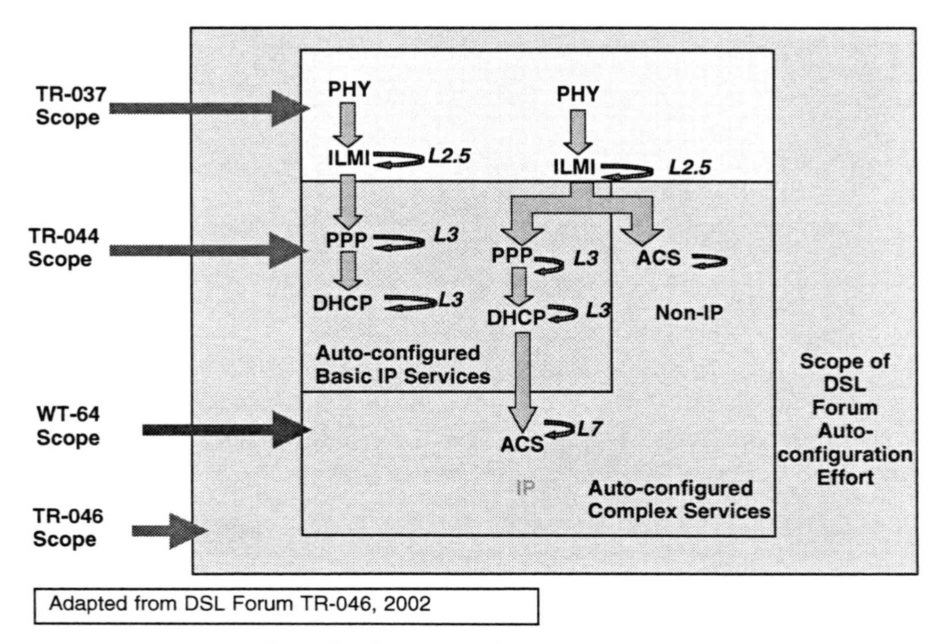

Figure 13.3 Autoconfiguration framework from DSL Forum TR-046.

3. *Non-IP Services.* Non-IP services cannot be configured using IP-based tools. Examples include non-IP voice-over DSL and certain video-over DSL services.

In all cases listed in Figure 13.3, the DSL PHY is assumed to provide tools within its own initialization protocol to automatically exchange the information required to allow the DSL path to initialize between the customer's premises and the provider's network. In those cases where ATM is used as the link layer, the ILMI (ATM Forum af-ilmi-065.0000 [1]) is used to transport the ATM configuration parameters between the network and the CPE. Extensions to the ILMI MIBs (management information bases) have been defined specifically to support DSL interfaces. These additions to the ILMI are defined in ATM Forum af-nm-0122.000 [2] af-nm-0165.0000 [3] and DSL Forum TR-37 [8].

Once the physical DSL connection and the ATM layer have been configured in the CPE, communications are possible to the equipment that supports standard IP stack tools to configure the IP parameters in the DSL CPE. In the case of "simple" IP environments, PPP and DHCP provided the necessary information to the CPE. In more complex environments, the two protocols are not sufficient to configure the CPE, and additional interfaces are required which are categorized in the DSL Forum's framework as ACS interfaces.

The non-IP architectures cannot use IP-based tools to configure the environment at the customer premises. Specialized ACS will be used to transport the service parameters to the CPE. An example of such an ACS is the configuration and management interfaces being defined for loop extension voice-over DSL (voice-over ATM) by the ATM Forum.

13.4.1 Use of ILMI to Configure the ATM Layer in a DSL Environment

TR-37 describes the use of the ATM ILMI, which provides the parameters that the CPE needs to use ATM virtual circuits to communicate over the DSL network. The ILMI [1] is an interface provided over the ATM protocols to allow configuration of the ATM interface on the link between two pieces of network equipment. The ILMI uses a specific ATM virtual circuit (VPI = 0, VCI = 9) to carry the SNMP management protocol between the equipment at each side of the link. The SNMP messages are directly carried on the ATM cell stream; no IP is required to support ILMI. SNMP MIBs, specifically designed for the configuration of ATM interfaces, are communicated over the SNMP interface provided by the ILMI. In addition to the ATM-specific MIBs defined in ATM Forum af-ilmi-065.0000, DSL-specific ILMI MIBs have been defined in ATM Forum Recommendations af-nm-0122.000 [2] and af-nm-0165.000 [3]. The following information must be provided to the CPE in order to allow it to utilize the ATM services provided by the network:

VPI/VCI: The virtual path identifier(s) and virtual circuit identifier(s) used over the interface. This information allows the DSL device to identify the virtual circuits (VCs) that are available on the interface.

PHY Identifier: This information indicates which latency channels are used for each VC. This is used with DSL interfaces such as ADSL that support multiple latency paths at the physical layer.

Encapsulation: This indicates the particular method used to carry the higher layer protocols over the particular ATM VC.

ATM Traffic Management Parameters: The ATM traffic parameters associated with each VC.

AAL: The ATM adaptation layer used over the particular VC.

L2 and L3 Client Protocols: Indicates the higher layer protocols carried over a particular VC.

The ILMI- and DSL-specific MIBs allow transport of this information to the DSL CPE. All information except the encapsulation and the L2 and L3 client protocols are known to the "owner" of the ATM network, the NAP (the "access network realm" in Figure 13.1). This information would be configured into the ATM network and DSLAM when the user's service is configured. The encapsulation and L2 and L3 protocols are, in general, determined by the service provider, the NSP (the "service provider realm" in Figure 13.1). The NSP must be able to communicate these to the NAP to allow the NAP to configure the ATM interface.

13.4.2 Configuring the Basic IP Environment

Once the initialization and configuration of the ATM layer is complete, the CPE is aware of which protocols are transported at Layer 3 of the stack on each ATM VC and the specific encapsulation used. For most IP environments, either PPP or DHCP or both may then be used to configure the appropriate IP parameters. DSL Forum TR-44 [9] discusses the use of these protocols in a DSL environment. Table 13.1 lists the appropriate configuration protocols for each of the commonly implemented IP architectures used over DSL. (These architectures are discussed in Chapter 14.)

PPP [4] allows the CPE to receive the following information from the network regarding the IP parameters:

- Assigned IP address
- Authentication mechanism used (AAA)
- Maximum transport unit size
- Domain name server addresses (DNS addresses)

PPP also supports the exchange of the authorization information used to identify and authorize the user's access to the NSP's network.

DHCP [6] allows the dynamic assignment of IP addresses and IP subnet masks by the network to the CPE.

Ethernet is a shared environment and the PPP protocol assumes a single PPP session is implemented on a point-to-point link, therefore an additional identifier is required to separate the multiple PPP sessions on LAN and ATM VC. This PPPoE session ID maps each PPPoE message to a particular end point on the Ethernet LAN. PPPoE utilizes a specialized initialization protocol to assign these IDs [7].

13.4.3 Complex IP Environments and Non-IP Environments

Neither DCP nor PPP allow for the configuration of features such as firewalls, complex routing rules, or network address translation (NAT) tables on the CPE. Additionally, in some service models the ability to configure arbitrary applications at the customer's site from the network is desirable. Proposals for interfaces to support these complex configurations include the development of specialized SNMP MIBs, the use of LDAP [5], and the development of specialized configuration

Table 13.1 IP Initialization Protocols Used with Common ATM-based End-to-End Architectures

Protocol Carried over ATM VC	Configuration Protocols for IP Layer
IP over Ethernet	DHCP
IP over AAL5	DHCP
PPP over ATM (PPPoA)	PPP, DHCP
PPP over Ethernet (PPPoE)	PPPoE initialization Protocol, PPP, DHCP

protocols. Similarly, non-IP–based services cannot be configured using tools such as PPP or DHCP. Certain applications such as voice-over DSL have their own configuration and management interfaces defined. In other cases, application-specific configuration interfaces may have to be developed. However, the same tools can be used for configuring the physical and ATM layers of all DSL CPE for both complex and simple higher-layer environments.

REFERENCES

[1] ATM Forum, *Integrated Local Management Interface 4.0, af-ilmi-0065.000,* September 1996.

[2] ATM Forum, *ILMI Auto-Configuration Extension,* af-nm-0122.000, ATM Forum, May 1999.

[3] ATM Forum, *Addendum to the ILMI Auto-Configuration Extension, af-nm-00165.000,* ATM Forum, 2001.

[4] IETF RFC 1661, *Point-to-Point Protocol,* W. Simpson, June 1994.

[5] IETF RFC 1777, *Lightweight Directory Access Protocol,* S. Kille, March 1995.

[6] IETF RFC 2131, *Dynamic Host Configuration Protocol,* R. Droms, March 1997.

[7] IETF RFC 2516, *Method for Transmitting PPP over Ethernet* (PPPoE), L. Mamakos, February 1999.

[8] DSL Forum TR-37, *Auto-Configuration for the Connection between the DSL Broadband Network Termination (B-NT) and the Network using ATM,* 2001.

[9] DSL Forum TR-44, *Auto-Configuration for Basic Internet (IP-based) Services,* 2001.

[10] DSL Forum TR-46, *Auto-configuration Architecture and Framework,* 2002.

[11] DSL Forum WT-64, *DSL Auto-Configuration Framework,* March 2001.

CHAPTER 14

NETWORK ASPECTS

14.1 EVALUATION OF PROTOCOL STACKS FOR DSL

The DSL Forum in TR-43 [1] discusses several ATM-based architectures that are currently deployed by carriers for DSL services. This document lists criteria that can be used for evaluating the appropriateness of a particular protocol stack for use in supporting a DSL service. Although TR-43 describes protocols that are currently deployed, these same criteria are also appropriate for the discussion of any protocol stack that might be used over a DSL system.

- Payload efficiency: How much overhead is required to support a particular protocol stack? What is the size of the protocol headers required, and what is the ratio of data required for these headers, compared with that available for carrying payload data?
- Error detection: Is the protocol stack capable of detecting errors introduced, as the data are transported end to end over the network? Is it capable of correcting these errors?
- Session multiplexing: Can the stack support multiple sessions from the customer's premises across the network? These separate sessions might

support multiple users from the premises, or multiple services to the same premises (e.g., voice and data). Also separate sessions might exist to separate end points at the far end of the network (e.g., simultaneous sessions from the same premises to different providers, such as an ISP and a corporate network).

- Multiprotocol support: What protocols can be supported over the stack? Given the same lower layers, what higher layer protocols can be supported on the stack?
- Autoconfiguration and management: From the viewpoint of equipment and software on the customer's premises, what information is required to configure this equipment to communicate over the network using a particular protocol stack? How much of this information is inherently carried in the information flows of the protocol? What information can be provided to the CPE over special management protocols? What are these management protocols? How much "bonding" (information exchange and network management cooperation) between the carrier and a service provider is required to allow the service to be configured? Is there any information that the end-user must configure manually to allow them to connect their equipment to the network?
- Service selection: Does the protocol stack allow the end-user to select the services that they use over the network? How does the protocol stack provide this function?
- Security: How secure are the communications? What features are inherent in the protocol to provide security to the user and to the network provider? What are the particular protocol stacks weaknesses?
- Authentication, authorization, and accounting (AAA): How does the stack provide support for these functions? How does it interact with existing tools to support these functions?
- Standards status: What parts of the stack are standardized? What is the status of the standardization for those that are not yet standards?

14.2 ATM-BASED ENVIRONMENTS

Most large-scale deployments of DSL by telecom carriers in North America and Europe use ATM to provide the end-to-end connections between the customer's premises and the service provider who provides access to the services used by the customer.

The use of ATM for the core of the access network, as shown in Figure 14.1, has several advantages for the access provider:

- Telephone companies have widely deployed ATM networks to provide other data services to their customers. They are thus familiar with their architecture, engineering, and management. The connection-based nature of ATM provides

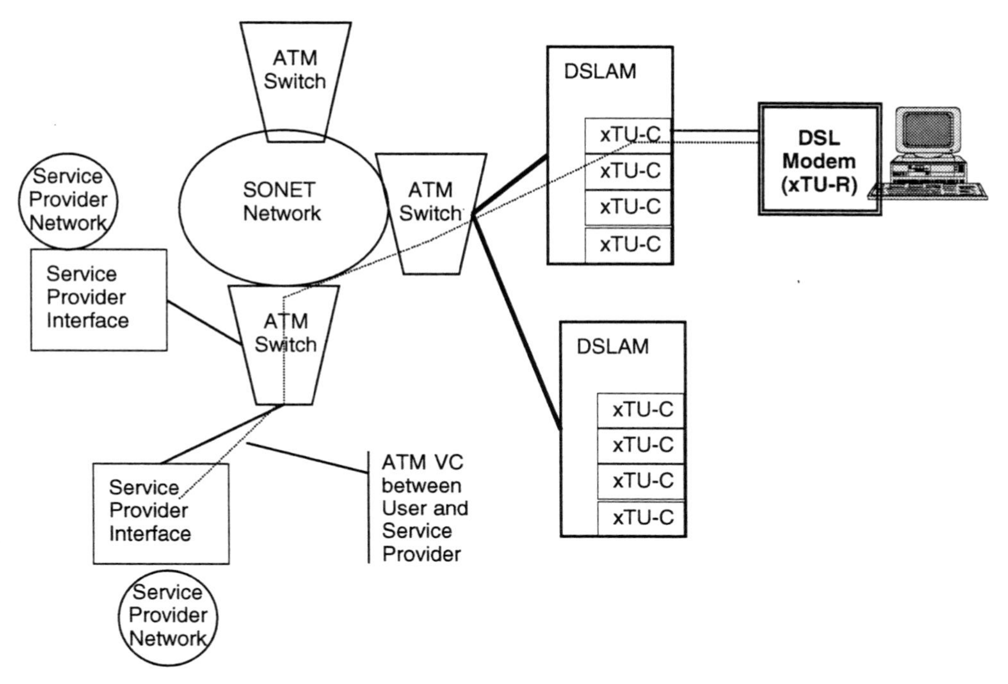

Figure 14.1 Overview of the use of ATM in DSL access architectures.

an architecture that the telephone carriers are comfortable with and know how to operate.

- The ATM-based core networks scale well as the number of customers supported by the network increase. The networks can be engineered so that the quality of service (QoS) to each customer is unaffected by the addition of new customers to the network.
- ATM protocols and networks support differentiated quality of service for different customers on the same network or for multiple services to the same customer. This gives the carriers the potential to offer multiple services to individual customers.

Table 14.1 lists the qualities of ATM-based architectures when compared with the criteria listed in DSL Forum TR-43.

14.2.1 AAL5

ATM adaptation Layer 5 (AAL5) defines the method for carrying datagram-based services (such as IP) over an ATM virtual circuit. An AAL5 PDU spans multiple cells and provides error detection for the transmitted data.

Table 14.2 lists the qualities of AAL5 when used to support DSL.

A stack based on ATM and AAL5 is used as the basis for most of the widely implemented DSL end-to-end protocols for data traffic. Figure 14.2 illustrates the stacks for these five protocols.

Table 14.1 Qualities of ATM-based Architectures

Criteria	ATM Qualities
Payload efficiency	ATM has a 5-byte header in every 53-byte cell resulting in a 10% "header tax" for all communications over ATM.
Session multiplexing	ATM provides intrinsic support for multiple sessions via the use of multiple virtual circuits between a network and the customer's premises.
Multiprotocol support	An ATM virtual circuit can support many different protocol stacks. Many IP-based environments can be supported using AAL5 [5]. AAL1 is appropriate for transport of services requiring emulation of a constant bit rate point-to-point circuit, and AAL2 was developed to carry digital telephony efficiently.
Autoconfiguration and management	The integrated local management interface (ILMI) provides a method of configuring the ATM service at the customer premises. DSL Forum TR-37 describes the use of ILMI for configuration of DSL services.
Service selection	Although the use of permanent virtual circuits (PVCs) requires preconfiguration, the use of switched virtual circuits (SVCs) allows communication between the users' premises and the service providers to be set up in real time as needed. SVC setup can be used as part of a service selection process or a user can select the appropriate preconfigured PVC to reach a particular service.
Security and AAA	The support of VCs end to end keeps communications for different users and services separated, which provides some security enhancements. The security functions of higher levels of the stack are unaffected by the use of ATM.
Error detection	Errors to cell headers introduced during transmission can be detected.
Standards status	ATM is defined in numerous standards developed by the ITU-T and ATM Forum.

14.2.2 Multiplexing Other Protocols on AAL5

IETF RFC 2684 [2] defines several methods for encapsulation of data protocols over ATM virtual circuits. It allows the definition interfaces that permit the ATM circuit to be used by a bridge to transport link level protocols (such as Ethernet) to the remote site. It also allows the definition of interfaces that permit routed protocols such as IP to be transported directly over the ATM virtual circuits. RFC 2684 supports two different methods for carrying connectionless network interconnect traffic, routed and bridged protocol data units (PDUs), over an ATM network. The first method allows multiplexing of multiple protocols over a single VCC (virtual circuit

Table 14.2 Qualities of AAL5

Criteria	AAL5 Qualities
Payload efficiency	AAL5 adds 8 bytes of trailer to each PDU. The effect on total payload efficiency is a function of the PDU length, which can be as high as 65,535 bytes.
Session multiplexing	AAL5 does not support session multiplexing within a particular VC. The higher levels of the stack supported by AAL5 may support multiple sessions.
Multiprotocol support	RFC 2684 describes methods for encapsulating multiple protocols over AAL5.
Autoconfiguration and management	The use of AAL5 on VCs on a DSL connection can be communicated to the user's CPE via the ILMI interface.
Service selection	This is a function of higher-layer protocols.
Security and AAA	AAL5 neither enhances nor degrades the security function of protocols above or below it in the stack.
Standards	AAL5 is defined in ITU-T Standards I363.5 [5].

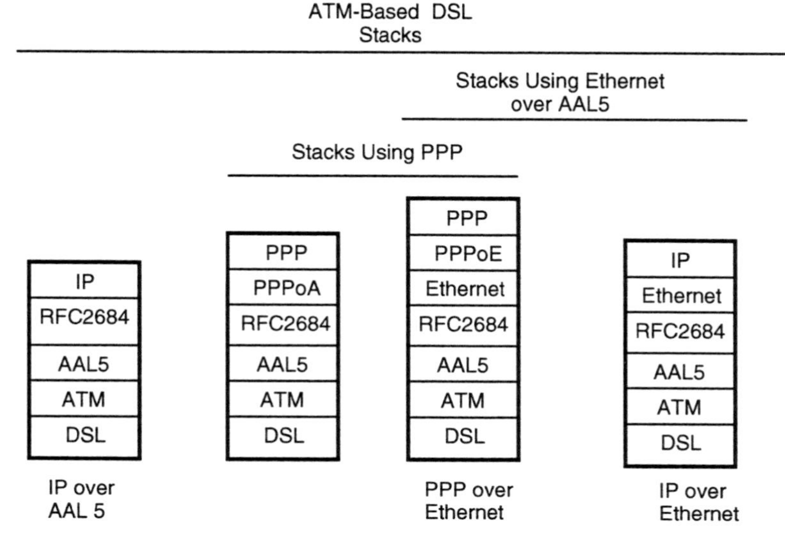

Figure 14.2 Stacks for four commonly implemented protocols over ATM/AAL5 at the "U" Interface. This Interface is the interface that carries the DSL physical layer between the customer's premises and the network. *Adapted from DSL Forum TR-43, 2001.*

connection). The protocol carried by a PDU is identified by prefixing the PDU with an IEEE 802.2 LLC header [9]. This method is called "LLC encapsulation." The second method does higher-layer protocol multiplexing implicitly by VC, and it is called "VC multiplexing." LLC encapsulation is used to carry Ethernet over ATM VC in a DSL environment, and VC multiplexing is used to carry IP directly over the AAL5 adaptation layer.

Bridged Ethernet

Figure 14.3 illustrates a DSL access architecture that uses IP over a bridged Ethernet. Using RFC 2684 LLC encapsulation allows the Ethernet packets to be carried over an ATM VC supporting AAL5 adaptation. If the Ethernet PDUs carry IP packets, then IP packets carried on the LAN will be transported over the Ethernet on the DSL connection. PPP over Ethernet discussed in Section 14.1 also uses LLC encapsulated Ethernet to bridge the premises LAN to the far end of the network.

Table 14.3 shows the IP over Ethernet qualities for the encapsulation of Ethernet over AAL5 using RFC 2684 encapsulation.

IP Directly on AAL5

IP can be directly transported on AAL5 using RFC 2684 VC multiplexing. In this case a router at the customer's premises terminates the DSL circuit and the ATM AAL5 VC. Figure 14.4 illustrates an access network supporting IP directly on AAL5.

Table 14.4 lists the qualities of a protocol stack supporting IP on AAL5.

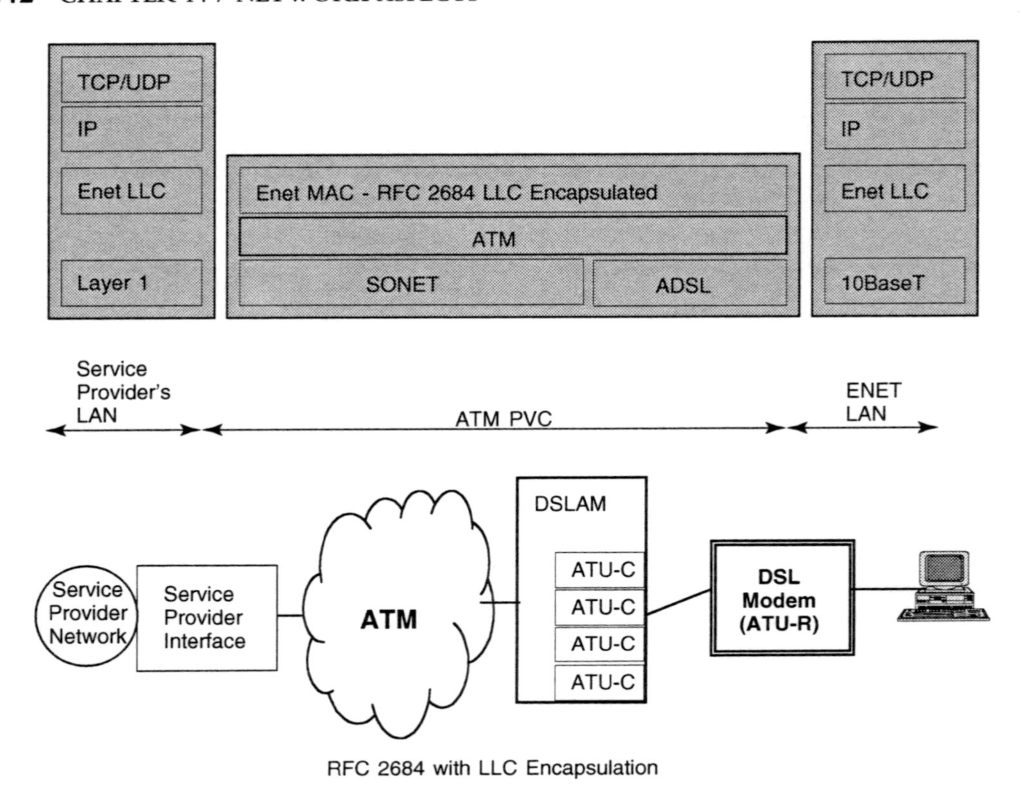

Figure 14.3 Bridged Ethernet directly transporting IP.

Table 14.3 IP over Ethernet Qualities

Criteria	IP over Ethernet Qualities
Payload efficiency	Ethernet requires an 8-byte header for each 1,492-byte Ethernet packet. The RFC 2684 LLC encapsulation requires an additional 8 bytes of header information.
Session multiplexing	RFC 2684 LLC encapsulation does not support sessions directly. However, higher lever protocols (such as PPPoE) can support sessions.
Multiprotocol support	RFC 2684 describes methods for encapsulating multiple protocols over AAL5. Ethernet is capable of supporting multiple protocols.
Autoconfiguration and management	The ATM ILMI can be used to indicate that LLC encapsulation is used. If IP is transported directly over Ethernet, standard IP tools such as DHCP can be used to configure the customer's devices over the network.
Service selection	This is a function of higher-layer protocols.
Security and AAA	This is not a function provided by RFC 2684. Higher-level protocols can be carried over LLC encapsulation, and can provide these functions.
Standards	IETF RFC 2684.

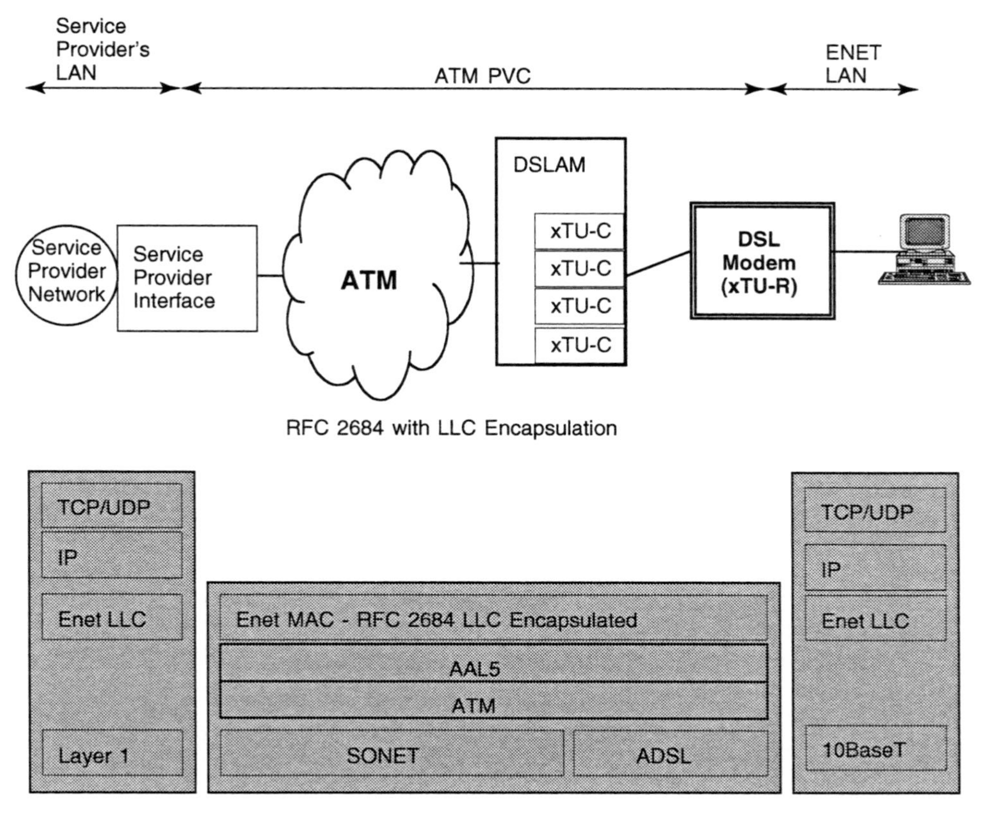

Figure 14.4 IP directly on AAL5.

Table 14.4 IP on AAL5 Qualities

Criteria	IP on AAL5 Qualities
Payload efficiency	No extra overhead is added beyond that required for ATM, AAL5, and IP components of the stack.
Session multiplexing	IP directly on AAL5 does not support session multiplexing.
Multiprotocol support	This protocol stack is meant to support IP and protocols that run over IP (other routed protocols such as IPX can also be supported).
Autoconfiguration and management	The use of IP over AAL5 can be communicated using the ILMI interface. The IP layers of the stack can be configured using standard IP tools.
Service selection	This is a not a function of this stack.
Security and AAA	This function is not inherent in this stack. However, security tools such IPsec can be supported over IP.
Standards	VC multiplexing is defined in RFC 2684.

Table 14.5 PPP Qualities

Criteria	PPP Qualities
Payload efficiency	PPP adds a 2-byte header to each PDU.
Session multiplexing	As PPP can support multiple simultaneous protocols, multiple sessions can be implemented.
Multiprotocol support	PPP allows the support of multiple protocols over each session. Each PDU contains a protocol ID to identify the payload type.
Autoconfiguration and management	PPP allows the negotiation of link capabilities such as compression and encryption using link control protocols (LCP) during session startup. Network control protocols (NCP) are used to support automatic address assignment. Additional configuration information can be provided using DHCP.
Service selection	It is possible to use the PPP negotiation process to select access to a particular service.
Security and AAA	The functionality provided by the PPP LCP protocols allows for the authentication and identification of users of a service. PPP allows the negotiation of the use of encryption on a link.
Standards	Defined in IETF RFC 1661 [6].

14.2.3 Use of PPP

Point-to-point protocol (PPP) was defined as the Internet standard for transmission of IP packets over serial lines. PPP supports the assignment and management of IP addresses, network protocol multiplexing, link configuration, error detection, and data-compression negotiation. The protocol is described in IETF RFC 1661 [6]. Although originally defined for use over point-to-point connection, such as those provided by analog modem over the dial telephone network, the protocol has been adapted for use over broadband connections such as those provided by DSL services. Two such adaptations of PPP have been defined:

- PPP over ATM (PPPoA)
- PPP over Ethernet (PPPoE)

PPP over ATM

PPP over ATM is described in IETF RFC 2364 [3] and DSL Forum TR-17 [8]. Figure 14.5 illustrates the protocol stack and network architecture for PPP over ATM.

PPP over ATM is specifically designed to support environments where the user's device supports the termination of both the ATM session and the PPP session. For example, a workstation or PC that supports an ATM stack directly and contains a DSL modem internally could support this protocol. Similarly, a router that used DSL for WAN access could terminate a PPP over ATM session to the service provider while routing traffic locally on the LAN.

Table 14.6 shows the qualities of PPP over ATM as a network protocol for DSL access.

PPP over Ethernet

PPP over Ethernet addresses the problem of connecting multiple devices over DSL access on the customer's premises. The protocol solves this problem by allow-

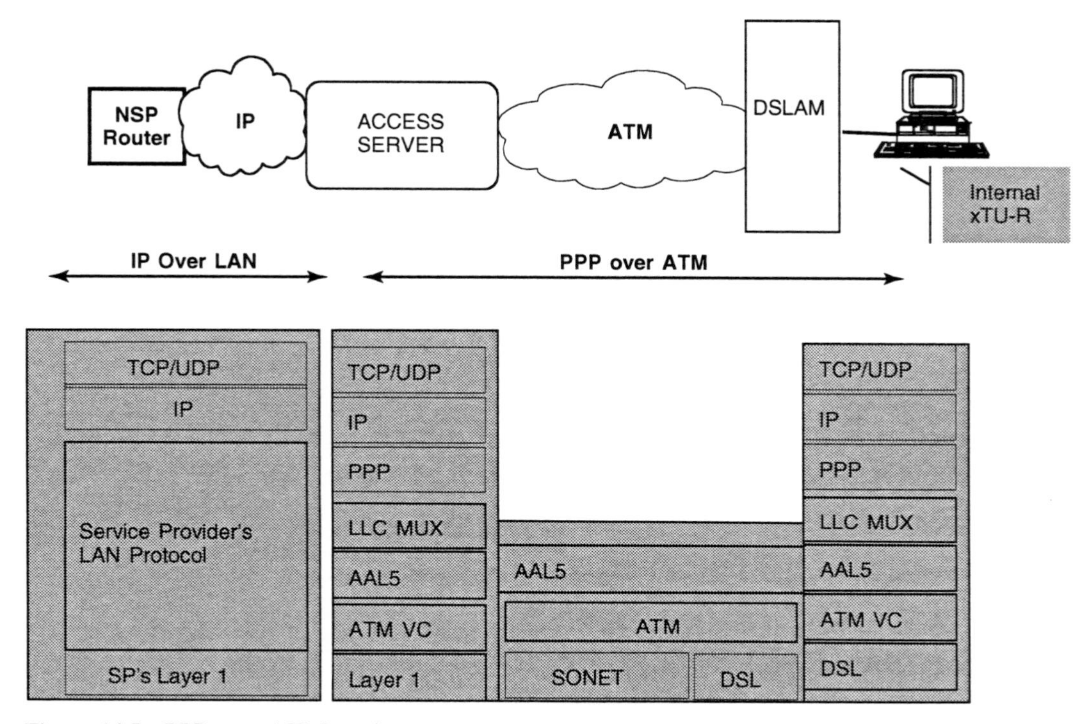

Figure 14.5 PPP over ATM stack.

ing the transport of multiple PPP sessions over an Ethernet LAN, which is multiplexed over AAL5 on the DSL WAN interface. As with PPP over AAL5, PPP over Ethernet can provide a PC user with an access environment that simulates the dial-up environment seen with analog modems on the PSTN. However, multiple PPP sessions can be multiplexed over one ATM VC, allowing relatively simple support for multiple devices over the DSL interfaces, and the ability for each user on the

Table 14.6 PPP over ATM Qualities

Criteria	PPP over ATM Qualities
Payload efficiency	No extra overhead is added beyond that required for the ATM, AAL5, and PPP components of the stack.
Session multiplexing	Only a single PPP session can be supported per ATM VC.
Multiprotocol support	PPP over ATM inherits the multiprotocol capabilities of PPP.
Autoconfiguration and management	PPP over ATM inherits the management tools that are provided by the ATM and PPP components of the protocol stack.
Service selection	It is possible to use the PPP negotiation process to select access to a particular service. Additionally, multiple ATM VCs could be used to access different services.
Security and AAA	The functionality provided by PPP provides the security and AAA functionality for PPP over ATM.
Standards	Defined in IETF RFC 2364.

LAN to select from multiple service providers over a single ATM VC. The configuration of the ATM network is therefore quite simple when PPP over Ethernet is supported; a single VC can support multiple simultaneous users of multiple services without the complexity or expense of a router on the customer's premises. Figure 14.6 illustrates the architecture of a PPP over Ethernet network.

PPP over Ethernet has the qualities shown in Table 14.7 as a protocol for DSL access.

14.3 ETHERNET DIRECTLY OVER DSL

Fiber-based Gigabit Ethernet capable of supporting public metropolitan area networks is an emerging technology for supporting a carrier's access network. As these networks are deployed, DSL interfaces become an obvious choice for providing

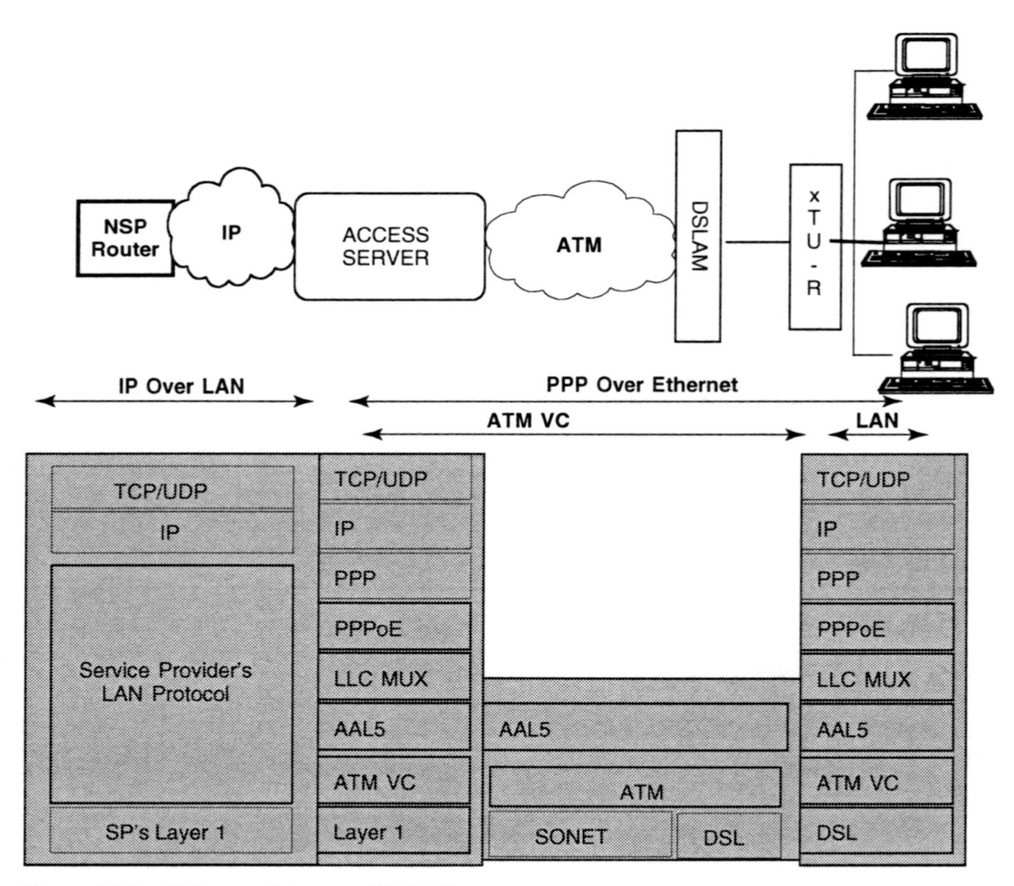

Figure 14.6 PPP over Ethernet (PPPoE).

Table 14.7 PPP over Ethernet Qualities

Criteria	PPPoE Qualities
Payload efficiency	PPPoE adds a 6-byte header beyond those that are required for AAL5, Ethernet, and PPP. As Ethernet is carried over the ATM network, the Ethernet overhead is added to the end-to-end communication.
Session multiplexing	Multiple PPP sessions can be multiplexed over the Ethernet LAN and over a single ATM VC.
Multiprotocol support	PPP allows the support of multiple protocols over each session. Each PDU contains a protocol ID to identify the payload type. Additionally, because Ethernet is extended over the WAN (using RFC 2684), protocols other than PPP can be supported on the same ATM VC simultaneously with multiple PPPoE sessions.
Autoconfiguration and management	The autoconfiguration functionality provided by ATM, PPP, and the IP stack are available to support autoconfiguration.
Service selection	The ability to negotiate an individual PPP session from each end point on the LAN allows each user to access a particular service on the far end of the network.
Security and AAA	The security functions provided by PPP are available when PPPoE is used.
Standards	Defined in IETF RFC 2516 [7]. This is an informational RFC that indicates common industry practice.

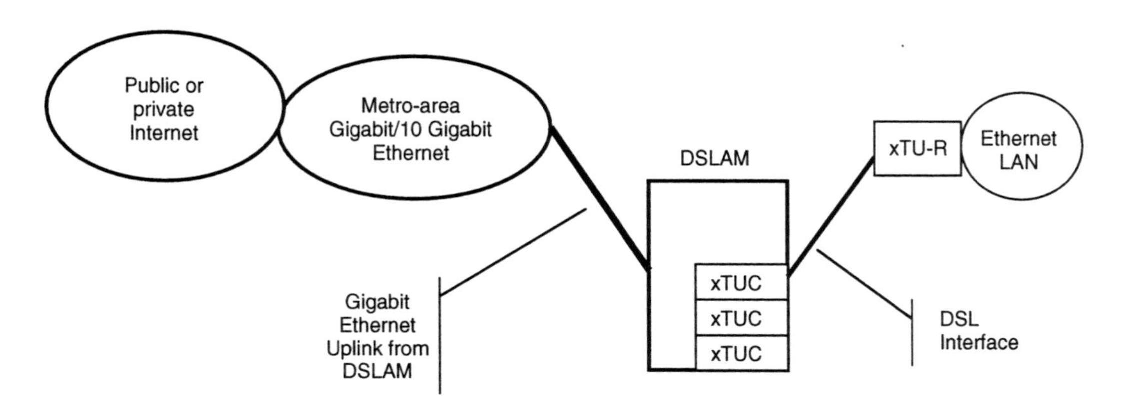

Protocol Stack for Ethernet Directly on DSL

Access Carriers Metropolitan Ethernet	DSL Interface	Customer Premises
IP	IP	IP
ETHERNET	ETHERNET	ETHERNET
Gigabit or 10 Gigabit Ethernet	DSL	100 BaseT

Figure 14.7 An overview of DSL used to access a metropolitan Ethernet.

customer access to these new networks. Figure 14.7 gives a high-level overview of the use of DSL to provide access to such a network.

Gigabit Ethernet has compelling attractions as a core transport for metropolitan access networks:

- Most LANs at the customer's premises use Ethernet. Extending these Ethernets LANs directly over the metropolitan access network provides a simple architecture.
- The development of routers capable of routing packets delivered at gigabit and higher rates (line speed routers) makes development of these networks relatively straightforward.
- The capital cost for metropolitan Ethernet architectures is likely to be lower than comparable architectures based on ATM or IP over SONET.
- It is claimed that management of these networks will be simpler than the alternatives.

As these networks are deployed, the use of DSL to provide the "last mile" of access to many customers becomes a virtual necessity.

- Fiber capable of supporting Gigabit Ethernet directly to a customer's site is economic only for the largest of customers of a metropolitan Ethernet. Other customers require the inexpensive but lower bandwidth interfaces provided by DSL technologies.
- Only DSL, with its use of twisted pair copper, provides the ubiquitous access required to reach all possible users.
- Although attempts are being made to standardize a specific Ethernet PHY (the IEEE "Ethernet in the last mile" effort) over telephone loop quality twisted pairs, DSL standards and implementations are already mature and capable of supporting Ethernet directly on the DSL PHY. Although additional standards work is required to define the direct support of Ethernet on DSL interfaces, this work is incremental and would be based on the foundation of existing DSL physical layer standards.

Table 14.8 Ethernet Directly over DSL

Criteria	Ethernet Directly over DSL
Payload efficiency	There should be little or no additional overhead beyond those already required for Ethernet and the protocols above it in the stack.
Session multiplexing	Ethernet is capable of supporting sessions via the use of PPPoE.
Multiprotocol support	Because the Ethernet is extended over the WAN, any stack capable of being carried over Ethernet can be transported.
Autoconfiguration and management	Ethernet is largely self-configuring.
Service selection	The service selection functions of protocols carried over Ethernet are available.
Security and AAA	Security is a function of the higher layers of the stack.
Standards	Ethernet is defined in the IEEE 802 series of standards. The specifics of adaptation of Ethernet to the various DSL technologies need to be standardized.

Ethernet directly over DSL has the qualities shown in Table 14.8 when compared with some of the criteria from DSL Forum TR-43.

REFERENCES

[1] DSL Forum TR-43, *Protocols at the U Interface for Accessing Data Networks Using ATM/DSL,* August 2001.
[2] D. Grossman and J. Heinanen, *Multiprotocol Encapsulation over ATM Adaptation Layer 5, IETF, RFC 2684,* September 1999.
[3] M. Kaycee, G. Gross, A. Lin, A. Malis, and J. Stephens, *PPP over AAL5, IETF RFC 2364,* July 1998.
[4] DSL Forum TR-037, *Auto-Configuration for the Connection between the DSL Broadband Network Termination (B-NT) and the Network Using ATM,* 2001.
[5] ITU-T, *I363.5, B-ISDN ATM Adaptation Layer (AAL) Type 5 Specification,* August 1996.
[6] W. Simpson, ed., *IETF RFC 1661, The Point-to-Point Protocol (PPP),* July 1994.
[7] L. Mamakos et al., *A Method for Transmitting PPP Over Ethernet, IETF RFC 2516,* February 1999.
[8] DSL Forum TR-017, *ATM over ADSL Recommendations,* 1999.
[9] IEEE 802.2 1998/ISO/IEC 8802-2, Information technology—Telecommunications and information exchange between systems—Local and metropolitan networks—Specific requirements, Part 2: Logical link control, 1998.

CHAPTER 15

DSL AND SECURITY

As DSL is implemented to support deployments of low-cost services to large masses of customers, network security issues that formerly were primarily the concerns of the larger enterprises and governments begin to effect the naive user in the home and small business. Many of these emergent issues are common to all forms of mass broadband access, that is, cable modem, fixed wireless, or DSL. Some of the security issues shared among all broadband access methods to mass retail users include:

- Problems raised by an always-connected service
- Naivete of the typical broadband accesses user
- Increasing complexity of the networks in homes and small businesses
- Complex interactions between the broadband users and the networks of others

In certain regards, DSL architectures provide an amount of inherent security that is not found in the other broadband access methods.

- DSL does not use a shared physical medium to reach individual customers.
- The use of end-to-end ATM virtual circuits in many DSL implementa-

tions keeps the communications of particular users separate from each other throughout much of the access network.

Even with these positive architectural features, the integrity of both the users data and systems and those of the access network and service providers must be considered in the design of any DSL system.

In ensuring the security of assets on any network or of the services supported by that network, the following areas must be considered:

- Unauthorized interception of data transmitted
- Masquerade of one user by another
- Malicious modification of data transmitted over the network
- Unauthorized access or modification to data or resources connected to the network
- Unauthorized modification of configuration of the networks
- Denial of use of the networks or resources to authorized users
- Theft of service
- The legitimate requirements of the government and owners of the networks to detect and prevent malicious and illegal activities

A secure environment has several important qualities. Those authorized to perform a particular action or to access particular resources can do so with relative ease, without being monitored by unauthorized "watchers." Those who are not authorized are blocked from accessing, modifying, or using data or resources that they are not entitled to. Attempts at malicious actions can be detected and prevented.

Although easy to describe, providing such a secure environment for any network is difficult. The security methods must prevent unauthorized activities while allowing users to perform their business with minimal interference. Security methods that are too difficult to use or administer will be ignored, or worse, disabled, by the authorized users who see themselves as being overly inconvenienced. Weak methods that may be easy to use or administer only produce a false sense of security. Those who may attack systems are creative, and will evolve their methods of attack. This evolution will usually occur more rapidly than the increases in security for the network and its resources. Unfortunately security is typically only improved after successful attacks. Security is an expense whose value, in a sense, is only seen when it fails. A successful security system on a network, or lack of loss on an unsecured system, tends to result in complacency among its users and administrators.

In a mass deployment of broadband access such as DSL, these issues are accentuated. The users may have few resources to devote to security, little knowledge, and little desire or time to increase that knowledge. The operators of the networks and the services on those networks are caught in a similar bind. They must keep costs low to be able to offer services at a price that their customers are willing to pay. If services are insecure and can be easily stolen from the carrier, or malicious action can easily bring down the network, the carrier may not be able to afford to offer the services. If the services are made too complex by security functions added by the carriers, the users may reject the services. On the other hand, if security for the user's resources is not sufficient, the users may avoid the DSL services out of fear of theft (of valuable information such as credit card data) or damage to their home and business computing resources.

15.1 GENERAL SECURITY ISSUES FOR BROADBAND SERVICES

Several issues exist for all mass-deployed broadband services:

- Services are always connected.
- Users can be naive.
- Networks in homes and businesses are becoming increasingly complex.
- Broadband users have more complex interactions with the networks of others.

15.1.1 Problems Raised by an Always-connected Service

Unlike dial-up connections to the Internet, DSL connections are likely to be always connected to the wide area resource. This opens the user from the home or small business to a number of attacks that are less likely in the traditional environment of intermittent connections.

In a dial-up environment, the IP address of the user's system is not static, as it is typically assigned anew from a pool of available addresses owned by the ISP each time the user establishes a connection. Therefore, there is no permanent relationship between a dial-up user and a particular IP address. This provides a certain level of security for the user. In this transient environment it is not possible to establish a relationship between a user and a particular address, which provides some anonymity for a user's communications on the public Internet. In the case of a DSL connection, which is permanently connected, the IP address is never reassigned and the user's communications can associated with a particular address. This opens the user to a number of security issues.

1. Potentially anonymous communications, such as individual access to particular Web sites, can be associated more easily with particular users, as the IP address for the user never changes.
2. Even in the case of encrypted communications such as those based on SHTML (secure Hypertext transfer protocol) and SSL (secure sockets layer), the TCP (or user datagram protocol, UDP) and IP headers to and from the user are not encrypted. The unchanging address potentially allows those not privy to the communication to still gather information about transactions from the user based on their IP address.

3. The fact that the user's resources are always connected to the network provides an opening to attack. The attacker has as much time as required to enter the user's system because the unchanging IP address and always "up" connection allows the attacker to continue the attack until the user's system is penetrated.

Once the attacker has gained access to the user's system, a number of types of mischief are possible. In addition to stealing the user's information or damaging the user's resources, the attacker can use the system for their own purposes by installing their own data or programs on the suborned system. The fact that the user's system is always connected makes the system potentially very desirable to the attacker. The attacked system can then be used for further network mischief, such as denial of service attacks on the network from the user's computer. The user's PC can be configured as a Web server without the user's knowledge. In addition to broadcasting information about the user's system to the outside world, the commandeered system can be used to distribute information that the hacker has installed on the user's computer. The high speed of the DSL connection makes the user's system even more desirable to a hacker than dial-up connections.

A number of relatively simple remedies reduce the danger of the attack on the permanent DSL connection.

1. Most PCs running the Microsoft® Windows™ operating system come with default configurations that are insecure for any computer in an environment that is connected to the Internet. Simply turning off the "file and printer sharing" option does much to increase the security of an always connected PC.
2. If the user turns off their computer when it is not in use, it reduces both the window of vulnerability to attack and the computer's usefulness to malicious users. However, this reduces some of the benefits of the broadband connection for the user. It may not be feasible for users such as small businesses or home offices where the computer is used nearly continuously.
3. Inexpensive personal firewall software can be installed on the users computer. These software packages can perform a number of security functions such as:
   a) Detecting suspicious activities by observing known communications patterns that are likely to be used by hackers.
   b) Blocking communications from suspicious addresses.
   c) Blocking communications from the Internet to TCP/UDP ports on the protected computer that could be used by an attacker.
   d) Auditing the configuration of the operating system and aiding the user in setting up their system to give them maximum protection.
   e) Notifying the user of potential intrusions.
4. A hardware-based firewall provides similar functions to the software based personal firewall tools; however, because it is hardware based, it can have additional functions, improved performance, and protect multiple devices on the home or small business local network. In many cases the DSL modem, a local router or switching hub, and a firewall can be integrated into a single device.

15.1.2 Naivete of the Typical Broadband Access User

Home and small business users are often ignorant of security issues, and even if they are interested in the problem, they are likely to have little time to manage security tools in their homes or offices. Unlike the large enterprise where a specialized staff can oversee the security systems on the network, the smaller users have very limited resources for this complex function. Large enterprises with their professional, full-time network security staffs often experience breaches. How can the home or small business user be expected to learn how to configure and monitor their environment with their much more limited resources? Assumptions cannot be made about the value of the information on any user's computer. Identity theft through hacking a home PC is a serious problem for a particular home user, and the information on a business computer may be vital to the continued existence of that small operation.

Although the vendors of tools such as personal firewalls and simple hardware-based security tools attempt to make their applications as self-configuring and intuitive as possible, some learning and resources are required on the part of the user. Additionally the continuing evolution of attacks requires periodic updates, reconfigurations, and replacements of even the most user-friendly tools.

Addressing this issue is difficult in the DSL environment. However, it also provides a means for access and service providers to offer additional services to their customers and thus differentiate themselves from other providers of broadband service or to obtain additional revenues from additional services to their customers.

1. A service provider could provide appropriate personal firewall and other security software to all users as part of the installation package for new users of the network. The hardware provided to support customers with home or small business LANs should be expected to contain firewall functionality.
2. Security consulting and "for fee" security help desk services to small business customers can be a source of additional revenue for the carriers and service providers.
3. The placing of security functions in common points of the access provider's network can provide protection to all users of the carrier's access network. Placing the firewall in an aggregation point in the carrier's network can support detection and filtering of suspicious communications and forbidden activities at a common point in the access network. However, this can only occur at cost to the carrier (that might be recoverable by making this an added value service to the end user). Also reductions in function, flexibility, and performance of the access system as seen by the end user would likely occur.
4. The carrier could manage the security functionality at the user's premises for the user. The software installed in the DSL access device could be installed, configured, and monitored by the carrier or service provider. Such a service is likely to be especially desirable for small businesses that may have critical resources to protect on their network connected systems. A particular advan-

tage of ATM-based DSL architectures for such a service is the possibility of using a dedicated virtual circuit to allow a provider to manage the security function remotely. This would provide some protection from access by hackers.

15.1.3 Increasing Complexity of the Networks in Homes and Small Businesses

Users of DSL in both the home and small business are likely to support more complex networks at their premises than have been typical of the such environments in the past. In addition to the security issues mentioned in the previous section, DSL environments create security issues that are specific to these environments.

In a home environment with multiple computers, one cannot assume that the users share identical security interests. For example, the PC used by a parent may access that person's employer over the DSL connection from the home office, while children may have PCs used solely for "entertainment" Internet access. In this case, incoming and outgoing access restrictions for the parent and his computer are obviously different from those for the children from their computers. A firewall function at the gateway to the home (or within the access or corporate network) is not sufficient to ensure that only the authorized user can access the parent's corporate network. Only security that extends from the parent's computer to the employer's network can provide such control. The support for a virtual private network from the parent's PC, through the gateway and through the Internet, is one solution to this issue. Additionally, the use of separate ATM virtual circuits (VCs) dedicated to each of the users can ensure that the traffic for these two environments is kept separate.

The existence of home networks with multiple resources connected to them itself adds considerably to the complexity of keeping the home or small business environment secure. Personal firewalls on a single PC are a suitable security solution for the home or small business with such a simple environment. However, in environments with multiple devices on a home LAN, managing separate firewalls on each PC can become an administrative nightmare. In a DSL implementation supporting a home network, hardware-based applications on the device supporting the DSL access to the home is an obvious solution. Not only does this centralize the security function in the home or business, but also it allows the use of enhancements such as multiple VCs, and network address translation to enhance the security to the entire home network.

15.1.4 Complex Interactions between the Broadband Users and the Networks of Others

A PC on a DSL-connected home network that accesses another secure network creates security issues for that network. Breaching of resources on a vulnerable home network may allow access by unauthorized users to the remote secure network. Thus, the protection for the home network becomes part of the security cordon for the remote network; one that is not under the control of the administrators of that network. The home network thus becomes a portal for hacking another network and a weak point for other networks accessed from that home.

15.2 DSL-SPECIFIC SECURITY ADVANTAGES

DSL access provides two specific security advantages over other mass broadband access technologies:

1. DSL provides a point-to-point connection over dedicated copper facilities between the users premises and the telephone company switching office or remote site. Other users in the neighborhood do not share transmission resources, and it is not possible for one DSL user to access the signal for a second user. This is contrasted with the situation for both cable modem and fixed wireless broadband access where other users, or malicious actors, have physical access to signals meant for others over the shared media of hybrid fiber COAX (for cable) or fixed wireless transmission. Unauthorized reception, masquerade, and theft of service are thus much more difficult in a DSL environment than either of the other mass broadband technologies.
2. ATM-based DSL architectures also provide one or more VCs dedicated to each customer over the DSL physical connection. These connections extend from the customer's premises through the access network to the edge of either the public Internet or the private networks which are accessed via the DSL connection. The existence of the separate VCs makes unauthorized monitoring, or access to the communications more difficult than in environments where only connectionless protocols such as TCP/IP are utilized. Additionally the use of VCs allow the creation of specific paths for security and management information sent between a service provider and particular user protecting any remote configuration of equipment at the user's premises. Dedicated VCs can also be used to provide secure communication between particular resources at the user premises (e.g., a particular PC used by one member of a household) and a particular access point in the network (e.g., the access point to a secure corporate network).

CHAPTER 16

VOICE-OVER DSL (VODSL)

In addition to supporting data or multimedia traffic, DSL can be used to convey voice simultaneously along with the data. Several methods are discussed in this chapter.

16.1 THE HISTORY OF DSL AND TELEPHONY

DSL technology and telephony have been intimately associated from the invention of the very first digital line protocols. The ISDN basic rate interface (ISDN BRI) was developed to carry both digital voice and data (at rates between 64 Kbps and 128 Kbps) over the same phone line to environments such as residences, small businesses, and larger business sites using Centrex-type services. T1 carrier systems were originally created to provide digital trunk lines between telephone company facilities. Their use in what may have been the first "digital subscriber line" service to business customers for providing both private voice and data services to business customer premises was originally seen as providing telephony services. The ISDN primary rate interface (ISDN PRI), which provides (in North America) twenty-three 64 Kbps channels for voice or data, is an extension of the T1, explicitly to provide combined voice and data to larger businesses. HDSL (high-speed digital subscriber line), the first of the technologies to be given the title DSL, was developed to provide improved and lower cost support for T1.

As illustrated in Figure 16.1, ADSL supports both traditional analog telephony and a DSL connection over the same loop. ADSL was specifically developed to support broadband access to a personal residence. By working on the same loop as a traditional analog POTS line, ADSL allows the conservation of loops in a telephone company's outside plant. This is often a scarce resource, especially in residential areas. Additionally use of the same loop for broadband access and traditional voice services means that in many cases the same wiring at the customer's premises can be used for both services. Current practice in North America is to use "in-line" filters (sometimes known as microfilters) to protect the voice and data services supported on the same inside wire pair from interfering with each other. Because these filters, which attach to the user's telephone equipment, can be installed by most residential users, operational costs for the telephone company are considerably reduced.

In Europe, especially Germany, ADSL service is provided on the same loop as basic rate ISDN. Using the functionality provided in ITU-T G.992.1 Annex B [3], the ADSL physical layer (PHY) is transported on the same loop as the ISDN BRI. As with "ADSL over POTS," the voice service and the ADSL service are separate; they share the same physical media in the outside plant but are provided by separate networks that are operated independently. In the case of both ADSL over POTS and ADSL over ISDN, the two services are frequency multiplexed on the same facilities. The POTS or ISDN operates at the same low frequencies (up to 4 kHz for POTS and up to about 100 kHz for ISDN) that they use when the loop does not also support ADSL. The ADSL operates at frequencies above 26 kHz in the case of POTS and 138 kHz in the case of ISDN. The carrier's networks supporting the broadband and voice services are entirely separate once the two frequency bands are separated in the central office (CO). Except for the addition of splitters or microfilters, the user continues to receive their phone service identically to how they received it before they had the ADSL service installed.

In the United States, the FCC has mandated line sharing since mid-2000. When line sharing is implemented, the incumbent telephone company provides the

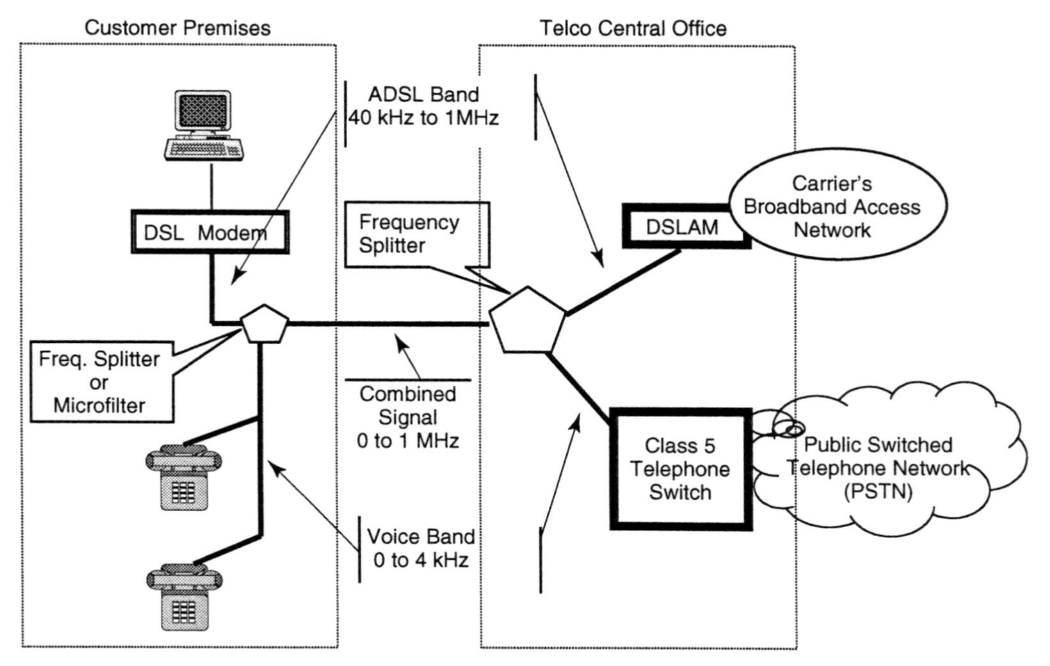

Figure 16.1 ADSL multiplexed with POTS.

loop for both the telephone service and DSL and also provides POTS service to the customer. The DSL service on the same loop can be provided by a competitive provider, who leases only the ability to carry DSL above the voice band. Thus, two separate services provided by separate companies can be provided over the same loop.

16.2 WHAT IS VOICE-OVER DSL?

In voice-over DSL (VoDSL), the voice and the telephony signaling is digitized and carried over the DSL PHY along with the data traffic. Unlike the frequency division multiplex provided with ADSL over POTS (or over ISDN), the voice channels are multiplexed into the DSL path. Digitally multiplexed lines are sometimes known as "derived lines." Although it is true that earlier digital technologies (e.g., T1, HDSL, and ISDN) were used to support telephony, VoDSL is distinguished from these voice environments in several ways. ADSL and SDSL (symmetric DSLs, both proprietary and the standard ITU-T G.991.2) are PHYs that were initially developed primarily to carry data. Unlike the telephony-oriented digital technologies that were developed primarily to transport voice traffic, carrying of voice traffic over the DSL protocols involves solving the problems carrying high-quality full-duplex voice traffic over protocols that were originally optimized for another purpose.

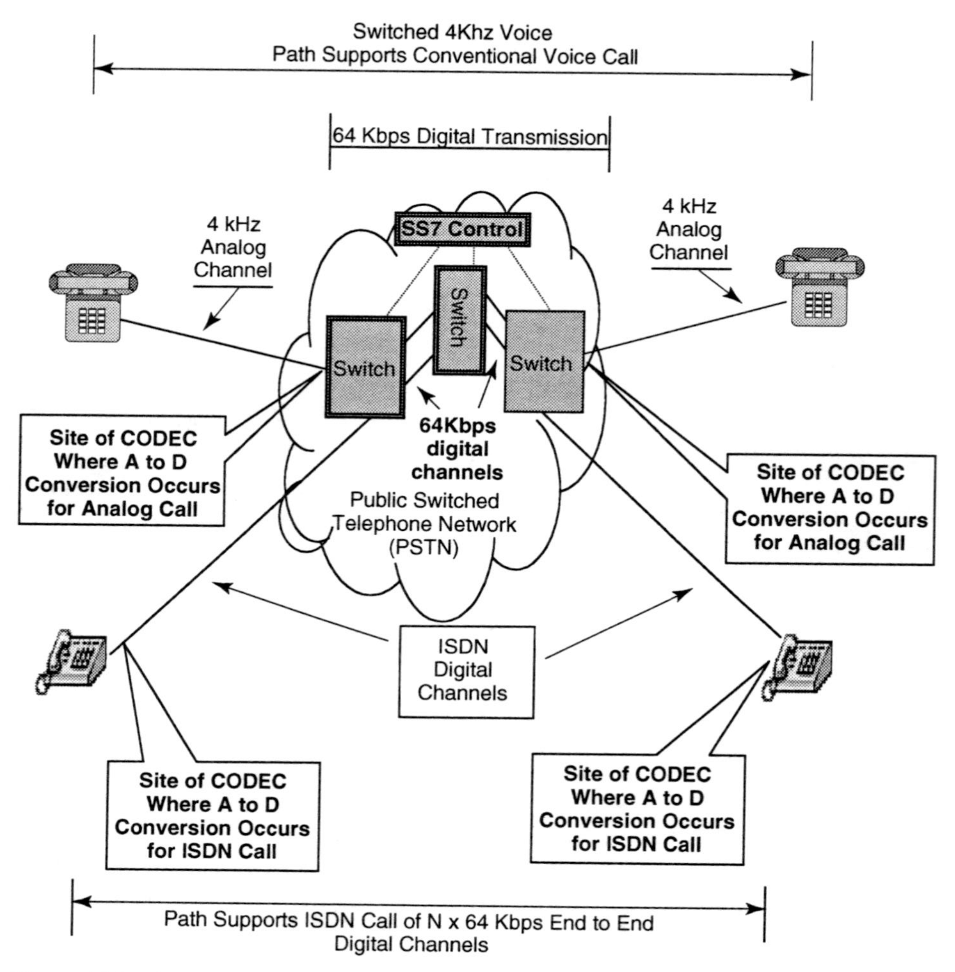

Figure 16.2 Overview of the PSTN carrying digitized telephony traffic.

The worldwide public switched telephone network (PSTN) typically transports digitized telephony in 64 Kbps connections that carry uncompressed digitized voice. The analog voice signal is digitized within the telephone network in the case of POTS, while the digitization occurs in the customer's telephone instrument in the case of ISDN services (Figure 16.2).

Thus, ISDN and T1 were designed from the onset for integration with the PSTN. They are easily channelized into 64 Kbps voice channels, have signaling channels and protocols defined which are compatible with those used in the PSTN, and are integrated with the existing tools used by carriers to manage and trouble shoot their networks. These telephony features are not inherently part of the features of the data-oriented DSLs. This is both a disadvantage and an advantage. Although features that support voice may need to be added to the DSL protocol stack either as enhancements to the PHY protocols or in the higher layers, the support for voice-over DSL is not linked to the legacy of existing PSTN protocols or architectures. Thus, VoDSL architectures can be integrated into both the legacy PSTN or into recently emerging telephony architectures. The PSTN is, and will remain for a considerable time, the main method worldwide for voice connectivity. However, new architectures are being developed that remove some of the existing limits on voice telephony. These architectural enhancements can include use of connectionless protocols that could reduce costs, use of digital compression to reduce bandwidth required to support voice, and novel control and management methods that could increase service flexibility and decrease time to market for new telephone services.

VoDSL supports voice services by multiplexing the voice traffic with the data traffic. Figure 16.3 illustrates this multiplex at the highest and most abstract level. Because the voice traffic is multiplexed within the data channel, it can be carried over the same network as the data traffic and separated from the data traffic at any of several appropriate points in the network. The voice traffic may be integrated with the PSTN at any of these places or may remain completely separated. The exact place where the voice traffic and the data traffic are demultiplexed is independent of the fact that DSL is used to carry both the voice and data traffic. There are several specific architectures proposed for VoDSL, which are discussed later in this chapter.

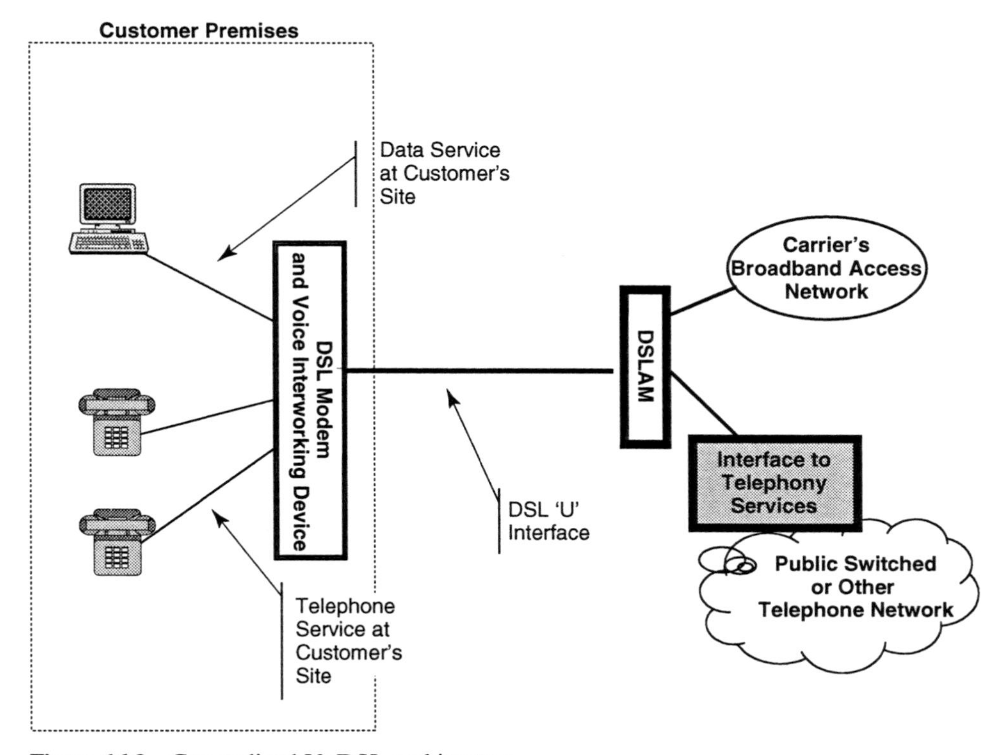

Figure 16.3 Generalized VoDSL architecture.

16.3 TELEPHONY AND ITS RAPID EVOLUTION

Figure 16.2 provides a schematic overview of the existing PSTN. This architecture provides a cost-effective method of delivering voice connectivity throughout the world. When combined with digital wireless (cellular) architectures, nearly universal access is available, certainly in the developed world and in large parts of the economically developing and even underdeveloped areas. However, there are a number of changes that are occurring that are causing rapid evolution to this architecture. These changes are among the important drivers for development and deployment of VoDSL:

- The deployment of fiber and "digital remotes" in the access network
- Support for voice-over packet and cell-based networks
- Novel signaling and control architectures that reduce cost and increase flexibility

16.3.1 Digital Loop Carriers (DLCs)

Traditionally POTS and ISDN services are provided over loops that terminate in the CO. "Line cards," equipment that is an integral part of the CO switch, provides the interface between the loop serving the customer and the remainder of the PSTN. Those loops that support DSL services are supported from DSLAMs, which are stand-alone devices that terminate the DSL services. In the case of ADSL, the POTS (or ISDN) and the ADSL signals are separated in the CO by a splitter, the high-frequency signal being sent to the DSLAM, and the low frequency signal going to the telephone switch.

For over forty years, remotes have been used to separate the location of the line card from the location of the telephone switch. In a remote, or DLC, the telephone services are multiplexed over a digital service from the remote to the switch. The line cards reside in the remote, while the signaling and switching functions remain on the switch.

In the first generation of DLCs (sometimes known as subscriber loop carrier or SLC, pronounced "slick"), one or more channalized T1s are used to connect the remote to the switch. The use of remotes has several advantages. The loop lengths are shorter, which saves copper and improves the quality of the voice signal for the user. The use of a carrier system such as T1 meant that long runs of copper from the CO to the users could be replaced by several long pairs to support the T1 carriers rather then one loop per user. Typically this first generation of remotes supports between 96 and 256 users per remote. The interface between the remote and the switch is defined in Telcordia TR-008 [9]. DLCs were originally used to serve relatively small aggregations of users that were as far from an existing CO as might occur in certain rural areas or a developing suburb.

Enhancements to this architecture include the use of a SONET/SDH fiber-optic connection between the remote and the CO and the use of enhanced protocols between the remote and the CO. These next generation digital loop carriers (NGDLCs) can support up to 2,000 customer lines per remote. The interfaces be-

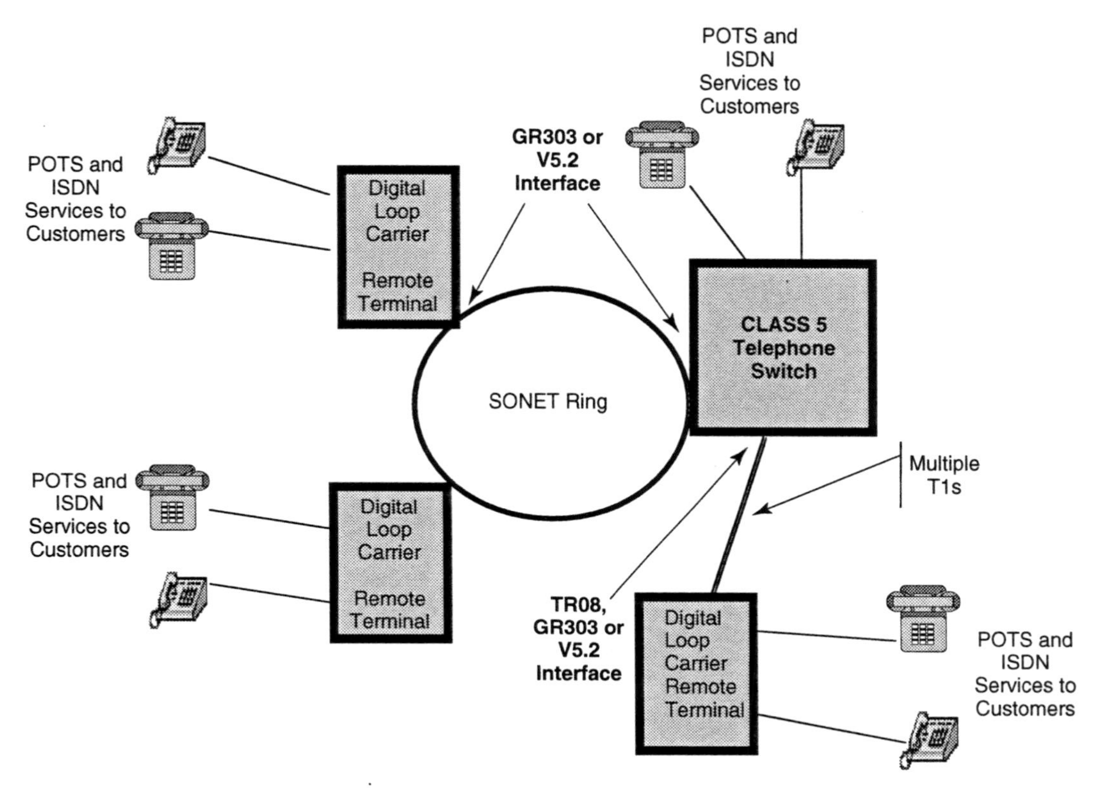

Figure 16.4 Digital loop carrier architectures.

tween the NGDLC and the CO switch are defined in Telcordia GR-303 [4] for North American–based systems and ITU-T V5.2 [11] for systems in Europe and the rest of the world. By the use of fiber, a very reliable high bandwidth interface between the remote and the CO is provided. The GR-303 and V5.2 interfaces support additional functionality and improved efficiencies in multiplexing. The use of SONET/SDH, which carries the traffic between remote and CO, means that highly efficient self-healing ring architectures can be used to connect multiple NGDLC to one CO. Figure 16.4 shows such an architecture.

Because of the development of DLC systems, new COs are rarely built in North America. Instead, new construction (such as new housing developments in suburbs) and rehabilitation of the loop plant (as when population density increases in an existing area) typically are served by adding a new remote to be connected to the existing CO. Because of this increasing deployment of DLC, approximately 30 percent of all customers of U.S. phone companies are currently served by remotes; this figure is steadily increasing.

The effect of DLC deployment on DSL deployment is discussed extensively in Chapter 9; however, the use of DLCs has several effects on the architectures and justifications for VoDSL.

1. The protocols used for supporting DLC and NGDLC from a class 5 telephone switch provide a method of connecting "gateways" that convert VoDSL protocols to a form native to the class 5 switch. If the gateway supports either the TR-08, GR-303 protocol, or V5.2 it is possible to make use of a VoDSL protocol to serve the customer premises completely transparent to the switch and the rest of the PSTN. The lines derived using VoDSL will appear to the PSTN as if they were conventional lines supported by NGDLC. Two of the more common VoDSL architectures, ATM-based broadband loop extension services (BLES) and channelized voice-over DSL (CVoDSL), are enabled by the existence of TR-08, GR-303, and V5.2 to hide the details of their implementation and operation from the PSTN. Figure 16.5 illustrates the use of a voice gateway and TR-08, GR-303, or V5.2 to provide VoDSL services.
2. The use of remotes typically means that loop lengths are relatively short between the customer's premises and the remote. In the United States, loop length between a CO and the customer premises may be almost any length (though rarely more than 30 kft), whereas loop length to remotes are almost always 12 kft or less. As providing VoDSL requires high-quality connections,

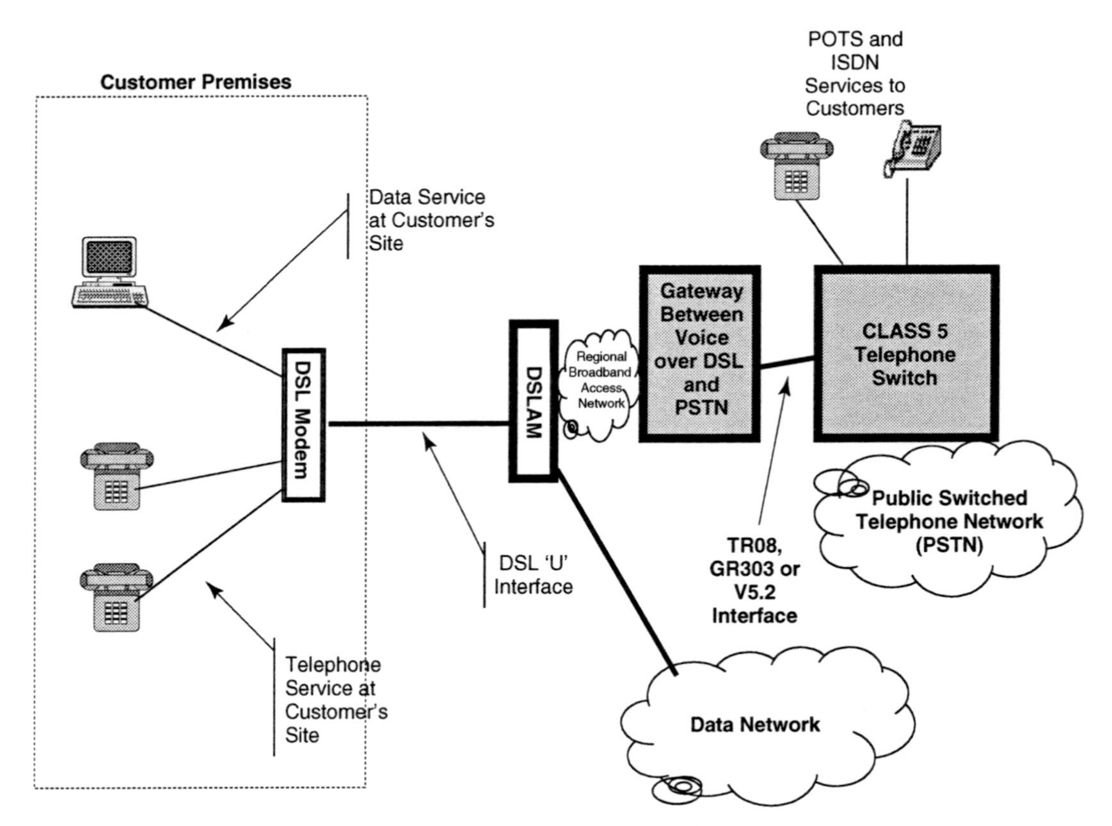

Figure 16.5 Use of TR-08, GR 303, and V5.2 to support VoDSL gateways.

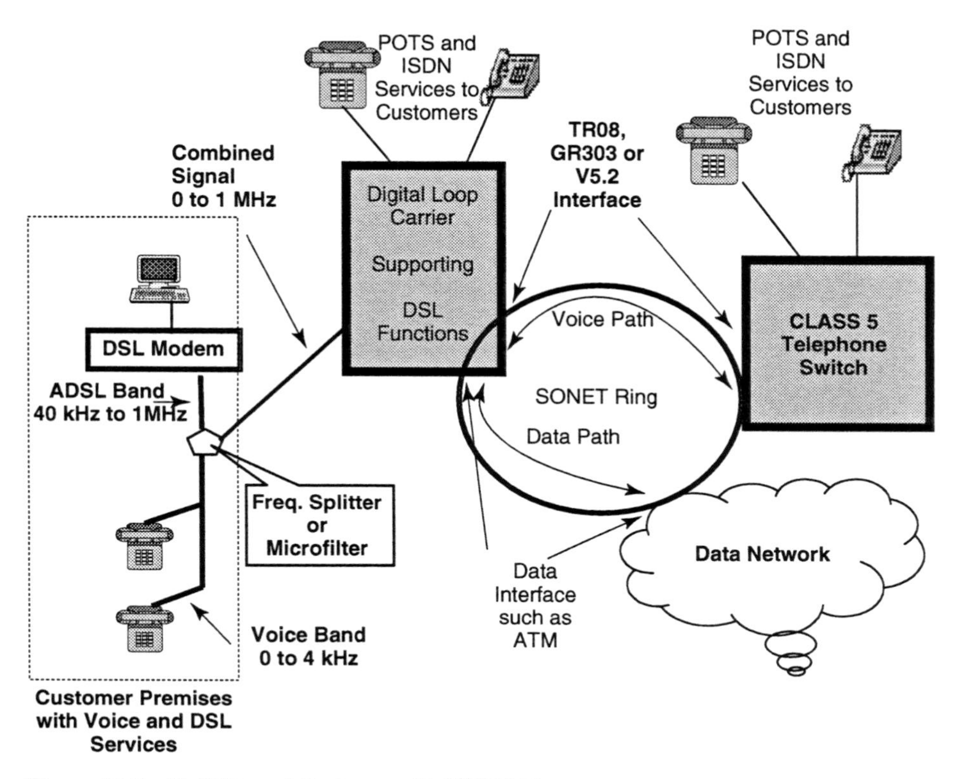

Figure 16.6 VoDSL architectures with NGDLCs.

the shorter loops help ensure that the DSL connections are suitable to carry both derived voice and data.

3. Because an NGDLC remote is served by SONET/SDH fiber, combining the DSLAM function for DSL and NGDLC function of telephony is aided by the deployment of the fiber served remotes. Additionally, placing the VoDSL "gateway" function described in item 1 above within an NGDLC allows for a very efficient integration of both the POTS and VoDSL functions. Figure 16.6 illustrates this combined architecture.

16.3.2 Support for Voice-over Packet and Cell-based Networks

The PSTN has been carrying voice for over 100 years, using a dedicated connection between the callers for each call. Originally these were physical connections, literally wire connections set up through the network between the two points. Although the PSTN no longer sets up copper connections between two callers, it still uses resources that are dedicated solely to each call. The analog POTS signal carried on the copper pairs is converted to a digital signal on the switch in the CO (or in the DLC). A series of dedicated 64 Kbps channels carries the signal to the terminating switch where the signal is converted back into an analog signal and transmitted to the receiver's telephone. Although these digital connections are multiplexed together on the transport systems that connect the switches, each call is allocated resources that are dedicated to the call for its duration.

This transport architecture has proven to be reliable and cost effective for many years. However, it is also inflexible and difficult to change; each call must use the allocated resources even if it is not the most efficient use of the bandwidth. The quality of voice that is received is limited by the analog bandwidth of the loops (4 khz) and digital encoding techniques and 64 kbps digital paths used in the network. A series of complex telephone switches must be coordinated by a dedicated and specialized signaling protocols. Because of this the services that can be offered to telephone users are often limited, and these services are difficult and expensive for carriers to change. Much of the operational expense of operating the PSTN is spent to ensure that the connections are stable and able to be constructed as needed.

The development of both the Internet protocols (IP) and asynchronous transfer mode (ATM) as data transport architectures allows for a fundamentally different structure for transporting voice.

The IP suite of protocols were originally developed to provide an extremely robust data networking environment. Communications are connectionless at the network layer. Rather than a dedicated connection for the data, each packet of data is routed separately and may take a unique path through the network. This routed architecture makes for a network that is very reliable and very extendable, as each router needs to know a limited amount of information about the entire network topology. Routers do not need to keep information about the state of connections set up through the network. Instead, such information is, if required at all, kept in devices connected to the periphery of the IP network.

Voice-over ATM conveys voice in an ATM virtual circuit (VC) and data in different ATM virtual circuits, at Layer 2 of the protocol stack. Unlike VoIP, VoATM uses the connections provided in the ATM network to ensure the proper quality to support voice adequately. However, unlike the TDM (time division multiplex) based PSTN, VoATM uses the 53-byte cells of ATM to make very efficient use of the bandwidth for transporting voice. VoATM architectures have been defined specifically for the DSL environment—DSL Forum TR 39 Annex A [TR 39] and ATM Forum af-vmoa-0145.0000 [1].

16.4 VoDSL ARCHITECTURES

There are three basic architectures for DSL, each of which makes use of a different layer of the communications stack for carrying the voice-over DSL connection. Figure 16.7 illustrates these three basic architectures:

1. Voice-over IP (VoIP) in which voice is encapsulated within IP data packets. In this case voice is transmitted at Layers 3 and 4 of the communications stack.

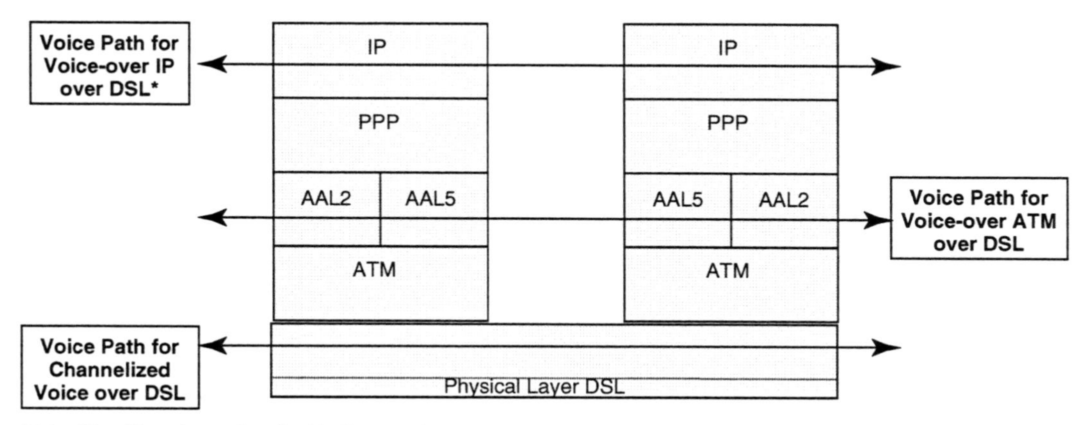

Figure 16.7 VoDSL architecture on the communications stack.

2. Voice-over ATM in which the voice is encapsulated within ATM cells. In this case the voice traffic is transmitted at layer 2 of the communication stack.
3. Channelized voice over DSL (CVoDSL), where the digitized voice is carried directly on the PHY of the DSL link.

16.4.1 VoIP and DSL

The Internet and its suite of protocols were not originally developed to support voice but rather data. However, the pervasiveness of the Internet and the flexibility and extendibility of its suite of protocols invites the use of IP to carry voice traffic. This, and the fact that digitized voice is just a form of data (albeit data with some very stringent delivery requirements) when seen by IP routers, has lead to the development of a large number of Voice-over IP (VoIP) architectures and protocols. Several of these are seeing rapid deployment in a number of telephony environments. Any telephony protocol must cover three basic areas: The specific methods of carrying the voice traffic (i.e., the "bearer" protocols); the methods used for setting up the communications between the two end points (i.e., the "signaling protocols"); and the quality of the service must be managed, the services configured, and user problems resolved. This requires the definition of management protocols.

Three Approaches to VoIP

H.323. One approach to providing VoIP is based on an existing ITU Recommendation H.323 [6]. This approach is currently widely deployed in ad hoc "voice from the PC implementations" which uses the public Internet for transmitting voice between PCs or between PCs and gateways to the PSTN. Because of its complex call processing, it is losing favor to other VoIP solutions. This architecture has four main components.

1. *Terminals.* A terminal is a computer or other device that supports the H.323 protocol and allows the termination of H.323 voice or multimedia sessions for use by the user. The H.323 protocols allow voice (or other multi-media sessions) to be supported between terminals.
2. *Gateways.* A gateway connects a network supporting H.323 to another network. In an H.323 network a gateway is required to connect that network to the PSTN or to a corporations internal legacy voice network. Gateways are not required to support communications between two H.323 terminals.
3. *Gatekeepers.* Gatekeepers provide information to terminals required for the setting up and management of H.323 voice sessions. This information includes address translation, admission control (that is, guaranteeing the network resources exist to set up a call before an attempt is made to allocate these resources), bandwidth management, and routing services. A gatekeeper is not required in an H.323 network but is highly desirable as it allows detailed network information to be hidden from terminals and instead be stewarded by the gatekeepers.
4. *Multi-point control units (MCUs).* MCUs allow the set up and management of calls such as conference calls between multiple terminals.

H.323 has the advantage of being widely deployed today, for example, in many PC-based "Net-to-Phone" services and PC-to-PC voice applications such as Microsoft's NetMeeting multimedia service. Because it makes few assumptions

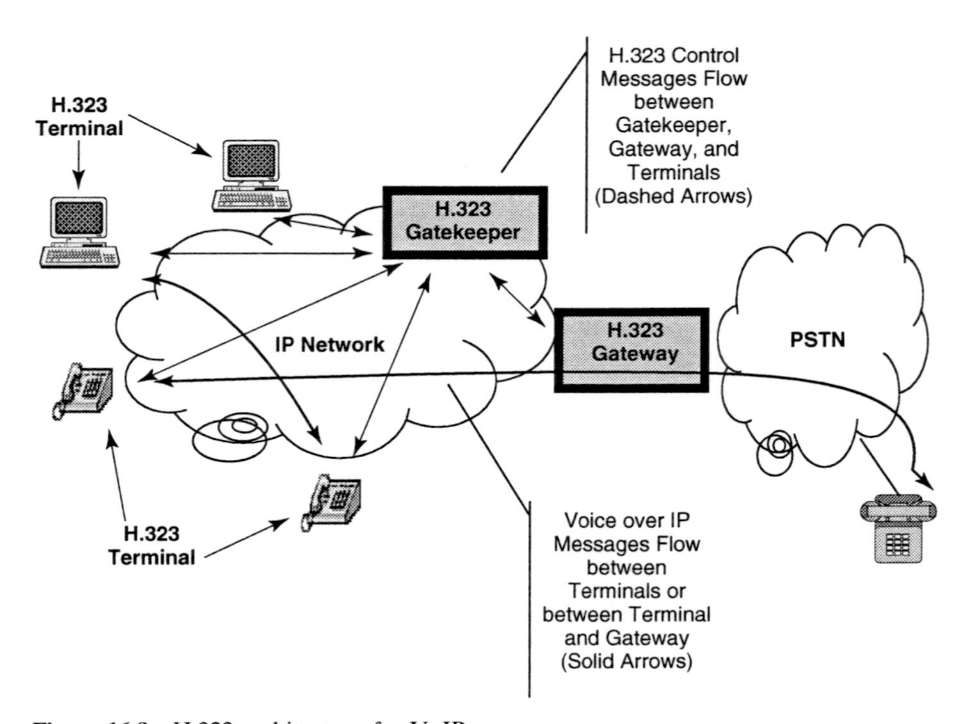

Figure 16.8 H.323 architecture for VoIP.

about the existing network and is quite tolerant of the fact that most IP networks do not guarantee quality of service (QoS), H.323 service is quite easy to set up on the existing public Internet or private intranets. It has several disadvantages: The call set-up model is very cumbersome and complex; it does not deal well with issues of stability and consistency that are vital in networks that claim to compete in quality with the PSTN; accounting and billing for services is difficult; it does not interact smoothly with those elements in a network that could guarantee quality of service.

In many cases, the use of H.323 is transparent to a DSL implementation. DSL is often used to provide a high bandwidth and always connected access to the Internet. This allows existing H.323 applications on the users' hosts (which are running software that makes then H.323 terminals) to make use of the improved service to enhance the H.323 VoIP performance. In a sense this enhancement comes for free; neither the access network nor the DSL CPE has been specifically enhanced to support H.323. Figure 16.8 illustrates a telephony system architecture based on H.323.

H.248/MEGACO (Media Gateway Control Protocol). MEGACO is defined in ITU Recommendation H.248 [5]. This recommendation is derived from work done in the IETF on the MEGACO (Media Gateway Control) specification. MEGACO is a rather centralized approach to the control of VoIP services. This

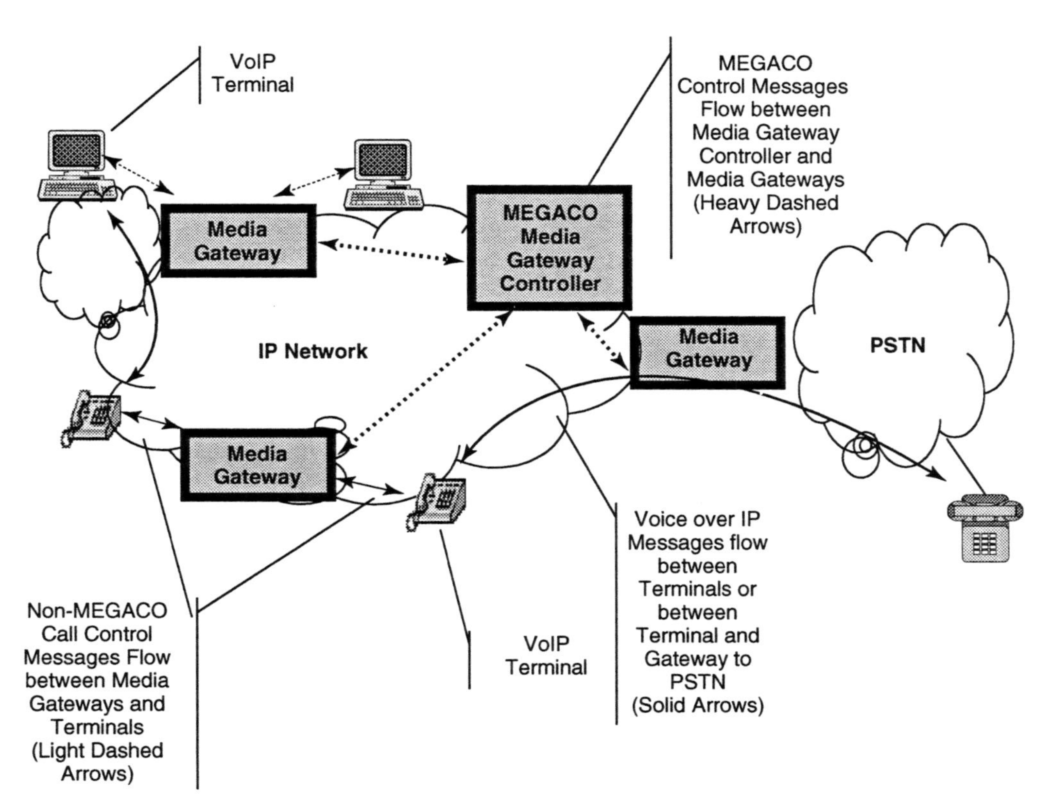

Figure 16.9 Overview of the MEGACO architecture.

approach uses a media gateway controller to set up and control connections between the users. The media gateway controller both mediates control and bearer setup with other networks, either the PSTN or other VoIP networks (such as H.323) and controls media gateways within the H.248 network. Media gateways are coordinated and controlled by the media gateway controllers which instruct the media gateways in setting up communications paths between themselves through the packet network. MEGACO resembles the centralized structure of signaling and control used in the PSTN. In the PSTN, SS7 is a specialized signaling protocol that instructs telephone switches to set up calls between each other. The media gateway controller serves a similar function for the media gateways in a H.248-based network. However, in the SS7-controlled PSTN, the SS7 network is a physically distinct network, whereas in the MEGACO environment the media gateway controller and the media gateways are typically supported over the same packet-based network.

MEGACO has a number of advantages: Its signaling protocols are easily extendable to deal with new environments. Its similarity in architectural philosophy to the existing SS7 controlled PSTN simplifies interworking between MEGACO-based environments and the legacy telephone network. It is specifically designed to support distributed control of complex telephony services in a packet environment, thus aiding in the development and deployment of softswitches, which are distributed and packet-based replacements for existing telephone switches. As H.248 was developed largely by engineers with a more telephony mindset than H.323, it deals with issues of accounting (billing) and network stability as an inherent aspect of the architecture.

The ATM forum is in the processes of defining H.248 interfaces for support of VoDSL loop extension services. These interfaces define MEGACO messages between a softswitch or gateway and a device supporting the traditional telephony interfaces found at a customer's premises. The use of MEGACO is shown in Figure 16.9.

SIP (Session Interconnect Protocol). SIP, based on IETF specifications RFC 2543, is a signaling protocol used for establishing sessions in an IP network [8]. A session could be a two-way telephone call or it could be a collaborative multimedia conference. SIP is a request-response protocol that closely resembles HTTP (hypertext transfer protocol, the communication protocol that underlies the World Wide Web) and SMTP (simple mail transport protocol, which is the basis of Internet email). The philosophy upon which SIP is designed is like that of many of the Internet applications in which a distributed family of servers provides addressing and location information while devices on the periphery of the network have the intelligence to actually support the applications. If MEGACO is the packet-voice environment that most resembles the PSTN, the SIP architecture is most Web-like. SIP provides the following services to control a session: Name translation and user location, which ensures that the call reaches the called party; feature negotiation, which allows the elements involved in a call to agree on the features that are supported; call participant management, which allows a user to establish a call or to add or delete additional users to a call; and call feature changes, which allow a user to modify the features of a call during the call.

There are two basic components to the SIP architecture: the User Agent and the SIP Network Server. The User Agent resides in the end systems that terminate the calls; the SIP Server resides in the network and coordinates calls. The main function of the SIP Server is to provide name resolution and user location, as the caller is unlikely to know the IP address or host name of the called party. The philosophy here is identical to that of Domain Name Servers that hide the need for a user to know an Internet address to connect to a service.

The advantage of SIP is that it is most converged of the various VoIP architectures with the structure of the Internet applications. Its developers thus expect that the great flexibility of this decentralized philosophy of application architecture will result in very low cost and very extendable voice applications that will be easily integrated with other IP-based applications. A SIP-based telephony architecture is shown in Figure 16.10.

All the VoIP architectures leverage the universal connectivity and growing efficiencies of the packet-switched Internet. It is expected by many that once the VoIP CPE elements are cost-reduced and the distributed call control feature matures, VoIP will enable voice transport at a cost less than the circuit-switched PSTN. Even more important it is hoped that VoIP will lower the development cost and speed the deployment of new voice services. VoIP maximizes the abilities to perform multimedia conferencing involving interactive combinations of voice, text,

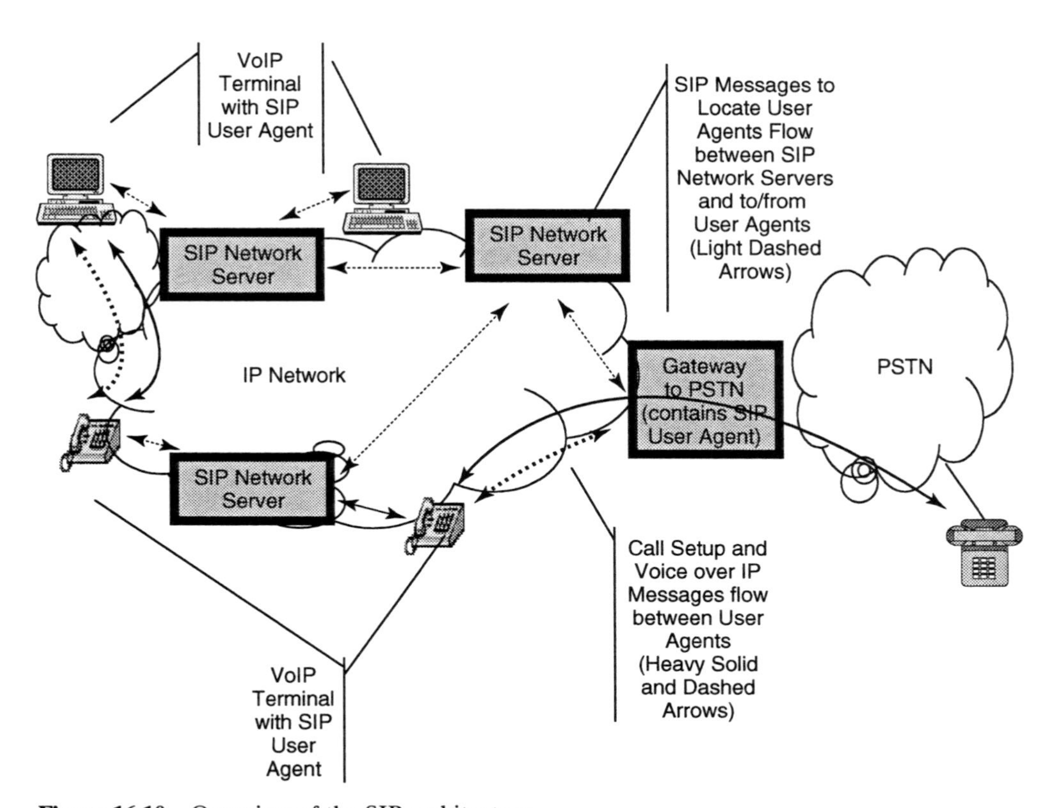

Figure 16.10 Overview of the SIP architecture.

data, and video; this is what H.323 is all about. Both MEGACO and SIP add enhanced call control, session stability, and the ability to efficiently bill for VoIP-based services. When used to support VoDSL, VoIP has the advantage of using the existing DSLAMs without hardware or firmware modifications. The IP packets flow through the DSLAM and ATM traffic aggregator the same way as data packets.

However, development and deployment of VoIP will require much effort to supplant the PSTN and its environment. The IP protocol stack was developed to provide a best effort data service. The basic IP protocol stack does not allow control of QoS, especially with regard to delay, guarantee of bandwidth to carry the messages, and control of jitter factor that are all vital to quality transmission of voice. In addition to the basic architectures that are mentioned above, successful deployment of VoIP requires enhancements to IP networks, especially the addition of QoS functionality provided by protocols such as RSVP, MPLS, and Diffserv. The deployment of these protocols is still in its earliest stages.

It is not clear whether H.248-MEGACO or SIP will ultimately be the favored approach for VoIP. The latency inherent in VoIP is a serious concern, especially for VoDSL. The variable packet delay in the routed IP network results in yet more delay in the voice IP conversion points where the packets must be buffered to avoid bit-stream underrun. Furthermore, lost packets can be a problem because there is not sufficient time to retransmit lost packets in a voice environment. Voice-band echo cancellation can reduce the perceived effects for a total one-way latency less than 200 ms. However, echo cancelers add some cost, and the voice quality is noticeable reduced when the total delay is above 200 ms.

16.4.2 VoATM on DSL—Loop Extension

Voice-over ATM conveys voice in an ATM virtual circuit (VC) and data in separate ATM VC. Unlike VoIP, VoATM uses the connections provided in the ATM network to ensure the proper quality to support voice adequately. However, unlike the TDM (time division multiplex)-based PSTN, VoATM uses the 53-byte cells of ATM to make very efficient use of the bandwidth for transporting voice. VoATM architectures have been defined specifically for the DSL environment—DSL Forum TR 39 Annex A [10] and ATM Forum af-vmoa-0145.0000 [1].

As most of the DSL implementations are based upon ATM networking, VoATM is a natural architecture for VoDSL. The greater efficiency of VoATM when compared with VoIP is very desirable for the DSL environment where line rates are relatively slow (often as little as 128 kbps) upstream when compared with the 10 Mbps or greater line rates found on Ethernet in corporate environments. Because voice is carried over its own ATM VC, QoS can be guaranteed for the voice traffic independently of that provided for the data traffic for a particular customer.

The loop extension architectures (BLES or LES) defined by the ATM Forum and DSL Forum, illustrated in Figure 16.11, allow telephony functions from existing telephony equipment at a customer's site to be connected to a telephone network over ATM connections. The loop extension service replaces the copper

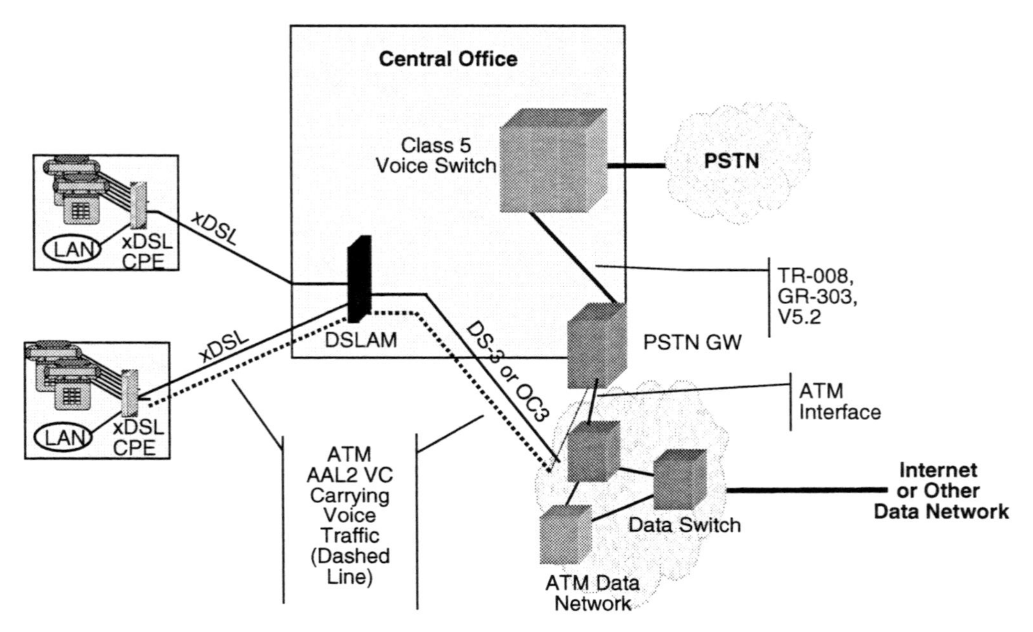

Figure 16.11 Overview of a VoATM architecture for DSL.

telephone loop with derived lines carried over an ATM VC. The basic elements in the architecture are:

1. The devices at the customer's premises that can support either POTS (plain old telephone service) or ISDN BRI.
2. A customer premises interworking function (CP-IWF) which converts the telephony interfaces into voice-over ATM and inserts the voice traffic on the ATM AAL2 VC.
3. The DSL link provides the interface to the carrier's ATM network.
4. The CO interworking function (CO-IWF or CO gateway) converts the voice-over ATM back into an interface that is supported by the telephone network. These interfaces include Telcordia GR-303 and TR-08 for North American implementations and ITU-T V5.2 for implementation elsewhere in the world.
5. The data traffic to and from the customer is carried on its own ATM VC, over the DSL link and through the carrier's ATM network to its own termination at the far end of the carrier's network.

The loop extension interface uses AAL2, defined in ITU I 366.2 [7] to carry the voice traffic. AAL2 allows multiple voice channels identified by a channel ID (CID) to be carried over a single ATM VC. Digitized voice from multiple conversations can thus be carried in a single ATM cell, resulting in a very efficient transport of voice traffic with little overhead. This is true even if the digitized voice is not compressed. By allowing efficient transport of uncompressed voice (pulse code

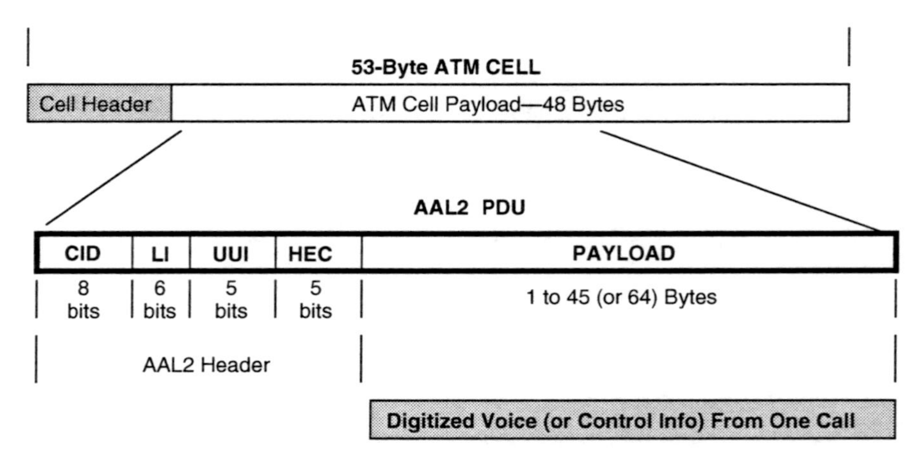

Figure 16.12 AAL2 cell structure.

modulated—PCM voice—as defined in ITU-T G.711 [2]), loop extension allows for a very simple conversion between analog and digital voice, which reduces transmission latency and simplifies the design of the CPE. Voice compression can also be supported.

Each ATM cell contains one or more AAL2 data structures. Each AAL2 data structure is composed of a 3-byte header and the data. The header contains:

1. An 8-bit channel identifier.
2. A 6-bit length indicator that indicates the length of the payload, which can be between 1 and 45 bytes.
3. A 5-bit UUI field. This is used to indicate the "profile" that indicates how the voice data is encoded and compressed.
4. A 5-bit error-detecting HEC code.

The digitized voice for a particular voice connection is placed in the payload. Thus, each call over the DSL interface is associated with a particular CID. The layout of the AAL2 cell is shown in Figure 16.12.

The signaling information for a particular call is carried in AAL2 messages. Loop extension supports channel associated signaling (CAS), used in North America, and common channel signaling (CCS), used in Europe:

- In CAS the signaling (such as the telephone is off-hook or on-hook or the ringing of a telephone) is encoded in a 4-bit code word. Each CAS message is carried in an AAL2 packet with the same CID as the voice traffic for the call.
- In CCS signaling specific messages are used for the signaling. In this case all messages for a particular interface are carried in CID 8.

Loop extension supports a management channel between the CO-IWF and the CP-IWF carried in CID 9.

16.4.3 VoSTM

Voice-over synchronous transfer mode (VoSTM, also known as channelized voice-over DSL [CVoDSL]) conveys voice calls over a separate bearer channel within the PHY. Voice in VoSTM is converted from an analog voice signal to digital: 64 kb/s G.711, 32 kb/s G.726, or 16 kb/s G.723. Several voice calls are placed in a DSL PHY bearer channel separate from data. At the DSLAM, the voice calls are combined with those from other lines and multiplexed into a TDM digital interface to the class 5 voice switch using an interface such as GR-303 or TR-008. Figure 16.13 illustrates this architecture.

The advantages claimed for VoSTM include low signal latency and reduced equipment complexity. The separate bearer channel would operate without data interleaving to achieve low signal latency, and thus would improve end-to-end voice quality by avoiding noticeable voice echo. The low latency would further eliminate the need for voice-band echo cancellation in the CPE, and the CPE would be further simplified by eliminating the complex AAL2 functionality. Furthermore, the voice is conveyed directly as TDM traffic, without ATM or IP chan-

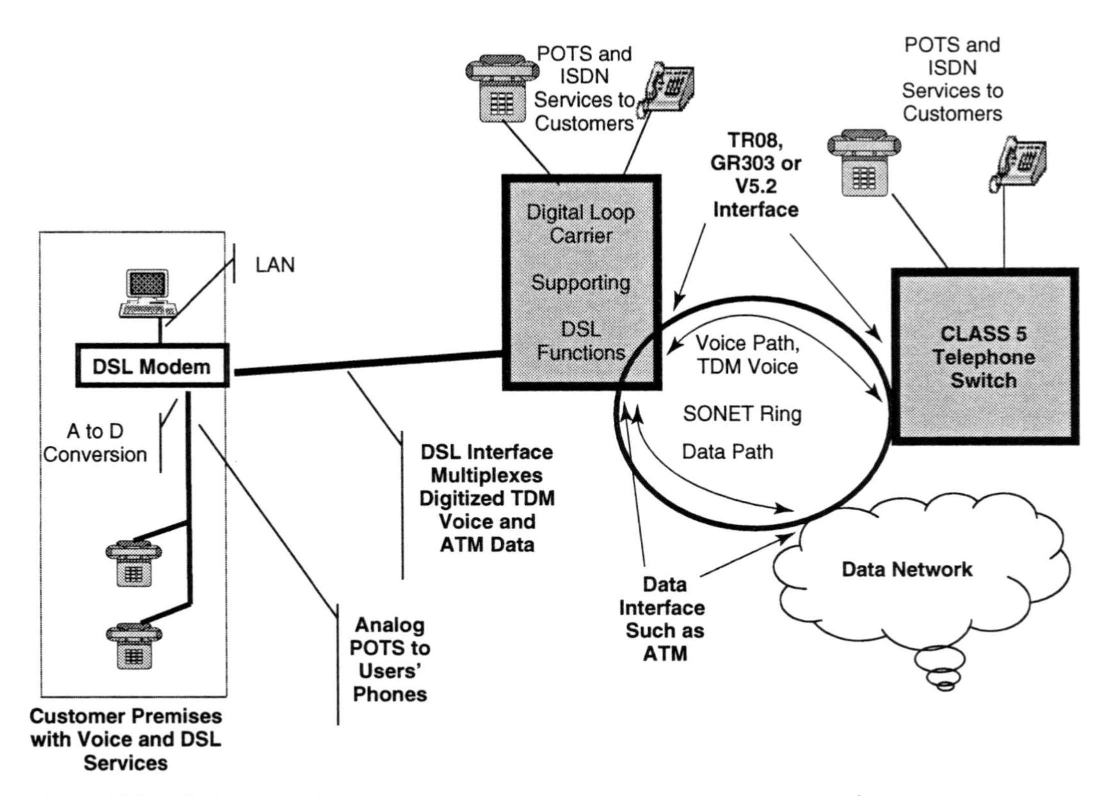

Figure 16.13 CVoDSL architecture.

nel overhead. The VoSTM method also avoids the need to administer the separate ATM virtual circuits needed for VoATM.

It has been claimed that VoSTM eliminates the cost and complexities of the voice gateway in the CO. Actually, the voice gateway would still be in the CO, but it is integrated within the DSLAM. This integration could provide some advantages. However, placing the gateway within the DSLAM precludes the efficiencies of pooling the traffic from several DSLAMs into one shared voice gateway.

VoSTM raises some complex issues. Would a system supporting up to several simultaneous voice calls (e.g., four calls) permanently reserve the full bandwidth needed for the maximum number of calls, and thereby waste this bandwidth the majority of the time when a smaller number of calls are active? Alternatively, would a system implement a complex dynamic PHY channel rate-repartitioning scheme to change the voice bearer channel rate in real time as calls begin and end. If the dynamic bearer bit rate approach is taken, then the bit rate of the data bearer channel will change dynamically, resulting in the need for the data multiplexers, switches, and routers to adapt to the changing bit rate.

The disadvantages of VoSTM include the difficulty of incorporating it into the embedded base of DSLAM and NGDLC equipment that now amounts to many millions of lines of network capacity. Whereas VoATM and VoIP would flow through the existing DSLAM and NGDLC equipment without replacement or upgrade, VoSTM would require at least a major in-service field upgrade to the hardware and software of every DSLAM or NGDLC that would need to support VoSTM. Because most ADSL systems in use operate in a low-latency mode already, the further latency reduction provided by VoSTM is of marginal value. It is doubtful that VoATM causes a need for voice-band echo cancellation in the IWD, because the VoATM latency for the DSL portion of the network can be about 2 ms. Furthermore, voice-band echo cancellation is usually performed at the network gateway, and is not needed elsewhere. The ATM cell overhead is more than recovered in the transport efficiency of silence suppression, where ATM cells are not sent during the silence between utterances. Silence suppression provides no benefit for data and fax type calls, but does provide an average traffic savings of over 50 percent for voice calls. Regarding the cost savings due to eliminating AAL2 processing, this would be true for implementations optimized to support only VoSTM. However, DSL chips often attempt to support all major options in an all-purpose chip; thus, AAL2 functions would likely not be eliminated.

16.5 THE APPEAL OF VOICE-OVER DSL

There are at least three distinct service opportunities provided by VoDSL: loop extension, PBX extension, and access to packetized voice networks. Although each of these opportunities will require different end-to-end architectures to implement them, they all will use the basic functionality of DSL to support simultaneous access for digitized voice and data. The last two of these opportunities can be supported by DSL access using enhancements to DSL functionality that are already planned for other purposes. However, loop extension requires CPE with capabilities specifically developed to support the service.

By use of the DSL physical access and a broadband network, home offices and small branch offices are given remote access to a corporate voice network. The goal is to extend these private voice networks in cost-effective ways to the smallest sites and to increase the integration available to telecommuters. Figure 16.14 is a schematic architecture for this service.

A PBX supporting a packet voice interface, either VoIP, VoFR (voice-over frame relay), or VoATM is connected to a remote site over a broadband network and a DSL access. The remote device may be a remote extension to the PBX to allow a teleworker to have seamless access to the corporate PBX, in which case a line-side interface is abstracted from the PBX. In other cases a branch PBX at the remote site is connected to the corporate voice network over the DSL access using the packetized voice as a trunk-side interface. The remote user may also access the corporate data network over a VPN.

Functions such as support for VPN and IPsec, QoS on the LAN (via protocols such as 801.2p/q), and QOS on the WAN (using protocols such as Diffserv) are required in the DSL CPE to support this service effectively. As these features are delivered, it will be possible to provide secure and high-quality remote PBX access as an integral part of a DSL data service to any telecommuter or small branch office.

16.5.1 Access to Packetized Voice Networks

New voice and multimedia architectures as discussed in Section 16.4 will provide backbones to support voice services based on VoIP and VoATM. The new broadband access methods will provide direct access to these networks from home LANs and small businesses. Interworking between these new networks and the traditional PSTN will likely occur on "trunk side" interfaces, while the function of today's class 5 voice switch will be distributed within an IP- or ATM-based network. DSL will provide broadband service to the home, multidwelling units, and small business to allow access to these networks, providing converged support for

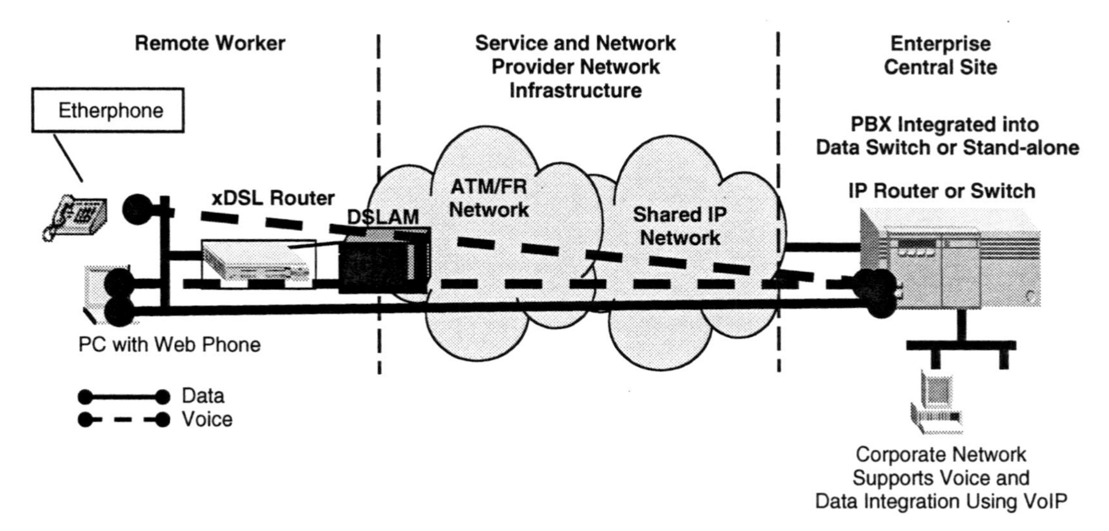

Figure 16.14 PBX extension architecture.

both data and voice. Figure 16.15 illustrates the architecture for DSL access to a network based on VoIP.

As with use of DSL to support PBX extension, much of the functionality required to support this type of service over DSL access will be developed as part of the natural planned enhancement of DSL CPE and network equipment.

16.5.2 Loop Extension

This service provides a high-quality emulation of POTS, or even ISDN service, to the end user, including the ability of the user to make use of their existing analog telephone equipment, small PBX, or key systems. All existing switched-based voice services such as CLASS and IN services will be supported transparently. The voice is digitized in an interworking device at the customer's premises and is multiplexed over ATM or packet with the user's data on the DSL physical access. The interworking with the PSTN occurs with a "line side" interface to a class 5 voice switch, such as GR-303, TR-08, or V5.2. The class 5 switch provides the telephony functionality and sees the gateway as a digital loop carrier system.

The broadband data network carries the voice traffic to a centralized line side gateway that interworks with a class 5 switch. The line side gateway appears to the class 5 voice switch as a if it were a digital loop carrier supported by either the TR-08 [9] or GR-303 (or outside North America V5.2) interfaces. Architectures like those shown in Figure 16.16, where the gateway functionality is integrated into the class 5 voice switch may be especially attractive to CLECs.

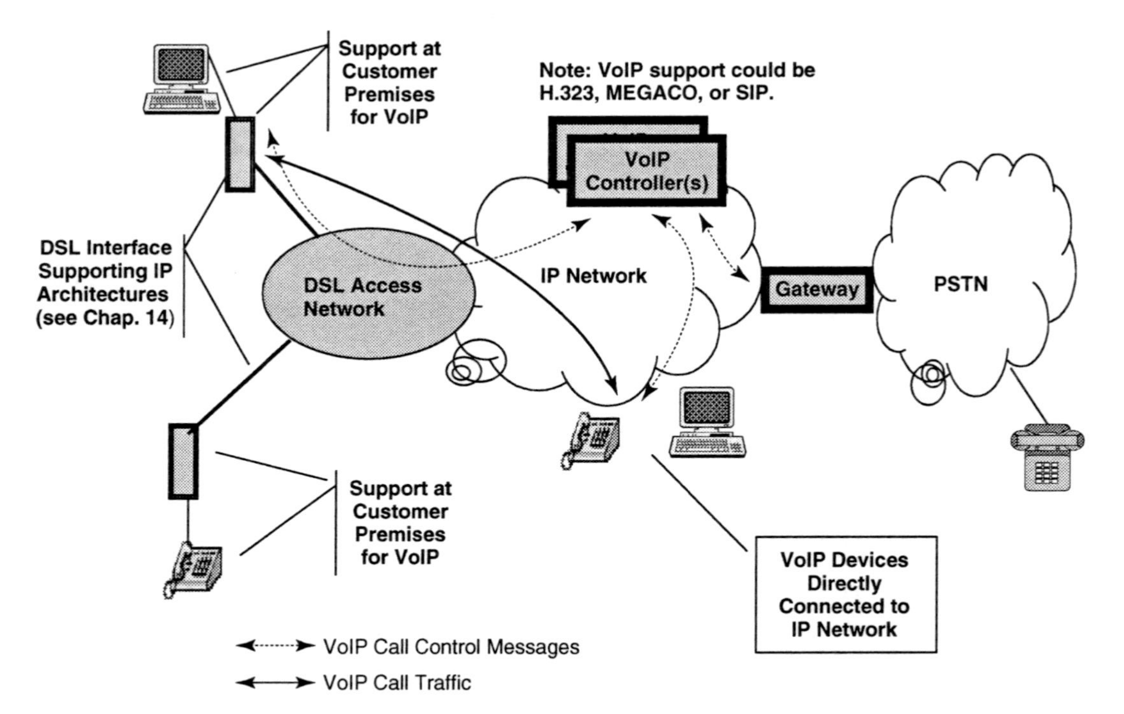

Figure 16.15 Access to packetized voice architectures.

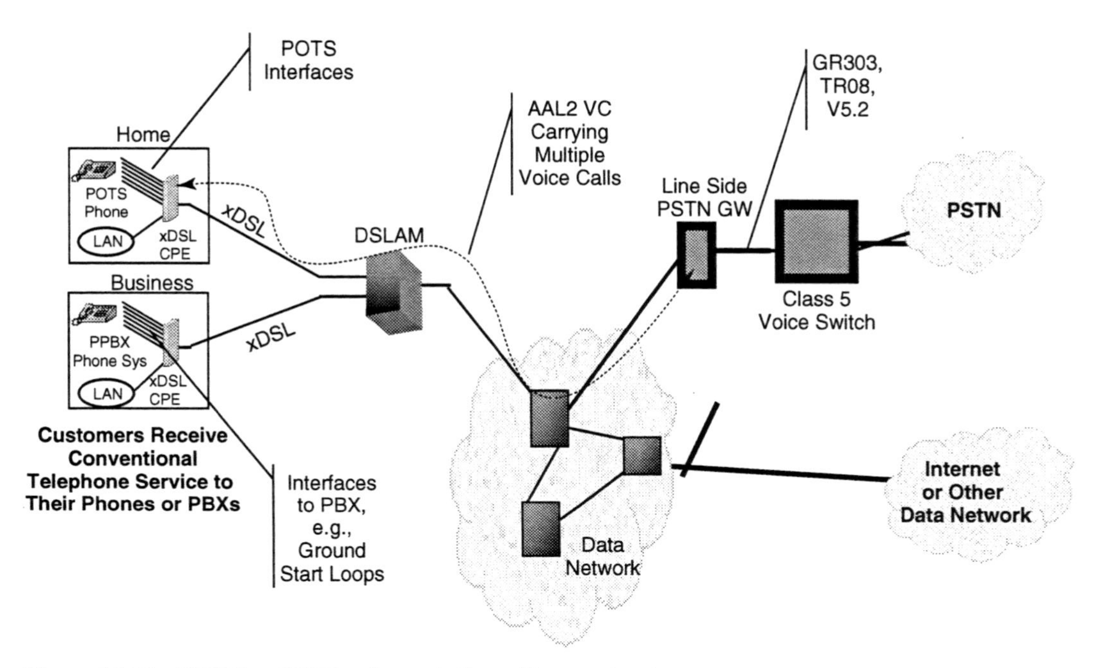

Figure 16.16 CLEC or IXE implementation of loop extension.

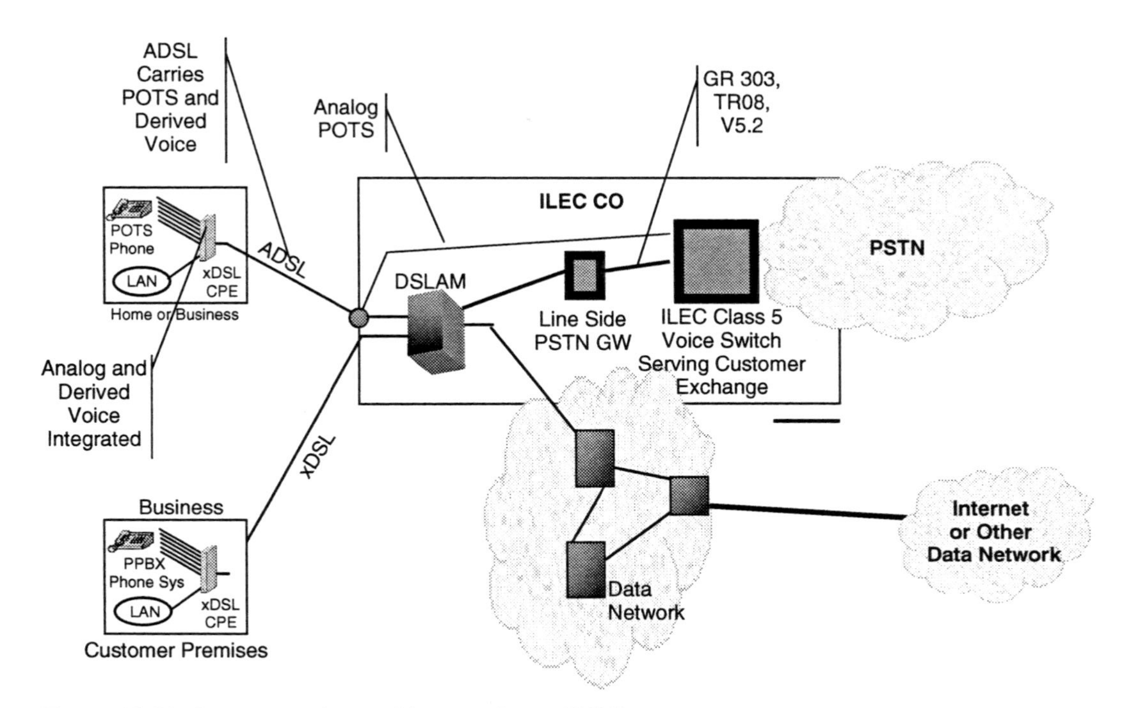

Figure 16.17 Loop extension architecture for an ILEC.

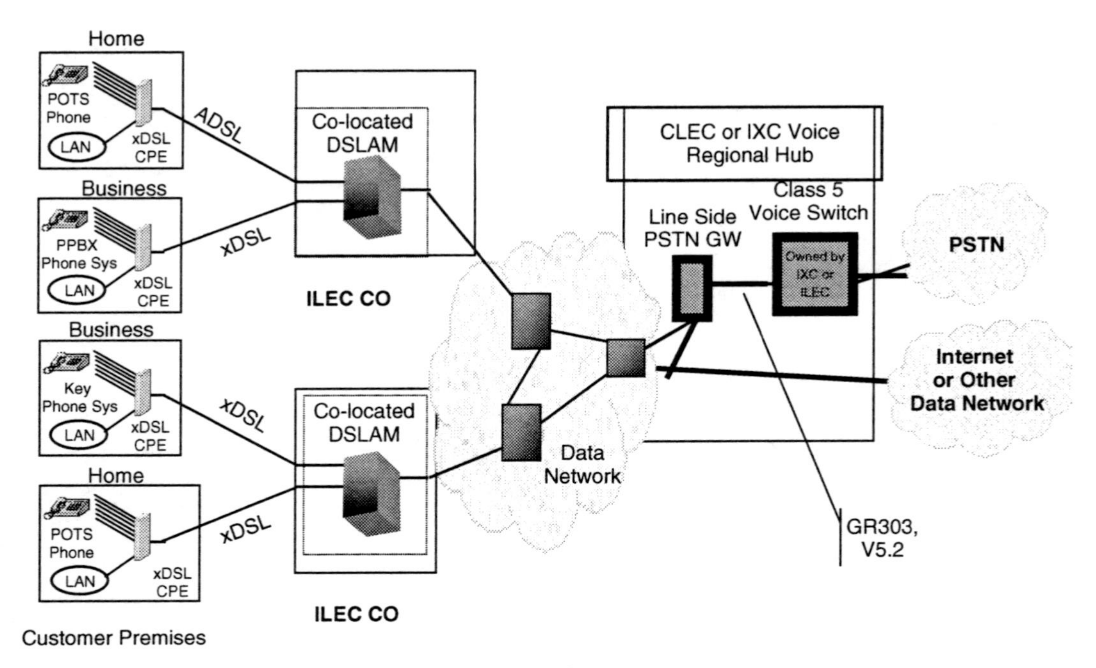

Figure 16.18 Loop extension architecture for CLEC.

If the case of an ILEC architecture like that in Figure 16.17 is possible, the incumbent telco already has a voice switch in every CO.

In the case of a CLEC, an architecture like that in Figure 16.18 may be desirable, as the CLECs can aggregate traffic to a centralized switch serving several COs. Architectures where the gateway is integrated into the DSLAM or both gateway and DSLAM functionality are integrated into a digital loop carrier system may be especially attractive to ILECs.

16.6 SERVICE MODELS BY CARRIERS

In addition to the general requirement for the VoDSL loop extension CPE listed above, the specific service model used by the carrier to provide VoDSL will effect the requirements for the CPE. Five potential service models that could be developed using loop extension over DSL are discussed. These five models are examples of how voice service might be offered over DSL by a carrier. Detailed requirements for CPE would of course be based on the specific filed requirements of carriers actually offering VoDSL loop extension services.

1. *Residential Supplemental Lines.* In this service model, the DSL access is used to provide additional voice lines as a relatively low-cost add-on to the DSL data service. The carrier offers low cost supplemental voice lines and does not attempt to replace the existing life line service provided by the ILEC with VoDSL provided lines. There may be no coordination between these supple-

mental lines and the existing lifeline service. These lines might be installed to serve a home office. Although the voice and service quality are similar to those for conventional analog POTS, no attempt is made to provide the same level of reliability. Instead the service is marketed as a low-cost enhancement bundled to a DSL data service. It is most likely that a CLEC would offer such a service to gain extra revenue from a residential data service. Requirements of CPE (customer premises equipment) supporting such a service model might include:

a) Support for two to four derived POTS lines.
b) CPE is placed on the user's desktop.
c) POTS interface integrated into DSL modem.
d) No power backup requirements.
e) Possible fail-over of supplemental extensions to the analog life line in case of power failure at modem.
f) Possible end-user self-installation.

2. *Residential Full Service Replacement.* In this service model, the carrier provides all of the home voice services using VoDSL. No analog life line service is provided. Instead the VoDSL equipment and network are designed to provide the reliability required for life line service to a home. The user's existing POTS home wiring and telephones continue to be used, and the existing connections to the analog NID are moved over the new VoDSL interworking device when this service is installed. This service is likely to be provided by a CLEC or IXC attempting to compete directly with the Incumbent Carrier. CPE requirements for this service might include:

   a) Support for two to four derived POTS lines.
   b) CPE is placed at the NID (network interface device—the entrance to the home), either inside or outside the home.
   c) CPE is protected against power failure with battery backup or a very-low-bandwidth line powered DSL backup mode.
   d) CPE meets network equipment levels of reliability.
   e) Data service is extended over Ethernet or Home LAN technology from NID location to various sites in home.
   f) CPE installed by carrier or service provider.

3. *Residential Full-Service Replacement Integrated with Existing Analog Service.* In this service model, the carrier provides additional voice lines using VoDSL, while the analog POTS service is maintained to support life line service. The analog and VoDSL supported lines are provided as an integrated service and are likely supported from the same class 5 switch. The ADSL service is carried on the same loop as the analog voice. In case of power failure, the extensions supported by VoDSL may be automatically switched to be served by the analog line. This service model may be attractive to an ILEC, especially to support homes that are currently served by pair gain technologies such as DAML. CPE requirements for this service might include:

   a) Support for 2 to 4 derived voice lines.
   b) CPE is placed at the NID, either inside of outside the home.

   c) All lines fail over to analog line in case of power failure.
   d) ADSL Splitter is integrated into CPE.
   e) Data Service is extended over Ethernet or Home LAN technology from NID location to various sites within the home.
   f) CPE is installed by carrier or service provider.

4. *POTS Replacement to Small Business.* In this service model, the carrier provides all of the small business voice services using VoDSL. Analog life line service is not necessarily provided, as CPE can be provided with batter backup or a UPS in a business environment. The user's existing POTS premises wiring, telephones, and PBXs continue to be used. The existing connections to the analog NID are moved over the new VoDSL interworking device when this service is installed. A CLEC, IXC, or ILEC might provide this service. CPE requirements for this service might include:
   a) Support for four to sixteen POTS interfaces.
   b) CPE placed in wiring closet.
   c) Battery backup or UPS.
   d) Optional support for an analog backup line.
   e) Data service extended into premises over 10/100 BaseT Ethernet.
   f) Network equipment level of reliability.
   g) CPE installed by carrier.
5. *Micro-DLC.* In this service model, the loop provider emulates multiple voice lines by placing a network-owned VoDSL unit on the line before the line enters the customer's premises. The customer-end unit is powered from the CO via a DC voltage applied to the line by VoDSL equipment at the CO. The VoDSL units at both ends of the line multiplex data and multiple digitized voice channels into the DSL payload. Multiple separate voice lines are provided at the demarcation point to the customer.
   a) Support for two to four POTS interfaces.
   b) Customer-end unit placed outside the customer's premises.
   c) Equipment powered by line-fed power from CO.
   d) No battery backup needed.
   e) No baseband analog voice transmission.
   f) Optional data service extended into premises using HomePNA (home phone network alliance) or wireless LAN.
   g) Installed by carrier.

16.7 STANDARDS EFFORTS FOR VoDSL

The DSL Forum has formed a voice working group and has defined requirements for both loop extension, which they have termed broadband loop emulation service (BLES), and for other packetized voice service architectures, which they have termed multi-service data networks (MDN). The existing technical reference

[10] define functional and OAM requirements for these services and defines detailed architectures for the support of voice for a BLES environment. Work continues on defining requirements for MDN (VoDSL for packetized voice networks, e.g., VoIP for DSL). Work also progresses to specify requirements for channelized voice over DSL in the DSL Forum. Free copies of approved specifications and lists of work in progress are available at the DSL Forum's Web site: www.dslforum.org.

The ATM Forum has defined detailed requirements for support of loop extension using AAL2 in their VMOA (voice and multimedia over ATM) working group. ATM Forum technical reference af-vmoa-0145.0000 [1] defines protocols for loop emulation services over AAL2. They based their service requirements on the work of the DSL Forum in TR-39. This effort defines bearer, control, and management planes for loop extension application over ATM using AAL2 encapsulation. The ATM Forum has continued to extend the architecture for voice-over AAL2 to use MEGACO signaling. They are also defining management interfaces for loop extension. Free copies of approved specifications and lists of work in progress are available at the ATM Forum's Web site: www.atmforum.com.

The International Telecommunications Union Telecom Section (ITU-T) defines international standards for the DSL PHYs, for ATM including voice-over ATM, and for packetized voice (H.323) and MEGACO (H.248). Their published recommendations can be purchased for download from their Web site: www.itu.int.

REFERENCES

[1] ATM Forum, *AF-VMOA-0145.0000, "Voice and Multimedia over ATM—Loop Emulation Service Using AAL2,"* 2000.

[2] ITU-T, *Recommendation G.711, Pulse Code Modulation of Voice Frequencies,* Geneva, 1988.

[3] ITU-T G.992.1 *Transmission Systems and Media Asymmetrical Digital Subscriber Line (ADSL) Transceivers,* Geneva, 1999.

[4] Telecordia, *GR-0303-Core Issue 3, IDLC Generic Requirements, Objectives and Interface,* 1999.

[5] ITU-T, *Recommendation H.248, Gateway Control Protocol,* Geneva, 2000.

[6] ITU-T, *Recommendation H.323, Packet-Based Multimedia Communications Systems,* Geneva, 2001.

[7] ITU-T Recommendation I.366.2, *AAL2 Service Specific Sublayer for Trunking,* Geneva, 1999.

[8] M. Handley, H. Schulzrinne, E. Schooler, and J. Rosenberg, *RFC 2543, SIP: Session Initiation Protocol,* IETF, March 1999.

[9] Telecordia, *TR-TSY-00008 Issue 2, Digital Interface between the SLC-96® Digital Loop Carrier and a Local Digital Switch,* 1987.

[10] DSL Forum, *DSL Forum TR-39 Requirements for Voice-over DSL,* March 2001.

[11] ETSI *EN 300 347-1, V5.2 Interface for the Support of Access Network (AN) Part 1: V5.2 Interface Specification,* 1999.

CHAPTER 17

STANDARDS

The development of standards was much simpler when telecommunications networks were solely provided by state-run phone companies (PTTs) and regulated monopolies such as the predivestiture AT&T. There was one central planning organization that dictated the standards. When a standard is developed privately by one or a few companies who have sufficient dominance to control market adoption, it is called a de facto standard. Today, most standards are developed by committees with hundreds of companies and diverse interests. The standards developed by recognized open and fair standards bodies are called de jure standards.

The standards arena contains international organizations such as the International Telecommunications Union (ITU), the International Standards Organization, and the International Electrotechnical Commission (ISO/IEC), which address the harmonization of standards for nearly all countries. The European Telecommunications Standards Institute (ETSI) addresses standards for its member countries. The IEEE, originally a U.S.–dominated organization, is now becoming more internationally focused. National standards for the United States are developed by the Electronics Industry Association/Telecommunications Industry Association (EIA/TIA) and the American National Standards Institute (ANSI) accredited Committees T1 and X3. TTC develops standards for Japan. Implementation forums have been formed to quickly address specific topics. The many standards organizations that do not have direct bearing on DSLs are not shown in Figure 17.1.

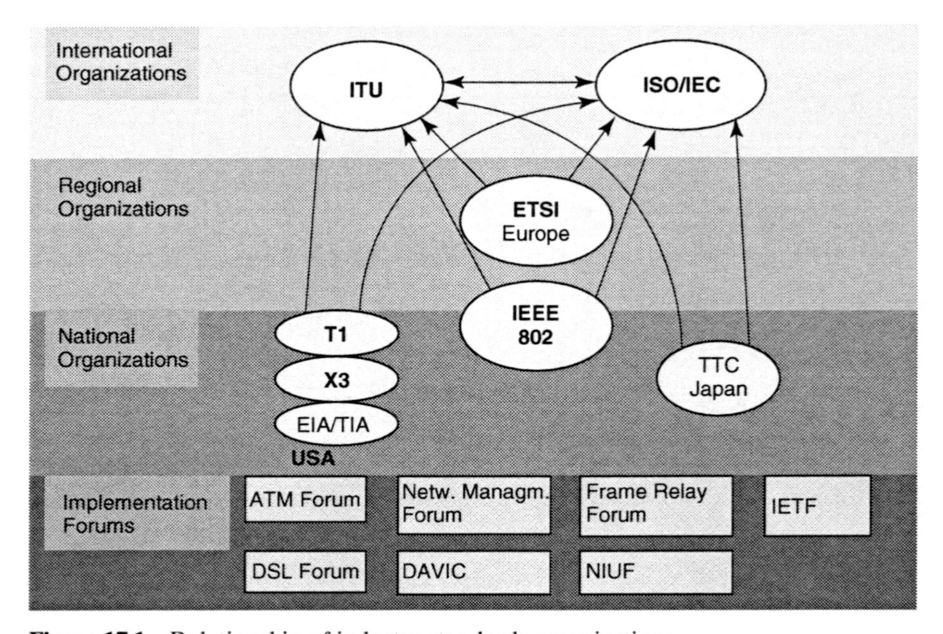

Figure 17.1 Relationship of industry standards organizations.

17.1 ITU

Founded in 1865 as the International Telegraph Union, the ITU later changed it name to the International Telecommunications Union in 1934 and became an agency of the United Nations in 1947 with its headquarters in Geneva, Switzerland. Approximately 190 countries are members of the ITU. The ITU consists of:

- The General Secretariat, which provides administrative and financial functions
- ITU-D, the Development Sector, which coordinates and assists the development of new telecommunications infrastructure, primarily in developing countries
- ITU-R, the Radio-Communications Sector, which ensures the fair, efficient, and economical use of radio-frequency spectrum for all uses
- ITU-T, the Telecommunications Standardization Sector, which studies technical, operating, and policy questions. Previously know as the CCITT, the ITU-T issues Recommendations (nonbinding standards) for the purpose of international standardization of systems and services. In 1997, the ITU-T consisted of:
  - World Telecommunications Standardization Conference (WTSC) which defines general ITU-T policy and approves "questions" (i.e., projects)
  - Telecommunications Standardization Advisory Group (TSAG) which provides advice regarding work priorities and strategies
  - Telecommunications Standardization Bureau (TSB) which provides secretariat (administrative) and coordination functions
- Study Group 1—Service definition
- Study Group 2—Network operation: network management, numbering, routing, service quality
- Study Group 3—Tariff and accounting principles
- Study Group 4—Network maintenance, TMN
- Study Group 5—Protection against electromagnetic environment effects: protection of equipment and humans from electromagnetic hazards
- Study Group 6—Outside plant: cables and associated structures
- Study Group 7—Data network and open system communications: data communications network aspects, frame relay
- Study Group 8—Terminals for telematic services: fax
- Study Group 9—Television and sound transmission
- Study Group 10—Languages for telecommunications applications
- Study Group 11—Switching and signaling: intelligent network
- Study Group 12—End-to-end transmission performance of networks and terminals
- Study Group 13—General network aspects: consideration new concepts, broadband ISDN (BISDN), global information infrastructure (GII)

- Study Group 14—Modems and transmission techniques for data, telegraph, and telematic services: modems, ISDN terminal adaptors
- Study Group 15—Transmission systems and equipment: subscriber access systems, DSLs, fiber optic transmission (DSL is addressed in Q4/15.)
- Study Group 16—Multimedia

Membership classes consist of:

- Administrations (countries)
- Recognized operating agencies (ROAs): network and service providers
- Manufacturers, scientific, and industrial organizations (SIOs)
- International and regional organizations

The ITU-T may be reached at ITU-T, Place des Nations, CH1211 Geneva 20 Switzerland, www.itu.int.

17.2 COMMITTEE T1

Committee T1, formed at the time of AT&T's divestiture, is the telecommunications standards body for the United States. Committee T1 is sponsored by the Alliance for Telecommunications Information Solutions (ATIS) and is accredited by the American National Standards Institute (ANSI). There were 70 voting members of Committee T1 in 2002.

In addition to Committee T1, which determines policy and direction, the committee contains an Advisory Group and the following technical subcommittees (TSCs):

- T1A1: Performance and Signal Processing
  - T1A1.2: Network Survivability
  - T1A1.3: Performance of Digital Networks and Services
  - T1A1.5: Multimedia Communications Coding and Performance
  - T1A1.7: Signal Processing and Network Performance for Voiceband Services
- T1E1: Interfaces, Power and Protection for Networks
  - T1E1.1: Physical Interfaces and Analog Access
  - T1E1.2: Wideband Access
  - T1E1.4: Digital Subscriber Loop Access
  - T1E1.5: Power Systems—Power Interfaces
  - T1E1.6: Power Systems—Human and Machine Interfaces
  - T1E1.7: Electrical Protection
  - T1E1.8: Physical Protection and Design

- T1M1: Internetwork Operations, Administrations, Maintenance and Provisioning
  - T1M1.3: Internetwork Operations, Testing, Operations Systems and Protocol
  - T1M1.5: OAM&P Architecture, Interface and Protocols
- T1P1: Systems Engineering, Standards Planning and Program Management
  - T1P1.1: NII/GII
  - T1P1.2: Personal Communications Service Descriptions and Network Architecture
  - T1P1.3: Personal Advanced Communications Systems
  - T1P1.5: PCS 1900
  - T1P1.6: CDMA/TDMA
  - T1P1.7: Wideband-CDMA
- T1S1: Services, Architectures, and Signaling
  - T1S1.1: Architecture and Services
  - T1S1.3: Common Channel Signaling
  - T1S1.5: Broadband ISDN
- T1X1: Digital Hierarchy and Synchronization
  - T1X1.3: Synchronization and Tributary Analysis Interfaces
  - T1X1.4: Metallic Hierarchical Interfaces
  - T1X1.5: Optical Hierarchical Interfaces

The TSC and its subtending working groups normally meet during the same week, four times per year. Additional interim meetings may be held when needed. All meetings are open to nonmembers.

The name of Technical Subcommittee T1E1 has often been confused with the committee's work on transmissions systems related to T1 and E1 carrier. As can be seen from the list of committee names above, the committee's name has nothing to do with T1 and E1 carrier.

Committee T1's offices are at 1200 G Street, NW Suite 500, Washington, DC 20005, Phone: 202-434-8845, Fax: 202-347-7125, Web: http://www.t1.org.

17.3 ETSI

The European Telecommunications Standards Institute develops standards and reports for its member countries. ETSI was formed in 1988 to carry on the work that had previously been done by CEPT. In 2001, ETSI membership consisted of 789 members from fifty-two countries within and outside of Europe.

ETSI Technical Committees consist of:

- TM—Transmission and Multiplexing
  - TM1—Transmission equipment, fibers, and cables
  - TM2—Transmission network management and performance

- TM3—Interfaces, architecture, and functional requirements of transport networks
- TM4—Radio relay systems
- TM6—Access transmission systems on metallic cables (DSLs)

- NA—Network Aspects
  - NA1—User interfaces, services, and charging
  - NA2—Numbering, addressing, routing, and Internet working
  - NA4—Network architecture, operations, maintenance, principle, and performance
  - NA5—Broadband networks
  - NA6—Intelligent networks
  - NA ECTM: European Coordination of TMN
- EE—Environmental engineering
- ERM—EMC and radio spectrum
- BRAN—Broadband radio access networks
- HF—Human factors
- ICC—Integrated circuit cards
- CN—Corporate networks
- CTM—Cordless terminal mobility
- MTS—Methods for testing and specification
- DECT—Digital enhanced cordless telecommunications
- DTA—Digital terminals and access
- RES—Radio equipment and systems
- SEC—Security
- MTA—Multimedia terminals and applications
- PTS—Pay terminals and systems
- SPS—Signaling protocols and switching
- STQ—Speech processing, transmission, and quality
- TMN—Telecommunications management networks

ETSI headquarters is in Sophia Antipolis in the south of France, and may be reached at http://www.etsi.org.

17.4 DSL FORUM

The DSL Forum is an international association of more than 350 companies (as of 2001) covering telecommunications, equipment, computing, networking, and service provider companies. Established in 1994, the Forum continues its drive for a mass market for DSL and delivers the benefits of this technology to end users around the world over existing copper telephone wire infrastructures.

Throughout its eight years, the DSL Forum has worked on defining the core technology as it develops, providing inputs to international standards bodies and establishing processes to deliver maximum effectiveness in the deployment and use of DSL. The Forum is focused on the complete portfolio of DSL technologies designed to deliver ubiquitous broadband services for a wide range of situations and applications that will continue the transformation of our day-to-day lives in an online world.

Best practices for autoconfiguration, flow through provisioning, and a range of other key facilitators of scaleable, global, mass-market deployment of DSL technology are fast-tracked by DSL Forum through its technical and marketing committees. This work takes place at quarterly, weeklong meetings and through continuous working group progress programs with formal technical reports developed from contributions and "Working Texts."

Technical Working Groups	**Marketing Working Groups**
Architecture and Transport	Ambassador Program
Autoconfiguration	Deployment Council
Emerging DSLs Study Group	Mindshare
Operations and Network Management	Summit and Best Practices
Testing and Interoperability	Tradeshows
VoDSL	Public Relations
	Web
	E-commerce

Further information on the DSL Forum, its work, members, and meeting schedule is available on www.dslforum.org. The DSL Forum's Web site dedicated to providing information to end-users is on www.dsllife.com. The DSL Forum offices are at 39355 California St. Suite 307, Fremont, CA 94538, phone: 1-510-608-5905.

17.5 ATM FORUM

The ATM Forum is an international forum founded in 1991 with the purpose of assisting the rapid and widespread implementation of ATM technology. The ATM Forum had more than 600 member companies in 2001. The ATM Forum consists of a managing Board, a Technical Committee, and Market Awareness Committees for Asia Pacific, Europe, and North America. The Technical Committee consists of the following working groups:

- ATM-IP Collaboration
- Control Signaling
- Frame-Based ATM
- Network Management

- Physical Layer
- Routing Addressing
- Residential Broadband
- Security
- Service Aspects and Applications
- Testing
- Traffic Management
- Voice and Telephony over ATM
- Wireless ATM

The ATM Forum may be reached at 1000 Executive Parkway #220, St. Louis, MO 63141 phone: 1-314-205-0200, www.atmforum.com.

17.6 BROADBAND CONTENT DELIVERY FORUM

The BCDF addresses the delivery of broadband content and applications via the Internet and all access methods: DSL, cable, and wireless. The BCDF contains the following groups: infrastructure, content, and applications; market development; market awareness; and European regional. Further information may be found at www.bcdforum.org.

17.7 TELEMANAGEMENT FORUM

Founded in 1988, the TMF addressed management of telecommunications networks in a open, multivendor environment with 250 members in the year 2001. The TMF work on the interworking of operational support systems (OSSs).

17.8 DAVIC

The Digital Audio-Visual Council (DAVIC) developed technical specifications to assure the end-to-end interoperability for digital audio-visual applications. DAVIC was an international forum that primarily addressed service provider and CPE aspects for video-on-demand and broadcast video services. DAVIC had over 200 member companies in 1997. In addition to the Forum Management committees, DAVIC consisted of working groups addressing applications, physical layers, information representation, security, subsystems, and systems integration. DAVIC produced a specification for TV distribution, video-on-demand, and teleshopping. DAVIC ceased operation in 1999.

17.9 IETF

The Internet Engineering Task Force (IETF) is comprised of users, researchers, service providers, and vendors of hardware and software; it focuses on Internet protocol (IP)–based solutions. The IETF is chartered by the Internet Society (ISOC), and receives guidance from the Internet Engineering Steering Group (IES) and the Internet Architecture Board (IAB). In 2001, the IETF consisted of the following subject areas:

- Applications
- General
- Internet
- Operations and management
- Routing
- Security
- Sub-IP
- Transport
- User services

Altogether, there are about 100 working groups. The IETF develops specifications called requests for comments (RFCs) which proceed in stages from less formal to more formal (proposed, draft, and full standard). To progress forward, evidence of multiple interoperable implementations must be provided. The IETF procedures are documented in RFC 2026. The IETF may be reached at phone: 703-620-8990, http://www.ietf.org.

17.10 EIA/TIA

The Electronics Industry Association and Telecommunications Industry Association (EIT/TIA) is ANSI accredited. EIA/TIA develops U.S. standards and technical reports pertaining to equipment, and develops U.S. positions to international standards bodies (ITU, ISO, IEC). EIA/TIA T41.9 has developed technical recommendations that have assisted the FCC with Part 68 of the U.S. telecommunications rules. The EIA/TIA had over 1,100 members in 2001, and may be reached at Suite 300, 2500 Wilson Boulevard, Arlington, VA 22201, phone: 703-907-7703, http://www.tiaonline.org.

17.11 IEEE

The Institute of Electrical and Electronic Engineers (IEEE) sponsors many standards committees. The most relevant to DSLs is IEEE Committee 802, founded in 1980 by the IEEE Computer Society. IEEE Committee 802 addresses ISO layer 1 and 2 aspects of local area network (LANs) and metropolitan area networks (MANs), and works closely with IEC/ISO JTC 1 SC6. In 1997, IEEE Committee 802 consisted of the following active working groups:

- P802.1: Overview, architecture, bridging, management
- P802.2: Logical link control
- P802.3: CSMA/CD (Ethernet)
- P802.4: Token bus
- P802.5: Token ring
- P802.6: DQDB (dual queue dual bus)
- P802.8: Broadband and fiber optics
- P802.9: Integrated services
- P802.10: Security
- P802.11: Wireless
- P802.12: Demand priority
- P802.14: Cable TV

IEEE P743 has developed techniques for measuring the analog characteristics of DSL signals.

The IEEE may be reached at 445 Hoes Lane, Piscataway, NJ 08855, phone: 908-562-3820, http://www.ieee.org.

17.12 THE VALUE OF STANDARDS AND PARTICIPATION IN THEIR DEVELOPMENT

Each version of DSL was nurtured and defined in an innovation incubator: standards committees. Members of the standards committees know that the standards process accomplishes much more than just writing the standard. Some managers fail to understand the value of sending top members of their technical staff to frequent standards meetings. Trying to explain the creative synergy and learning opportunities gained from participating in the standards process can be rather like trying to explain why a rose is beautiful. It is hard to explain the obvious; standards set the future course of the industry.

Standardization enables the provision of interoperable elements and systems from multiple independent sources. This helps to ensure that compatible elements/systems will continue to be available even if the original source of supply ends. The feasibility of replacing one source of supply with another enables a competitive environment that stimulates low prices, good support, and continuing improvements. One may buy a mixture of interchangeable elements from different suppliers. Standards specify performance, reliability, safety characteristics, and methods for measurement and test. The buyer has the assurance that the standard elements comply to a comprehensive set of requirements. Thus, even if the buyer is not an expert, he or she can be assured that the standard element has the necessary performance and features. Standard systems provide the user (who may be a network technician) with a common look-and-feel; this makes the system quick to learn. Standard elements may be reused elsewhere. Standards improve manufacturing and vendor efficiencies by reducing the number of unique versions of product that must be developed and supported.

What is less obvious is that the standardization process is a stimulus to innovation. Vendors are more likely to introduce new products when they see industry-wide buy-in to the new products. Through interactions in the standards committees, the vendors see the interest of the potential buyers. To assure success, the vendors and operators prepare for the associated systems to be in place. Through the personal relationships developed among the representatives of many industry sectors, the committee members explore what could be possible, as well as each other's intentions, needs, and concerns. To develop a full system view, an open interchange of ideas is needed among the system vendors, unit vendors, IC vendors, software providers, customer equipment vendors, test equipment vendors, network operators, content providers, end users, university researchers, and consultants. The lively review and criticism of proposals by peers in the same and other industry sectors is a rapid and effective means to find the best overall solution and ensure that nothing is overlooked. A well-managed standards committee exhibits both competition and cooperation. The members compete with each other in an effort to find a better solution. Yet the members also know that they must cooperate to produce results in a timely manner. With top technical talent supplied by the member companies and strong committee leadership, extraordinary results can be attained by the multidisciplinary interactions. Standards committees are often criticized for being slow, but it does take time to develop industry-wide buy-in and to develop a high-quality system definition. Once the standard is in place, one can proceed quickly and with confidence.

The best standards specify only what is necessary to assure interoperability, performance, reliability, and safety while not overly constraining implementation, optimization, and improvement. A well-managed standards process does not hinder competition, innovation, availability, and cost reduction. Suppliers can differentiate their products with supplemental features, cost efficiencies, versatility, ease of use, customer service, and other aspects.

Without a standard, a market leader may delay the entry of competitors into a market. However, if the market is attractive, competitors will eventually find a way to produce similar and perhaps compatible products. In most cases, the market leader would have been more successful from the larger market stimulated by a standard.

Involvement in the development of standards is an excellent means to learn what other companies (customers and competitors) have done and plan to do. Members of the standards committees are in the ideal position to learn of the latest development in technology and business plans. Standards set the future course of the industry, and those present during the development of the standard understand its implications and implementation far better than an outside reader of the final document. There are opportunities to help steer the directions dictated by the standards to benefit one's own company. Lastly, participation in the committees can help establish a company's credibility. The personal relationships with fellow committee members can be very useful building business relationships. The members who are willing to share some information, provide high-quality contributions, and can compromise when necessary can have a far larger influence over the final standard than other members. The chairperson of the standards committee can have significant influence over the outcome; however, there are limits.

A company cannot participate in every standards committee. Each company must individually select those committees that have direct impact on their company's interests. Furthermore, the opportunity to stimulate progress, the ability to influence the outcome, the prognosis for timely results, the value of the knowledge gained, and the relationship built must be considered for each committee. Once a company decides which committees it should attend, then it must decide if it will only monitor the meetings or actively participate and contribute. Attending the meetings and only taking notes (monitoring) has nearly the same costs to the company as active participation. Only active participation provides the benefits of speeding progress, establishing company credibility, influencing the outcome to benefit the company, and learning the lessons that only come from being fully engaged in the process. Members must work to understand the needs and concerns of the other members and be flexible whenever and wherever possible.

17.13 THE STANDARDS PROCESS

The process used by Committee T1 is typical of processes used by many standards committees. The process begins with a proposal for a new standard or technical report. This may first require the approval of a new project and possibly a new committee or working group. Project approval authorizes a specific committee to perform certain activities. Contributions (technical papers that may make a proposal or offer pertinent information) are provided to the working group by its members. The working group chairperson may appoint one or more editors to prepare a working document based on the contributions that have gained the working group's agreement and the material developed during the meetings. Most standards committees use a consensus process whereby extensive efforts are made to devise a solution to which all (or nearly all) members can agree to. The consensus process can take longer than a simple voting process where the majority wins. However, the consensus process often generates superior industry support. Also, in-depth consideration of minority opinions can often add value to the end result. When a final-draft proposed standard is ready, it is sent with a voting ballot to all voting members of the committee. Members may vote YES, NO, or ABSTAIN. Members may also attach detailed comments to their ballot response. The committee reviews the ballot comments (regardless of whether the comment was on a YES or NO vote) and attempts to resolve all comments by making revisions to the draft document. If necessary, the revised document is sent out for a second ballot (sometimes called a default ballot). For most committees, the standard is approved when a two-thirds majority of the votes are YES. Upon approval, the document is published. Many committees will review an approved standard five years after its issue. To expedite the issuance of a standard, it may be released with some aspects to be addressed in a future issue of the standard. In this case, work on the Issue 2 standard will begin upon completion of the Issue 1 standard.

In the United States, Section 273 of the Telecommunication Act of 1996 requires Non-Accredited Standards Developments Organizations (NASDOs) to follow rules of due process which assure open and fair operation. Further, U.S. antitrust laws require standard development organizations to avoid anticompetitive behavior. As a result, standard activities must avoid discussion of cost, price, product availability, market allocation, and other topics that might lead to anticompetitive behavior. Most standards development organizations require their members to identify any intellectual property (e.g., patents) that may be necessary for implementation of a proposed standard and to agree to license the intellectual property at fair and nondiscriminatory terms to an unlimited number of parties. This may be interpreted as offering a license to all parties at a cost that does not prevent them from offering products that are truly competitive.

17.13.1 When to Develop a Standard

The success of a standard depends on timing. If work starts too soon, the committee will waste time due to poor requirements and constantly changing technology fundamentals. If work starts too late, much of the market window may be lost or proprietary products may form a de facto standard. Work on a standard may begin while a technology is in the laboratory stage. However, before completion of the standard, it is best to have completed some field trials of prestandard systems to provide the benefit of practical experience learned from operating in the field. The committee leadership must keep a close watch on the interests of the committee members, the marketplace, and technology trends. New projects should be started quickly once several committee members express a strong interest in contributing work to a new project. In the case of ISDN, strong arguments can be made that standards were too late to address the market and too early for maturity of the technology.

17.13.2 Is a Standard Needed?

There is no definitive rule for when to start work on a standard or if a standard should be developed at all. Standards are necessary when equipment from multiple vendors must interwork. In many cases, a standard can help create a market that attracts many competitive suppliers. Often the end-user's interests drive the need for standards. End-users desire standards to assure service uniformity across many service providers. Network operators desire standards to encourage multiple sources for the systems they will buy. Equipment vendors desire standards to avoid the need to develop many redundant versions of their products. Regulatory and legal developments can create the need for new standards.

Compliance with consensus standards are voluntary. Market dynamics may drive suppliers to comply with the standards or to provide proprietary products for differentiation. A lack of standards can stimulate the introduction of government regulation where noncompliance results in penalties. Even the harshest critic of the standards process will admit that the regulatory process is inflexible and much slower.

So, if a new standard is needed, where will it be developed? Selection of the most suitable committee should be based largely on the expertise of the committee membership and the committee charter. Several committees may have an interest in the new topic. Sometimes certain companies will attempt to place the development of a standard in a committee where they have strong influence. In the event that more than one committee must be involved in the standard's development, it is vital that the committees establish and maintain a cooperative and coordinated relationship. This may be accomplished by one committee having lead responsibility and the other committee(s) providing input in defined areas. Another method is for two committees to jointly develop the standard as peers; this often involves a series of joint meetings. It is essential that conflicts between committees be quickly resolved to avoid an intercommittee war which can waste an enormous amount of time.

In some cases, there is no existing committee with an interest in developing the new standard. The creation of a new committee may be necessary. The creation of a new subcommittee within an existing standards development organization requires the approval of the parent organization and the election or appointment of subcommittee leadership. The creation of a new standards development organization, which is not a subpart of an existing standards development organization, requires a massive administrative effort to begin and also to maintain. Thus, an exhaustive effort should first be made to find a way for the standard to be developed within an existing standards development organization.

17.13.3 Standard or Standards?

Some have suggested that multiple standards for the same item can help stimulate competition. However, the fundamental value of a standard lies in the uniformity it creates. All systems defined to the standard will interwork, and as future enhancements are introduced, there is single base for which compatibility must be maintained. Lively competition is often seen with standardized products (e.g., V.34 modems). A market with multiple standards represents a high-risk venture and may discourage the entry of suppliers and confuse customers. There are some cases where multiple standards are warranted. Different standards may be needed to address different applications (e.g., a low-cost, low-performance consumer grade version, and a higher-cost, higher-performance professional grade version). For political reasons different standards may apply for different areas (for example, different standards for different countries). International standards bodies such as the ITU-T try to minimize the extent of country-specific standards. In a voluntary standards environment, a company or an alliance of companies can create and publish their own specification. Thus, it is not necessary to produce redundant standards.

APPENDIX A

OVERVIEW OF TELCO OPERATIONS SUPPORT SYSTEMS

Operations support systems (OSSs) are the brains of the network. Telecommunications networks are too vast and complex for a network operator to keep all the aspects in a persons head, or even on the pages of a notebook. OSSs are computer programs that manipulate specialized databases and communicate with many pieces of equipment in the network. The OSSs keep track of what equipment is in the network, how the equipment is configured, what equipment is associated with each customer's service, problems in the equipment operation, tracking of service provisioning and trouble repair, and billing the customer.

THE ITU'S TMN (TELECOMMUNICATIONS MANAGEMENT NETWORK) RECOMMENDATIONS

M.3000 through M.3600 describe network management in terms of FCAPS:

Fault: Alarm surveillance and testing

Configuration: Provisioning of equipment and facilities

Accounting: Billing and customer service records

Performance monitoring: Records of traffic and errors

Security: Restricted access to operations systems

The Web-Based Enterprise Management (WBEM) group of the Distributed Management Task Force (DMTF) created the common information model (CIM), which uses the universal modeling language (UML) to define the data that may be used by the CMIP, CORBA, and XML systems described below.

The TeleManagement Forum also addresses operations systems with a top-down perspective.

Many major network operators use an ad hoc collection of network operations systems that have little or nothing in common with ITU (TMN) or TM Forum standards. Many of the legacy OSSs predate TMN. For some major network operators, this consists of network elements (NEs) with associated element management systems (EMSs) where the NE and EMS are provided by the same equipment vendor, and the EMS usually manages many NEs of the same type. The EMS usually performs the following functions with the NE: alarm surveillance, testing, provisioning, and performance monitoring. The EMS is considered to be a tier-1 management system, and may connect to "tier-1.5" service management system (SMS) or a tier-2 network management system (NMS). The TMN standards defines Q3 as the interface between the NMS and the EMS.

The International Standard Organization (ISO—www.iso.ch) defined a TMN-based OSS software structure that includes a service layer protocol (CMISE: common management information service element) and presentation layer protocol (CMIP: common management information protocol). CMIP-CMISE has seen little adoption by the industry. Instead, CORBA (common object request broker architecture) from the Object Management Group (www.omg.org) has been widely adopted by the industry. CORBA is an open object-oriented software architecture for distributed OSS systems that enables applications from different vendors to interwork via networks. CORBA uses the standard IIOP protocol and the interface description language (IDL) to define interfaces.

Three principal languages are used for communications between operations systems: TL1, SNMP, and XML. TL1 (transaction language one) has been used for many years by the legacy telco operations systems such as TIRKS, NMA, and LMOS. Additional functions, such as managing new types of NEs, usually require new TL1 messages and updates to the necessary operations systems. Updates to the operations systems can be expensive and time consuming. Updates to Telcordia (formerly Bellcore) operations systems use a process called OSMINE.

Simple network management protocol (SNMP) management systems originated in the world data networks. HP-Openview was one of the first SNMP based management systems. SNMP is based upon access to standardized information records termed management information bases (MIBs). SNMP-v1 is widely used for alarm surveillance and monitoring. SNMP-v2 added much needed security features, but its complexity caused SNMP-v2 to be unpopular. Recently, SNMP-v3 has added provisioning capabilities and the security functions have been simplified. SNMP-v3 is expected to gain widespread use. SNMP management systems are object oriented and thus are comparatively easier to update and expand.

Extensible markup language (XML) has recently been introduced for electronic data interchange (EDI, also known as electronic bonding), and was derived from hypertext markup language (HTML). XML is primarily used for automated service negotiation between CLECs and ILECs.

Network management architecture (NMA) is the principal OSS used for telephone network alarm surveillance and performance monitoring. NMA was developed by Telcordia. TL1 interfaces from the network elements to convey alarms: critical loss of service to many lines, major loss of service to a few lines, and minor impaired service. NMA also collects E2A relay contact closures that indicate environment conditions: fire, high water, door open, loss of primary power, low battery. A fault in one network element can result in the trouble being detected in associated network elements and at several layers in each element. With TL1, each layer of each element can report faults, and thus a phenomenon know as an alarm storm can result where one fault can cause many alarms. NMA contains a root cause analysis function that determines the origin for related alarms.

Special services (not locally circuit switched, e.g., HICAP, DDS) are tested using digital test systems such as SARTS (special access remote test system—Telcordia), the Teradyne test system, and Hekimian-REACT that perform digital loop-backs and measure bit error rates. The digital transport test systems often gain access to the circuit via a digital cross-connect system (DCS) such as Titan (Tellabs) and DACS (Lucent). Subscriber lines for circuit switched services (POTS and ISDN) are tested by the loop maintenance operations system (LMOS—Lucent) that controls the metallic loop test (MLT—Lucent) system. MLT measures the analog properties of the lines: resistance tip-ring, conductor-ground, line voltage, and loop capacitance. Line test systems gain access to the line via the metallic test trunk of a local digital switch (LDS) such as the 5ESS (Lucent), DMS-100 (Nortel), or EWSD (Siemens).

Configuration consists of equipment preprovisioning and service provisioning. Equipment preprovisioning (aka "inventory creation") occurs when a network element is placed in the network and then entered into the inventory of network equipment that is ready for use. As a rule, information is entered during equipment preprovisioning whenever possible, so that minimizes the per-service-order effort. An OSS such as TMM (technology management module—Telcordia) is used to enter new network elements into the SWITCH, LFACS, and TIRKS databases. Upon the receipt of a service order from a customer representative, the service provisioning (aka service activation) is performed to allocate equipment to the customer's service and to configure the equipment for the indicated type of service. Flow-through provisioning is a term used to describe service provisioning that requires no manual intervention issue work orders following the service order entry.

The primary telephone network configuration operations system is TIRKS (trunk integrated record keeping system), developed by Telcordia. TIRKS keeps track of virtually all central office (CO) switching and transmission equipment, synchronous optical network (SONET) equipment, and HICAP circuits (including cable records for HICAPs). TIRKS interfaces with SWITCH (Telcordia) to manage GR303 interfaces to switches and main distributing frame (MDF) wire jumpers that connect outside line to CO equipment. Prior to SWITCH, MDF wire jumpers were managed by COSMOS (Telcordia). Recent change memory administration center (RCMAC) manages customer specific information in the local voice switch. TIRKSs also interfaces with LFACS (loop facilities assignment system—Telcordia) to keep track of what subscriber lines are available and what line is associated with each customer service. DLESA (digital loop electronics service activation) adds capabilities to TIRKS, SWITCH, LFACS, and several OSSs to enable flow-through service provisioning for next generation digital loop carrier (NGDLC) systems. Traditional DLC (SLC-96, SLC-5, Fujitsu DLC) equipment and subtending subscriber loops are inventoried and assigned in LFACS. Network service database

(NSDB—Telcordia) provides unified access to the TIRKS database for systems, including OPS/INE and WFA. OPS/INE (Telcordia) generates TL1 messages to NGDLC equipment to configure the time slot assignments in the NGDLC. OPS/INE is being superceded by TEMS (transport element management system). WFA (work force administration—Telcordia) generates work orders that inform technicians of the specific service provisioning and repair work that is needed and keeps track of the progress of the work. WFA (Telcordia) consists of WFA-C (controls the flow of orders), WFA-DI (dispatch-in to CO technicians), and WFA-DO (dispatch-out to outside plant technicians). The loop equipment inventory system (LEIS—Telcordia) keeps track of equipment in the outside plant, including DLC. Customer service representatives use the service order retrieval and distribution systems (SORD) that forwards the service order to the service order activation and control (SOAC—Telcordia) system interfaces with TIRKS to generate the service order and keep track of the completion of service provisioning. SOAC parses the service order and forwards the appropriate information to LFACS, SWITCH, and TIRKS.

Accounting consists of service negotiation and billing. Bulk orders from major customers and carriers are entered into EXACT via electronic-bonded interfaces. Most customer service orders are entered by a telco service representative by a front-end service order entry system that feeds into SOAC. Customer billing is generated based on the universal service order code (USOC) and feature identifier (FID) combined with out-of-service records and automatic message accounting (AMA) records of feature usage and called number and call duration. Examples of billing systems include CABS (for carrier billing) and CRIS (for retail billing).

Having reached this point, the reader should not be surprised to hear that even a simple feature addition can require millions of dollars and about two years for the necessary OSS development.

APPENDIX B

A Photographic Tour of the Telephone Company Network

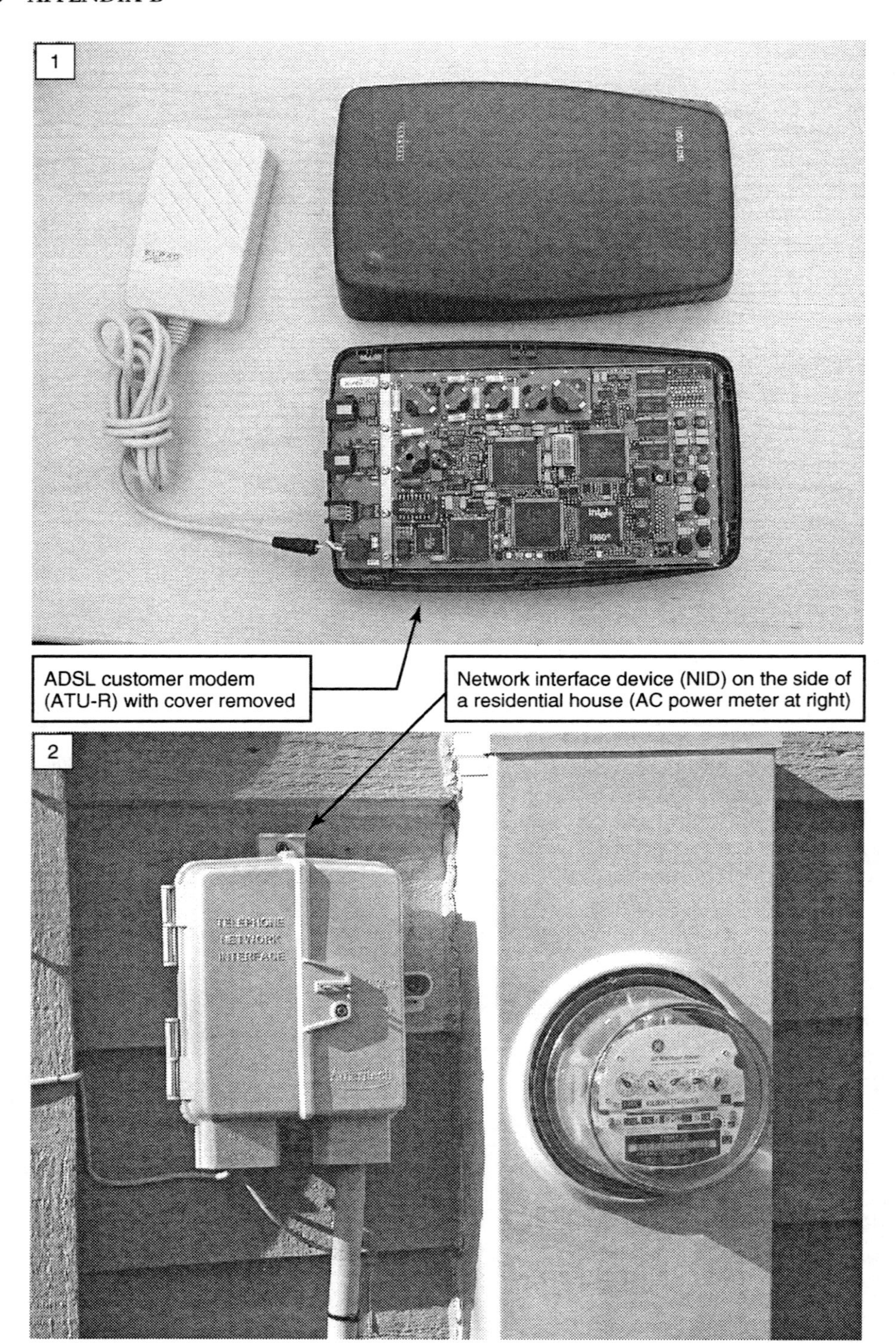
1
ADSL customer modem (ATU-R) with cover removed
Network interface device (NID) on the side of a residential house (AC power meter at right)
2
TELEPHONE NETWORK INTERFACE

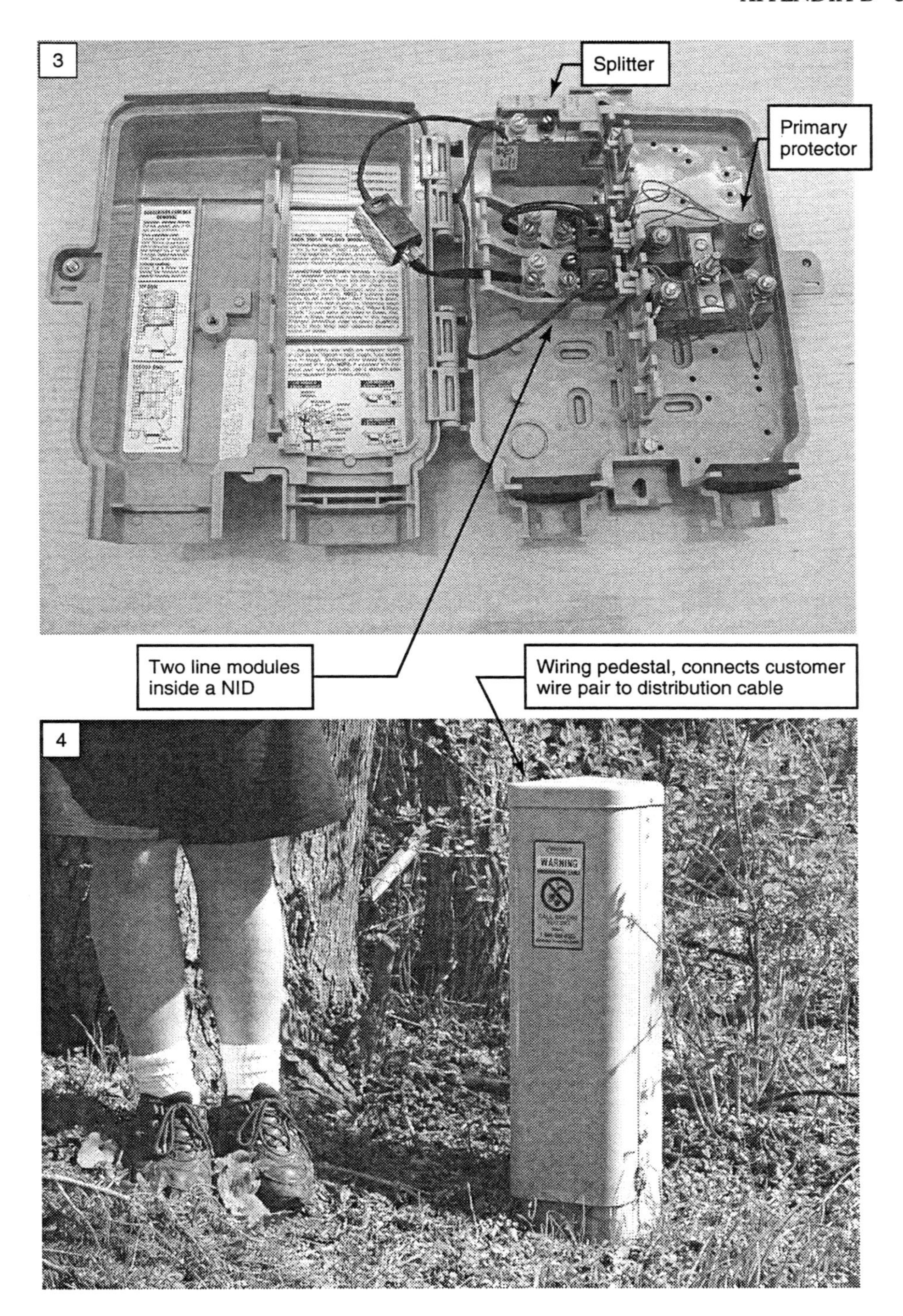
3
Splitter
Primary protector
Two line modules inside a NID
Wiring pedestal, connects customer wire pair to distribution cable
4
WARNING

5
Splice case
Drop wire to customer premises
Ready access distribution terminal, connects drop wires
Distribution terminal, connects drop wires to distribution cable
6

An aerial splice case used for copper and fiber.

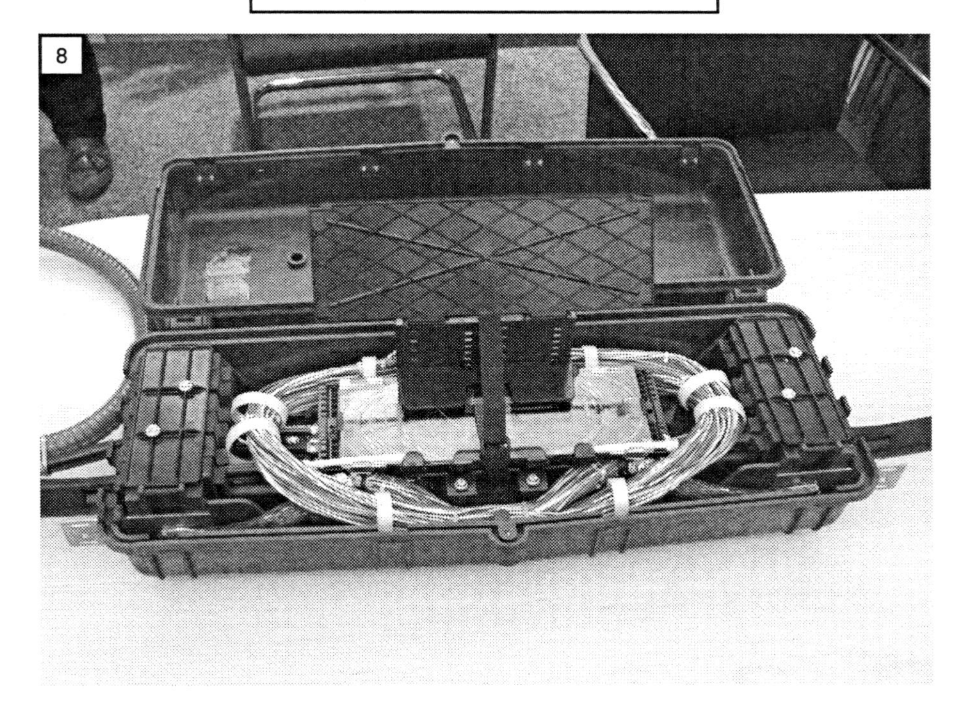

9
Feeder Distribution Interface (FDI), also know as SAC box and crossbox. Connects feeder cable to distribution cables.
Open FDI
10

11
Ameritech
A smaller FDI
A 1500 pair aircore PIC cable with
metal sheath. Colored plastic ribbons
are wrapped around binder groups.
12

13
Cable binder groups
(photo courtesy of
Craig Valenti

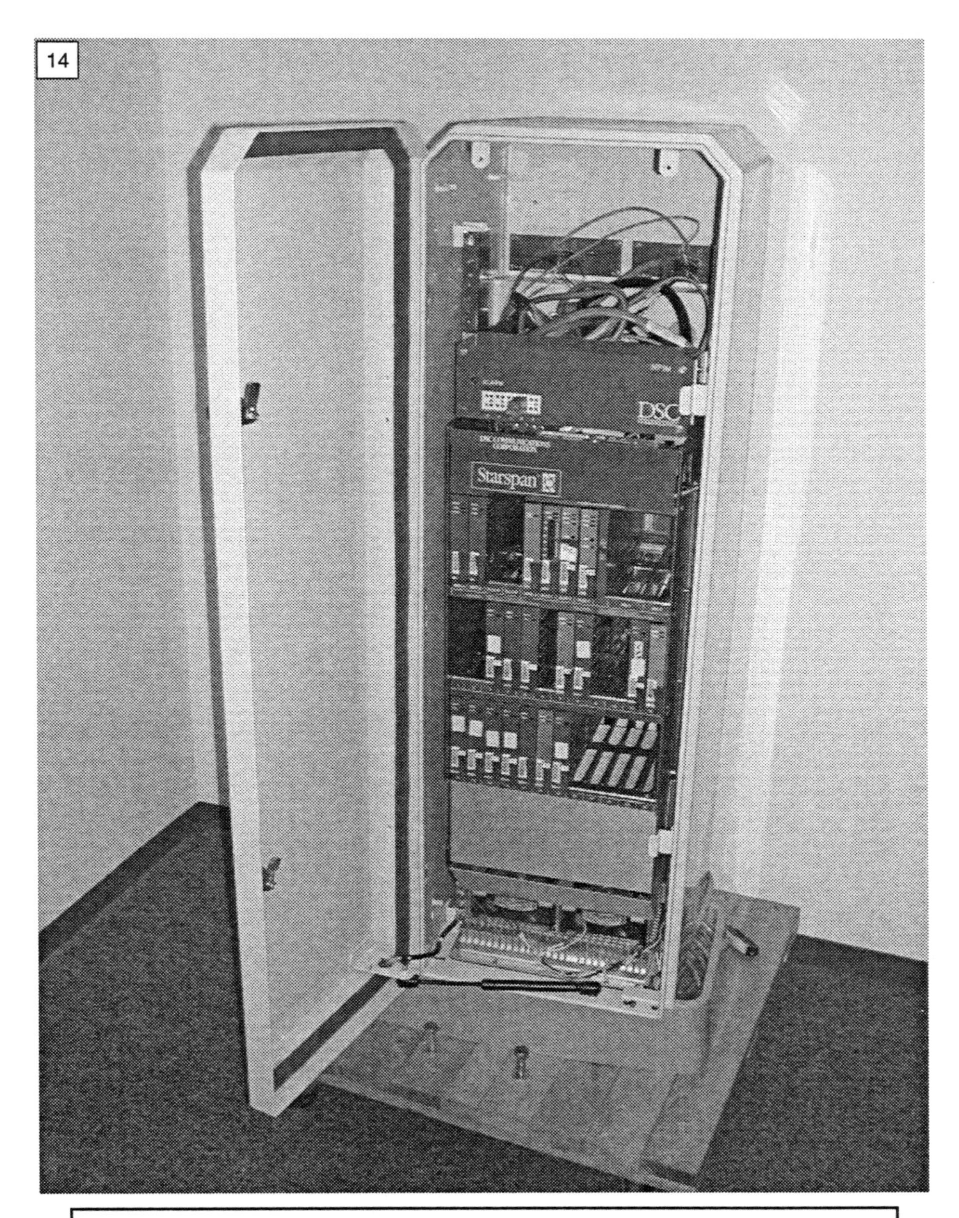

An opened optical network unit (ONU) able to serve 96 lines (laboratory model), about 1 meter tall.

15
Hand hole, contains cable splice
Next generation digital loop carrier remote terminal (NGDLC-RT) cabinet
16

Above ground portion of a buried controlled environment vault (CEV). A portable generator is attached to a power box in the background

Underground CO cable vault. 3,600-pair feeder cable is spliced into many 100-pair cables going to MDF

Feeder cables in underground conduits enter the CO cable vault

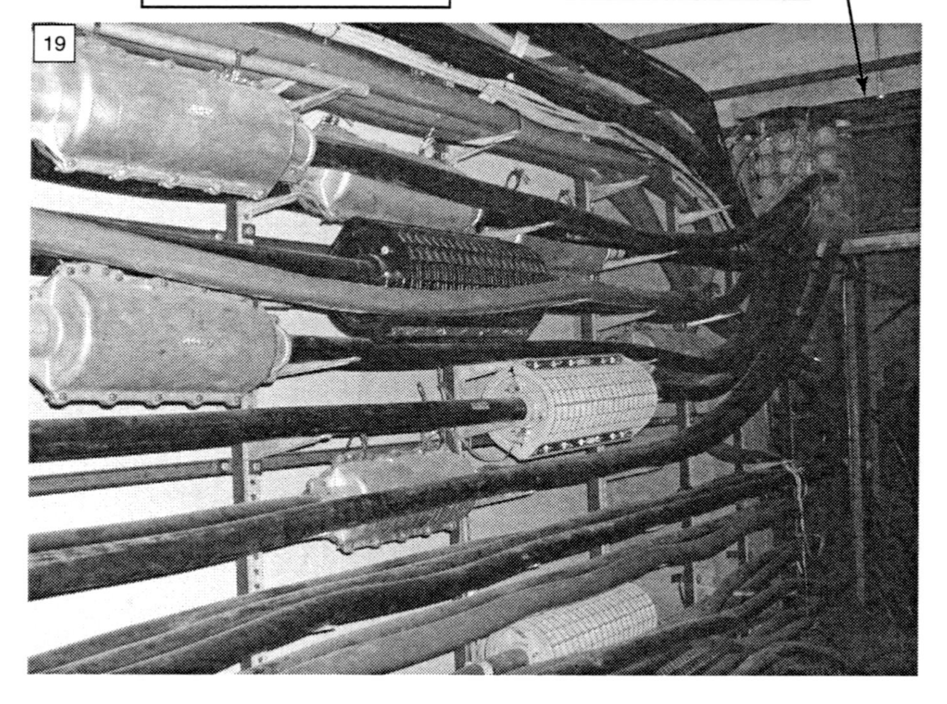

20
COSMIC ™ main
distribution frame (MDF)
MDF wiring blocks
21
3
4
5
6

22
Fiber cable rack
Copper cable rack
Overhead cable racks
23

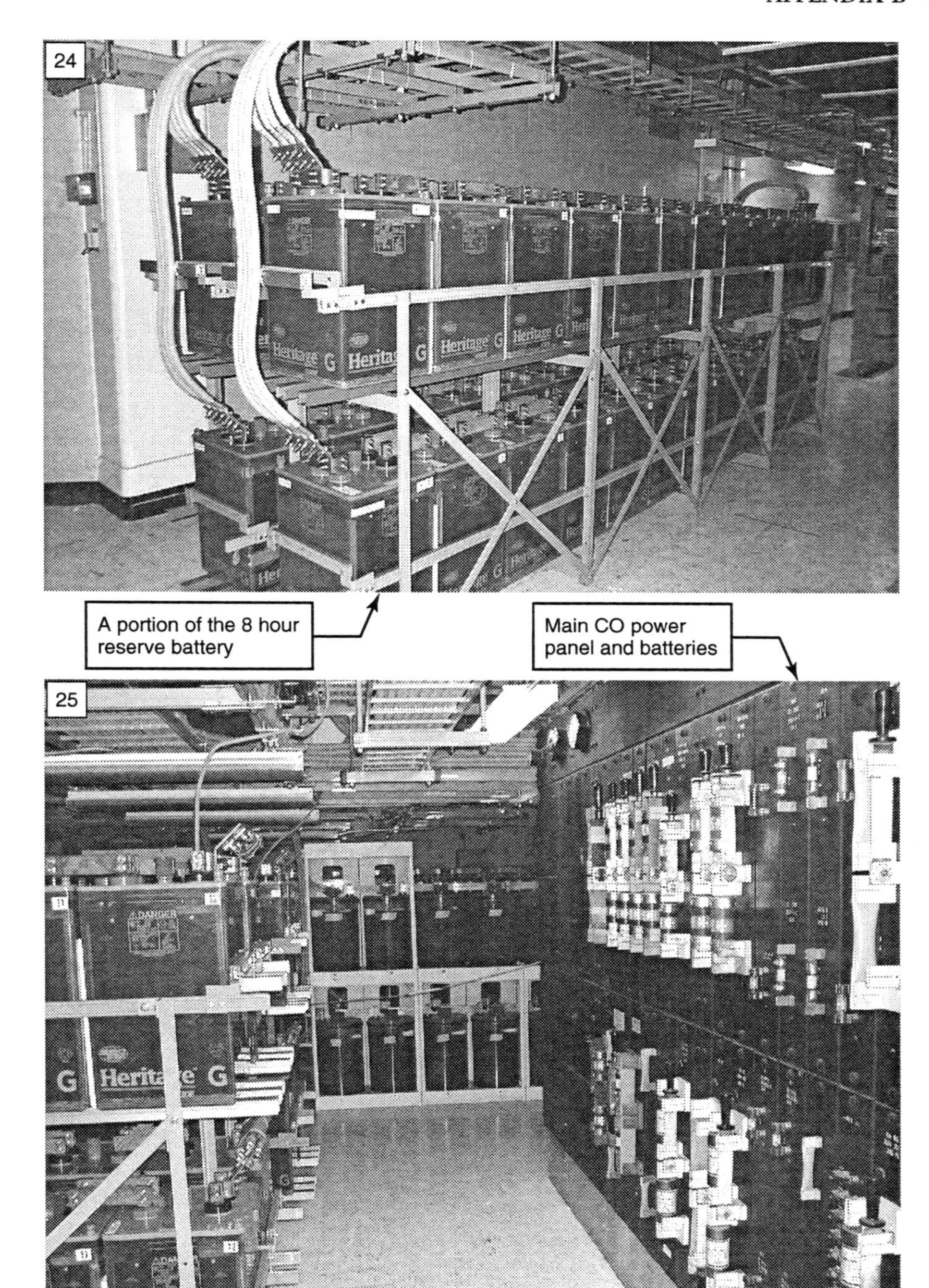
24
A portion of the 8 hour reserve battery
Main CO power panel and batteries
25

9-foot high transmission equipment aisle, 1970-era D4 equipment at left
26

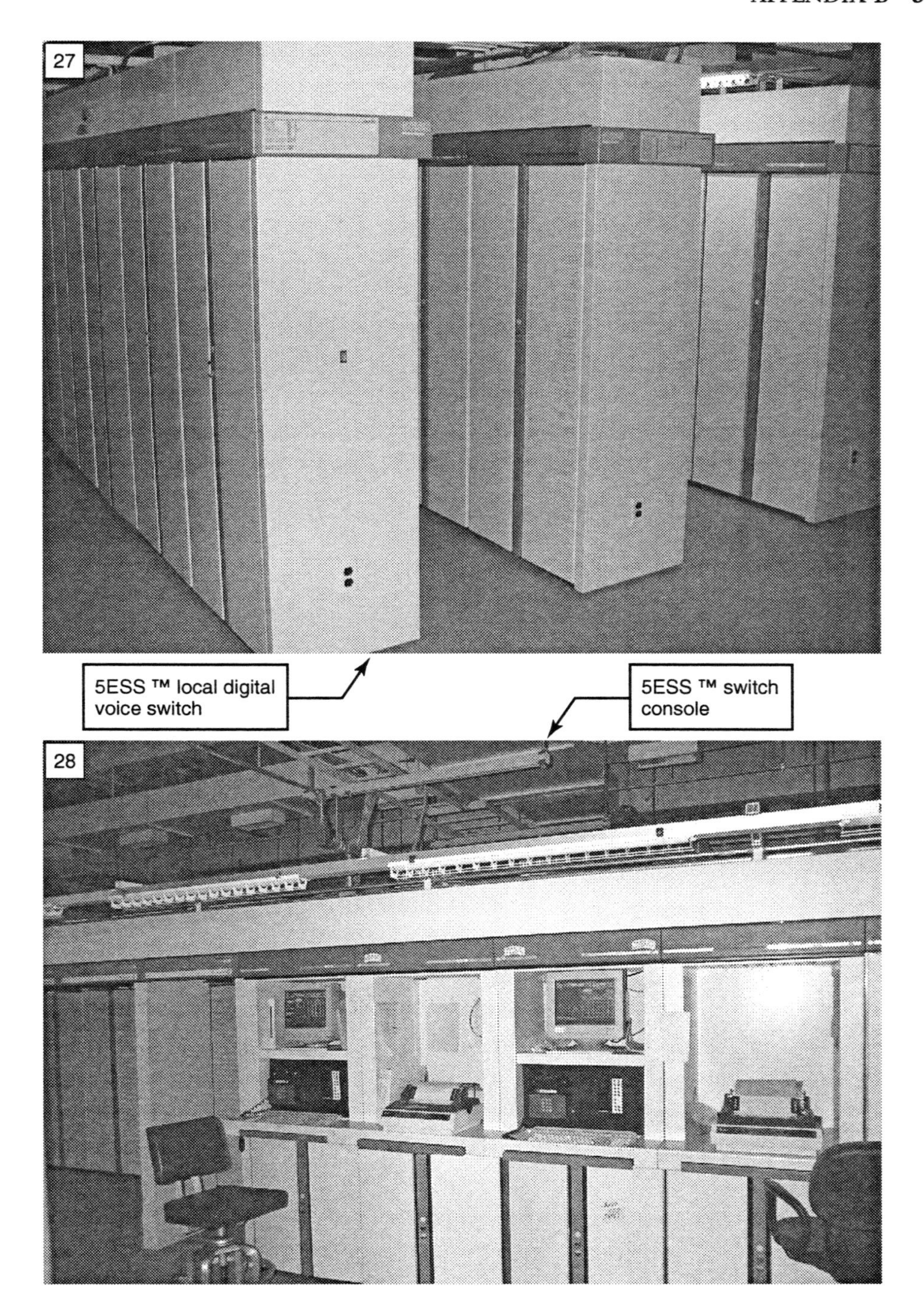
27
5ESS ™ local digital voice switch
5ESS ™ switch console
28

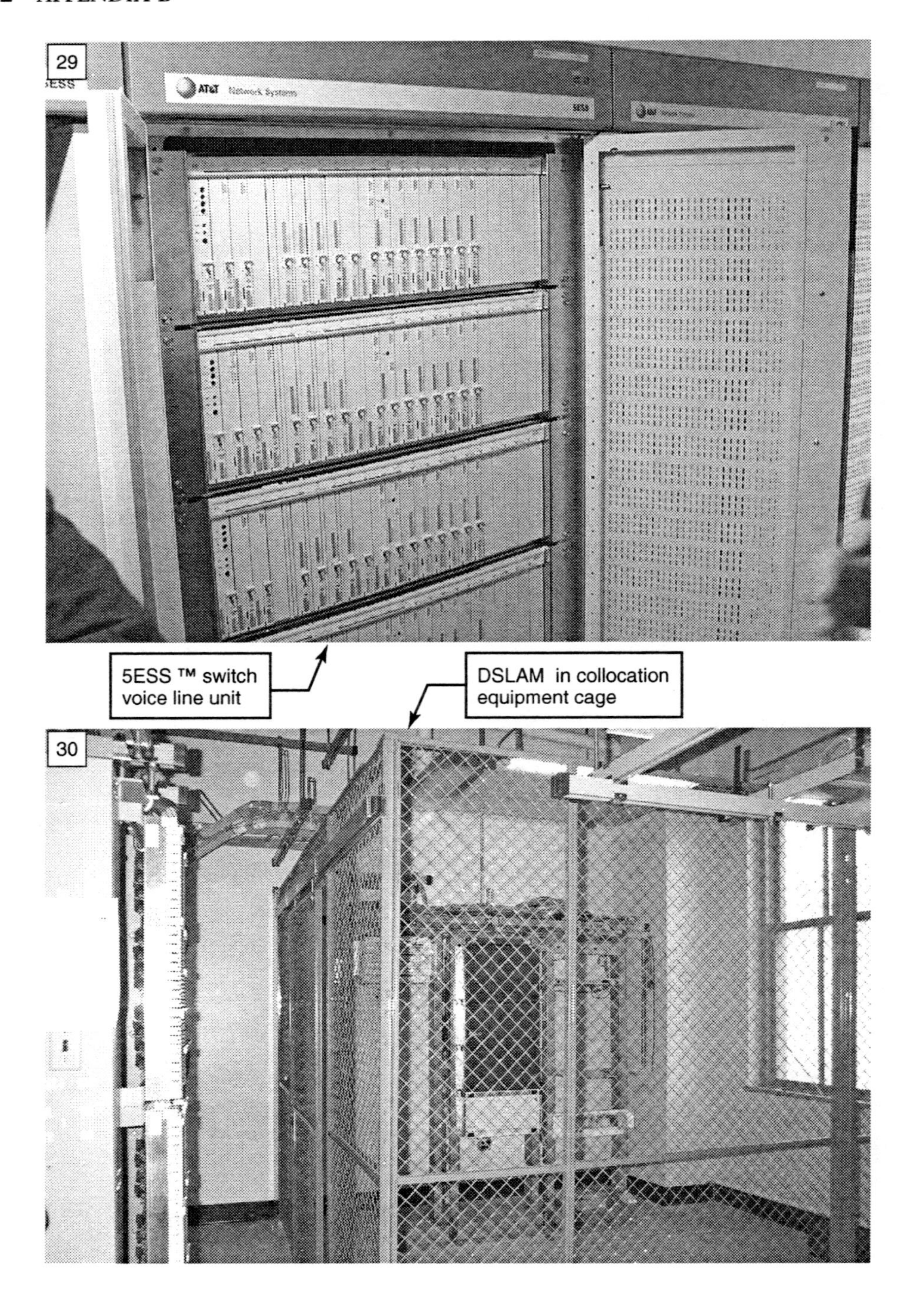
29
5ESS ™ switch
voice line unit
DSLAM in collocation
equipment cage
30

Fiber multiplexers

HDSL CO shelf

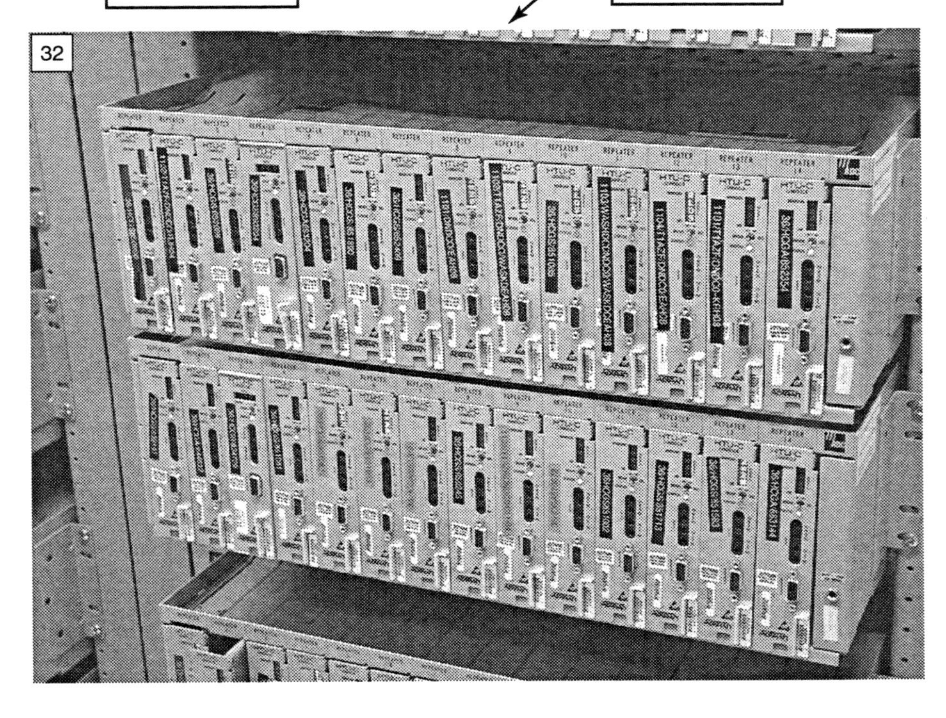

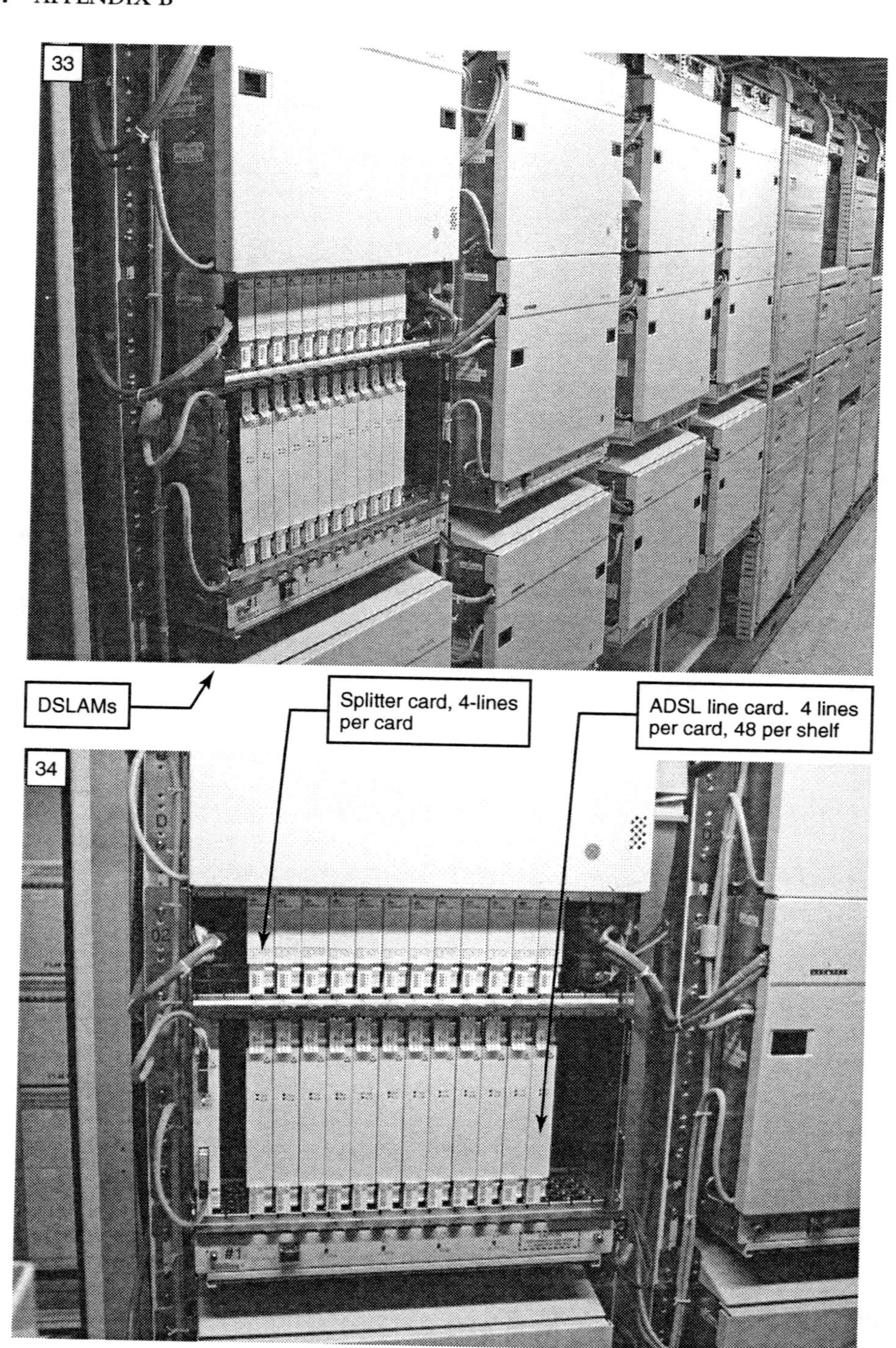
33
DSLAMs
Splitter card, 4-lines per card
ADSL line card. 4 lines per card, 48 per shelf
34

35
SIECOR
CO splitter
DSLAM in CO virtual collocation space

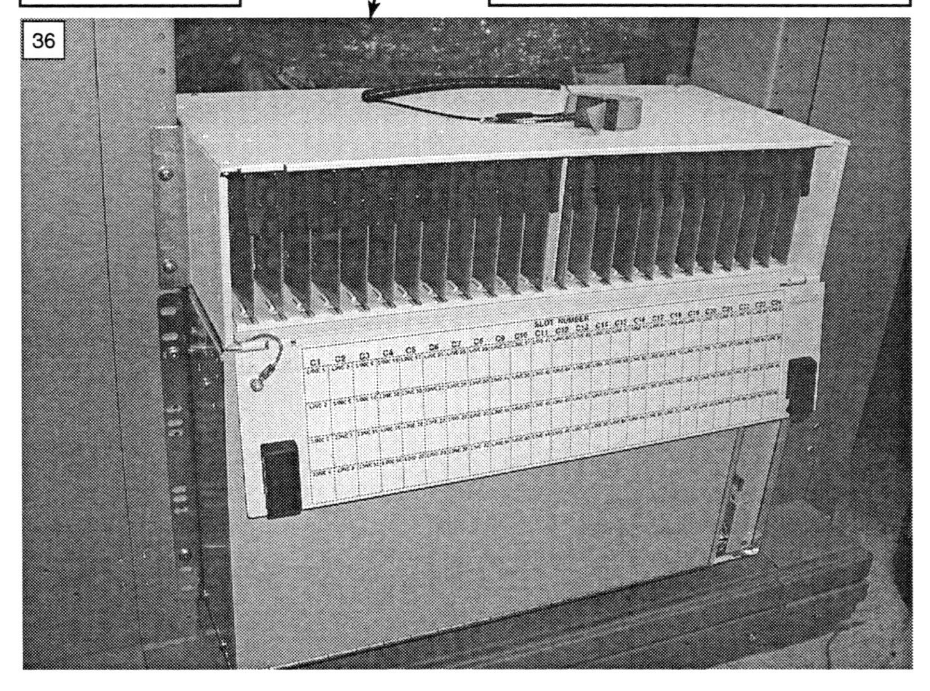
36
CO splitter in CO virtual collocation space. Note ESD wrist strap on top of shelf.
SLOT NUMBER

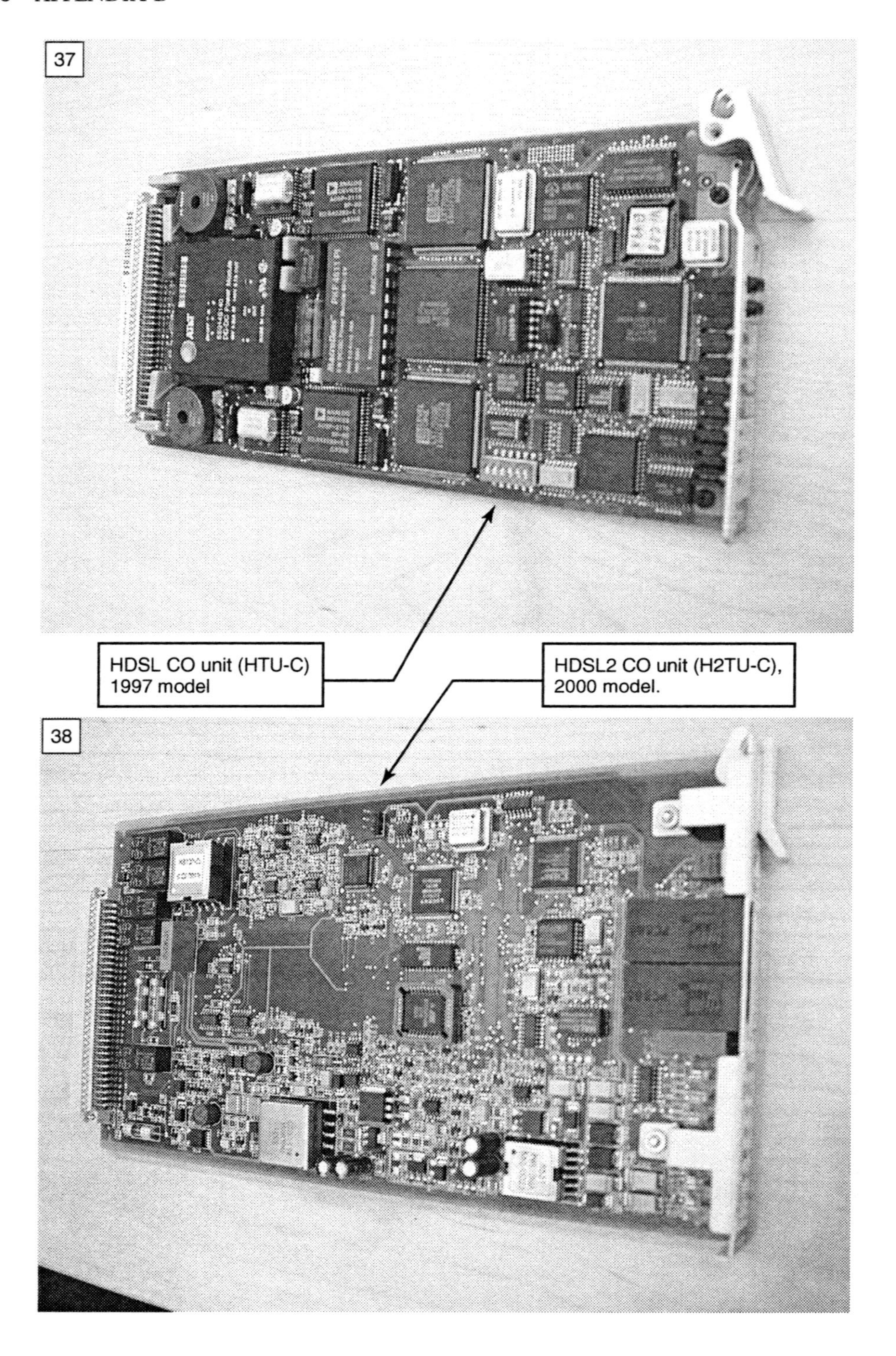
37
HDSL CO unit (HTU-C)
1997 model
HDSL2 CO unit (H2TU-C),
2000 model.
38

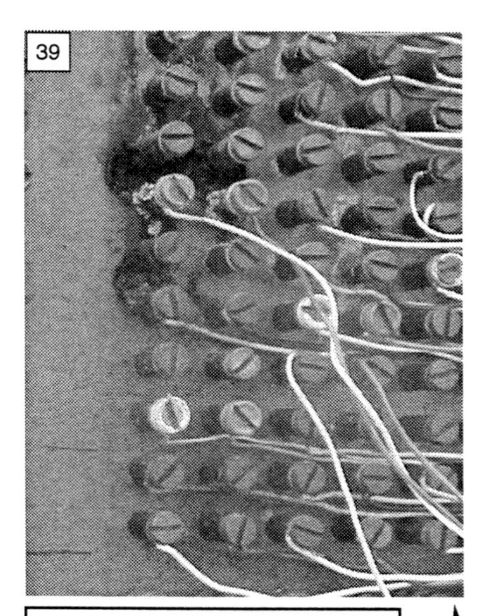

Corrosion on outside plant cross-connect panel. Positive power feeding voltage is suspected to have caused the corrosion.

New and corroded cable brackets. Positive power feeding voltage is suspected to have caused the corrosion.

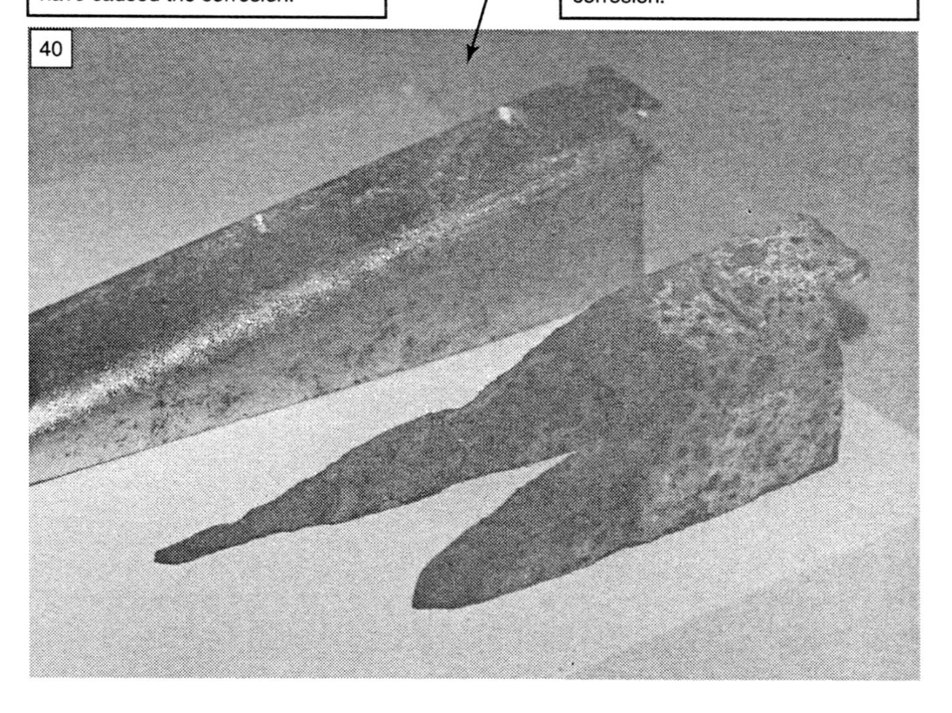

Glossary

1000BaseT: Gigabit Ethernet on unshielded twisted pair wire.

100BaseT: 100 Mb/s Ethernet on unshielded twisted pair wire.

10BaseT: 10 Mb/s Ethernet on unshielded twisted pair wire.

2B+D: Two 64 kb/s B channels and one 16 kb/s D channel are transported by basic rate ISDN.

2B1Q: 2-Binary 1-Quaternary line code is used for BRI and HDSL. Binary information is represented by four amplitude levels.

3B2T: A baseband line code where three binary bits are encoded into two ternary symbols.

4B3T: A baseband line code where four binary bits are encoded into three ternary symbols.

AAA: Authentication, authorization, and accounting—three security functions that occur when a user gains access to a networked system.

AAL2: ATM Adaptation Layer 2—The ATM adaptation layer that is optimized for the transport of multiple simultaneous voice sessions.

AAL5: ATM Adaptation Layer 5—The ATM adaptation layer that is optimized for the transport of packet-based protocols such as IP.

ABCD parameters: The transfer characteristics of a two-port network describing the input voltage and current to the output voltage and current.

ABR: Available bit rate, an ATM quality of service class where the network dynamically controls that cell rate from the source.

ac: Alternating current.

ADPCM: Adaptive differential pulse code modulation, 32 kb/s voice coding using ITU Rec. G.726.

ADSL: Asymmetric digital subscriber line simultaneously transports high-bit-rate digital information towards the subscriber, lower rate data from the customer, and analog voice via one twisted-wire-pair. Relevant standards: T1.413, ITU G.992.1, ITU G.992.2.

AFE: Analog front end, functions include the analog-digital conversion, analog filter, and line driver.

AGC: Automatic gain control, receiver adaptation to the received signal level so as to reduce dynamic range of the signal input to the analog-to-digital converter.

Aliasing: Sampling a signal at intervals greater than Nyquist rate, resulting in the inability to accurately reproduce the signal.

AMI: Alternate mark inversion line code, aka "bipolar," used for T1 transmission systems. Binary information is represented by pulses with three possible amplitudes.

ANSI: American National Standards Institute; accredits standards bodies, for example: committee T1 for telecommunications.

AP: Access point, wireless LAN hub or gateway.

API: Application program interface. A defined set of parameters for communication with a software application.

ARP: Address resolution protocol translates physical layer address to IP address.

ASCII: American Standard Code for Information Interchange: Binary coding of 128 alphanumeric characters as 7 bits plus an optional eighth parity bit. Extended ASCII, which uses 8-bit coding, is more widely used today.

ASP: Application service provider, generally hosts software on a server or gateway function.

ATIS: Alliance for Telecommunications Industry Solutions sponsors Standards Committee T1.

ATM: Asynchronous transfer mode. Link layer using 53 byte cells.

Attenuation: The signal loss resulting from traversing the transmission medium.

ATU-C: ADSL transmission unit at the central office side of the subscriber line. The ATU-C performs transceiver functions, including modulation, coding, and equalization.

ATU-R: ADSL transmission unit at the remote side of the subscriber line. The ATU-R performs transceiver functions, including modulation, coding, and equalization.

AWG: American wire gauge measure of conductor diameter for wires.

AWGN: Additive white Gaussian noise.

B8ZS: Bipolar with 8 zero substitution line code use for T1 transmission. Ones are encoded as pulses of alternating polarity and eight consecutive zeros are represented by a pulse of the same polarity as the previous pulse.

Baseband: Transmission using a technique that places signal energy in a frequency band starting at approximately zero.

Baud: A symbol or element modulated for transmission.

B Channel: A 64 kb/s bearer channel used for ISDN.

BER: Bit error ratio, the proportion of bits that are altered from their correct value.

Binder Group: A bundle of twisted wire pairs that maintain close proximity throughout a section of telephone cable.

Biphase: A baseband line code, also known as the Manchester line code.

Bit: Binary digit.

Bit Stuffing: Extra bit(s) that are conditionally inserted into the frame to adjust the transmitted bit rate.

BLEC: Building local exchange carrier which provides access service within a multi-tenant building.

BLES: Broadband loop extension service.

BRI: Basic rate ISDN provides 2B+D transport. See 2B+D, ISDN.

Bridged Tap: A wire stub connected to the transmission path.

Broadband: Digital communication at bit rates above 128 kb/s.

Byte: A group of eight bits.

CAC: Connection admission control, which means to deny a connection if network capacity does not exist.

CAP: Carrierless AM/PM is a line code similar to QAM used for some DSL systems.

CAS: Channel associated signaling. Call signaling is conveyed within the same bearer channel as the call.

CAT5: ANSI/EIA/TIA-568 standard category 5 unshielded twisted pair cable for LAN bit rates up to 100 Mb/s.

CBR: Constant bit rate, a quality of service to assure a fixed bit rate and delay.

CCS: Common channel signaling.

CLASS: Telephone services such as call forwarding and caller name ID.

Class 5 Switch: A central office–based telephone switch that directly serves an end-user.

CLEC: Competitive local exchange carrier. An access service provider other than the incumbent carrier in a region.

CO: Central office, aka local exchange. A building where telephone lines terminate and connect to transmission and switching equipment.

Codec: Coder-decoder.

Companding: Compressing the dynamic range of a signal prior to transmission, with matching expansion at the receiver to regain the original signal.

Convolutional code: A code that depends on the current bit sequence and the encoder state.

CRC: Cyclic redundancy check. An error detecting or correcting code.

Crest factor: The ratio of peak to RMS signal voltage, also called peak-to-average-ratio, PAR.

Crosstalk (XTLK): The unintentional electromagnetic coupling of a signal to other wire pairs in a cable. *See* NEXT *and* FEXT.

CSA: Carrier Serving Area loop design rules specify the characteristics of loops served by digital loop carrier sites.

CSMA/CA: Carrier sense multiple access with collision avoidance, a media access control used by IEEE 802.11 wireless LAN.

CSMA/CD: Carrier sense multiple access with collision detection, a media access control method used by Ethernet.

Cyclic Code: An error correcting code implemented with a feedback shift register, for example, CRC.

CVoDSL: Channelized voice-over DSL.

DA: Distribution area is a loop serving area for a feeder distribution interface (FDI).

dB: Decibel; ten times of the common logarithm of the ratio of relative powers.

dBm: dB relative to one milliwatt.

dBrnc. The logarithmic power ratio of a C-message weighted filtered signal with respect to 1 picowatt.

dc: Direct current.

D Channel: A 16 kb/s channel used by basic rate ISDN lines to carry signaling and packet information.

DDS: Digital data service. A tarried private line data service using AMI-type transmission of symmetric rates from 1.2 to 64 kb/s.

DFE: Decision feedback equalizer.

DFT: Discrete Fourier transform. A signal transformation that is often implemented as a fast Fourier transformation on a digital signal processor.

DHCP: Dynamic host configuration protocol. A protocol for assigning an IP address for the duration of a session.

Diffserv: A protocol for providing control of quality of service on an IP network.

Digital Duplexing: a highly flexible and high-performance duplexing system used by DMT VDSL.

Digital Loop Carrier: *See* DLC

Discrete Time Domain: Signal values that are defined at periodic time intervals.

Distribution cable: The portion of the telephone loop plant that connects the feeder cable to the drop wires.

DLC: Digital loop carrier. A system that is often located remotely from the central office to multiplex the service from many customer lines on to a high-speed line between the central office and the DLC remote terminal.

DMT: Discrete multitone is a multicarrier transmission technique that uses a fast Fourier transform (FFT) and inverse FFT to allocate the transmitted bits among many narrowband QAM modulated tones, depending on the transport capacity of each tone.

Downstream: Information flowing from the network to the customer.

Drop Wire: The section of the local loop connecting the distribution cable to the customer premises.

DS0: A 64 kb/s bit rate.

DS1: A 1.544 Mb/s bit rate.

DS3: A 44.736 Mb/s bit rate.

DSL: Digital subscriber line may be used to indicated the basic rate ISDN transceivers and the local loop to which they are connected. May also be used to indicate all types of DSL.

DSM: Dynamic spectrum management can be used to allow DSL systems to self-improve as a function of binder situation.

DSP: Digital signal processing performs filtering and other signal modifications in the digital domain by first converting analog signals to a digital representation.

DSSS: Direct sequence spread spectrum. Modulation method used by IEEE 802.11 wireless LAN.

DSX-1: DS1 Cross-connect. A 1.544 Mb/s AMI signal used for short distances to interconnect equipment within a CO.

DTMF: Dual tone multifrequency: Signaling used by touch-tone telephone.

DWMT: Discrete wavelet multitone. A version of DMT that uses a wavelet type transform in place of the FFT. *See* DMT.

E1: A 2.048 Mb/s bit rate.

E3: A 34 Mb/s bit rate.

EC: Echo cancellation is a DSP-time domain technique for removing echos.

ECH: Echo-canceled hybrid is a 2-to-4 wire conversion with echo cancellation. A hybrid transform is often used to interface to the line.

EMC: Electromagnetic compatibility prevents unintended radio frequency interference.

EMI: Electromagnetic interference.

EMS: Element management system.

Encapsulation: The addition of a protocol header to a protocol data unit

ES: Errored second. An interval of one second containing one or more bit errors.

Ethernet: A link layer local access network protocol using CSMA/CD media access control.

ETSI: European Telecommunications Standards Institute.

FCC: Federal Communications Commission is the U.S. government agency that regulates the radio, television, and telecommunications industries.

FDI: Feeder distribution interface is a cabinet that contains a cross-connect field used to connect feeder cables to distribution cables.

FDM: Frequency division multiplexing permits more than one information stream to be sent over a line by subdividing the line into separate frequency bands.

FEC: Forward error control. Errors are corrected by the receiver using redundant information sent by the transmitter.

Feeder: Cable between the central office and the FDI point.

FEXT: Far end crosstalk results from transmitted signals being coupled to another wire pair interfering with the reception at the far end of the line.

FFT: Fast Fourier transform. An algorithm for efficiently implementing via digital signal processors the conversion from the time domain to the frequency domain.

FHSS: Frequency hopping spread spectrum.

FIR: Finite impulse response. A FIR filter utilizes a limited number of delay and multiplication elements.

FM: Frequency modulation uses changes in frequency of a carrier signal to represent information.

Fourier Transform: Using a sinusoidal expansion to represent a signal.

Fractionally Spaced Equalizer: An equalizer using multiples of the symbol rate.

Frame Relay: A low-complexity wide area network packet protocol.

FTP: File transfer protocol.

Full duplex: Simultaneous transmission in upstream and downstream directions.

Galois field: A closed algebra field used to describe an encoder or decoder.

Gateway: A device that enables different types of networks to interwork.

Gaussian distribution: A bell shaped distribution function, aka a normal distribution.

GDFE: Generalized decision feedback equalization is a canonical design method for DSL systems with ISI and/or crosstalk.

G.lite: ITU Recommendation G.992.2 for ADSL operating up to 1.5 Mb/s in the downstream direction.

GR-303: A Telcordia developed interface between a NGDLC/DLC and a class 5 switch.

Half-Duplex: Transmission that alternates the upstream and downstream information flow.

Hamming Distance: The number of bits having a different value for a pair of code words.

HDLC: High level data link control, a layer 2 protocol.

HDSL: High-bit rate digital subscriber line permits 1.544 or 2.048 Mb/s symmetric transmission over two or three wire pairs which meet CSA design rules.

HDSL2: Second-generation HDSL uses one pair of wires for 1.544 Mb/s symmetric transmission over CSA loops.

HEC: Header error correction.

HTML: Hypertext markup language, used to describe World Wide Web information.

HTTP: Hypertext transfer protocol: protocol for World Wide Web messages.

HTU-C: HDSL transmission unit at the central office side of the line. The HTU-C performs transceiver functions, including modulation, coding, and equalization.

HTU-R: HDSL transmission unit at the remote side of the line. The HTU-R performs transceiver functions, including modulation, coding, and equalization.

IAD: Integrated access device, a customer premises device that includes a DSL modem with voice and data functions.

IDFT: Inverse discrete Fourier transform. A discrete form of the IFFT.

IDSL: A DSL that used BRI transceivers to convey 128 kb/s packet-mode transport via a local loop.

IEEE: Institute of Electrical and Electronics Engineers.

IETF: Internet Engineering Task Force, develops Internet specifications.

IFFT: Inverse fast Fourier transform. An algorithm for efficiently implementing via digital signal processors the conversion from the frequency domain to the time domain.

ILEC: Incumbent local exchange carrier, the telephone service provider that was originally the sole service provider in a region.

ILMI: Integrated network management protocol.

Impedance: The relationship between the voltage applied to the resulting current.

Impulse Noise: An unwanted signal of short duration often resulting from the coupling of energy from an electrical transient from a nearby source.

Impulse Response: The time-domain response of a network to a impulse input.

IN: Intelligent network—advanced telephone services that make use of the SS7 network.

IP: Internet protocol. Layer 3 protocol used as part of TCP/IP; used in the Internet, as well as private networks.

IPSec: Internet protocol security.

ISDN: Integrated services digital network, provides end-to-end digital circuit switched and packet switched transport. *See* BRI.

ISI: Intersymbol interference. Signal energy of a symbol that transfers into nearby symbols due to channel dispersion.

ISP: Internet service provider.

ITU: The International Telecommunications Union develops international standards. The telecommunications sector, ITU-T, was previously called the CCITT.

IWD: Interworking device.

JPEG: Joint photographic experts group, method for compressed digital still images.

kft: Kilofeet, a thousand feet.

L2TP: Layer 2 tunneling protocol, used with PPP to convey many packet streams via one connection.

LAN: Local access network, a means to interconnect many digital devices within a few hundred meters.

Latency: The time for information to flow between two points of a network.

LES: Loop extension service.

Lifeline POTS: Commonly used to describe voice service that is powered from the network, so that service does not depend on customer premises power. This is distinct from the meaning of "lifeline" for regulators, meaning voice service for low-income persons.

Line Code: A modulation method.

Loaded Loop: A loaded loop contains series inductors, typically spaced every 6 kft, for the purpose of improving the voice-band performance of long loops. However, DSL operation over loaded loops is not possible due to excessive loss at higher frequencies.

LMS: Least mean square. An algorithm used for adaptive filters.

Loop, also local loop: Another term for the telephone line connecting the customer premises to the telephone network.

Loopback: A diagnostic test method whereby the signal received by an element is replicated and retransmitted to the source of the signal. In some cases, the system is taken out of service to perform the loopback.

LT: Line termination. The LT is the DSL transceiver at the network end of the line.

MAC: Media access control, the means to control access to the physical medium.

Margin: Signal-to-noise ratio in excess of that necessary to attain a reference bit error rate.

MCM: Multicarrier modulation, for example, DMT.

MDB: Modified duobinary. A baseband line code.

MDF: Main distributing frame. A wiring cross-connect field in the central office used to connect outside lines to CO equipment.

MEGACO: Media gateway control protocol.

MIB: Management information base, an SNMP object used to define the characteristics of a network element.

ML: Maximum likelihood is an optimum method by which to detect transmitted messages in a receiver.

Modem: Modulator/demodulator, a device that converts digital signals for efficient transmission and reception over a certain physical media.

MPEG: Motion picture experts group of ISO, develops video compression standards.

MPLS: Multiprotocol label switching

NAP: Network access provider. The entity that provides a broadband access network which allows a user to connect to a service network.

NAT: Network address translation.

NEXT: Near-end crosstalk results in the signal transmitted from a wire pair coupling into another wire pair and interfering with the reception of signals at a receiver located at the same end of the line as the disturbing transmitter.

NGDLC: Next generation digital loop carrier.

NIC: Network interface card, a communications interface circuit card for a PC.

NID: Network interface device, a connection point located at the demarcation between the network and the customer installation.

NSP: Network service provider. The entity that provides a network that support services that can be accessed by a user via an access network.

NT: Network termination. The NT is the DSL transceiver at the customer end of the line.

Nyquist Interval: The maximum time between equally spaced samples necessary for accurate representation of a signal.

OCn: Optical carrier of *n* times 51.84 Mb/s.

OSP: Outside plant, telephone cable, supporting infrastructure (poles, ducts), midspan equipment (loading coils, repeaters, DLC), and FDI

Overhead Information: Information other than end-user data.

PAM: Pulse amplitude modulation.

Passband: The modulation of signals which permits the placement of the modulated signal in a specified frequency band, for example, CAP, DMT, QAM.

PC: Personal computer.

PCM: Pulse code modulation, used for voice coding.

PDH: Plesiochronous digital hierarchy, used for asynchronous multiplexed transmission.

PDU: Protocol data unit.

PHY: The physical layer of a communications stack.

Ping: TCP/IP message used to measure packet transport latency.

PLL: Phase locked loop. A feedback loop with narrow bandwidth used to recover signal timing.

PM: Phase modulation use changes in the phase of a carrier signal to represent information.

POTS: Plain old telephone service, circuit switched analog voice.

POTS splitter: *See* splitter.

PPP: Point-to-point protocol, used for session control.

PPPoE: PPP over Ethernet. A protocol that extends multiple PPP sessions over an Ethernet interface.

PSTN: Public switched telephone network. The traditional circuit switched voiceband network.

PVC: Permanent virtual circuit. A provisioned ATM or frame relay connection.

QAM: Quadrature amplitude modulation is a passband modulation technique that represents information as changes in carrier phase and amplitude.

QPSK: Quadrature phase shift keying. A passband modulation method.

QoS: Quality of service. Virtual circuit characterization in terms of throughput, delay, and delay variation.

Quad Cable: Cables where four wires are twisted as a unit. High crosstalk is experienced among the wires within a quad unit.

Quantization Noise: The noise resulting from analog to digital conversion.

Quat: A quaternary symbol representing two bits with four-level symbol.

RADSL: Rate adaptive digital subscriber line is an ADSL, which adjusts its transmission rate to the capacity of the line.

Remote Terminal: Digital loop carrier at a site remote from a CO.

RFI: Radio frequency interference.

RJ-11: A six-position modular plug/jack with four conductors commonly used to connect to two-line phones.

RJ-14: A six-position modular plug/jack with two conductors commonly used to connect to single-line phones.

RJ-45: An eight-position and conductor plug/jack commonly used for 10BaseT.

RS Code: Reed-Solomon code. A code used for error detection or correction.

RT: Remote terminal, equipment located distant from the central office. Often located in an outdoor cabinet or underground vault.

SDSL: Variously used to describe symmetric DSLs and single-pair DSLs. May or may not carry POTS (voice) on the same loop.

SES: Severely errored second, a one second interval with a bit error ratio greater than 10^{-2}.

SIP: Session initiation protocol, IETF RFC 2543 method to initiate and manage Internet sessions.

SNMP: Simple network management protocol. A network management protocol using TCP/IP to access management information base (MIB) information.

SNR: Signal-to-noise ratio. A signal quality measure.

SONET: Synchronous optical network.

Spectral Balancing: Generally describes methods for adaptively determining the spectra of users in a common binder or cable of DSL systems.

Splitter: Filter(s) used to combine voice-band and higher-band data signals to coexist in different frequency bands on the same pair of wires.

SS7: Signaling system 7.

STP: Shielded twisted pair. A metallic sheath is provided to reduce RFI.

SVC: Switched virtual circuit. An ATM or frame relay connection established in real time via signaling messages.

T1 (Committee): Telecommunications standards committee for the United States.

T1 (Carrier): 1.544 Mb/s symmetric AMI transmission.

T1E1.4: The U.S. standards working group responsible for DSL standards. T1E1.4 is one of several Working Groups within Technical Subcommittee T1E1.

TCM: Trellis coded modulation. A convolutional code that provides coding gain without increasing bandwidth.

TCM: Time compression multiplexing, aka Ping-Pong, permits two-way transmission by the use of alternating short one-way transmission bursts.

TCP: Transmission control protocol.

TDM: Time division multiplexing.

Throughput: The effective rate of information flow though a system.

TPS-TC: Transport protocol specific transmission convergence.

TR-08: A Telcordia developed Interface between a DLC and a class 5 Switch.

Unavailable Seconds: A duration beginning with ten consecutive severely errored seconds and ending with ten seconds that are not severely errored.

UBR: Unspecified bit rate, a best effort quality of service.

UNE: Unbundled network element.

UNI: User to network interface.

Upstream: Information flowing from the customer to the network.

USB: Universal serial bus, an interface supporting up to 12 Mb/s.

UTP: Unshielded twisted pair.

V5.2: An ITU-T specified protocol for connecting a digital loop carrier (remote terminal) to a telephone switch.

VC: virtual circuit.

VCI: virtual channel identifier used for ATM.

VDSL: Very-high bit rate digital subscriber line systems transmit information at rates from 12 to 52 Mb/s.

Vectoring: An optimum means of co-generating and/or co-receiving multiple DSL signals.

Viterbi: An algorithm used for reception of trellis coded modulation.

VLSI: Very large scale integration permits millions of transistors in one integrated circuit.

VoATM: Voice-over ATM.

VoDSL: Voice-over DSL.

VoIP: Voice-over Internet protocol. Voice conveyed via IP packets.

VoSTM: Voice-over synchronous transfer mode (CVoDSL).

VPI: Virtual path identifier used by ATMVPN: virtual private network.

WAN: Wide area network, covering more than one city.

Water-filling: A procedure for determining best-transmitted spectra.

White Noise: Noise with equal power at all frequencies.

XDSL: A generic term that applies to all DSL technologies.

XML: Extensible markup language, a more flexible version of HTML.

Zeptosecond: 10^{-21} seconds, the time interval between a traffic signal turning green and a New York taxi driver honking the horn.

Zettasecond: 10^{21} seconds, about 300 years.

INDEX

B

C

D

L

M

N

O

P

Q

R

S

W

X

Y

Z

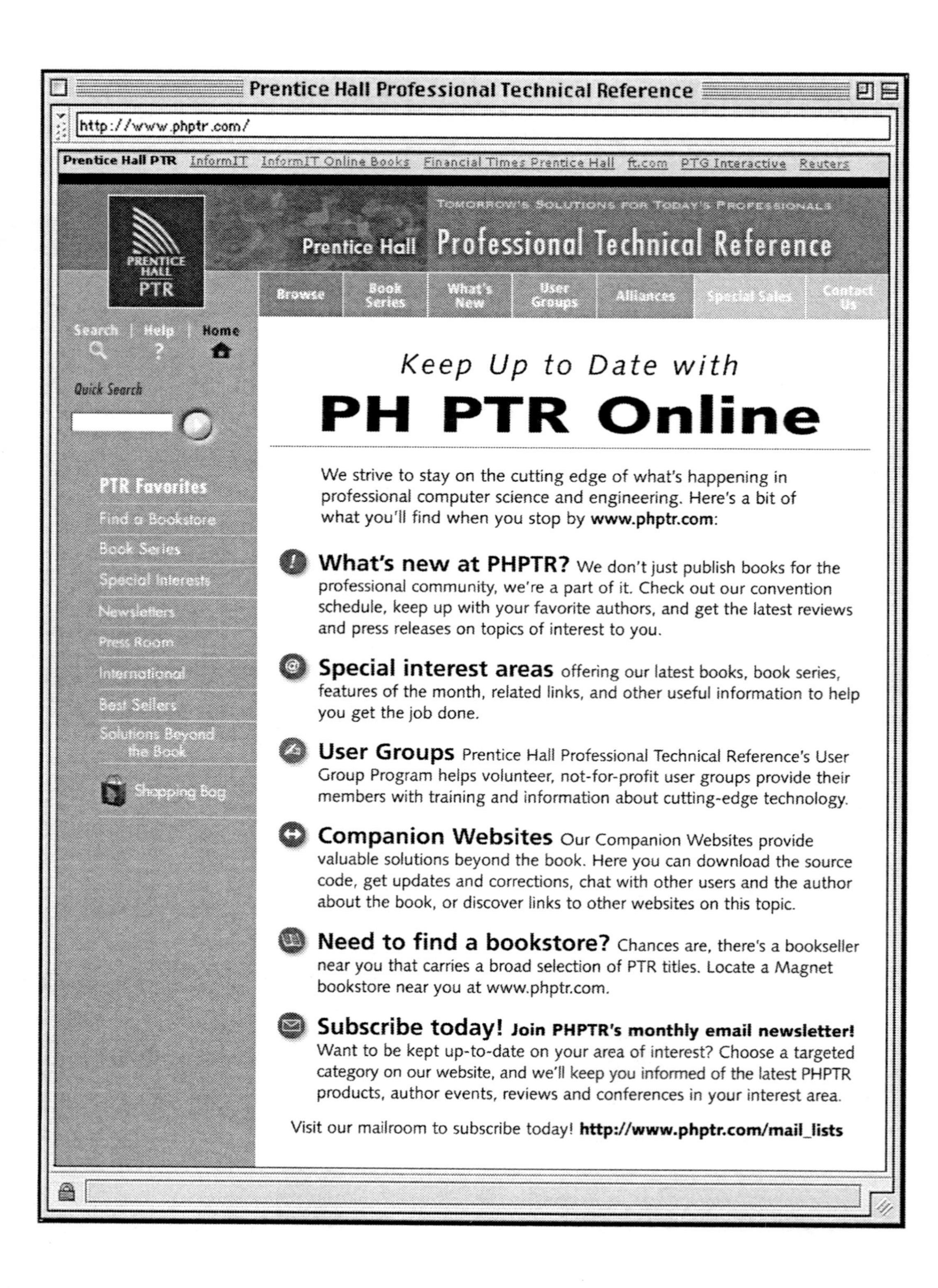
Prentice Hall Professional Technical Reference
http://www.phptr.com/
Prentice Hall PTR InformIT InformIT Online Books Financial Times Prentice Hall ft.com PTG Interactive Reuters
Tomorrow's Solutions for Today's Professionals
Prentice Hall Professional Technical Reference
PRENTICE HALL PTR
Browse
Book Series
What's New
User Groups
Alliances
Special Sales
Contact Us
Search | Help | Home
Quick Search
PTR Favorites
Find a Bookstore
Book Series
Special Interests
Newsletters
Press Room
International
Best Sellers
Solutions Beyond the Book
Shopping Bag
Keep Up to Date with
PH PTR Online
We strive to stay on the cutting edge of what's happening in professional computer science and engineering. Here's a bit of what you'll find when you stop by www.phptr.com:
What's new at PHPTR? We don't just publish books for the professional community, we're a part of it. Check out our convention schedule, keep up with your favorite authors, and get the latest reviews and press releases on topics of interest to you.
Special interest areas offering our latest books, book series, features of the month, related links, and other useful information to help you get the job done.
User Groups Prentice Hall Professional Technical Reference's User Group Program helps volunteer, not-for-profit user groups provide their members with training and information about cutting-edge technology.
Companion Websites Our Companion Websites provide valuable solutions beyond the book. Here you can download the source code, get updates and corrections, chat with other users and the author about the book, or discover links to other websites on this topic.
Need to find a bookstore? Chances are, there's a bookseller near you that carries a broad selection of PTR titles. Locate a Magnet bookstore near you at www.phptr.com.
Subscribe today! Join PHPTR's monthly email newsletter!
Want to be kept up-to-date on your area of interest? Choose a targeted category on our website, and we'll keep you informed of the latest PHPTR products, author events, reviews and conferences in your interest area.
Visit our mailroom to subscribe today! http://www.phptr.com/mail_lists